PCR P

A LABORATORY MANUAL

Second Edition

PCR Primer

A LABORATORY MANUAL

Second Edition

EDITED BY

Carl W. Dieffenbach
National Institute of Allergy and Infectious Diseases

Gabriela S. Dveksler
Uniformed Services University of the Health Sciences

COLD SPRING HARBOR LABORATORY PRESS
Cold Spring Harbor, New York

PCR Primer: A Laboratory Manual
Second Edition

Printed in the United States of America.

**Acquisitions Editor
  and Publisher:**   John Inglis
Project Manager: Judy Cuddihy
Developmental Editors: Siân Curtis and Tracy Kuhlman
Project Coordinator: Inez Sialiano
Production Editor: Patricia Barker
Desktop Editor: Danny deBruin
Book Designer: Denise Weiss
Cover Designer: Ed Atkeson

Front cover artwork: Illustrated model of DNA courtesy of Slim Films (http://www.slimfilms.com; ©Slim Films).

Library of Congress Cataloging-in-Publication Data

PCR Primer : a laboratory manual / edited by Carl W. Dieffenbach, Gabriela S.
    Dveksler.— 2nd ed.
        p.cm.
    Includes bibliographical references and index.
    ISBN 0-87969-653-2 (cloth : alk. paper) — ISBN 0-87969-654-0 (pbk. : alk. paper)
    1. Polymerase chain reaction—Laboratory manuals. I. Dieffenbach, Carl W. II.
    Dveksler, Gabriela S.

QP606.D46P359 2003
572'.43—dc21   2003053085

10 9 8 7 6 5 4 3 2 1

All Cold Spring Harbor Laboratory Press publications may be ordered directly from Cold Spring Harbor Laboratory Press, 500 Sunnyside Blvd., Woodbury, N.Y. 11797-2924. Phone: 1-800-843-4388 in Continental U.S. and Canada. All other locations: (516) 422-4100. FAX: (516) 422-4097. E-mail: cshpress@cshl.edu. For a complete catalog of all Cold Spring Harbor Laboratory Press publications, visit our World Wide Web Site http://www.cshlpress.com./

This book is dedicated to
Sara,
Rebecca,
Ethan,
and Ari

Contents

Preface

THE POLYMERASE CHAIN REACTION HAS BECOME one of the most widely used techniques in molecular biology, and it is very rare nowadays to find a molecular laboratory without a PCR machine. This technique has facilitated many traditional cloning experiments and has made possible procedures that were previously impossible, usually because of a lack of adequate amounts of nucleic acid. Coupling the information available from genome sequencing with the ability to synthesize DNA at will, PCR provides investigators with tools to address nearly every question in molecular biology. Since the first edition of *PCR Primer: A Laboratory Manual*, there have been some significant advances, as well as incremental improvements in technology and methods. This second edition has captured these advances and integrated them with the tried-and-true methods, which have been updated.

Our thanks and appreciation go to the authors of the chapters, who provided us with manuscripts and were patient with our edits to ensure continuity and consistency throughout the book. We thank the wonderful staff at the Cold Spring Harbor Laboratory Press that has been an essential part of this process: Inez Sialiano has cheerfully kept us on track and on schedule, and Patricia Barker has smoothed out the production process. As with the first edition, this book would not have been possible without Judy Cuddihy, whose advice, guidance, and assistance were essential for the completion of this project. Now it is up to you, the reader of this book, to apply what you know to advance knowledge and make the world a better place.

C.W.D.
G.S.D.

INTRODUCTION TO PCR

The polymerase chain reaction (PCR) is a deceptively simple technique. The process of repetitive bidirectional DNA synthesis via primer extension of a specific region of nucleic acid is so fundamental in design that it can be, and has been, applied in seemingly endless ways to solve problems in nearly every field of science. This basic description implies several critical processes: (1) that the target DNA is single stranded, and that conditions occur for target sequence-specific annealing of the primers; (2) that the primers are extended, converting the target DNA from single stranded to double stranded; and (3) that the double-stranded DNA denatures, beginning the process anew.

These processes are functionally programmed into the PCR machine so that each reaction proceeds through each step in an orderly manner. DNA denaturation occurs when the reaction is heated to 92–96°C. The time required to denature the DNA depends on its complexity, the geometry of the tube, the thermal cycler, and the volume of the reaction. For DNA sequences that have a high G+C content, the addition of betaine as described in "PCR Amplification of Highly GC-rich Regions" (p. 43) has been reported to improve the yield of the PCR.

After denaturation, the oligonucleotide primers hybridize to their complementary single-stranded target sequences. The temperature of this step depends on the length and base composition of the primers. In "PCR Primer Design" (p. 61), methods are described to select primers that have the maximum likelihood of success. Because primers hybridize nearly instantaneously to their target sequence at the proper annealing temperature, this step in the cycle can be manipulated to improve the yield and specificity of the PCR, as described in "Optimization and Troubleshooting in PCR" (p. 35).

The last step is the extension of the oligonucleotide primer by a thermostable polymerase. Traditionally, this portion of the cycle is carried out at 72°C. The time required to copy the template fully depends on the length of the PCR product. The time considerations as related to the length of the PCR product are described in "Tips for Long and Accurate PCR" (p. 53).

The major strength of PCR is its ability to specifically detect a vanishingly small amount of target nucleic acid sequence, even in the presence of a vast excess of other nucleic acid. Whether one is a novice or an expert, the amplification capacity of PCR can also cause tremendous problems.

CONTAMINATION ISSUES

The most serious issue with the widespread use of PCR is the contamination of new reactions with amplified product from old reactions or target nucleic acids. This can occur during several steps prior to the actual amplification reaction. The chapters "Dealing with Carryover Contamination in PCR: An Enzymatic Strategy" (p. 15) and "Setting up a PCR Laboratory" (p. 5) provide specific protocols to reduce contamination, and the latter chapter discusses how to integrate these methods into good laboratory practices and improved laboratory design.

To start at the beginning with laboratory design is a luxury that most of us cannot afford. In this regard, it is never too late to implement some basic procedures to minimize contamination. These are (1) physical separation of pre- and post-PCR activities; (2) separation of standard molecular biology from pre-PCR; (3) dedication of equipment and supplies solely to the pre-PCR area; and (4) judicious use of positive and negative controls.

Another critical step in contamination control is the sensible handling of reagents in the pre-PCR area. Rules for the safe handling of reagents and supplies in the PCR laboratory include:

1. Prepare solutions free from contaminating nucleic acids and/or nucleases (both DNases and RNases).

2. Use the highest-quality water in all PCR reagents. Water should be freshly distilled/deionized, filtered using a 0.22-micron filter, and autoclaved. USP-certified water that has been filtered and autoclaved is also sufficient for use in PCR. Aliquot the water so that a new bottle is used for each set of experiments.

3. Sodium azide, 0.025%, should be added to any reagent stored at room temperature. This does not inhibit the amplification reaction.

4. For consistency and reproducibility, all reagents should be made up in large volumes, then tested and, if acceptable, dispensed into single-usage volumes for storage.

5. Use only disposable, sterile bottles and tubes for all sample and reagent preparation.

6. Test all new sample preparation reagents before using on new, valuable specimens.

7. Carefully store all pipettes used in sample preparation and pre-PCR when not in use. Ziploc bags are good for this purpose.

8. Wear gloves at all times when preparing reagents, handling samples, and setting up reactions. The handling of the reaction after amplification must be in a different area of the laboratory.

9. Dedicated lab coats are required for both the pre-PCR and post-PCR areas. Keep the experimental flow from sample preparation, to pre-PCR, into post-PCR analysis.

STANDARD PCR

With all the improvements in PCR, what is currently considered standard PCR? PCR amplification of a template requires two oligonucleotide primers, the four deoxynucleotide triphosphates (dNTPs), magnesium ions in molar excess of the dNTPs, and a thermostable DNA polymerase to perform DNA synthesis. The target region to be amplified should be between 150 and 400 nucleotides. The quantities of oligonucleotide primers, dNTPs, and magnesium, and the final volume, may vary for each specific application.

The components of a standard PCR protocol using *Taq* DNA polymerase are as follows:

10x Enzyme-specific reaction buffer:
10–50 mM Tris-HCl, pH 7.5–9.0
6–50 mM KCl or $(NH_4)_2SO_4$
1.5–5.0 mM $MgCl_2$ or $MgSO_4$

0.2 mM of each dATP, dGTP, dCTP, and dTTP

0.1–1.0 μM of each oligonucleotide primer

2.0–2.5 units of a thermostable DNA polymerase

Nucleic acid template, 10^2–10^5 copies

Distilled water to 50 μl

CHOICE OF REAGENTS AND PROCEDURES

Depending on the application, *Taq* polymerase may not be the best choice. In the chapters "High-Fidelity PCR Enzymes" (p. 21) and "Tips for Long and Accurate PCR" (p. 53), the case is made for using enzymes from other thermophiles or mixtures of enzymes to deliver consistent results without risk of the appearance of mutation in the product. The methods proposed by Barnes in "Tips for Long and Accurate PCR" are sufficiently robust to easily amplify target sequences from 300 nucleotides up to very long products in the kilobase range.

For any of these reactions to work, there is a need to establish conditions for selective amplification of the target sequence. The determination of the temperatures to be used at each step in the PCR cycle and the length of the primer extension step are dependent on the base composition of the target sequence, the length of amplicon, and the annealing temperatures of the primers. Methods to define conditions for amplification are covered in "Optimization and Troubleshooting in PCR" (p. 35).

Another method that has proven very useful in the optimization of PCR is the use of the hot-start procedure. Hot-start PCR is defined as any method that prevents the primer extension of inappropriately bound primers. Although mispriming can be avoided in some situations by paying close attention to the primer design and by selecting the correct annealing temperature for the amplification reaction, it is also necessary to avoid having the components of the reaction at a temperature lower than the annealing temperature chosen for the specific amplification. This can be easily accomplished with hot-start methods. There are a number of recommended methods for this, including uracil-*N*-glycosylase and dUTP, a wax barrier, a wax bead impregnated with magnesium, an anti-*Taq* DNA polymerase monoclonal antibody, or the purchase of enzymes precomplexed with a heat-labile inhibitor. With all of these methods, after heating the reaction to 92°C for the first time, all the reaction components mix, and DNA synthesis occurs only from accurately hybridized primers. Hot-start methods are described in three chapters: "Dealing with Carryover Contamination in PCR: An Enzymatic Strategy" (p. 15); "High-Fidelity PCR Enzymes" (p. 21); and "Optimization and Troubleshooting in PCR" (p. 35); and "Amplification of RNA: High-temperature Reverse Transcription and DNA Amplification with a Magnesium-activated Thermostable DNA Polymerase (p. 211). Another important consideration in the optimization of PCR is the overall GC content of the target DNA. PCR depends on the complete denaturation of the target DNA each cycle. As discussed in "PCR Amplification of Highly GC-rich Regions" (p. 43), special methods are often needed to get the DNA to amplify.

With the best enzymes and buffers, PCR cannot proceed without quality template nucleic acid or PCR primers. Nucleic acid purification is such an important topic that it has a separate section. The "PCR Primer Design" chapter (p. 61) covers all aspects of design including the construction of Taqman and Scorpion probes for real-time PCR.

CONTROLS

Even with the best reagents, primers, and supplies, proper controls must be performed to test for PCR performance and presence of DNA contamination. As a rule, there should be a panel of negative, weak, and strong positive control samples to assay the efficiency and cleanliness of the sample preparation and pre-PCR process. The negative controls that are run alongside each set of samples should be constructed to assay for sample-to-sample contamination as well as for contamination with PCR products. The negative controls should include all reagents except the template DNA.

The amount of target DNA in positive controls should be minimized for two reasons. To test for robust amplification, only 10^1–10^4 copies of the target sequence should be included in the reaction to serve as a valid control. Second, by limiting the amount of DNA, there is less chance of this material serving as a source of contamination. When setting up the reactions, always pipette the template in the tubes designated as positive controls last.

PCR-BASED SEQUENCING

Finally, PCR has helped to revolutionize DNA sequencing. The ability of PCR-based methods to selectively amplify and sequence in the absence of radioactivity, with limited amounts of template, which has or has not been cloned into a vector, helped provide the systems to accomplish the Human Genome Project. A robust method is provided in "Nonradioactive Cycle Sequencing of PCR-amplified DNA" (p. 75).

The chapters in this section lay the foundation for harnessing the full capacity of PCR. Careful attention to laboratory practices and incorporation of contamination control methods will make PCR a seamless part of your laboratory.

1 Setting Up a PCR Laboratory

Theodore E. Mifflin

Department of Pathology, University of Virginia, Charlottesville, Virginia 22908

Development of the polymerase chain reaction (PCR) as a basic component of the molecular biology laboratory has occurred very rapidly from its inception in 1985. Since then, more than 15,500 articles have been published in which this technique was used. (See Table 1 for additional information sources for PCR.) As PCR became more widely used, scientists rapidly learned more about it and, as a result, learned that the PCR had its strong points and its deficiencies. Very quickly, PCR demonstrated its power to amplify very small amounts (e.g., a single copy) of template nucleic acid and to amplify different nucleic acids (e.g., DNA and RNA). At the same time, laboratory personnel learned that this biochemical reaction had a unique deficiency; namely, a strong susceptibility to contamination from its own product. Early experience with the PCR soon showed that additional precautions were needed (Lo et al. 1988; Kwok and Higuchi 1989). This chapter is devoted to establishing a PCR laboratory whose operations will give reliable and contamination-free results.

CONTAMINATION ISSUES

PCR contamination remains an issue for laboratories performing forensic procedures and detection of infectious agents (Pellett et al. 1999; Scherczinger et al. 1999). There are a number of approaches to control of PCR contamination, and the degree of stringency that is required in a laboratory is often determined by the assay being performed.

TABLE 1. Listing of Internet Web sites and URLs for PCR information

#	Web site (URL)	Comment
1.	http://highveld.com/pcr.html	PCR Jump station (Web portal)
2.	http://www.dnalc.org/resources/BiologyAnimationLibrary.htm	Downloadable animated video of PCR (PC or MAC)
3.	http://ncbi.nlm.nih.gov	GenBank site for checking amplimer (primer) sequences
4.	http://www.mbpinc.com (then select "Tech Reports" for monograph)	PCR contamination monograph
5.	http://info.med.yale.edu/genetics/ward/tavi/PCR.html	Guide to multiplex PCR

Amplicon Aerosol

The single most important source of PCR product contamination is the generation of aerosols of PCR amplicons that is associated with the post-PCR analysis. Methods for eliminating this aerosol range from physical design of laboratories and use of specific pipettes to chemical and enzymatic approaches. The choice of method is often dependent on the frequency of amplification of a target amplicon and the relative amounts and concentrations of the amplicons created by the PCR.

Target Template Contaminants

In addition to post-PCR contamination, the target template itself can be a source of contamination. For example, DNA templates are typically more troublesome as contaminants because they are more stable than RNA targets. Detection of infectious agents typically demands the most stringent contamination efforts, whereas detection of other targets, such as those from inherited disease, may require less contamination control. Regardless of the template to be detected, good laboratory practices should be followed (Kwok and Higuchi 1989) (see details below).

Real-time PCR Systems

PCR systems exist that provide direct measurement of amplicon accumulation during the reaction. These real-time PCR systems offer an alternative approach to the traditional post-PCR analysis methods. From a contamination control perspective, the collection of data during the amplification reaction by using a fluorescence-based detection system eliminates the need to handle the sample. Thus, when these PCRs are completed, the detection and analysis are complete, the reaction tubes remain sealed, and there is no amplicon escape. Arranging the PCR laboratory to perform these homogeneous (or real-time) PCRs requires a different approach, which is addressed later in this chapter. Because the major use of PCR as a laboratory tool still depends on a separate post-amplification manipulation, this approach is the one primarily addressed.

CONTAMINATION PREVENTION APPROACHES IN THE PCR LABORATORY

The PCR laboratory typically is involved with activities that include sample preparation, PCR reaction assembly, PCR execution, and post-PCR analysis. These activities are summarized in Figure 1. When arranged in this linear fashion, these activities can be collected into two major groups, the pre-PCR activities (sample preparation and PCR preparation) and the post-PCR activities (PCR execution and analysis).

Use of the PCR for research and diagnostic purposes requires that some additional procedural limitations be observed so that the reaction yields valid results. As awareness of the PCR's susceptibility to contamination became known, Kwok and Higuchi (1989) presented some additional guidelines for researchers using the PCR. Consistently observing these guidelines is essential for successfully operating a PCR laboratory on a long-term basis. They form part of a network of protocols focused on maintaining a PCR laboratory in a contamination-free condition.

Contamination can arise from several different sources, such as previous amplification and purification of plasmid clones, repeated isolation of template (genomic) nucleic acids, and previously amplified molecules ("amplicons"). Although most attention in a PCR lab-

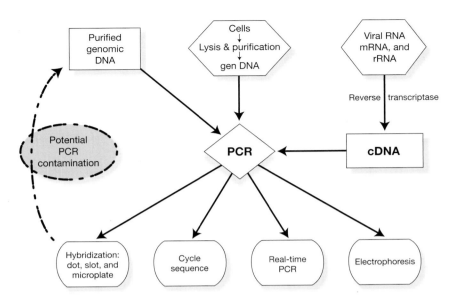

FIGURE 1. Outline of sample processing and analysis in a PCR laboratory.

oratory is focused on the last category of contamination, the other two sources should not be ignored. A prudent approach to controlling their contamination is to segregate the more standard recombinant DNA activities into separate areas of the laboratory and, in particular, to limit the performance of PCR activities to its own area.

For PCR amplicon contamination, it is the control and removal of the PCR amplicons that form the basis for the contamination control program. When PCR is used in research laboratories, either a greater variety of templates or amplifications of a specific template will be studied or manipulated and, thus, controlling amplicon contamination may be less challenging. In a diagnostic laboratory, there can be more opportunities for PCR contamination due to the repeated analysis of selected templates and the fact that PCR assays may be performed at or near the detection limit of PCR. The last possibility is especially demanding and thus requires a much more rigorous approach to controlling PCR contamination.

The essential parts of this contamination control program include space and time separation of pre- and post-PCR activities, use of physical aids, use of ultraviolet (UV) light, use of aliquoted PCR reagents, incorporation of numerous positive and negative or blank PCRs (H_2O substituted for template), and use of one or more various contamination control methods that use chemical and biochemical reactions. The underlying theme in these actions is the recognition that amplicon contamination cannot be seen, felt, or a priori detected before it happens. Use of consistent, careful technique coupled with liberal incorporation and monitoring of PCR blanks will ensure a vigilant, proactive approach to PCR contamination.

Space and Time Separation

As illustrated in Figure 1, the main source of the feedback contamination is the amplicons generated by the previous PCR. By separating the source of the amplicons' (e.g., post-PCR) activities from the pre-PCR activities, the potential for contamination is significantly reduced. This separation is best illustrated by separating the facilities in space, so that there are two rooms where these activities occur (Fig. 2). If this is not achievable, different areas designated for sample preparation and PCR setup can be located away from the area for

PCR Laboratory Organization

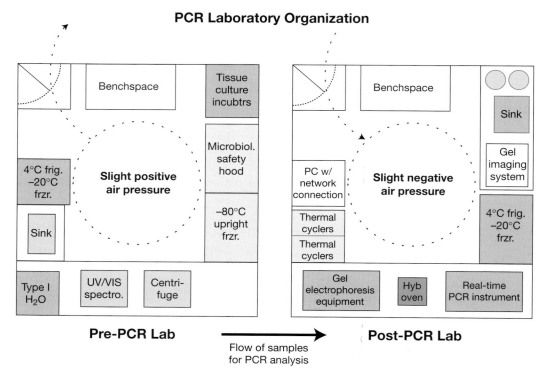

FIGURE 2. Organization of a PCR laboratory with separate pre- and post-PCR rooms.

post-PCR analysis. If all activities are to be performed in a single room, sample preparation should occur inside a laminar flow hood, preferably equipped with a UV light. The walls of the hood should be wiped with a fresh (or freshly made) 10% bleach solution (1 part regular bleach: 9 parts water) before processing samples or preparing PCR samples. Waste materials that contain PCR amplicons should not be allowed to accumulate in an area that is also frequented by other personnel who may be eventually involved with template isolation and purification. Additionally, the laboratory should consider establishing a daily schedule for performing PCR. Sample preparation and pre-PCR should be morning activities. Once completed, the pre-PCR supplies and equipment can be stored, and the afternoon can be devoted to the post-PCR analysis.

Laboratory Space Arrangement

As mentioned above, the ideal arrangement of the PCR facility is to have the pre- and post-PCR areas located in separate rooms (see Fig. 2), each with dedicated resources. A source of deionized water needs to be present in both rooms, as well as dedicated centrifuges, storage freezers/refrigerators, and storage of supplies. Even telephones, computers, and other electronic communications should also be dedicated.

Rarely is laboratory space allocated strictly for PCR; it often must be shared with other procedures. In a research laboratory, cell culture and other biochemical protocols such as analysis for enzymes and other biochemical species, protein and nucleic acid isolation, cloning, transfection, and purification may also occur. Thus, PCR protocols may likely be integrated into the laboratory's operation with consequent need to share facilities and bench space.

In the case of the diagnostic laboratory, a much more limited menu of activities occurs that is focused primarily on sample isolation and processing, although other possible activities such as cloning or sequencing can happen as well. There is usually a greater demand to detect and quantify low amounts of template in biological samples in the diagnostic laboratory, and this places a greater need on preventing amplicon contamination proactively.

A research institution may be unable to have separate facilities unless an arrangement can be developed for a "core" PCR setup facility, with each laboratory performing its own PCR and subsequent amplification. Where there is a requirement for detection of very low amounts of template nucleic acid (DNA or RNA), separate facilities offer the greatest likelihood that contamination-free results can be obtained over a long time frame.

An alternative to two-facility arrangements is found in real-time PCR methods. Because the results of each PCR are provided throughout the reaction, there is no need to open the PCR containers, and these can be discarded while still sealed. At least one instrument manufacturer is developing a completely self-contained automated system (the AmpliPrep: Taqman from Roche Molecular Systems, Alameda, CA) that would eliminate the need for two separate rooms. However, there may be situations where the contents of individual tubes and capillaries created during real-time PCR analysis need to be analyzed independently, and so these containers would need to be opened. This should happen in a different room from where the real-time PCRs were set up and performed.

Equipment in PCR Laboratories

To ensure that pre-PCR and post-PCR events remain separated, each room must have its own separate set of equipment, including pipettors, reagents, pipettor tips, racks, and so forth. Moreover, these items should not leave the area to which they are assigned. Each should be labeled as to location and used in that location only. Lab coats should be dedicated for both areas as well. Because pipetting forms the basis for most PCR analysis, each area needs its own dedicated pipettors that are never exchanged between work areas. To assist with this, color-coded pipettors (e.g., green for pre-PCR work, red for post-PCR work) can be used. When pre-PCR pipettors and tips are not in use, they should be stored in airtight bags to keep them clean. Reactions should be constructed using master mixes, and the template should always be added last using positive displacement tips to prevent pipettors from becoming cross-contaminated while pipetting samples that contain template. These types of pipettors and tips are available from several sources and can be purchased in sterilized packs. It is important to remember that barrier tips cannot be autoclaved.

Pre-PCR Activities

The definition of pre-PCR is the protocols and equipment required for the isolation of nucleic acid and the assembly of the reaction to amplify the samples. During the last 10 years, there has been much progress in developing devices that perform these activities in an automated fashion. Most PCR laboratories still perform these tasks using manual procedures.

What is the minimum needed to equip a PCR laboratory for sample preparation, PCR reagent preparation, and PCR assay setup? Because most of the activities revolve around pipetting of liquids, these activities should be examined most closely; in particular, the manual pipettors and pipette tips. As discussed previously, positive displacement tips or barrier methods should be used to pipette the template into the reactions as the last step.

There is a risk of creating aerosols in the preparation for RNA and DNA templates. If a large number of specimens of one type are processed on a routine basis, the laboratory may wish to treat this method with care and perform it in a hood or biosafety cabinet (Fig. 2). Because of the effectiveness of ultraviolet light (UV) for amplicon control, use of UV inside the cabinet prior to sample preparation or PCR reagent preparation is advisable. Alternatively, any one of a number of small, benchtop-size cabinets that use UV irradiation can also be utilized. These are dedicated to PCR use and are large enough to contain several pipettors, racks, and some reagents.

Environmental Considerations

- *Air handling.* For extremely high-performance PCR laboratories that will be involved with detecting very-low-prevalence DNA or RNA molecules (e.g., infectious disease agents in clinical samples), additional measures may be necessary to prevent contamination from the air being recirculated between the pre- and post-PCR laboratories. In this case, the air handlers need to be separate and the air pressure individually adjusted in each laboratory. In the pre-PCR laboratory, there should be a slight positive pressure compared to the air in the connecting hallway. The post-PCR laboratory, in contrast, should be at slightly reduced pressure to pull air in from the outside and thereby prevent escape of amplicons from the completed PCR samples being analyzed inside the lab (Fig. 2). Finally, the air handlers for the pre- and post-PCR laboratories need to be connected to separate air ducts, and each must lead to a separate location for exhaust.

- *UV irradiation.* It is possible to exploit further the sensitivity of nucleic acid to UV by using UV to sterilize the entire pre-PCR laboratory. This can be done by having UV lights placed in the ceiling fixtures and connecting their activation to a lock-out mechanism on the exit door so they only illuminate when the last person in the lab closes and locks the external lab door. If this type of hardware is installed, it must be accompanied by a ventilation system to eliminate the UV-generated ozone and a rigidly enforced schedule of monitoring the performance of the UV bulbs. These light fixtures accumulate a residue arising from the precipitation of oxidation products on the glass of the bulb. If this is not removed monthly, the UV system is not effective.

- *Protective clothing.* To further prevent PCR amplicons from leaving the post-PCR lab, each investigator should have a dedicated post-PCR lab coat. Additionally, each investigator should have a general molecular biology lab coat and a separate coat for pre-PCR. In extreme cases, a disposable gown and booties should be worn.

- *Adhesive paper at lab entrances.* This approach effectively prevents trace amounts of dust and debris from entering the laboratory. It is a rather expensive approach to controlling contamination, but may be worth the expense for selected applications.

Sterilization of Reagents

Because PCR laboratories perform some molecular biology methods that require sterile reagents, some may need to be autoclaved. The single most critical reagent is water. Sterile USP water can be quickly converted to PCR water by filtering it through two 0.45-micron nitrocellulose filters. These filters have a very high binding capacity for nucleic acid and proteins. If the laboratory is involved in amplification of very small quantities of bacterial DNA, the USP water should be autoclaved separately from all other reagents before filtration. In general, reagents and solid items destined for the pre-PCR lab should be autoclaved separately from other supplies. It is important to note that spent tissue culture fluids, bacterial culture supernatants, bacterial media plates with recombinant cultures or plasmids, and samples from the post-PCR lab represent a large potential reservoir of contaminating DNA and should also be autoclaved separately from any material that will enter the pre-PCR lab.

Contamination Control

As mentioned earlier, a variety of approaches can be used to control PCR amplicon contamination. They can be grouped into two broad methods: (1) methods that use physical means to prevent dispersion of PCR amplicons and (2) methods that exploit some type of chemical reaction to render the amplicons incapable of serving as templates in a new round of PCR. Each of these has a place in the PCR laboratory, and most successful PCR laboratories use a spectrum of these methods to effectively control contamination.

Physical Methods

This category includes the physical barrier approaches. The most popular is the use of either positive displacement or barrier pipette tips to prevent aerosols. These barrier methods prevent the reintroduction of small amounts of a contaminating aerosolized sample into the next sample that is pipetted (Fig. 3). Use of these tips is generally recommended in the pre-PCR areas of the laboratory where samples are being processed and template nucleic acids (DNA and RNA) are being isolated and purified. Use of these tips is necessary and cost-effective in the post-PCR laboratory because there is already a large amount of amplicon present.

An adjunct to these tips is the use of a laminar flow hood or biological safety cabinet to facilitate preparation of PCR samples and reagents. When they are prepared in such an enclosure, there is much less chance of an external source of PCR amplicon contaminating the samples and reagents being manipulated for the subsequent PCRs.

Chemical Methods

A number of chemical approaches have been developed during the last 20 years. However, only a few have seen any real success in becoming routine in their use and application for controlling PCR contamination.

- *UV photolinking.* This approach can be used in both a pre-PCR and post-PCR setting. The basis for this reaction is that adjacent pyrimidines on a DNA strand can be cross-linked when exposed to UV light of 254 nm (Gordon and Haseltine 1982). The reaction is very fast and can be effective for bigger amplicons; i.e., those greater than 700 bp. Smaller amplicons are harder to inactivate because there are fewer adjacent pyrimidines. There is also a role for interstrand cross-linking in UV inactivation. Once cross-linked, the pyrimidine dimers cannot be excised and so the DNA polymerase is sterically blocked, or the DNA cannot completely denature, and the synthesis reaction is effectively halted.

    UV photolinking is most often used in a pre-PCR setting in which the equipment is installed in a small tabletop cabinet that is used for sample preparation. All of the items to create a PCR are placed inside the cabinet and then illuminated before the PCRs are assembled. There are some caveats with this approach, however, that are worth mentioning. First, there is a safety concern about exposure to UV light, and this must be addressed. Second, the photoreaction favors thymidine over cytidine by about 10:1.

Use of Open Pipet Tips Leads to Pipettor Contamination

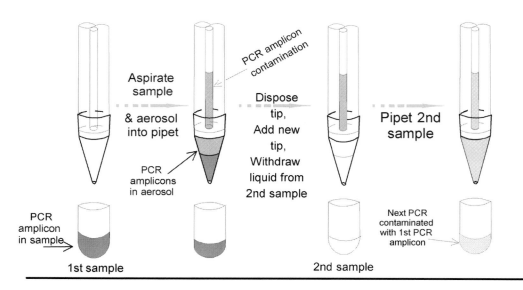

Use of Barrier Pipet Tips Prevents Pipettor Contamination

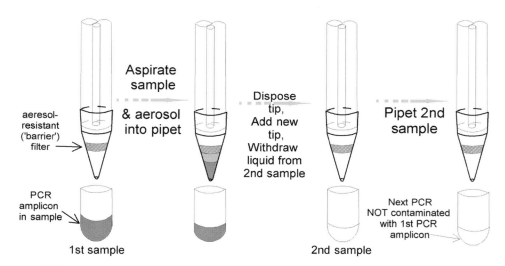

FIGURE 3. Use of barrier tips to prevent amplicon contamination in the PCR laboratory.

Therefore, amplicons that are AT-rich are more efficiently disabled than AT-poor sequences. Third, decreasing length of the amplicon usually gives a lower rate of protection; therefore, short amplicons are not well controlled. The approach is nevertheless effective and should be used when possible.

- *Uracil-DNA-glycosylase.* This enzyme (also known as UDG) is very effective at destroying PCR amplicons when vigorously used for sample preparation (Longo et al. 1990; Thornton et al. 1992). The method is so important in contamination control that it is covered separately in Chapter 2. Briefly, at the pre-PCR step, dTTP is substituted with

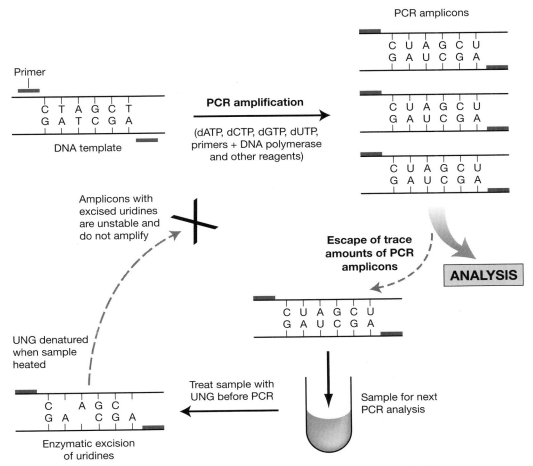

FIGURE 4. Use of the UDG reaction to prevent PCR contamination.

dUTP, and UDG is included in the reaction mix. All other reaction components remain the same. During the PCR, DNA polymerase substitutes dU for dT in the growing DNA strand. In the final product, there is now dU instead of dT in the DNA sequence. Before any new sample is processed, it first is exposed to UDG enzyme. If UDG comes across any U-containing DNA strands, the U's are cleaved, leaving the strand with gaps. Following heating in the next PCR, the abasic strands fall apart and cannot be amplified. The use of UDG provides the added advantage of a hot start by degrading all PCR products made prior to the first full cycle. This is discussed in Chapters 2, 3, 4, and 16.

Most PCR laboratories use one or more of these reactions in conjunction with laboratory operation and design to control PCR amplicon contamination effectively (Newton 1995). Selection of the best combination of these techniques is driven by factors such as types of templates, their prevalence, and others, as discussed earlier in this chapter.

PATENTS AND LICENSES

The use of PCR is covered by U.S. patents that were issued in the mid-1980s (Mullis 1987; Mullis et al. 1987). If results from PCR are to be used for diagnostic purposes and/or a fee for service will be arranged, then an additional license(s) will need to be negotiated for the use of the PCR with Roche Molecular Systems (Alameda, CA). This arrangement may also be necessary for other in vitro amplification reactions such as nucleic acid sequence-based amplification (NASBA; BioMerieux, Durham, NC), transcription-mediated amplification (TMA; GenProbe, San Diego, CA), and strand displacement amplification (SDA; Becton Dickinson, Franklin Lakes, NJ) and their corresponding licensing organizations.

Not only are licenses required for the use of the amplification reaction(s), but there are also some PCR-based diagnostic tests for specific genes that are patented and therefore require an additional use license to be negotiated. In each case, laboratory personnel should clarify the license issue(s) prior to performing and offering test results.

SUMMARY

This chapter has provided a brief overview of essential items that need to be considered when creating a PCR laboratory from the "bare walls." Many times, some of the needed resources are already present, making the conversion process easier. However, it should still be recognized that a PCR laboratory has its own unique requirements, and those should be carefully considered when evaluating what is needed to start and operate this kind of laboratory successfully.

REFERENCES

Gordon L.K. and Haseltine W.A. 1982. Quantitation of cyclobutane pyrimidine dimer formation in double- and single-stranded DNA fragments of defined sequence. *Radiat. Res.* **89:** 99–112.

Kwok S. and Higuchi R. 1989. Avoiding false positives with PCR (erratum *Nature* [1989] **339:** 490). *Nature* **339:** 237–238.

Lo Y.M., Mehal W.Z., and Fleming K.A. 1988. False-positive results and the polymerase chain reaction. *Lancet* **2:** 679.

Longo M.C., Berninger M.S., and Hartley J.L. 1990. Use of uracil DNA glycosylase to control carry-over contamination in polymerase chain reactions. *Gene* **93:** 125–128.

Mullis K.B. 1987. Process for amplifying nucleic acid sequences. U.S. Patent #4,683,202.

Mullis K.B., Erlich H.A., Arnheim N., Horn G.T., Saiki R.K., and Scharf S.J. 1987. Process for amplifying, detecting, and/or cloning nucleic acid sequences. U.S. Patent #4,683,195.

Newton C.R. 1995. Setting up a PCR laboratory. In PCR: Essential data. (ed. C.R. Newton), p. 216. Wiley, New York.

Pellett P.E., Spira T.J., Bagasra O., Boshoff C., Corey L., de Lellis L., Huang M.L., Lin J.C., Matthews S., Monini P., Rimessi P., Sosa C., Wood C., and Stewart J.A. 1999. Multicenter comparison of PCR assays for detection of human herpesvirus 8 DNA in semen. *J. Clin. Microbiol.* **37:** 1298–1301.

Scherczinger C.A., Ladd C., Bourke M.T., Adamowicz M.S., Johannes P.M., Scherczinger R., Beesley T., and Lee H.C. 1999. A systematic analysis of PCR contamination. *J. Forensic Sci.* **44:** 1042–1045.

Thornton C.G., Hartley J.L., and Rashtchian A. 1992. Utilizing uracil DNA glycosylase to control carryover contamination in PCR: Characterization of residual UDG activity following thermal cycling. *BioTechniques* **13:** 180–184.

2 Dealing with Carryover Contamination in PCR: An Enzymatic Strategy

Kyusung Park and Jun Lee

Research and Development, Invitrogen Corporation, Carlsbad, California 92008

Since its introduction, the polymerase chain reaction has found a solid niche in broad areas of the academic, industrial, forensic, and medical communities where people routinely handle nucleic acids. Applications in these areas have developed mainly as a result of PCR's high sensitivity. This sensitivity, however, has proven time and time again to be a double-edged sword. The same high sensitivity that allows a minute quantity of valuable DNA to be amplified also facilitates the amplification of minute quantities of contaminating DNA. When PCR is used for diagnostic or forensic applications, the presence of such contaminants can be disastrous. Repeated use of PCR and manipulation of the products can cause the reaction mixture to form an aerosol, a mist of droplets that provides an easy means of contaminating neighboring samples and work areas. Contrary to popular belief, the production of aerosols is not rare. Such contamination arising from aerosols is referred to as "carryover contamination."

IRRADIATION AND ADDITION OF PHOTOREACTIVE CHEMICALS

A few general precautions in the laboratory, such as physical isolation of a space and equipment for sample preparation from the rest of the laboratory, can reduce the incidence of carryover contamination (Victor et al. 1993); however, more practical and effective ways for its prevention have been sought. The most effective way to eliminate the contaminants would be by destroying the foreign DNA without affecting the genuine target sequences, or the PCR product itself, while minimizing manipulation of the sample. Irradiation of PCR products with UV or gamma rays is a relatively simple way to inactivate DNA in solution or in a dry state (Deragon et al. 1990; Sarkar and Sommer 1990; Fox et al. 1991). However, the process is inefficient and, although it can be used to treat some PCR reagents (e.g., PCR buffers and water), it destroys target DNA, PCR enzymes, and primers. These reagents must be subjected to an alternative decontamination procedure.

The addition of certain photoreactive chemicals while irradiating the PCR product can render DNA unavailable for further amplification, or "sterile," by cross-linking the complementary strands together or forming monoadducts (Cimino et al. 1991; Isaacs et al. 1991). The most widely used chemicals for this purpose are psoralen and its derivatives, e.g., isopsoralen (IP-10). Photochemical inactivation methods are more effective than UV irradiation (Isaacs et al. 1991). However, high concentrations of psoralen compounds (>25 µg/ml) inhibit PCR and necessitate the inclusion of glycerol, bovine serum albumin (BSA), or dimethyl sulfoxide (DMSO) in the PCR mixture (Isaacs et al. 1991; Rys and Persing 1993).

To avoid this inhibitory effect, post-PCR addition of IP-10 was tested. To prevent the formation of aerosols, the reactions were compartmentalized using AmpliWax. After amplification, the reactions were cooled, and IP-10 was inoculated on top of the solid AmpliWax. The tubes were then reheated, melting the AmpliWax and allowing the IP-10 to enter the PCR. This treatment effectively "sterilized" the PCR products without inhibiting PCR (De la Viuda et al. 1996).

NUCLEASE TREATMENT

Another way to avoid carryover contamination is to treat the reaction mixtures with nucleases. The most widely used enzymes for this are restriction endonucleases (Furrer et al. 1990; DeFillipes 1991), DNase I (Furrer et al. 1990), and exonuclease III (Zhu et al. 1991). Multiple restriction enzymes can be used to enhance the effectiveness, but again, the target DNA cannot be subjected to this type of decontamination. Primers, as single-stranded oligonucleotides, are resistant to restriction endonuclease digestion (unless partial annealing creates the consensus sequence for the enzyme), but they are still susceptible to DNase I and exonuclease III. Additional drawbacks to this method include the need for extensive manipulation. For example, the nuclease reaction buffer is unlikely to be compatible with PCR. For optimum results, multistep buffer exchanges must be performed prior to PCR. Increased handling of reagents will inevitably lead to greater risk of contamination.

CHEMICAL DECONTAMINATION

Modifying PCR products with mutagenic agents, such as hydroxylamine hydrochloride (Aslanzadeh 1993), or degrading with hydrochloric acid or sodium hypochlorite (bleach), has been shown to decontaminate PCR reagents (Prince and Andrus 1992). Decontamination of work areas with bleach also reduces carryover contamination. However, because of the corrosive nature of these chemicals, they are of limited use. Highly reactive agents, such as hydrogen peroxide, can also be used to decontaminate by nicking or cleaving DNA, but removal of such chemicals after decontamination is not a trivial matter.

DISTINGUISHING TEMPLATE DNA AND PCR PRODUCTS

The ideal method of preventing carryover contamination would distinguish template DNA and PCR products. Such a method would drastically reduce or even remove the possibility of carryover contamination. One way of differentiating product from template is to include dUTP in place of dTTP in the PCR (Longo et al. 1990). This would result in amplified DNA that contains uracil residues, which can be selectively destroyed by the action of uracil DNA glycosylase (UDG). This enzyme hydrolyzes uracil glycosidic bonds, thus rendering the amplicon unamplifiable by DNA polymerases. Unfortunately, it has been shown that the proofreading DNA polymerases tested are inhibited by the presence of dUTP in the PCR (Barnes 1994; Cheng et al. 1994; Lasken et al. 1996; see also Chapter 3, this volume). The presence of uracil residues in DNA may or may not affect post-PCR procedures using the amplified products. Several restriction endonucleases do not digest DNA segments containing uracil residues (Beebe et al. 1992). It is necessary to use an *ung⁻* strain of the *Escherichia coli* host for cloning of a uracil-containing DNA fragment because endogenous UDG in a wild-type host would inactivate the incoming DNA. Otherwise, uracil-containing DNA should be considered normal in most aspects (Fig. 1) (Bodnar et al. 1983; Wang et al. 1992).

First round of PCR

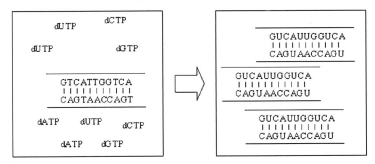

PCR in presence of dUTP

Decontamination prior to next round of PCR

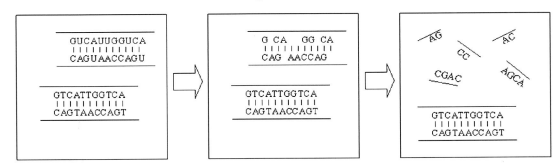

UDG treatment before PCR   94°C preincubation before PCR

FIGURE 1. Schematic representation of representative uracil-UDG decontamination steps in PCR. Uracil residues are incorporated into PCR-amplified DNA in a reaction mixture with dUTP in place of dTTP. In subsequent reactions where a uracil-containing amplicon might be carried over, preincubation at 37°C for 10 minutes followed by a 94°C incubation for 10 minutes selectively degrades the carryover contaminants while template DNA and primer remain intact.

There are two ways to utilize uracil-UDG methodology. PCR can be carried out using uracil-containing primers, or dUTP can be incorporated into the amplicon during PCR. If uracil residues are present in the primer, each DNA strand of the amplicon will contain uracil residues at its 5′ end. UDG treatment of the amplicon results in truncation of these 5′ sequences, thus destroying the primer-binding site and preventing further amplification (Longo et al. 1990). Similar results can be obtained by modifying primers with ribonucleotides close to the 3′ end (Rys and Persing 1993; Walder et al. 1993). Alkali treatment of PCR products would truncate the 5′ region, thus inhibiting further amplification. In these methods, complete inactivation of UDG or neutralization of alkali is required before the addition of primers. Note that the primers could be a source of contamination.

PROTOCOL

Incorporation of dUTP into the amplicon during PCR is the most effective and widely used method of decontaminating PCR products (Longo et al. 1990; Hartley and Rashtchian 1995). The following protocol describes the procedure for the incorporation of dUTP in PCR amplicons and subsequent treatment with UDG.

MATERIALS

BUFFERS AND SOLUTIONS

dNTP solution (containing dATP, dCTP, dGTP, and dUTP, each at 10 mM)

> For most applications, direct substitution of dUTP should suffice (Rys and Persing 1993; Kox et al. 1994). If necessary, to compensate for inhibition of the polymerase, the final concentration of dUTP may be raised to 0.6 to 1 mM (Wang et al. 1992; Hohlfeld et al. 1994).

MgCl$_2$ solution<!>, 50 mM

> A final Mg^{++} concentration of 1.5 mM should suffice in most cases. However, this concentration may require optimization, depending on the nucleotide (e.g., dUTP) concentrations and the primer specificity. Increased nucleotide concentration should be matched by increased Mg^{++}.

PCR buffer, 10x (100 mM Tris-HCl<!>, pH 8.4, 500 mM KCl<!>)

NUCLEIC ACIDS AND OLIGONUCLEOTIDES

Forward primer (10 μM) in H$_2$O
Reverse primer (10 μM) in H$_2$O

ENZYMES AND ENZYME BUFFERS

Uracil DNA glycosylase, 1 unit/μl (Invitrogen)

> One unit of UDG will remove up to 5 ng of carryover contaminants using this protocol. This can be reduced to 0.01 unit depending on the extent of decontamination required (Kox et al. 1994).

Taq DNA polymerase, 5 units/μl

> Hot-start Platinum Taq DNA polymerase, 5 units/μl (Invitrogen 10 966-018) allows reactions to be prepared at room temperature.

SPECIAL EQUIPMENT

Thermal cycler programmed with desired amplification protocol (note modifications listed in steps 3 and 4)
Microcentrifuge tubes (thin-walled for DNA amplification)

METHOD

1. In a sterile microfuge tube, on ice, assemble the following PCR components for each reaction:

PCR buffer, 10x	5 μl
dATP, dCTP, dGTP, dUTP solution (each at 10 mM)	1 μl
Forward and reverse primers (10 μM each)	1 μl
Uracil DNA glycosylase, 1 unit/μl	1 unit
Taq DNA polymerase, 5 units/μl	0.5 μl
MgCl$_2$, 50 mM	1.5 μl
H$_2$O	40 μl

2. Incubate at 37°C for 10 minutes prior to temperature cycling.

> UDG removes uracil bases from carryover contamination during this step. It is reported that decontamination can also be achieved at room temperature (de Wit et al. 1993; Wang et al. 1992).

3. Increase the pre-PCR denaturation step at 94°C to 10 minutes to inactivate UDG and to hydrolyze the contaminant DNA (Hohlfeld et al. 1994; Kox et al. 1994).

4. Add two additional cycles to the number of cycles in the normal PCR program.

   These additional cycles compensate for the reduced efficiency of amplification in the presence of dUTP (Longo et al. 1990).

5. Maintain the final temperature at 72°C upon completion of temperature cycling until further usage/storage.

   UDG from *E. coli* has been reported to recover a small fraction of its catalytic activity when it is returned to lower temperatures after temperature cycling (Thornton et al. 1992).

6. Store reaction mixtures with uracil-containing DNA and residual UDG activity at –20°C until use.

DISCUSSION

Carryover contamination is a major cause of false-positive results when primer pairs are used repeatedly in DNA amplification. The UDG decontamination procedure presented here has been used successfully by several groups and applied in several PCR-based diagnostic tests: for example, the detection of *Mycobacterium tuberculosis* (Nolte et al. 1993), *Mycobacterium leprae* (de Wit et al. 1993), human immunodeficiency virus (HIV) (Butcher and Spadoro 1992), and *Borrelia burgdorferi* (Dodge et al. 1992). All of Roche's PCR diagnostic tests utilize a decontamination protocol based on UDG application.

DNA polymerases with a 3´ exonuclease (proofreading) activity, such as *Pfu*, Vent, and DeepVent DNA polymerases, however, are inhibited by the presence of dUTP in the reaction (Barnes 1994; Cheng et al. 1994; Lasken et al. 1996). This inhibition is thought to explain why UDG decontamination could not be applied to long PCR using a proofreading DNA polymerase in combination with *Taq* DNA polymerase. Proofreading DNA polymerases are extremely sensitive to even small amounts of dUTP. Even the small amount of dUTP present in a PCR mixture arising from deamination of dCTP during temperature cycling is reported to reduce performance (Hogrefe et al. 2002).

The high efficacy of the UDG decontamination protocol cannot substitute for good laboratory practices. However, a combination of careful working practices and an effective decontamination protocol can control the problem of carryover contamination.

REFERENCES

Aslanzadeh J. 1993. Application of hydroxylamine hydrochloride for post-PCR sterilization. *Mol. Cell. Probes* **7:** 145–150.

Barnes W.M. 1994. PCR amplification of up to 35-kb DNA with high fidelity and high yield from lambda bacteriophage templates. *Proc. Natl. Acad. Sci.* **91:** 2216–2220.

Beebe R.L., Thornton C.G., Hartley J.L., and Rashtchian A. 1992. Contamination-free polymerase chain reaction: Endonuclease cleavage and cloning of dU-PCR products. *Focus* **14:** 53–56.

Bodnar J.W., Zempsky W., Warder D., Bergson C., and Ward D.C. 1983. Effect of nucleotide analogs on the cleavage of DNA by the restriction enzymes AluI, DdeI, HinfI, RsaI, and TaqI. *J. Biol. Chem.* **258:** 15206–15213.

Butcher A. and Spadoro J. 1992. Using PCR detection of HIV-1 infection. *Clin. Immunol. Newslett.* **12:** 73–76.

Cheng S., Fockler C., Barnes W.M., and Higuchi R. 1994. Effective amplification of long targets from cloned inserts and human genomic DNA. *Proc. Natl. Acad. Sci.* **91:** 5695–5699.

Cimino G.D., Metchette K.C., Tessman J.W., Hearst J.E., and Isaacs S.T. 1991. Post-PCR sterilization: A method to control carryover contamination for the polymerase chain reaction. *Nucleic Acids Res.* **19:** 99–107.

DeFillipes F.M. 1991. Decontaminating the polymerase chain reaction. *BioTechniques* **10**: 26–29.

De la Viuda M., Fille M., Ruiz J., and Aslanzadeh J. 1996. Use of AmpliWax to optimize amplicon sterilization by Isopsoralen. *J. Clin. Microbiol.* **34**: 3115–3119.

Deragon J.M., Sinnett D., Mitchell G., Potier M., and Labuda D. 1990. Use of gamma irradiation to eliminate DNA contamination for PCR. *Nucleic Acids Res.* **18**: 6149.

de Wit M.Y., Douglas J.T., McFadden J., and Klatser P.R. 1993. Polymerase chain reaction for detection of *Mycobacterium leprae* in nasal swab specimens. *J. Clin. Microbiol.* **31**: 502–506.

Dodge D.E., Nersesian R., and Sun R. 1992. Diagnosis of the Lyme disease spirochete *Borrelia burgdorferi. Clin. Immunol. Newsl.* **12**: 69–73.

Fox J.C., Mouir A.K., Webster A., and Emery V.C. 1991. Eliminating PCR contamination: Is UV irradiation the answer? *J. Virol. Methods* **33**: 375–382.

Furrer B., Candrian U., Wieland P., and Luthy J. 1990. Improving PCR efficiency. *Nature* **346**: 324.

Hartley J.L. and Rashtchian A. 1995. Enzymatic control of carryover contamination in PCR. In *PCR primer* (ed. C.W. Dieffenbach and G.S. Dveksler), pp. 23–29. Cold Spring Harbor Laboratory Press, Cold Spirng Harbor, New York.

Hogrefe H.H., Hansen C.J., Scott B.R., and Nielson K.B. 2002. Archaeal dUTPase enhances PCR amplifications with archaeal DNA polymerases by preventing dUTP incorporation. *Proc. Natl. Acad. Sci.* **99**: 596–601.

Hohlfeld P., Daffos F., Costa J.M., Thulliez P., Forestier F., and Vidaud M. 1994. Prenatal diagnosis of congenital toxoplasmosis with a polymerase-chain-reaction test on amniotic fluid. *N. Engl. J. Med.* **331**: 695–699.

Isaacs S.T., Tessman J.W., Metchette K.C., Hearst J.E., and Cimino G.D. 1991. Post-PCR sterilization: Development and application to an HIV-1 diagnostic assay. *Nucleic Acids Res.* **19**: 109–116.

Kox L.F., Rhienthong D., Miranda A.M., Udomsantisuk N., Ellis K., van Leeuwen J., van Heusden S., Kuijper S., and Kolk A.H. 1994. A more reliable PCR for detection of *Mycobacterium tuberculosis* in clinical samples. *J. Clin. Microbiol.* **32**: 672–678.

Lasken R.S., Schuster D.M., and Rashtchian A. 1996. Archaebacterial DNA polymerases tightly bind uracil-containing DNA. *J. Biol. Chem.* **271**: 17692–17696.

Longo M.C., Berninger M.S., and Hartley J.L. 1990. Use of uracil DNA glycosylase to control carryover contamination in polymerase chain reactions. *Gene* **93**: 125–128.

Nolte F.S., Metchock B., McGowan J.E., Jr., Edwards A., Okwumabua O., Thurmond C., Mitchell P.S., Plikaytis B., and Shinnick T. 1993. Direct detection of *Mycobacterium tuberculosis* in sputum by polymerase chain reaction and DNA hybridization. *J. Clin. Microbiol.* **31**: 1777–1782.

Prince A.M. and Andrus L. 1992. PCR: How to kill unwanted DNA. *BioTechniques* **12**: 358–360.

Rys P.N. and Persing D.H. 1993. Preventing false positives: Quantitative evaluation of three protocols for inactivation of polymerase chain reaction amplification products. *J. Clin. Microbiol.* **31**: 2356–2360.

Sarkar G. and Sommer S.S. 1990. Shedding light on PCR contamination. *Nature* **343**: 27.

Thornton C.G., Hartley J.L., and Rashtchian A. 1992. Utilizing uracil DNA glycosylase to control carryover contamination in PCR: Characterization of residual UDG activity following thermal cycling. *BioTechniques* **13**: 180–184.

Victor T., Jordan A., du Toit R., and Van Helden P.D. 1993. Laboratory experience and guidelines for avoiding false positive chain reaction results. *Eur. J. Clin. Chem. Clin. Biochem.* **31**: 531–535.

Walder R.Y., Hayes J.R., and Walder J.A. 1993. Use of PCR primers containing a 3´-terminal ribose residue to prevent cross-contamination of amplified sequences. *Nucleic Acids Res.* **21**: 4339–4343.

Wang X., Chen T., Kim D., and Piomelli S. 1992. Prevention of carryover contamination in the detection of beta S and beta C genes by polymerase chain reaction. *Am. J. Hematol.* **40**: 146–148.

Zhu Y.S., Isaacs S.T., Cimino C.D., and Hearst J.E. 1991. The use of exonuclease III for polymerase chain reaction sterilization. *Nucleic Acids Res.* **19**: 2511.

3 High-Fidelity PCR Enzymes

Holly H. Hogrefe and Michael C. Borns

Stratagene Cloning Systems, La Jolla, California 92037

High-fidelity PCR enzymes are valuable for minimizing the introduction of amplification errors in products that will be cloned, sequenced, and expressed. Significant time and effort can be saved by employing high-fidelity amplification procedures that eliminate the need for downstream error-correction steps and minimize the number of clones that must be sequenced to obtain error-free constructs or accurate consensus sequences. Moreover, the use of high-fidelity amplification conditions is essential when analyzing very small amounts of template DNA or rare molecules in heterogeneous populations (Cha and Thilly 1995). Amplifications employing small amounts of template DNA are especially prone to high mutant frequencies, due to PCR-generated errors in early cycles ("jackpot" artifacts) and high target doublings (Cha and Thilly 1995). When analyzing rare sequences, such as allelic polymorphisms in individual mRNA transcripts (Frohman et al. 1988), allelic stages of single cells (Li et al. 1990), or rare mutations in human cells (Andre et al. 1997), it is essential that polymerase-generated errors (PCR-induced noise) be minimized to prevent masking of rare DNA sequences.

PCR fidelity is largely determined by the intrinsic error rate of DNA polymerases under the reaction conditions employed. Parameters contributing to DNA polymerase fidelity have been reviewed previously (Kunkel 1992; Goodman et al. 1993; Goodman and Fygenson 1998; Kunkel and Bebenek 2000) and include the tendency of a polymerase to incorporate incorrect nucleotides, the rate at which the enzyme can extend from mispaired 3′ primer termini, and the presence of an integral 3′→5′ exonuclease activity, which can remove mispaired bases (proofreading activity). The importance of proofreading is evident in comparisons of base substitution error rates between non-proofreading (10^{-2} to $\geq 10^{-6}$) and proofreading (10^{-6} to 10^{-7}) DNA polymerases (Kunkel 1992; Cline et al. 1996). DNA polymerase error rates are influenced by PCR conditions and can be minimized by optimizing pH, Mg^{++} concentration, and nucleotide concentrations (Eckert and Kunkel 1990, 1991; Ling et al. 1991; Cline et al. 1996).

Taq DNA polymerase is suitable for a number of PCR applications and is still considered by many to be the industry standard. However, the performance of *Taq* is limited in more challenging applications, such as those requiring high fidelity, synthesis of long (>2 kb) amplicons, and/or amplification of GC-rich sequences. *Taq* DNA polymerase lacks proofreading activity and, as a result, exhibits relatively poor fidelity. Over the past decade, a number of PCR enzymes have been introduced that provide higher fidelity and, in many cases, improved performance compared to *Taq*. This chapter surveys commercial PCR enzymes developed for high-fidelity PCR applications, such as cloning, mutation detection, and site-directed mutagenesis. We provide detailed information regarding the composition, PCR characteristics, and intrinsic error rates of high-fidelity PCR enzymes, and include additional suggestions for minimizing PCR mutation frequency.

HIGH-FIDELITY PCR ENZYMES

Proofreading Archaeal DNA Polymerases

High-fidelity PCR enzymes include proofreading archaeal DNA polymerases (Table 1) and DNA polymerase blends (Table 2). Commercial proofreading DNA polymerases have been obtained from *Thermococcus* and *Pyrococcus* species of hyperthermophilic archaea and are classified as Family B-type DNA polymerases (Perler et al. 1996). Unlike thermophilic eubacterial DNA polymerases (e.g., *Taq*), which may or may not possess 3´→5´ exonuclease activity, all archaeal B-type DNA polymerases possess proofreading activity and lack an associated 5´→3´ exonuclease activity.

The kinetic properties of several thermostable DNA polymerases have been reported previously (Perler et al. 1996; Takagi et al. 1997; Hogrefe et al. 2001). Comparisons of steady-state kinetic parameters indicate that archaeal proofreading DNA polymerases exhibit lower K_m [DNA] values (0.01–0.7 nM) and similar K_m [dNTPs] values (16–57 μM) compared to those reported for *Taq* (1–4 nM, K_m [DNA]; 16–24 μM, K_m [dNTPs]). Most archaeal proofreading DNA polymerases (*Pfu*, Vent, Deep Vent) exhibit limited processivities in vitro (<20 bases)(Table 1). The only known exception is KOD DNA polymerase, which is reported to be 10- to 15-fold more processive than *Pfu* and DeepVent DNA polymerases (Takagi et al. 1997). Polymerization rates determined for PCR enzymes range from 9 to 25 nucleotides s^{-1} (*Pfu*) up to 47–61 nucleotides s^{-1} (*Taq*) and 106–138 nucleotides s^{-1} (KOD) (Tagaki et al. 1997; Hogrefe et al. 2001).

Unlike *Taq*, which possesses a structure-specific 5´→3´ endonuclease activity that cleaves 5´ flap structures (Lyamichev et al. 1993), archaeal DNA polymerases exhibit temperature-dependent strand-displacement activity (e.g., detectable at ≥70°C for *Pfu* and ≥63°C for Vent) (Kong et al. 1993; Hogrefe et al. 2001). *Taq* DNA polymerase also adds extra non-templated nucleotide(s) to the 3´ ends of PCR fragments, and, as a result, *Taq*-generated PCR products can be directly cloned into vectors containing 3´ T overhangs (Hu 1993; Zhou and Gomez-Sanchez 2000). In contrast, archaeal DNA polymerases lack terminal extendase activity, and hence, produce blunt fragments that can be cloned directly into blunt-ended vectors (Hu 1993; Costa and Weiner 1994).

Uracil Poisoning of Archaeal DNA Polymerases

Unlike *Taq*, archaeal DNA polymerases possess a "read-ahead" function that detects uracil (dU) residues in the template strand and stalls synthesis (Greagg et al. 1999). Uracil detection is unique to archaeal DNA polymerases (e.g., *Pfu*, Vent) and is thought to represent the first step in a pathway to repair DNA cytosine deamination (dCMP→dUMP) in archaea (Greagg et al. 1999). Stalling of DNA synthesis opposite uracil has significant implications for high-fidelity amplification with archaeal DNA polymerases. Techniques requiring dUTP (e.g., dUTP/UDG decontamination methods, Longo et al. 1990) or uracil-containing oligonucleotides cannot be performed with proofreading DNA polymerases (Slupphaug et al. 1993; Sakaguchi et al. 1996). Even more importantly, uracil stalling has been shown to compromise the performance of archaeal DNA polymerases under standard PCR conditions (Hogrefe et al. 2002).

We found that during PCR amplification, a small amount of dCTP undergoes deamination to dUTP (% dUTP varies with cycling time) and is subsequently incorporated by archaeal DNA polymerases. Once incorporated, uracil-containing DNA inhibits archaeal DNA polymerases, limiting their efficiency. We found that adding a thermostable dUTPase (dUTP→dUMP + PP$_i$) to amplification reactions carried out with *Pfu*, Vent, and Deep Vent DNA polymerases significantly increases PCR product yields by preventing dUTP incorporation (Hogrefe et al. 2002). Moreover, the target-length capability of *Pfu* DNA polymerase is dramatically improved in the presence of dUTPase (e.g., increased from <2 kb to 14 kb; Hogrefe et al. 2002). Long-range PCR is particularly susceptible to dUTP poisoning due to the use of prolonged extension times (1–2 min per kb at 72°C) that promote dUTP formation. It is presently unknown to what extent cytosine deamination in template DNA (as compared to dCTP) limits the performance of archaeal DNA polymerases.

PCR Characteristics of Proofreading DNA Polymerases

The source, composition, and PCR characteristics of commercial proofreading enzymes are provided in Table 1. *PfuUltra* DNA polymerase is formulated with a proprietary *Pfu* mutant that provides 3-fold higher fidelity than *Pfu*. In addition, the *PfuTurbo* and *PfuUltra* enzymes contain *P. furiosus* dUTPase (ArchaeMaxx Polymerase Enhancing Factor) to minimize uracil poisoning. As a result, both yield and target-length capability are vastly improved, and genomic targets up to 19 kb in length have been amplified (Hogrefe et al. 1997; Borns et al. 2000). According to the manufacturer, Platinum Pfx DNA polymerase is recommended for amplifying genomic targets up to 12 kb, whereas *Pfu*, *Tgo*, KOD HiFi, and ProofStart DNA polymerases are suitable for amplifying shorter fragments up to 3.5–6 kb in length (Table 1).

Several proofreading DNA polymerases are available as hot-start formulations. Heat-reversible inactivation is achieved by adding monoclonal antibodies that neutralize polymerase and 3´→5´ exonuclease activities (*PfuTurbo* Hot Start, Platinum Pfx, KOD Hot Start; no preactivation required) or by chemically modifying the DNA polymerase (ProofStart; requires 95°C preactivation step). With proofreading DNA polymerases, high background and/or low product yield may result from extension of nonspecifically annealed primers at ambient temperatures (common with *Taq*; Kellogg et al. 1994) or from degradation of primers and DNA template during room-temperature reaction assembly (unique to proof-readers). In our experience, hot-start formulations provide improved yield and/or specificity when amplifying low-copy-number targets in complex backgrounds (Borns et al. 2001) or when using KOD DNA polymerase (B. Arezi and W. Xing, Stratagene, pers. comm.).

Each manufacturer recommends somewhat different PCR conditions for optimal performance (Table 1). All manufacturers of proofreading enzymes recommend taking measures to minimize nonspecific degradation of PCR primers or products, including using relatively high nucleotide concentrations (200–300 μM each), adding proofreading enzymes last to PCRs (post dNTPs), titrating the amount of enzyme, and using sufficient PCR primer concentrations. When testing different proofreading PCR enzymes, researchers are strongly encouraged to follow each manufacturer's recommendation for enzyme quantity and extension time. In general, lower yields are obtained with *Tgo*, Platinum Pfx, and KOD HiFi DNA polymerases when excess enzyme (>1–1.25 units; amplicons up to 2–3 kb) or longer extension times (>1 min/kb) are used, presumably due to exonucleolytic degradation of PCR primers and/or products (Takagi et al. 1997). In our experience, *Pfu* amplifications are less sensitive to the use of excess (>2.5 units) enzyme, perhaps because *Pfu* exhibits less 3´→5´ exonuclease activity compared to KOD DNA polymerase (~24-fold slower degradation rate per unit; Takagi et al. 1997).

TABLE 1. Characteristics of high-fidelity PCR enzymes

DNA polymerase	Exonuclease activity 3′→5′	5′→3′	Processivity (bases)	Polymerization rate (s⁻¹)	Uracil stalling	Product name (manufacturer)	Notes and recommendations for use	Recommended target length	Hot start
P. furiosus	yes	no	10[6], <20[8]	9.3[6], 25[8]	yes[1] (dU-DNA minimized by ArchaeMaxx in PfuTurbo and PfuUltra)	Pfu DNA polymerase (Stratagene)	cloned Pfu PCR buffer optimized for fidelity[9]; use 2.5 units/50 µl for ≤2 kb; use 200 µM each dNTPs and 2 min/kb at 72°C extensions	up to 4 kb genomic	no
						PfuTurbo DNA polymerase (Stratagene)	formulated with ArchaeMaxx factor; genomic <10 kb: use 2.5 units/50 µl, 200 µM each dNTPs and either 1 min/kb (≤6 kb) or 2 min/kb (>6 kb) at 72°C extensions; genomic >10 kb: use 5 units/50 µl, 500 µM each dNTPs and 2 min/kb at 68°C extensions	up to 19 kb genomic[2]	yes[a]
						PfuUltra DNA polymerase (Stratagene)	formulated with ArchaeMaxx factor and Pfu mutant that improves fidelity; see PfuTurbo recommendations	up to 17 kb genomic	yes[a]
T. litoralis	yes	no	11[6], 7[7]	16.7[7]	yes[1]	Vent DNA polymerase (New England BioLabs)		N.R.	no
P. sp. GB-D	yes	no	<20[8]	23[8]	yes[1]	DeepVent DNA polymerase (New England BioLabs)		N.R.	no
T. kodakaraensis KOD1	yes	no	>300[8]	106–138[8]	yes[1]	Platinum Pfx DNA polymerase[b] (Invitrogen)	only sold as hot-start version[a]; includes PCRₓ Enhancer Solution for problematic/high-G/C templates; use fewer units (1–1.25 units/50 µl) for ≤ 3 kb and 2.5 units for >3 kb targets; use 300 µM each dNTPs and 1 min/kb at 68°C extensions	up to 12 kb genomic and 20 kb vector[3]	yes[a]

						KOD HiFi (Toyobo/Novagen)	includes dNTPs and 2 buffers; use fewer units (1 unit/50 µl) for ≤2-kb targets; use 200 µM each dNTPs and 1.5 min at 72°C extensions; can use shorter cycling times for cDNA/vector targets; use buffer #2 to improve PCR of long (5–6 kb) or genomic targets	up to 6 kb	yes[a]
T. gorgonarius	yes	no			yes	*Tgo* (Roche)	titrate *Tgo*; use 0.4–1.25 units/50 µl for ≤3.5 kb; use 200 µM each dNTPs and 40 s/kb at 72°C extensions; combine 2 separate master mixes to minimize DNA degradation	up to 3.5 kb	no
P. spp.	yes	no			yes	ProofStart (Qiagen)	only sold as hot-start version; reversibly modified DNA polymerase; preactivate for 5 min at 95°C; includes Q-Solution additive to enhance PCRs of difficult templates; use 300 µM each dNTPs, and 2.5 units/50 µl and 1 min/kb at 72°C extensions (≤2 kb) or 5 units/50 µl and 2 min/kb at 72°C extensions (>2 kb targets)	up to 4 kb[4]	yes
Thermus aquaticus	no	yes	10[6], 42[7]	46.7[6], 61[8]	no	numerous	see manufacturers' recommendations	up to 5 kb	yes

Information from product manuals, unless otherwise specified; low-fidelity *Taq* included for comparative purposes.

N.R., No recommendations provided by manufacturer; (*P.*) *Pyrococcus*, (*T.*) *Thermococcus*.

References: [1]Hogrefe et al. 2002; [2]Borns et al. 2000; [3]Expressions 8.5:13 (Invitrogen); [4]Qiagen Technical Services; [5]Nishioka et al. 2001; [6]Hogrefe et al. 2001; [7]Kong et al. 1993; [8]Takagi et al. 1997; [9]Cline et al. 1996.

[a]Hot-start formulation contains polymerase- and exonuclease-neutralizing monoclonal antibodies.

[b]Source identified by manufacturer as *Pyrococcus sp.* strain KOD, but reclassified as *T. kodakaraensis* KOD1[5].

TABLE 2. Characteristics of high-fidelity DNA polymerase blends

DNA polymerase (manufacturer)	Blend composition			Hot start	Recommended target length	Standard PCR conditions		Recommended applications
	major polymerase	minor polymerase	additives			units; [each dNTP]; extension conditions	additional notes	
Herculase (Stratagene)	Pfu	Taq	ArchaeMaxx factor	yes[a]	up to 37 kb genomic and 48 kb vector	2.5 units/50 μl; 200 μM dNTPs; 1 min/kb at 72°C (<10 kb) or 5 units/50 μl; 500 μM dNTPs; 1 min/kb at 68°C (>10 kb)	includes DMSO to enhance long or GC-rich PCR (titrate: 0–7%); manual has optimized protocol for GC-rich targets; to further increase yield can add 10 s per cycle for cycles 11–30	applications requiring high fidelity, yield, and length challenging applications long and/or high GC-templates
TaqPlus Precision (Stratagene)	Taq	Pfu	none	no	up to 10 kb genomic and 15 kb vector	2.5 units/50 μl; 200 μM dNTPs; 1 min/kb at 72°C		applications requiring high fidelity and yield
Expand High Fidelity (Roche)	Taq	Tgo	none	no	up to 5 kb genomic	2.6 units/50 μl; 200 μM dNTPs; 40 s/kb at 72°C (≤3 kb) or 68°C (>3 kb) plus 5 s per cycle for cycles 11–30	Mg++-containing and Mg++-free buffers provided; combine 2 separate master mixes to minimize DNA degradation	high-fidelity applications use instead of Taq for higher fidelity, yield, and sensitivity
Platinum Taq High Fidelity (Invitrogen)	Taq	DeepVent	Taq-neutralizing MAb	yes (only version available)	up to 12 kb; up to 20 kb with optimization	1 unit/50 μl; 200 μM dNTPs; 1 min/kb at 68°C	up to 2.5 U of enzyme may be required for some systems	high-fidelity applications hot start PCR long PCR
Advantage-HF (Clontech)	KlenTaq	Proofreading DNA polymerase	Taq-neutralizing MAb	yes (only version available)	up to 2.5 kb	1 μl/50 μl (1x); 200 μM dNTPs; 2-step protocol with 4 min at 68°C for anneal/extend	includes dNTPs, 2 buffers, and PCR control; mix HF buffer with cDNA buffer to improve long PCR; shorter extension times may be used for shorter targets	high-fidelity applications

Information from manufacturers' catalog or product manual, unless otherwise specified. MAb, monoclonal antibody.
[a] Hot-start formulation contains Pfu- and Taq-neutralizing monoclonal antibodies.

With all proofreading enzymes, synthesizing longer amplicons or amplifying GC-rich (>70%) sequences typically requires additional optimization. In general, amplification of longer targets requires more enzyme units, higher nucleotide concentrations, and/or longer extension times. To enhance amplifications of problematic or GC-rich templates, researchers can add DMSO to *Pfu* formulations (3–10%; titrate in 1% increments) or use the proprietary PCR additives that are provided with Platinum Pfx (PCRx Solution) and ProofStart (Q-Solution) DNA polymerases (Table 1).

High-Fidelity DNA Polymerase Blends

In addition to proofreading DNA polymerases, several DNA polymerase blends have been introduced for high-fidelity PCR (Table 2). With the exception of Herculase DNA polymerase (*Pfu*-based mixture, see below), commercial DNA polymerase blends consist predominantly of *Taq* plus a lesser amount of a proofreading DNA polymerase (e.g., *Pfu*, Deep Vent, *Tgo*) to enhance PCR product yields, amplification of long targets, and fidelity (Barnes 1994). The fidelity of *Taq*-based blends is typically improved by increasing the proportion of proofreading to nonproofreading DNA polymerase and by modifying the PCR buffer to optimize yield. Since product yield and target-length capability decrease with increasing proofreading:nonproofreading polymerase ratios (Barnes 1994), higher-fidelity *Taq*-based blends typically exhibit reduced performance compared to blends optimized for yield and length (i.e., blends with lower proofreading:nonproofreading polymerase ratios). In general, high-fidelity *Taq*-based blends provide superior performance compared to *Taq* alone with respect to fidelity, yield, and target-length capability (Table 2).

In contrast to other commercial blends, Herculase DNA polymerase consists predominantly of *Pfu* DNA polymerase and a lesser proportion of *Taq* DNA polymerase. Herculase also contains the ArchaeMaxx enhancer to prevent dUTP incorporation and to enhance *Pfu*'s contribution to blend performance and fidelity (Hogrefe et al. 2002). We have found that higher *Pfu:Taq* ratios can be employed successfully in the presence of dUTPase, but not in the absence of dUTPase. Thus, reduced performance of blends containing a higher proportion of proofreading activity (Barnes 1994) can most likely be attributed to sensitivity of the proofreading component to uracil poisoning. As a result of its unique composition (high *Pfu:Taq* ratio, ArcheaMaxx factor), Herculase DNA polymerase exhibits several distinct characteristics compared to *Taq*-based blends, including higher fidelity (see below) and very broad target-length capability (0–37 kb genomic, Table 2). Herculase DNA polymerase is recommended for high-fidelity applications that are too difficult for proofreading DNA polymerases alone, such as amplifications of extra-long (>10 kb) genomic targets or highly GC-rich sequences (Borns and Hogrefe 2000a,b).

PCR FIDELITY

Error Rate Calculations

DNA polymerase fidelity is expressed in terms of error rate, i.e., the number of misincorporated nucleotides per base synthesized. In PCR-based fidelity assays, error rates (ER) are calculated as:

$$\text{ER} = \frac{\text{number of mutations per bp}}{\text{number of amplicon doublings}}$$

where number of amplicon doublings (*d*) is quantified from the amount of input target DNA and amplicon yield, as

$$2^d = \frac{\text{amplicon yield}}{\text{input target DNA}}$$

The error rates of *Pfu* and *Taq* DNA polymerases have been measured using several different fidelity assays, including denaturing gradient gel electrophoresis (DGGE) analyses, reversion mutation assays, and forward mutation assays (for review, see Cha and Thilly 1995). *Pfu* DNA polymerase exhibits an error rate of 0.7×10^{-6} to 1.8×10^{-6} mutations per bp per doubling (Lundberg et al. 1991; Cline et al. 1996; Andre et al. 1997; Li-Sucholeiki and Thilly 2000), which means that the probability of a base being mutated in a single round of replication is ~1–3 per 1,500,000 nucleotides. At this rate, after 20 doublings (10^6-fold amplification), ~1–2.5% of 1-kb amplification products will contain mutations. In comparison, published error rates for *Taq* range from 0.5×10^{-5} to 21×10^{-5} mutations per bp per doubling, and include 7.2×10^{-5} to 21×10^{-5} using DGGE (Keohavong and Thilly 1989; Ling et al. 1991), 0.8×10^{-5} to 1.0×10^{-5} (*lacI*) and 1.8×10^{-5} (p53) using PCR-based phenotypic assays (Lundberg et al. 1991; Flaman et al. 1994; Cline et al. 1996), 2×10^{-5} using a gap-filling *lacZ* assay (Eckert and Kunkel 1990), and 0.5×10^{-5} to 2.7×10^{-5} by DNA sequencing of PCR products (Martell et al. 1992; Bracho et al. 1998). At these rates, anywhere from 10% to 100% of 1-kb products amplified with *Taq* will contain one or more mutations (doublings = 20; percent of mutation-containing products, 10–420%).

Variation in published error rates reflects differences in the reaction conditions (e.g., pH, dNTP concentration, Mg^{++} concentration, DNA template sequence) and types of fidelity assays employed (Eckert and Kunkel 1991; Ling et al. 1991; Cha and Thilly 1995). Because different assays are likely to measure different parameters, error rates should only be compared among PCR enzymes tested in the same assay (Perler et al. 1996) and, preferably, according to manufacturers' recommendations.

Error Rates of High-Fidelity PCR Enzymes

Our laboratory routinely employs a PCR-based forward mutation assay that utilizes the well-characterized *lacI* target gene (Lundberg et al. 1991; Cline et al. 1996). In this assay, a 1.9-kb sequence encoding *lacIOZα* is amplified and cloned, and the percentage of clones containing a mutation in *lacI* (% blue) is determined in a color-screening assay (Fig. 1). When performing fidelity assays based on phenotypic changes, it is essential that the number of base changes producing a scorable mutant phenotype is known. Sequence analyses of over 30,000 *lacI* mutants indicate that single-base substitutions can be identified at 349 sites in the 1080-bp *lacI* gene by color screening (Provost et al. 1993). Using the *lacI* phenotypic assay, error rates are calculated as

$$\frac{lacI^- \text{ mutant frequency}}{(349 \text{ bases})(d)}$$

where *d* = number of amplicon doublings.

We have measured the error rates of several DNA polymerases using the *lacI* assay (Table 3). Error rates were measured in each enzyme's recommended PCR buffer, and whenever possible, identical PCR conditions were used, including DNA template concentration, PCR cycling parameters, and number of PCR cycles performed. The only exceptions

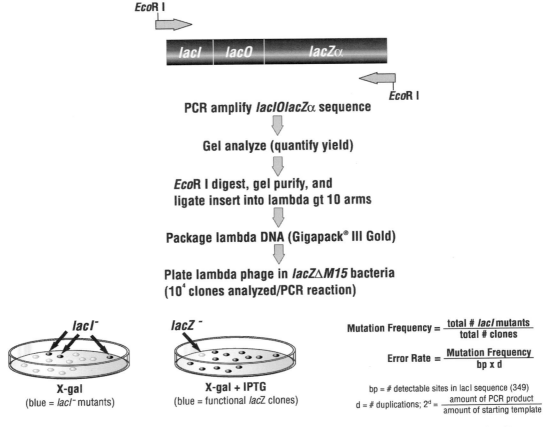

Mutation Frequency = $\dfrac{\text{total \# } lacI \text{ mutants}}{\text{total \# clones}}$

Error Rate = $\dfrac{\text{Mutation Frequency}}{\text{bp} \times \text{d}}$

bp = # detectable sites in lacI sequence (349)

d = # duplications; $2^d = \dfrac{\text{amount of PCR product}}{\text{amount of starting template}}$

FIGURE 1. *lac*I fidelity assay. A 1.9-kb sequence encoding *lacIOZα* is PCR-amplified with oligonucleotide primers containing 5′ *Eco*RI restriction sites. The fragments are digested with *Eco*RI, quantified and purified by gel electrophoresis, and ligated to lambda gt10 arms. The ligation reactions are packaged and the lambda phage is used to infect an α-complementing *E. coli* host strain. Aliquots of infected cells are plated on LB plates with top agar containing either X-gal or X-gal plus IPTG. Mutant frequencies are determined by dividing the total number of blue plaques, shown here in black (*lacI*⁻ mutants), on X-gal plates by the total number of plaques containing a functional *lacZα* sequence (blue plaques, shown here in black, on X-gal plus IPTG plates). Template doubling and error rate are calculated using the equations shown.

were that each manufacturer's recommendations were followed with respect to number of enzyme units, nucleotide concentration, primer concentration, and extension temperature (see Table 3). To allow assay-to-assay comparisons, *Pfu* (or *PfuTurbo*) DNA polymerase was run in every assay, and error rates were normalized relative to the mean value of 1.3×10^{-6} mutations per bp per doubling, as determined for *Pfu* in study #1 (Cline et al. 1996). *Taq* DNA polymerase, serving as a second internal control, exhibited mean error rates of 8.0×10^{-6} (study #1; 11 PCRs) and 9.1×10^{-6} (mean of studies #2–5; 14 PCRs) mutations per bp per doubling.

As expected, proofreading DNA polymerases exhibited significantly lower error rates (1–3 errors per 1,000,000 bases) compared to *Taq* DNA polymerase (8–9 errors per 1,000,000 bases). The *PfuUltra* mutant DNA polymerase formulation exhibited an error rate (4×10^{-7} mutations per bp per duplication) that was 3- to 8-fold lower than the error rates of native proofreading DNA polymerases. Relative differences in error rate observed with the *lac*I assay (Table 3) are consistent with those obtained using a p53-based forward mutation assay (e.g., *Pfu* < DeepVent < *Taq*; Flaman et al. 1994) and DGGE (e.g., *Pfu* < Vent <

TABLE 3. Error rates of high-fidelity PCR enzymes

DNA polymerase	Study #	No. of PCRs	Error rate[a] ($\times 10^{-6} \pm$ S.D.)	Accuracy (error rate^{-1} in bases)	Percentage of clones with mutations (10^6-fold amplification)		
					1-kb amplicon	5-kb amplicon	10-kb amplicon
Proofreading DNA polymerases							
Pfu, PfuTurbo[1,2]	1	10	1.3 ± 0.2	770,000	2.6	13	26
PfuUltra	5	12	0.4 ± 0.04	2,500,000	0.8	4	8
Tgo	5	2	2.2 ± 0.1	450,000	4.4	N.R.	N.R.
DeepVent[1]	1	4	2.7 ± 0.2	370,000	5.4	N.R.	N.R.
Vent[1]	1	6	2.8 ± 0.9	360,000	5.6	N.R.	N.R.
Platinum Pfx[4]	4	4	3.5 ± 1.0	290,000	7	35	70
High-fidelity blends							
Herculase[3]	3	4	2.8 ± 0.5	360,000	5.6	28	56
TaqPlus Precision[3]	2–3	13	4.0 ± 1.3	250,000	8	40	80
Platinum Taq High Fidelity[3]	3	2	5.8 ± 0.3	170,000	11.6	58	100
Advantage HF[3]	3	2	6.1 ± 0.07	160,000	12.2	N.R.	N.R.
Taq[1]	1	11	8.0 ± 3.9	125,000	16	80	N.R.
	2–5	14	9.1 ± 2.4	110,000	18.2	91	N.R.

N.R., Not recommended for 5- to 10-kb target sizes.

[a]Error rates were measured in each enzyme's recommended PCR buffer using the cycling conditions described in Cline et al. (1996). PCRs (50 μl) contained 0.2 μM each primer, 205 ng or 200 μM each nucleotide, 2.5 ng of target DNA, and 2.5 units of DNA polymerase, with the following manufacturer-recommended exceptions: Platinum Pfx: 300 μM each nucleotide, 1.25 units of enzyme, and 68°C extension temperature; Tgo: 1 unit of enzyme and 0.4 μM each primer; Vent/Deep Vent: 1 unit of enzyme.

References: [1]Cline et al. 1996; [2]Hogrefe et al. 1997; [3]Borns and Hogrefe 2000a; [4]Borns and Hogrefe 2000b.

Taq; Cha and Thilly 1995). In general, the error rates of high-fidelity DNA polymerase blends (3–6 errors per 1,000,000 bases) are intermediate between proofreading DNA polymerases and *Taq* (Table 3).

The use of high-fidelity DNA polymerases becomes increasingly important as the size of the amplicon increases (Table 3). With *Taq*, the percentage of clones (or amplicons) expected to contain mutations in a 10^6-fold amplification reaction increases from 4% (250 bp) to 16% (1 kb) to 80% (5 kb), and the number of clones that should be sequenced to obtain an error-free clone (95% confidence) increases from 1 to 2 to 14, respectively ($0.95 = 1 - [1 - f]^n$, where f = frequency of error-free clones and n = number of clones sequenced; Jeltsch and Lanio 2002). When amplifying a broader range of targets with high-fidelity blends (0.25 to 10 kb; ER = $2.8 - 5.8 \times 10^{-6}$), the percentage of clones likely to contain mutations increases from 1–3% (250 bp) to 5–11% (1 kb) to 28–58% (5 kb) to 56–100% (10 kb), and the number of clones that should be sequenced increases from 1–2 (up to 1 kb) to 3–5 (5 kb) to >>6 (10 kb). When amplifying similarly sized targets with *PfuUltra* (ER = 4×10^{-7}), the frequency of error-containing clones is <1% (up to 1 kb), 4% (5 kb), and 8% (10 kb), and sequencing 1 (up to 6 kb) or 2 (6–10 kb) clones should be sufficient for identifying an error-free clone. Therefore, PCR enzyme fidelity is of paramount importance when amplifying longer DNA sequences that will be cloned, sequenced, and expressed.

OPTIMIZING PCR FIDELITY

PCR error rate can be minimized by employing the highest-fidelity PCR enzyme available for the desired application. As discussed above, commercial high-fidelity DNA polymerases show considerable variation in error rates, ranging from 0.4×10^{-6} to 3.5×10^{-6} for proofreading DNA polymerases, up to 2.8×10^{-6} to 6.1×10^{-6} for DNA polymerase blends (Table 3). However, when selecting a PCR enzyme, parameters other than fidelity may have to be considered. As shown in Table 4, current high-fidelity PCR enzymes are incompatible with dUTP/UDG decontamination (Longo et al. 1990; Slupphaug et al. 1993) and direct TA cloning methods (Zhou and Gomez-Sanchez 2000). However, postamplification addition of 3′ A-overhangs with *Taq* (8–10 minutes at 72°C, TA Cloning Kit manual, Invitrogen) improves the TA cloning efficiency of blunt-ended fragments amplified with proofreading enzymes. Thus, suitable high-fidelity enzyme formulations are available for nearly every PCR application (Table 4).

In addition to enzyme choice, researchers should also consider optimizing reaction conditions to reduce PCR mutation frequency further. Although error rate is an intrinsic prop-

TABLE 4. PCR enzyme applications

Enzyme group	Fidelity	Product yield	Target length	Cloning strategy		dUTP/UDG method	GC-rich targets
				TA[a]	blunt		
Proofreading	****	*/**	*/**	N.R.	****	N.R.	***
Blends							
Herculase	***	****	****	*	*	N.R.	****
High fidelity	**	**	**	*	*	N.R.	**
Taq-based							
Long *Taq*-based	*	***	***	**	N.R.	N.R.	**
Taq	N.R.	*	*	****	N.R.	****	*

Relative performance shown on scale from * (low/poor) to **** (high/best); */** = variation within an enzyme group. N.R., not recommended.
[a]Assumes no post-PCR steps; TA cloning efficiency can be improved by post-PCR incubation with *Taq* DNA polymerase.

erty of DNA polymerases (under defined reaction conditions), observed mutation frequencies can vary from PCR to PCR, depending on the number of amplicon doublings. For example, assuming we amplify a 1-kb fragment using Taq (ER, 8×10^{-6} mutations per bp per doubling), a PCR generating 5 μg of amplicon from 5 pg of target DNA has undergone 20 target doublings and produced 1.6 mutations per 10,000 bases (~3/20 clones with mutations). In comparison, a PCR generating 5 μg of amplicon from 75 ng of target DNA has undergone only 6 target doublings (67-fold amplification) and introduced 0.5 mutations per 10,000 bases (~1/20 clones with mutations). Therefore, researchers can minimize mutation frequency by limiting the number of target duplications; for example, by increasing the amount of input DNA template or reducing the number of PCR cycles.

Additional reduction in mutation frequency may be achieved by optimizing buffer composition, nucleotide and polymerase concentration, and/or cycling conditions for a particular PCR enzyme. As discussed above, the error rates shown in Table 3 were obtained using the PCR buffer and nucleotide concentration recommended by each manufacturer, which may or may not be optimal with respect to fidelity. High-fidelity PCR conditions have been developed for Taq, Vent, and Pfu DNA polymerases (Eckert and Kunkel 1990, 1991; Ling et al. 1991; Matilla et al. 1991; Cline et al. 1996). For example, the error rate of Pfu decreases from 2.6×10^{-6} to 1.1×10^{-6} as the nucleotide concentration is lowered from 1 mM to 100 μM each (Cline et al. 1996). Even greater changes in Pfu's error rate were observed as the Mg^{++} concentration was increased from 1 mM (4.9×10^{-6}) to 2 mM $MgSO_4$ (1.3×10^{-6})(at 200 μM each dNTP, pH 8.8) and the pH was increased from pH 7.5 (8.2×10^{-6}) to pH 8.8 (1.3×10^{-6})(at 200 μM each dNTP, 2 mM $MgSO_4$) (Cline et al. 1996). For enzymes whose pH and Mg^{++} optima are unknown, researchers can expect to achieve lower mutation frequencies by using the lowest balanced nucleotide concentration compatible with yield (e.g., 25–150 μM each). In addition, using shorter synthesis times and lower enzyme concentrations is likely to minimize polymerase extension from mispaired or misaligned primer termini (Eckert and Kunkel 1991).

DISCUSSION

Since the introduction of Taq DNA polymerase in the late 1980s, significant progress has been made in developing PCR enzyme formulations with improved fidelity and PCR performance. Proofreading DNA polymerases offer significantly higher fidelity compared to Taq, and initial problems associated with their use (low yield, poor reliability) have been largely overcome by reducing uracil poisoning (Stratagene's Pfu formulations), preparing blends with Taq DNA polymerase, and developing faster, more processive proofreading DNA polymerases. In fact, many of the new high-fidelity enzyme formulations provide significantly improved yield and target-length capability compared to Taq.

REFERENCES

Andre P., Kim A., Khrapko K., and Thilly W.G. 1997. Fidelity and mutational spectrum of Pfu DNA polymerase on a human mitochondrial DNA sequence. *Genome Res.* **7:** 843–852.

Barnes W.M. 1994. PCR amplification of up to 35-kb DNA with high fidelity and high yield from λ bacteriophage templates. *Proc. Natl. Acad. Sci.* **91:** 2216–2220.

Borns M. and Hogrefe H. 2000a. Unique DNA polymerase formulation excels in a broad range of PCR application. *Strategies* **13:** 1–3.

———. 2000b. Unique enhanced DNA polymerase delivers high fidelity and great PCR performance. *Strategies* **13:** 76–79.

Borns M., Cline J., and Hogrefe H. 2000. Comparing fidelity and performance of proofreading PCR enzymes. *Strategies* **13:** 27–30.

Borns M., Scott B., and Hogrefe H. 2001. Most accurate PCR enzyme improved with hot start feature. *Strategies* **14:** 5–8.

Bracho M.A., Moya A., and Barrio E. 1998. Contribution of Taq polymerase-induced errors to the estimation of RNA virus diversity. *J. Gen. Virol.* **79:** 2921–2928.

Cha R.S. and Thilly W.G. 1995. Specificity, efficiency, and fidelity of PCR. In *PCR primer: A laboratory manual* (ed. C.W. Dieffenbach and G.S. Dveksler), pp. 37–51. Cold Spring Harbor Laboratory Press, Cold Spring Harbor, New York.

Cline J., Braman J.C., and Hogrefe H.H. 1996. PCR fidelity of *Pfu* DNA polymerase and other thermostable DNA polymerases. *Nucleic Acids Res.* **24:** 3546–3551.

Costa G. L. and Weiner M.P. 1994. Protocols for cloning and analysis of blunt-ended PCR-generated DNA fragments. *PCR Methods Appl.* **3:** S95–106.

Eckert K.A. and Kunkel T.A. 1990. High fidelity DNA synthesis by the *Thermus aquaticus* DNA polymerase. *Nucleic Acids Res.* **18:** 3739–3744.

———. 1991. DNA polymerase fidelity and the polymerase chain reaction. *PCR Methods Appl.* **1:** 17–24.

Flaman J.-M., Frebourg T., Moreau V., Charbonnier F., Martin C., Ishioka C., Friend S.H., and Iggo R. 1994. A rapid PCR fidelity assay. *Nucleic Acids Res.* **22:** 3259–3260.

Frohman M.A., Dush M.K., and Martin G.R. 1988. Rapid production of full-length cDNAs from rare transcripts: Amplification using a single gene specific oligonucleotide primer. *Proc. Natl. Acad. Sci.* **85:** 8998–9002.

Goodman M.F. and Fygenson K.D. 1998. DNA polymerase fidelity: From genetics toward a biochemical understanding. *Genetics* **148:** 1475–1482.

Goodman M.F., Creighton S., Bloom L.B., and Petruska J. 1993. Biochemical basis of DNA replication fidelity. *Crit. Rev. Biochem. Mol. Biol.* **28:** 83–126.

Greagg M.A., Fogg M.J., Panayotou G., Evans S.J., Connolly B.A., and Pearl L.H. 1999. A read-ahead function in archaeal DNA polymerases detects promutagenic template-strand uracil. *Proc. Natl. Acad. Sci.* **96:** 9045–9050.

Hogrefe H.H., Cline J., Lovejoy A.E., and Nielson K.B. 2001. DNA polymerases from hyperthermophiles. *Methods Enzymol.* **334:** 91–116.

Hogrefe H.H., Hansen C.J., Scott B.R., and Nielson K.B. 2002. Archaeal dUTPase enhances PCR amplifications with archaeal DNA polymerase by preventing dUTP incorporation. *Proc. Natl. Acad. Sci.* **99:** 596–601.

Hogrefe H., Scott B., Nielson K., Hedden V., Hansen C., Cline J., Bai F., Amberg J., Allen R., and Madden M. 1997. Novel PCR enhancing factor improves performance of Pfu DNA polymerase. *Strategies* **10:** 93–96.

Hu G. 1993. DNA polymerase-catalyzed addition of nontemplated extra nucleotides to the 3′ end of a DNA fragment. *DNA Cell Biol.* **12:** 763–770.

Jeltsch A. and Lanio T. 2002. Site-directed mutagenesis by polymerase chain reaction. In *In vitro mutagenesis protocol*, 2nd edition (ed. J. Braman), pp 85–94. Humana Press, Totowa, New Jersey.

Kellogg D.E., Rybalkin I., Chen S., Mukhamedova N., Vlaski T., Siebert P.D., and Chenchik A. 1994. TaqStart antibody: "Hot start" PCR facilitated by a neutralizing monoclonal antibody directed against Taq DNA polymerase. *BioTechniques* **16:** 1134–1137.

Keohavong P. and Thilly W.G. 1989. Fidelity of DNA polymerases in DNA amplification. *Proc. Natl. Acad. Sci.* **86:** 9253–9257.

Kong H., Kucera R.B., and Jack W.E. 1993. Characterization of a DNA polymerase from the hyperthermophile archaea *Thermococcus litoralis*. Vent DNA polymerase, steady state kinetics, thermal stability, processivity, strand displacement, and exonuclease activities. *J. Biol. Chem.* **268:** 1965–1975.

Kunkel T.A. 1992. DNA replication fidelity. *J. Biol. Chem.* **267:** 18251–18254.

Kunkel T.A. and Bebenek K. 2000. DNA replication fidelity. *Annu. Rev. Biochem.* **69:** 497–529.

Li H., Cui X., and Arnheim N. 1990. Direct electrophoretic detection of the allelic state of a single DNA molecule in human sperm by using the polymerase chain reaction. *Proc. Natl. Acad. Sci.* **87:** 4580–4584.

Ling L.L., Keohavong P., Dias C., and Thilly W.G. 1991. Optimization of the polymerase chain reaction with regard to fidelity: Modified T7, Taq, and Vent DNA polymerases. *PCR Methods Appl.* **1:** 63–69.

Li-Sucholeiki X.C. and Thilly W.G. 2000. A sensitive scanning technology for low frequency nuclear point mutations in human genomic DNA. *Nucleic Acids Res.* **28:** E44.

Longo M.C., Berninger M.S., and Hartley J.L. 1990. Use of uracil DNA glycosylase to control carryover contamination in polymerase chain reactions. *Gene* **93:** 125–128.

Lundberg K.S., Shoemaker D.D., Adams M.W.W., Short J.M., Sorge J.A., and Mathur E.J. 1991. High-fidelity amplification using a thermostable DNA polymerase isolated from *Pyrococcus furiosus. Gene* **108:** 1–6.

Lyamichev V., Brow M.A., and Dahlberg J.E. 1993. Structure-specific endonucleolytic cleavage of nucleic acids by eubacterial DNA polymerase. *Science* **260:** 778–783.

Martell M., Esteban J.I., Quer J., Genesca J., Weiner A., Esteban R., Guardia J., and Gomez J. 1992. Hepatitis C virus (HCV) circulates as a population of different but closely related genomes: Quasispecies nature of HCV genome distribution. *J. Virol.* **66:** 3225–3229.

Mattila P., Korpela J., Tenkanen T., and Pitkanen K. 1991. Fidelity of DNA synthesis by the *Thermococcus litoralis* DNA polymerase—an extremely heat stable enzyme with proofreading activity. *Nucleic Acids Res.* **19:** 4967–4973.

Nishioka M., Mizuguchi H., Fujiwara S., Komatsubara S., Kitabayashi M., Uemura H., Takagi M., and Imanaka T. 2001. Long and accurate PCR with a mixture of KOD DNA polymerase and its exonuclease deficient mutant enzyme. *J. Biotechnol.* **88:** 141–149.

Perler F.B., Kumar S., and Kong H. 1996. Thermostable DNA polymerases. *Adv. Protein Chem.* **48:** 377–435.

Provost G.S., Kretz P.L., Hamner R.T., Matthews C.D., Rogers B.J., Lundberg K.S., Dycaico M.J., and Short J.M. 1993. Transgenic systems for in vivo mutation analysis. *Mutat. Res.* **288:** 133–149.

Sakaguchi A.Y., Sedlak M., Harris J.M., and Sarosdy M.F. 1996. Cautionary note on the use of dUMP-containing PCR primers with Pfu and Vent$_R$™ DNA polymerases. *BioTechniques* **21:** 368–369.

Slupphaug G., Alseth I., Eftedal I., Volden G., and Krokan H.E. 1993. Low incorporation of dUMP by some thermostable DNA polymerases may limit their use in PCR amplifications. *Anal. Biochem.* **211:** 164–169.

Takagi M., Nishioka M., Kakihara H., Kitabayashi M., Inoue H., Kawakami B., Oka M., and Imanaka T. 1997. Characterization of DNA polymerase for *Pyrococcus* sp. strain KOD1 and its application to PCR. *Appl. Environ. Microbiol.* **63:** 4504–4510.

Zhou M.Y. and Gomez-Sanchez C.E. 2000. Universal TA cloning. *Curr. Issues Mol. Biol.* **2:** 1–7.

Optimization and Troubleshooting in PCR

4

Kenneth H. Roux

Department of Biological Science, Florida State University, Tallahassee, Florida 32306-4370

The use of PCR to generate large amounts of a desired product can be a double-edged sword. Failure to amplify under optimum conditions can lead to the generation of multiple undefined and unwanted products, even to the exclusion of the desired product. At the other extreme, no product may be produced. A typical response at this point is to vary one or more of the many parameters that are known to contribute to primer-template fidelity and primer extension. High on the list of optimization variables are Mg^{++} concentrations, buffer pH, and cycling conditions. With regard to the last, the annealing temperature is most important. The situation is further complicated by the fact that some of the variables are quite interdependent. For example, because dNTPs directly chelate a proportional number of Mg^{++} ions, an increase in the concentration of dNTPs decreases the concentration of free Mg^{++} available to influence polymerase function.

TOUCHDOWN PCR

Touchdown (TD) PCR represents a fundamentally different approach to PCR optimization (Don et al. 1991). Rather than multiple reaction tubes, each with different reagent concentration and/or set of cycling parameters, a single tube, or a small set of tubes, is run under cycling conditions that inherently favor amplification of the desired amplicon, often to the exclusion of artifactual amplicons and "primer–dimers." Multiple cycles are programmed such that the annealing segments in sequential cycles are run at incrementally lower temperatures (see below). As cycling progresses, the annealing-segment temperature, which was selected to be initially above the suspected T_m, gradually declines to, and falls below, this level. This strategy helps ensure that the first primer-template hybridization events involve only those reactants with the greatest complementarity; i.e., those yielding the target amplicon. Even though the annealing temperature may eventually drop down to the T_m of nonspecific hybridizations, the target amplicon will have already begun its geometric amplification and is thus in a position to outcompete any lagging (nonspecific) PCR products during the remaining cycles. Because the aim is to avoid low-T_m priming during the earlier cycles, it is imperative that the "hot start" modification (D'Aquila et al. 1991; Erlich et al. 1991; Ruano et al. 1992) (see below) be utilized with TD PCR. TD PCR should be viewed not so much as a method of determining the optimum cycling conditions for a specific PCR, but as a potential one-step method for approaching optimal amplification. We have found that a variety of otherwise satisfactory single-amplicon-yielding reactions are rendered more robust (i.e., yield more product) when subjected to TD PCR (Hecker and Roux 1996).

TD PCR is of particular value when the degree of identity between the primer and template is unknown (Roux 1994; Hecker and Roux 1996). This situation often arises when primers are designed on the basis of amino acid sequences, when members of a multigene family are amplified, or when evolutionary PCR is attempted; i.e., amplification of DNA from one species using primers with identity to a homologous segment of another species. In such cases, the mismatches between the primers and template may result in T_m values that are so low that they approach the T_m values of the spurious priming sites. Degenerate primers with multiple base variation or inosine residues are often used in such situations (Knoth et al. 1988; Lee et al. 1988; Patil and Dekker 1990; Batzer et al. 1991; Peterson et al. 1991), but the greater variety of sequences in the former case and the relaxed stringency in the latter case might tend to increase the chances of nonspecific priming. Moreover, in some cases, the locations of potential base mismatches will be unknown. Although TD PCR can be used with degenerate primers (Batzer et al. 1991), we have shown that nondegenerate primers displaying a significant degree of template-sequence mismatch can yield single-target amplicons of single-copy genes from genomic DNA under standard buffer conditions (Roux 1994). Even mismatches clustered near the 3′ end of the primer are tolerated. TD PCR has the added benefit of compensating for suboptimal buffer composition (e.g., Mg^{++} concentration) as well (Hecker and Roux 1996).

PROCEDURE

Programming the Thermal Cycler for Touchdown PCR

The goal in programming for TD PCR is to produce a series of cycles with progressively lower annealing temperatures. The annealing temperature range should span about 15°C and extend from at least a few degrees above the estimated T_m to 10 or so degrees below. For example, for a calculated primer-template T_m of 62°C with no degeneracy, program the thermal cycler to decrease the annealing temperature 1°C every second cycle (i.e., run two cycles per degree) from 65°C to 50°C, followed by 15 additional cycles at 50°C.

Some thermal cyclers (e.g., PCR Express, Hybaid) readily accommodate TD PCR and are easily programmed to decrease the temperature of a segment automatically by a fixed amount per cycle (e.g., 0.5°C/cycle). For others, a long series of files must be linked or extensive strings of commands entered. In these latter cases, it may be more convenient to create a "generic" TD PCR program covering a broader temperature range (~20°C) than to reprogram every time the range needs to be modified by a few degrees. Another alternative to programming restrictions and inconvenience is to use stepdown PCR, in which fewer but more abrupt steps (e.g., seven 2°C steps or five 3°C steps) are used (Hecker and Roux 1996).

The continued presence of spurious bands following TD PCR indicates that the initial annealing temperature was too low, that there is a relatively small gap between the T_m values of the target and unwanted amplicons, and/or that the unwanted amplicons are being more efficiently amplified. Raising the number of cycles per 1°C-descending step to three or four will give the target amplicon an added competitive advantage before the initiation of the spurious amplification. A proportional number of cycles should be removed from the end of the program to prevent excess cycling and the concomitant degradation of the amplicon and generation of high-molecular-weight smears (Bell and DeMarini 1991).

Modifications of TD PCR for use with degenerate and mismatched primers include lowering the annealing temperature range (e.g., 50°C declining to 35°C) while keeping the last

15 cycles at 50°C or more (once priming has begun, the primers are fully complementary to the newly formed amplicons, have a much higher T_m, and do not benefit from excessively low annealing temperatures).

Optimization Strategy

The example presented here is for TD PCR, but the same principles apply to conventional PCR.

1. Design optimal primer pairs that are closely matched in T_m. For additional discussion of primer design, see Chapter 7.

2. Calculate or estimate approximate T_m. Program the thermal cycler for TD PCR as described above.

3. Set up several standard hot-start PCR mixes incorporating a range of Mg^{++} concentrations and including appropriate positive and negative controls. Use 10^4 to 10^5 copies of the template.

4. Amplify as above and analyze products by agarose or acrylamide gel electrophoresis.

    a. If weak or no product is detected:

        i. Subject reaction tubes to 10 additional cycles at a constant annealing temperature (i.e., 55°C) and recheck.

        ii. Reamplify 10-fold dilutions (1:100 to 1:10,000) of initial TD PCR at a fixed annealing temperature for 30 cycles.

        iii. Use more template and check for the presence of inhibitors in the template preparation by spiking the original PCR mix with dilutions of a known positive (demonstrably amplifiable) template.

        iv. Add, extend, or increase the temperature of the initial template denaturation step prior to cycling (5 minutes at 95°C is standard).

        v. Vary concentrations of other buffer components (pH, *Taq* polymerase, dNTPs, primers).

        vi. Add enhancers to the PCR mix.

        vii. Reamplify dilutions (1:100 to 1:10,000) of first reaction using nested primers.

        viii. Abandon the chosen primer set, design new primers, and begin again. Depending on the experimenter's degree of impatience and tolerance for frustration, this step might supercede any of the above.

    b. If multiple products or a high-molecular-weight smear is observed:

        i. Raise the maximum and minimum annealing temperatures (i.e., shift the range upward) in the TD PCR program.

        ii. Decrease the total number of cycles by eliminating some cycles from the bottom of the range and/or from the terminal constant temperature cycles.

        iii. Increase the number of cycles per degree annealing temperature by one cycle, i.e., to three cycles/degree. Doing so may necessitate removing some lower-end and/or terminal cycles to prevent smearing due to excess cycling.

        iv. Vary concentrations of other buffer components (pH, *Taq* polymerase, dNTPs, primers).

        v. Attempt band purification followed by reamplification. Target bands can be cut from agarose gels and allowed to diffuse out or be liberated by freeze/thaw cycles or enzymatic gel digestion. Alternatively, a small plug of gel can be removed with a micropipette tip or, most simply, by stabbing the band directly in the gel with an autoclaved toothpick and inoculating a fresh reaction tube.

   vi. Reamplify $1:10^4$ and $1:10^5$ dilutions of first reaction using nested primers.

  vii. If all else fails, abandon primer set, design new primers, and begin again.

Other Optimization Strategies

Several other optimization strategies have been developed for standard PCR, and most are applicable to TD PCR as well. Each is discussed briefly below. Variables that affect PCR product specificity and yield are listed in Table 1. Additional discussion of PCR optimization and contamination avoidance strategies can be found in Newton and Graham (1994) and McPherson and Møller (2000).

Enhancing Agents

Various additives such as dimethylsulfoxide (DMSO) (1–10%), polyethylene glycol (PEG) 6000 (5–15%), glycerol (5–20%), nonionic detergents, formamide (1.25–10%), and bovine serum albumin (10 to 100 μg/ml) can also be incorporated into the reaction to increase specificity and yield (Pomp and Medrano 1991; Newton and Graham 1994). In fact, some reactions may amplify only in the presence of such additives (Pomp and Medrano 1991). Several optimization kits incorporating these and other enhancing agents, and a variety of buffers, are currently marketed. See Chapters 5 and 12 for discussions of enhancing agents.

Matrix Analyses

The basic challenge is to devise an optimization protocol that is efficient in both time and cost. A full matrix analysis, in which several values for each of the variables are tested in

TABLE 1. Conditions favoring enhanced specificity

- Use hot start
- Use TD PCR (favors enhances specificity *and* sensitivity)
- Optimize primer design
- $\downarrow Mg^{++}$
- $\downarrow$dNTP (also favors higher fidelity)
- Optimize pH
- $\downarrow Taq$ polymerase
- $\downarrow$ Cycle segment lengths
- $\downarrow$ Number of cycles
- $\uparrow$ Annealing temp
- $\downarrow$ Inhibitors
- $\uparrow$ Ramp speed
- Add and optimize enhancer(s)
- $\downarrow$ Primer concentration
- $\downarrow$ Primer degeneracy
- $\uparrow$ Template denaturation efficiency

Adjusting conditions in the direction opposite that listed above usually favors increased sensitivity (i.e., more product) and the concomitant risk of nonspecific amplification. The aim is to strike a balance between these two opposing tendencies. $\uparrow$ and $\downarrow$ signify increase and decrease, respectively.

combination with each of the other variables, can quickly become overwhelmingly cumbersome and costly. The size of the matrix can be significantly pared down by applying the Taguchi method (Taguchi 1986), in which several key variables are simultaneously altered (Cobb and Clarkson 1994). A more typical strategy is to run a simple matrix analysis focused on those parameters most likely to have the greatest impact on PCR primer hybridization and enzyme fidelity; i.e., Mg^{++} concentration and annealing temperature.

Mg^{++} Concentration

Mg^{++} concentration is the easiest to manipulate because all concentration variations can be run simultaneously in separate tubes. Suppliers of *Taq* polymerase now provide $MgCl_2$ solution separate from the rest of the standard reaction buffer to simplify its adjustment. A typical two-step optimization series might first include Mg^{++} at 0.5-mM increments from 0.5 to 5.0-mM and, after the range is narrowed, a second round covered by several 0.2- or 0.3-mM increments.

Annealing Temperature

Optimization of annealing temperature begins with calculation of the T_m values of the primer-template pairs by one of several methods, the simplest being $T_m = 4(G + C) + 2(A + T)$ for primers less than 21 bases long. A single-base mismatch lowers the T_m by about 5°C. More complex formulas can also be used (Sambrook et al. 1989; Sharrocks 1994), but in practice, because the T_m is variously affected by the individual buffer components and even the primer and template concentrations, any calculated T_m value should be regarded as an approximation. Several reactions run at temperature increments (2–5°C) straddling a point 5°C below the calculated T_m will give a first approximation of the optimum annealing temperature for a given set of reaction conditions. It should be noted that some primers, for reasons that are not entirely apparent, are refractory to optimization (He et al. 1994). One possible explanation may be that unique characteristics of the target amplicon give a T_m above the temperature of the denaturation cycle segment (Sharrocks 1994). If permissible, it may be more time- and cost-efficient simply to design a second set of primers that hybridize to neighboring DNA. Thermal cyclers that generate a uniform temperature gradient across the heating block can greatly simplify determination of the optimum annealing temperature, but the precise thermal characteristics yielding the optimum amplification may be difficult to determine for some models.

Cycle Number, Reamplification, and Product Smearing

Increasing the number of cycles may enhance an anemic reaction, but this modification can also lead to the generation of spurious bands and to smears composed of high-molecular-weight products rich in single-stranded DNA (Bell and DeMarini 1991). Similar smearing can occur under normal conditions if the quantity of starting template is too great, as often occurs in attempts to reamplify from a previous PCR. A general rule of thumb for reamplification of a product that has been detected on an agarose gel is to use 1 μl of a 1:10^4 to 10^5 dilution of the PCR.

HOT-START PCR

Even brief incubations of a PCR mix at temperatures significantly below the T_m can result in primer–dimer formation and nonspecific priming. Hot-start PCR methods (D'Aquila et al. 1991; Erlich et al. 1991; Ruano et al. 1992) can dramatically reduce these problems. The aim is to withhold at least one of the critical components from participating in the reaction until the temperature in the first cycle rises above the T_m of the reactants. For example, in smaller assays incorporating an oil overlay, one of the components common to all tubes (e.g., *Taq* polymerase) can be initially withheld and added only after the temperature rises above 85°C during the first denaturing stage. Alternatively, a wax bead can be melted over the bulk of the reaction mix in each tube and allowed to solidify, and the withheld component can be pipetted on top of the wax cap. These beads can be made in the laboratory (Bassam and Caetano-Anolles 1993; Wainwright and Seifert 1993) or purchased (Ampliwax PCR Gems, Perkin Elmer). During the temperature ramp into the first denaturation segment, the wax will melt and the final component will become incorporated and mixed by convection in each tube, a great convenience when dealing with large numbers of tubes. Another variation on this theme is the use of antibody to *Taq* (TaqStart Antibodies, Clontech) that binds to and prevents the function of the enzyme until the antibody is denatured by high heat in the first cycle segment. One can also buy TaqStart Antibodies preassociated with *Taq* DNA polymerase (JumpStart Taq, Sigma) or forms of *Taq* that are inherently inactive at lower temperature (AmpliTaq Gold, P E Biosystems; HotStarTaq, Qiagen). Additional information on these enzymes is provided in Chapter 3. These modifications are compatible with techniques that seek to avoid the extra handling and purification steps accompanying oil and wax addition and sample recovery. Wax-encapsulated magnesium (Start Spheres, Stratagene) or encapsulated *Taq* polymerase (TaqBeads, Promega) is also available. The amount of wax in each is minimal and does not form a vapor barrier.

▶ TROUBLESHOOTING

- *Little or no detectable product.* When the Mg^{++} concentration, buffer pH, and cycling parameters have all been adjusted; when extra cycles have been added; and when lower annealing temperatures and TD PCR have been tried, and still no product is seen on ethidium bromide-stained gels (acrylamide gels are considerably more sensitive than agarose gels), what should be the next step? Lengthening the initial denaturation step and/or increasing temperature will increase the likelihood that the template DNA is fully denatured to provide the maximal number of priming sites. The addition of betaine, as recommended in Chapters 5 and 6, should also be tried. Standard conditions for this optional step are 5 minutes at 95°C. An in-tube thermocouple can be used to predetermine that the indicated temperature will correspond to the actual sample temperature. Amplification may have occurred but have been inefficient. If so, the amplicons can be revealed by a probe of the dried gel or a blot. A secondary amplification using the same primers or, preferably, nested primers may be all that is needed to generate a specific product. Serial 10-fold dilutions ranging from 1:100 to 1:10,000 should be used.

  Little or no product may indicate the presence of inhibitors in the DNA sample. Numerous inhibitors of PCR have been described. These include ionic detergents (e.g., SDS and Sarkosyl; Weyant et al. 1990), phenol, heparin (Beutler et al. 1990), xylene cyanol, and bromophenol blue (Hoppe et al. 1992). First try reamplification of a 100-fold dilution of the starting template, because it is often possible to dilute out the inhibitor before the DNA template. Alternatively, test for inhibitor in the template preparation by spiking the original PCR mix with dilutions of known positive (demonstrably amplifiable) template. Reextraction, ethanol precipitation, and/or centrifugal ultrafiltration may resolve the problem. Proteinase K carryover can serve to digest the *Taq* polymerase but is readily denatured by a 5-minute incubation at 95°C.

ACKNOWLEDGMENTS

I thank Rani Dhanarajan, Dan Garza, and Karl Hecker for their valuable comments.

REFERENCES

Bassam B.J. and Caetano-Anolles G. 1993. Automated "hot start" PCR using mineral oil and paraffin wax. *BioTechniques* **14:** 30–34.

Batzer M.A., Carlton J.E., and Deininger P.L. 1991. Enhanced evolutionary PCR using oligonucleotides with inosine at the 3´-terminus. *Nucleic Acids Res.* **19:** 5081.

Bell D.A. and DeMarini D. 1991. Excessive cycling converts PCR products to random-length higher molecular weight fragments. *Nucleic Acids Res.* **19:** 5079.

Beutler E., Gelbart T., and Kuhl W. 1990. Interference of heparin with the polymerase chain reaction. *BioTechniques* **9:** 166.

Cobb B.D. and Clarkson J.M. 1994. A simple procedure for optimizing the polymerase chain reaction (PCR) using modified Taguchi methods. *Nucleic Acids Res.* **22:** 3801–3805.

D'Aquila R.T., Bechtel L.J., Videler J.A., Eron J.J., Gorczyca P., and Kaplan J.C. 1991. Maximizing sensitivity and specificity of PCR by pre-amplification heating. *Nucleic Acids Res.* **19:** 3749.

Don R.H., Cox P.T., Wainwright B.J., Baker K., and Mattick J.S. 1991. 'Touchdown' PCR to circumvent spurious priming during gene amplification. *Nucleic Acids Res.* **19:** 4008.

Erlich H.A., Gelfand D., Sninsky J.J. 1991. Recent advances in the polymerase chain reaction. *Science* **252:** 1643–1651.

He Q., Marjamaki M., Soini H., Mertsola J., and Viljanen M.K. 1994. Primers are decisive for sensitivity of PCR. *BioTechniques* **17:** 82–87.

Hecker K.H. and Roux K.H. 1996. High and low annealing temperatures increase both specificity and yield in touchdown and stepdown PCR. *BioTechniques* **20:** 478–485.

Hoppe B.L., Conti-Tronconi B.M., and Horton R.M. 1992. Gel-loading dyes compatible with PCR. *BioTechniques* **12:** 679–680.

Knoth K., Roberds S., Poteet C., and Tamkun M. 1988. Highly degenerate, inosine-containing primers specifically amplify rare cDNA using the polymerase chain reaction. *Nucleic Acids Res.* **16:** 10932.

Lee C.C., Wu X.W., Gibbs R.A., Cook R.G., Muzny D.M., and Caskey C.T. 1988. Generation of cDNA probes directed by amino acid sequence: Cloning of urate oxidase. *Science* **239:** 1288–1291.

McPherson M.J. and Møller S.G. 2000. *PCR basics: From background to bench*. Springer-Verlag, New York.

Newton C.R. and Graham A. 1994. *Introduction to biotechniques*. Bios Scientific, Oxford, United Kingdom.

Patil R.V. and E.E. Dekker. 1990. PCR amplification of an *Escherichia coli* gene using mixed primers containing deoxyinosine at ambiguous positions in degenerate amino acid codons. *Nucleic Acids Res.* **18:** 3080.

Peterson M.G., Inostroza J., Maxon M.E., Flores O., Adomon A., Reinberg D., and Tjian R. 1991. Structure and functional properties of human general transcription factor IIE. *Nature* **354:** 369–373.

Pomp D. and Medrano J.F. 1991. Organic solvents as facilitators of polymerase chain reaction. *BioTechniques* 10: 58–59.

Roux K.H. 1994. Using mismatched primer-template pairs in touchdown PCR. *BioTechniques* **16:** 812–814.

Ruano G., Pagliaro E.M., Schwartz T.R., Lamy K., Messina D., Gaensslen R.E., and Lee H.C. 1992. Heat-soaked PCR: An efficient method for DNA amplification with applications to forensic analysis. *Biotechniques.* 13: 266–274.

Sambrook J., Fritsch E.F., and Maniatis T. 1989. *Molecular cloning: A laboratory manual*, 2nd edition. Cold Spring Harbor Laboratory Press, Cold Spring Harbor, New York.

Sharrocks A.D. 1994. The design of primers for PCR. In *PCR technology: Current innovations* (ed. H.G. Griffin and A.M. Griffin), pp. 5-11. CRC Press, Boca Raton, Florida.

Taguchi G. 1986. *Introduction to quality engineering*: Designing quality intro products and processes. Asian Productivity Organization, UNIPUB, New York and The Organization, Tokyo, Japan.

Wainwright L.A. and Seifert H.S. 1993. Paraffin beads can replace mineral oil as an evaporation barrier in PCR. *BioTechniques* **14:** 34–36.

Weyant R.S., Edmonds P., Swaminathan B. 1990. Effect of ionic and nonionic detergents on the *Taq* polymerase. *BioTechniques* **9:** 308–309.

5 PCR Amplification of Highly GC-rich Regions

Lise Lotte Hansen[1] and Just Justesen[2]

[1]Department of Human Genetics, [2]Department of Molecular Biology, University of Aarhus, DK-8000, Aarhus C, Denmark

DNA segments with very high GC content have proved difficult to handle in a wide range of molecular analyses. GC-rich regions may form rigid, constrained secondary structures that are difficult or impossible for the DNA polymerases to enter under standard PCR conditions. Highly GC-rich regions can obstruct PCR-dependent analyses in the following situations:

1. *Standard PCR amplification.* Highly GC-rich regions prevent template denaturation, and hence product synthesis.

2. *Multiplex PCR amplification.* The preferential amplification of low-GC-content sequences results in misinterpretation of the balance between the two sequences.

3. *Quantitative PCR amplification.* GC-rich targets may form heteroduplexes with internal standards, which are designed to be very similar to target sequences, leading to misinterpretation of results.

4. *DNA sequencing.* DNA polymerase tends to stall in GC-rich regions, which results in compression and difficulty in interpreting these parts of the sequence.

5. *cDNA synthesis.* cDNA synthesis of GC-rich templates may create shorter fragments that lack the GC-rich portions of the sequence. The result is a skewed representation of the original mRNA molecules.

Over the years, different approaches have been used to overcome the problems caused by secondary structure. Hot-start *Taq* polymerases have been especially designed to function only after an extended initial denaturing step at high temperature. The specificity of these polymerases is higher than that of standard *Taq* polymerases. To overcome the problem of DNA sequence compressions, the template can be denatured using either sodium hydroxide or the nucleotide analogs deoxyinositol triphosphate (dITP), and 7-deaza GTP could partly substitute the dGTP (Agarwal and Perl 1993; Mutter and Boynton 1995; Turner and Jenkins 1995). Organic additives such as dimethylsulfoxide (DMSO) (Winship 1989; Pomp and Medrano 1991), formamide (Sarkar et al. 1990), betaine (Baskaran et al. 1996; Weissensteiner and Lanchbury 1996; Henke et al. 1997), glycine (Sarkar et al. 1990), low-molecular-weight sulfones that are chemically related to DMSO (e.g., tetramethylene sulfoxide, tetramethyl sulfone [sulfolane] and methyl sulfone) (Chakrabarti and Schutt 2001a, 2002), and, more recently, low-molecular-weight amides (Chakrabarti and Schutt 2001b) have all proved successful, to some degree, in solving the problems associated with highly constricted DNA and RNA structures.

BETAINE

Betaine (*N,N,N,*-trimethylglycine) is an amino acid analog and a major metabolite of choline metabolism, which is present in liver and kidney cells (Mar et al. 1995). Betaine is the major regulator of osmotic pressure in plants, a role that serves to protect proteins in vivo and facilitates refolding of proteins in vitro (Tieman et al. 2001). Addition of betaine to a PCR mix protects the *Taq* polymerase from denaturation during the high-temperature steps of the cycles, thereby maintaining the efficiency of the enzyme during amplification (Hengen 1997). Betaine is a zwitterion at neutral pH and, even at high concentrations (> 5 M), it has only a minimal effect on the electrostatic interactions between DNA and proteins. This property facilitates the maintenance of stable DNA–protein complexes (Rees et al. 1993).

High concentrations of betaine eliminate the differences in melting temperature between AT and GC domains, without changing the double-stranded DNA conformation of the B form. It has been proposed that betaine preferentially binds to AT pairs in a weak and noncooperative manner in the major groove of the B-form double helix, which may explain the ability of betaine to eliminate the discrepancy in the melting properties of the different base pairs (Rees et al. 1993). It has also been suggested that betaine changes the hydration of AT regions, thereby affecting the local structure of the DNA molecule, and that a relative increase in hydration of GC-rich regions may increase the flexibility of these rigid sequences (Mytelka and Chamberlin 1996).

LOW-MOLECULAR-WEIGHT SULFOXIDES, AMIDES, AND GLYCINES

In the search for more potent amplification enhancers for highly GC-rich templates, several classes of low-molecular-weight compounds, sulfoxides and amides, have been analyzed. Among the sulfoxides, tetramethylene sulfoxide and tetramethylene sulfolane were found to efficiently enhance the amplification of templates with GC contents from 52% to 73%. (Chakrabarti and Schutt 2001a, 2002). In a separate study, acetamide had the highest specificity of the analyzed chemicals, which were pyrrolidones, formamides, acetamides, and high-chain primary amides such as propionamide and isobutylamide (Chakrabarti and Schutt 2001b). It should be noted, however, that all these comparisons were carried out using a number of different cDNA templates.

Chemicals, such as *N,N*-dimethylglycine, *N*-monomethylglycine (sarcosine), and trimethylamine *N*-oxide (TMANO), which are structurally similar to betaine, have been analyzed for their effect on T7 polymerase sequencing of supercoiled DNA. Betaine proved to be the most efficient facilitator, followed by TMANO, and then by *N,N*-dimethylglycine, which was intermediate in activity. Sarcosine showed only a minor ability to overcome the secondary structures of the DNA template (Mytelka and Chamberlin 1996).

BETAINE APPLICATIONS

DNA Sequencing

DNA regions with high melting temperatures, or with the consensus sequence of Py-G-C, may cause DNA polymerases to stall, resulting in compression of the DNA sequencing ladder and difficulty in the correct interpretation of the sequence (Mytelka and Chamberlin 1996). The addition of 2 M betaine to these difficult sequencing reactions can eliminate the problem more efficiently than the structurally similar chemicals dimethylglycine and sarcosine. The efficiency of TMANO, used at 0.25–2.0 M, was similar to that of betaine in these sequencing reactions.

Multiplex PCR Amplification

GC-rich fragments may be strongly underrepresented after standard multiplex PCR amplifications. The addition of 1 M betaine and 5% (v/v) DMSO may equalize the amplification efficiency of different DNA fragments (Baskaran et al. 1996).

Reverse Transcription PCR

Addition of 2 M betaine alone, or accompanied by 0.6 M trehalose, can dramatically enhance reverse transcriptase reactions. For example, in the presence of these additives, the yield of 12.5-kb cDNA fragments was increased 9-fold, as compared with reverse transcriptase (RT)-PCR amplifications without additives (Spiess and Ivell 2002).

Quantitative PCR

Internal standards for use in quantitative PCR amplifications are designed to be very similar to the target molecule. Shammas et al. observed that the target and the internal standard formed heteroduplexes/recombinants, even after the first amplification cycle. Addition of 2 M betaine, 5% formamide, and 10% DMSO to the PCR mix reduced the amount of recombinants to below a detectable level (Shammas et al. 2001; McDowell et al. 1998).

Microarray Studies

In addition to its use in enhancing amplification of GC-rich sequences, betaine proved useful in microarray studies. The addition of 1.5 M betaine to the DNA solution results in homogeneous spots and a significant reduction of the nonspecific background signal. Betaine reduces the evaporation from the DNA samples in the microtiter plates during the manufacturing of the slides. After the DNA has been applied to the glass slide, the diminished evaporation may provide enough time for the DNA to be distributed uniformly within each spot, thereby inhibiting the "doughnut effect" that disturbs the automatic interpretation of the results (Diehl et al. 2001).

PROTOCOL 1 ESTABLISHING AN AMPLIFICATION PROCEDURE FOR HIGHLY GC-RICH REGIONS

The following procedure details the establishment of an amplification procedure for GC-rich sequences. In this example, three amplicons from two different genes containing GC-rich sequences were used. Fragment 1 comprised the initial part of exon 1 from the human release factor 3 (GSPT1/hRF3), a 387-bp sequence that contains two trinucleotide repeats: a (GGC) followed by an (AGC), and a total GC content of 75%. The second and third fragments come from the Klotho gene (LocusLink number for the genes Klotho1: 9365 and GSPT1/hRF3: 2935). Fragment 2 comprised the 5′ end of Klotho exon 1 (375 bp, 81% GC) and fragment 3, the 3′ part of Klotho exon 1 (350 bp, 70% GC).

The DNA fragments were each amplified in the presence of either 5% DMSO, 1 M betaine, 2 M betaine, 1 M betaine and 5% DMSO, 2 M betaine and 5% DMSO, or 0.4 M tetramethylene sulfone, or without any of the enhancers.

MATERIALS

BUFFERS, SOLUTIONS, AND REAGENTS
Betaine (Sigma-Aldrich)
dNTP solution (containing all four dNTPs, each at 25 mM)
Dimethyl sulfoxide (DMSO)<!> (Sigma-Aldrich)
Tetramethylene sulfone (Sulfolane)<!> (Sigma-Aldrich)

ENZYMES AND ENZYME BUFFERS
Taq polymerase
Polymerase buffer (as supplied by enzyme manufacturer)

NUCLEIC ACIDS AND OLIGONUCLEOTIDES
Human genomic DNA, 20 ng per amplification reaction

hRF3 (hGSPT1) primers:	hRF3f: CCGCCTCTGTCGTCGTCGC
	hRF3r: CCGCGCTGAGGTTCTCCC
Klotho exon 1a primers:	Klo1af: CTCGCAGGTAATTATTGCCAG
	Klo1ar: GATGGACGCACCCTTG
Klotho exon 1c primers:	Klo1cf: GTGCAGCCCGTGGTCAC
	Klo1cr: GACTCAGTTCCCACACTTC

SPECIAL EQUIPMENT
Thermal cycler

ADDITIONAL ITEM
Equipment and reagents for agarose gel electrophoresis

METHOD

1. In a sterile 0.2-ml microfuge tube, mix the following reagents;

Genomic DNA	20 ng
Primers	15 pmoles of each
Taq polymerase	0.5 units
10x buffer	2.5 µl
dNTP solution (containing all four dNTPs)	250 µM
H$_2$O	25 µl

   Depending on the chosen conditions, include either DMSO (5%); betaine (1 M); betaine (2 M); betaine and DMSO (1 M and 5%, respectively); betaine and DMSO (2 M and 5%, respectively); tetramethylene sulfone (0.4 M); or no enhancers (final concentrations) in the PCR mix.

2. Carry out 30 cycles of amplification using a step program listed in program A or B as follows:

Program A: This program was used for the Klotho exon 1a and 1c fragments.

Cycle number	Denaturation	Annealing	Polymerization/Extension
1	40 sec at 96°C	36 sec at specified annealing temp.	40 sec at 72°C
36	30 sec at 95°C	36 sec at specified annealing temp.	40 sec at 72°C
1	30 sec at 95°C	36 sec at specified annealing temp.	10 min at 72°C

Alternatively, a longer denaturing step may be introduced. We have experienced that a short initial denaturing is sufficient for amplification of genomic fragments of up to 80% GC content.

Program B: This program was used for amplification of GSPT1/hRF3.

Cycle number	Denaturation	Annealing	Polymerization/Extension
1	5 min at 95°C	36 sec at specified annealing temp.	40 sec at 72°C
30	30 sec at 95°C	36 sec at specified annealing temp.	40 sec at 72°C
1	30 sec at 95°C	36 sec at specified annealing temp.	10 min at 72°C

3. Analyze an aliquot of the amplification product on an agarose gel.

Results

The results are shown in Figures 1–3. As shown in Figure 1, nearly all of the enhancers had a positive effect of the yield of GSPT1. However, different sequences have different responses to the additives. In Figures 2 and 3, the Klotho exon 1a fragment and exon 1c fragment were maximally enhanced by the combination of DMSO and betaine (Fig. 2, lane 5; Fig. 3, lane 5).

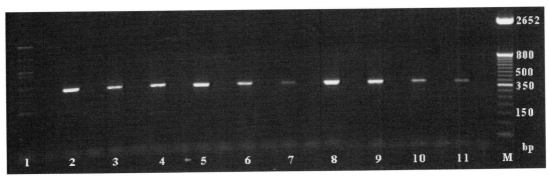

FIGURE 1. PCR amplification of the 5′ part of human release factor 3 (GSPT1). The fragment contains two trinucleotide repeats with 75% GC content. (Lane *1*) No additives; (lane *2*) 5% DMSO; (lane *3*) 1 M betaine; (lane *4*) 2 M betaine; (lane *5*) 1 M betaine and 5% DMSO; (lane *6*) 2 M betaine and 5% DMSO; (lane *7*) 0.4 M tetramethylene sulfolane; (lane *8*) 1 M betaine and 10% DMSO; (lane *9*) 10% DMSO; (lane *10*) 2 M betaine and 10% DMSO; (lane *11*) 0.3 M tetramethylene sulfolane; (lane *M*) 50-bp molecular-weight marker.

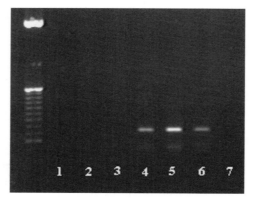

FIGURE 2. PCR amplification of Klotho exon 1a, a fragment with 81% GC content. (Lane *1*) No additives; (lane *2*) 5% DMSO; (lane *3*) 1 M betaine; (lane *4*) 2 M betaine; (lane *5*) 1 M betaine and 5% DMSO; (lane *6*) 2 M betaine and 5% DMSO; (lane *7*) 0.4 M tetramethylene sulfolane. The size marker is a 50-bp ladder.

The results in Figures 2 and 4 show that Klotho exon 1a (81% GC content) was PCR-amplified according to program A in the presence of increasing concentrations of betaine. The optimal yield of the PCR product was seen with 5% DMSO and 1.5 M betaine (program A) as shown in Figure 4.

▶ TROUBLESHOOTING

It is important to evaluate the GC content of the template sequences. The following sections offer some advice on solving the problems associated with GC-rich templates (>50% GC).

- *No PCR product.* Lower the annealing temperature to 2°C below the lowest primer melting temperature (T_m). As a rule of thumb, AT adds 2°C and GC 4°C to the T_m of a sequence. Then add 1.5 M betaine and 5% (v/v) DMSO to the PCR amplification.

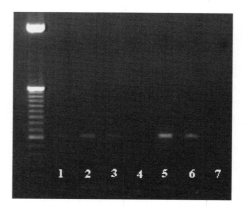

FIGURE 3. PCR amplification of the 3′ part of Klotho exon 1c with a GC content of 70%. (Lane *1*) No additives; (lane *2*) 5% DMSO; (lane *3*) 1 M betaine; (lane *4*) 2 M betaine; (lane *5*) 1 M betaine and 5% DMSO; (lane *6*) 2 M betaine and 5% DMSO; (lane *7*) 0.4 M tetramethylene sulfolane. The size marker is a 50-bp ladder.

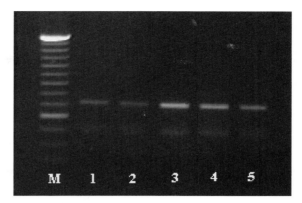

FIGURE 4. Klotho exon 1a amplified with 5% DMSO and different amounts of betaine. (Lane *1*) 0.5 M betaine; (lane *2*) 1 M betaine; (lane *3*) 1.5 M betaine; (lane *4*) 2 M betaine; (lane *5*) 2.5 M betaine. M is a 50-bp molecular-weight marker.

If no effect is seen, increase the betaine concentration to 2.5 M, add 10% (v/v) DMSO, and double the amount of *Taq* polymerase (DMSO inhibits *Taq* polymerase activity). The annealing temperature can be further decreased. Under normal conditions, lower annealing temperature leads to generation of nonspecific products, but the addition of betaine diminishes this.

- *Many unspecific PCR products.* Increase the annealing temperature and add 1.5 M betaine and 5% (v/v) DMSO. See Figure 1, lane 1, which shows a number of nonspecific PCR products generated from a PCR amplification of two adjacent trinucleotide repeated sequences with no additives. If no effect is seen, increase the betaine and DMSO concentrations further.

DISCUSSION

In this chapter, we have focused on betaine as the most successful PCR amplification enhancer of the wide spectrum of additives used to overcome constrained DNA structures. Furthermore, betaine has proved very efficient in a variety of PCR techniques, with both RNA and DNA as templates (Table 1).

We found that highly GC-rich regions are very efficiently amplified in the presence of 1.5–2.5 M betaine, and the yield for some fragments is further increased by addition of 5–10% DMSO. The amplification of a fragment with 81% GC content was only possible in the presence of either 2 M betaine or 1–2 M betaine and 5% DMSO (Fig. 2).

DMSO was seen to further enhance the amplification of a fragment with 70% GC, with the highest yield in the presence of 1 M betaine and 5% DMSO (Fig. 3). For the GSPT1 fragment (75% GC), many unspecific PCR products were present in the reaction without any additives. DMSO or betaine, alone or together, enhanced the amplification, and the product was a single clear and specific band on the agarose gel (Fig. 1).

Importantly, betaine does not interfere with further processing of the PCR product such as mutation detection, single-nucleotide polymorphism (SNP) analysis, and real-time PCR. This is not true for amplicons modified by the incorporation of 7-deaza GTP or dITP instead of dGTP.

TABLE 1. PCR applications that can be improved by addition of betaine

PCR applications	Additional reagents	Advantages	References
cDNA synthesis and 5′ RACE		overcome the secondary structures or loops in RNA, which may inhibit the first strand synthesis, thereby resulting in non-full length transcript	Spiess and Ivell (2002)
DNA cycle sequencing	Nucleotide analogs (dITP, 7-deaza GTP)	overcome the compression of sequences due to stalling of the polymerase	Mytelka and Chamberlin (1996)
Long and accurate PCR amplification	DMSO	increased yield of long PCR products (~5 kilo base pair)	Baskaran et al. (1996); Hengen (1997)
Long-chain cDNA synthesis	Trehalose	increased yield of 12.5 kb synthesis fragments from cDNA (~9 fold)	Spiess and Ivell (2002)
Multiplex PCR amplification	DMSO	equal representation of GC-rich and non GC-rich alleles in the PCR product	Baskaran et al. (1996)
Quantitative PCR amplification		high reproducibility; suppression of recombination leading to equal proportional amplification of the fragments	Shammas et al. (2001)

CONCLUSION

The optimal conditions for PCR amplification of a DNA template vary according to the GC content, composition of the template, and even the PCR machine. Comparison of the available information has led us to suggest the following scheme for solving PCR/sequencing problems due to GC-rich templates:

When optimizing the reaction conditions, add 1.5 M betaine (± 0.5 M) or 5% DMSO. If this does not give the desired result, a combination of the two can be tried. If this fails, or if the result is disappointing, then the addition of 0.8 M of either pyrrilidone or acetamide, 0.7 M tetramethylene sulfoxide or 0.4 M tetramethylene sulfone (sulfonal) may help.

ACKNOWLEDGMENTS

E. Hein and T. Kjeldsen are acknowledged for excellent technical assistance and Professor L.L. Kisselev for introducing us to betaine. This work was supported by The Novo Nordisk Foundation.

REFERENCES

Agarwal R.K. and Perl A. 1993. PCR amplification of highly GC-rich DNA template after denaturation by NaOH. *Nucleic Acids Res*. 21: 5283-5284.

Baskaran N., Kandpal R.P., Bhargava A.K., Glynn M.W., Bale A., and Weissman S.M. 1996. Uniform amplification of a mixture of deoxyribonucleic acids with varying GC content. *Genome Res*. **6:** 633–638.

Chakrabarti R. and Schutt C.E. 2001a. The enhancement of PCR amplification by low molecular-weight sulfones. *Gene* **274:** 293–298.

———. 2001b. The enhancement of PCR amplification by low molecular weight amides. *Nucleic Acids Res*. **29:** 2377–2381.

———. 2002. Novel sulfoxides facilitate GC-rich template amplification. *BioTechniques*. **32:** 866–874.

Diehl F., Grahlmann S., Beier M., and Hoheisel J.D. 2001. Manufacturing DNA microarrays of high spot homogeneity and reduced background signal. *Nucleic Acids Res.* **29:** E38.

Hengen P.N. 1997. Optimizing multiplex and LA-PCR with betaine. *Trends Biochem. Sci.* **22:** 225–226.

Henke W., Herdel K., Jung K., Schnorr D., and Loening S.A. 1997. Betaine improves the PCR amplification of GC-rich DNA sequences. *Nucleic Acids Res.* **25:** 3957–3958.

Mar M.H., Ridky T.W., Garner S.C., and Zeisel S.H. 1995. A method for the determination of betaine in tissues using high performance liquid chromatography. *J. Nutr. Biochem.* **6:** 392–398.

McDowell D.G., Burns N.A., and Parkes H.C. 1998. Localised sequence regions possessing high melting temperatures prevent the amplification of a DNA mimic in competitive PCR. *Nucleic Acids Res.* **26:** 3340–3347.

Mutter G.L. and Boynton K.A. 1995. PCR bias in amplification of androgen receptor alleles, a trinucleotide repeat marker used in clonality studies. *Nucleic Acids Res.* **23:** 1411–1418.

Mytelka D.S. and Chamberlin M.J. 1996. Analysis and suppression of DNA polymerase pauses associated with a trinucleotide consensus. *Nucleic Acids Res.* **24:** 2774–2781.

Pomp D. and Medrano J.F. 1991. Organic solvents as facilitators of polymerase chain reaction. *BioTechniques.* **10:** 58–59.

Rees W.A., Yager T.D., Korte J., and Von Hippel P.H. 1993. Betaine can eliminate the base pair composition dependence of DNA melting. *Biochemistry* **32:** 137–144.

Sarkar G., Kapelner S., and Sommer S.S. 1990. Formamide can dramatically improve the specificity of PCR. *Nucleic Acids Res.* **18:** 7465.

Shammas F.V., Heikkila R., and Osland A. 2001. Fluorescence-based method for measuring and determining the mechanisms of recombination in quantitative PCR. *Clin. Chim. Acta* **304:** 19–28.

Spiess A.N. and Ivell R. 2002. A highly efficient method for long-chain cDNA synthesis using trehalose and betaine. *Anal. Biochem.* **301:** 168–174.

Tieman B.C., Johnston M.F., and Fisher M.T. 2001. A comparison of the GroE chaperonin requirements for sequentially and structurally homologous malate dehydrogenases: The importance of folding kinetics and solution environment. *J. Biol. Chem.* **276:** 44541–44550.

Turner S.L. and Jenkins F.J. 1995. Use of deoxyinosine in PCR to improve amplification of GC-rich DNA. *BioTechniques.* **19:** 48–52.

Weissensteiner T. and Lanchbury J.S. 1996. Strategy for controlling preferential amplification and avoiding false negatives in PCR typing. *BioTechniques.* **21:** 1102–1108.

Winship P.R. 1989. An improved method for directly sequencing PCR amplified material using dimethyl sulphoxide. *Nucleic Acids Res.* **17:** 1266.

6 Tips for Long and Accurate PCR

Wayne M. Barnes

Department of Biochemistry and Molecular Biophysics, Washington University School of Medicine, St. Louis, Missouri 63110, and DNA Polymerase Technology, Inc., St. Louis, Missouri, 63108

In 1992, the author, as well as others, was working on mixing and fusing various domains of different DNA polymerases to achieve longer amplicons, combined with the robust reliability of *Taq* DNA polymerase. It was assumed that processivity (long "hang time," i.e., staying on the DNA for long periods and long distances) was key to long PCR. That may yet prove true, but so far, the opposite is true. Klentaq1, an amino-terminal deletion of *Taq* DNA polymerase, is even less processive than the full-length *Taq*, yet it was used to set the distance record of 35 kb in 1993. Ironically, it was found that less hang time allowed other components of an enzyme mixture to access the growing chain to repair or edit problem DNA molecules that were limiting extension—those with a mismatched base at the 3´ end.

In this chapter, the acronym LA means "long and accurate"—"long" being more than the ~3-kb limit of efficient PCR without a mixture of enzymes and "accurate" meaning high fidelity, because the long PCR mixtures were found to simultaneously provide some 10-fold fewer mutations in the PCR product (Barnes 1994a). LA used as a suffix means a mixture of DNA polymerases, the major one usually being *Taq* or Klentaq1 (which have no 3´-exonuclease proofreading activity) and the minor one an archaebacterial DNA polymerase such as DeepVent, Vent, or *Pfu* (Barnes 1994a). TaqLA means *Taq* DNA polymerase mixed with a small amount of DeepVent or *Pfu*. TaqLA is sold under various names, such as Expand, Elongase, ExTaq, TaqPlus, and Accutaq. KlentaqLA means Klentaq1 with a similar low-level partner of proofreading DNA polymerase.

Other components are being introduced to improve LA PCR. Among these are one enzyme component and one chemical component as follows:

1. **dUTPase.** Stratagene (Hogrefe et al. 2002) has discovered that a buildup of deaminated dCTP, namely dUTP, was causing the incorporation of uracil into DNA. Uracil in DNA is a disaster for the archaebacterial component of LA PCR mixtures of DNA polymerase, because the archeabacterial DNA polymerases (such as *Pfu*) bind almost irreversiby to dU-containing DNA under in vitro conditions (Lasken et al. 1996). A cure introduced by Stratagene is thermostable dUTPase. This enzyme is now included in some of their PCR enzyme mixtures, in addition to the DNA polymerase components with low and high 3´ exonuclease.

2. **Betaine.** Although introduced for high-GC targets, betaine never hurts, and usually helps, even for long PCR up to at least 20 kb. It is included at surprisingly high levels, such as 2 M final (Baskaran et al. 1996; see tip 4 below).

Since the introduction of mixtures of DNA polymerases (Barnes 1994a), most PCRs of any length have improved in reliability and in yield of product. Nevertheless, some amplifications do fail. When detailed conditions for a failing PCR are examined, usually the experimenter has broken about half of the following "rules" that the author has learned the hard way.

Using these components may improve PCRs so much that they will over-cycle, causing the appearance of a very fat product band and visible side products that only become visible when a (finished) reaction continues to cycle after the main product band is fully formed. For a maximally clean signal, add less template and/or use 5 fewer cycles than used for previous, less efficient amplification.

PROTOCOL

LONG AND ACCURATE PCR

The following protocol is consistent with the rules listed in the left column of Table 1. Suppose 10 various primer pairs will be used for ~5 kb, the template DNA is constant, and the reaction volume is 50 μl. This protocol only has one no-template control, with one of the primer pairs. For high numbers of cycles (25–40) a no-template and/or a one-primer control will also be advisable for each target.

MATERIALS

BUFFERS, SOLUTIONS, AND REAGENTS

Assemble the following reagents with autoclaved water and store them at the recommended temperature.

5 M Betaine (Sigma B-2629 or B-2754)

Filter-sterilize and store at room temperature.

10x KLA PCR reaction buffer, pH 9.2

500 mM Tris base<!> (Sigma; Trizma Base)

160 mM $(NH_4)_2SO_4$ <!>

25 mM $MgCl_2$ <!>

1% Tween 20

> ▼ CAUTION
>
> See Appendix for appropriate handling of materials marked with <!>.

The pH comes out to 9.2 (the optimal for long PCR) without adjustment. For some plasmid replicons, such as colE1, pH 7.9 is superior (see Chapter 29). Add dilute (2 molar, dropwise) HCl to adjust the pH of the 10x KLA to 7.9. Do not stick a pH meter probe into the buffer, as this could spread DNA contamination. Instead, test small aliquots separately. Do not freeze 10x PCR buffer. Filter-sterilize and store at 4°C (don't worry if it becomes cloudy; just shake well before use.) The level of magnesium in KLA is optimal for KlentaqLA. For mixtures that use full-length *Taq*, use 10x TLA, which has only 7.5 mM $MgCl_2$.

TEN+BSA

10 mM Tris-HCl<!>, pH 7.9

10 mM NaCl

0.1 mM EDTA

100 μg/ml bovine serum albumin (BSA)

Store at –20°C.

Always dilute the template DNA in TEN + BSA at or below 10 ng/ml. This increases the reproducibility, especially after freezing.

ENZYMES AND ENZYME BUFFERS

KlentaqLA, 50–55 units/μl (For recommendations on enzyme concentrations as a function of amplicon length, see tip 2, below.)

NUCLEIC ACIDS AND OLIGONUCLEOTIDES

10/40 dNTP mix

10 mM each dNTP

40 mM $MgCl_2$<!>

Each dNTP may be purchased in the dry form from Pharmacia Biotech. Dissolve to 100 mM stock solutions in water and store at –80°C. Store the 10/40 mix in 500-µl aliquots at –80°C, as well, except for the aliquot in current use, which may be stored at –20°C. Surprisingly, higher yields of PCR product are observed using this dNTP mix rather than the (more expensive) ones made up from purchased liquid (100 mM) dNTPs. The author has not done this comparison with the newly available dUTPase, so it remains possible that the slight problem with liquid stocks is dUTP.

10 µM primers

Dilute each primer to 10 pmoles per µl, and use 1 µl per 50-µl reaction. Store frozen at –20°C.

SPECIAL EQUIPMENT

Thermal cycler

METHOD

1. Assemble the following master mix on ice:

10x KLA, pH 9.2. Use pH 7.9 for some plasmid vectors.	60 µl
10/40 dNTP mix (final 100 µM each)	6 µl
5 M Betaine (final 1.2 M) Use final 2 M for some plasmid vectors.	156 µl
H₂O to make 6 x 95 µl.	345 µl
KlentaqLA (1 µl if target is under 2 kb).	3 µl

   Mix thoroughly.

2. Remove the no-template control reaction (48 µl), and add 2 µl of primer set #1 for the no-template control.

3. Add 5.5 µl of genomic DNA at 10 ng/µl to the master mix, and mix thoroughly.

4. Distribute 48 µl to each reaction tube.

5. Add 1 µl of each primer (or 2 µl of each primer pair) to each reaction. Do not worry about mixing, because convection during the cycling will take care of that.

6. Program the thermal cycler to achieve no more than 5–10 seconds at temperature for the heat step of each cycle. This may be a program instruction of 5 seconds to 50 seconds, depending on the brand of thermal cycler.

   Each cycle is thus 50 seconds at 92°C, 10 minutes at 63°C. Use the following table to estimate reasonable cycle numbers.

Template DNA	PCR cycle number
Plasmid	18–22
Bacterial	25–30
Mammalian	35–40

TABLE 1. Tips for reliable long and accurate PCR

	DO		DON'T
1.	Use a maximum of about 1 ng of DNA template per µl of PCR.		Use too much DNA template. 1 µg is way too much.
2.	Per 100 µl of PCR volume, use about 0.1 µl of KlentaqLA per kb of target size. Above 10 kb target size, use a range of 1 to 1.3 µl of enzyme.		Use plain *Taq*. Instead, use a mixture of enzymes licensed under US Patent 5,436,149 (Barnes 1995). These catalyze better PCR even for short targets.
3.	Consider Mg^{++} concentration as the *excess* over the level of dNTPs. After each dNTP has chelated a Mg, the amount excess optimal for KlentaqLA is 2.5 mM. For TaqLA, it's only 0.75 mM.		Try less Mg than dNTPs. Don't add dNTPs without their "own" Mg in their stock.
4.	Include 1.3 M betaine, final, in the PCRs, and set the melt temperature of each cycle to 92–93°C. which is about 2–3°C lower than without betaine.		Use betaine over 20 kb. (Please E-mail exceptions with gel pictures to author.)
5.	Preheat for 5 minutes at 68°C, so that the 2-second window for the melting step can be hit accurately at the first cycle.		Depurinate the template by preheating for 1–10 minutes at 95°C. DNA polymerase cannot cross an abasic site. This is the main current limit to target size. *Except:* Genomic DNA can benefit from a short preheat.
6.	Set the length of each 95°C (93°C with betaine) melt step at the minimum time, say 2 seconds; no more than 5–10 seconds at temperature in the reaction. Robo-cycler needs 30–50 seconds to provide this 5 seconds at temperature.		Heat more than 10 seconds for the melt step of each cycle. Longer will depurinate the DNA template, and decrease yields of targets, especially over 8 kb, since known enzymes cannot cross abasic sites.
7.	Use filter barrier pipette tips for setting up PCRs. Otherwise aerosols can contaminate stock solutions with DNA and "bad seed" from a previous PCR.		Use unfilter barrier tips for any pre-PCR manipulations. Unfiltered tips are OK only for loading gels.
8.	Use separate sets of pipettors for pre- and post-PCR steps. Pre-PCR is setting up PCR. Post-PCR is loading gels or cloning PCR products, etc.		Use standard pipettors for any solution that could go into a PCR. If it is affordable, do not even run gels and/or PCR cycling in the same room with, or upwind of, rooms with PCR setup benches.

TABLE 1. (*continued*)

9. **DO** Use only 1–10 ug of proteinase K per 100 µl of mouse cell suspension.	**DON'T** Use too much proteinase K. Use too much salt, because proteinase K is inhibited by salt around 0.3 M.
10. **DO** Use nice long extension steps. 10 minutes for 2–5 kb, up to 22 minutes for 20–35 kb.	**DON'T** Insist on speed and 30-second to 4-minute extensions.
11. **DO** Extend at 68°C, or even lower, such as 62–65°C.	**DON'T** Extend at 72°C. This may be the optimum for a DNA polymerase assay using nicked calf-thymus DNA as template, but that is not PCR.
12. **DO** Add enzyme from full-strength stock, to a mix for multiple reactions.	**DON'T** Dilute the enzyme mixture in anything but reaction mix with KLA buffer in there first, except for very short periods.
13. **DO** Assemble the RT reaction on ice, add reverse transcriptase, and then warm to 45°C.	**DON'T** Heat the RNA to 90°C or so. Unless there is excess EDTA, Mg and other metals will cause RNA breakdown.

▶ TROUBLESHOOTING

The following tips should be kept in mind in order to establish and maintain good LA PCR performance in the laboratory. Some of these may seem counterintuitive, but this is what works.

- *Tip 1. Use a maximum of about 1 ng of DNA template per microliter of PCR reaction.* More is not always better. Something else must be wrong if you are tempted to use more than 1 ng template per microliter of reaction. For plasmid as template, much less is best: Never use more than 0.1 ng/µl, and usually use only 1 ng in the whole reaction.

- *Tip 2. Plain Taq is old technology.* Use a mixture of enzymes optimized for efficiency, fidelity, and length. The author favors his original mixture used to discover long and accurate PCR (Barnes 1994a), namely KlentaqLA, which is a mixture of Klentaq1 and DeepVent. This mixture gives consistently more product than mixtures with full-length *Taq* as the major component. The increased yield with Klentaq1 occurs because the 5´-exonuclease (deleted in Klentaq1, an amino-terminal deletion of *Taq* with increased thermal stability) can negatively affect the product (although only at high concentrations because the 5´-exonuclease actually has poor activity on the end of double-stranded DNA; it is optimized instead for forked structures, such as the structure encountered during nick-translation.)

  A disadvantage of KlentaqLA is that a different amount is optimal for different target lengths, although 0.5 µl will usually work well. For 100 µl of reaction volume, only 1/32 µl (1.5 units of enzyme) is enough for 300 bp, and 1 µl can be too much. 1.3 µl (65–70 units of enzyme) is necessary for 35 kb. For TaqLA, one amount (0.5 µl) seems to fit all. For some cases, such as long RT-PCR, a TaqLA (mixture) product such as Expand sometimes works better.

- *Tip 3. Keep Mg⁺⁺ concentration at molar excess over the level of dNTPs.* TaqLA mixtures seem problematic until one realizes that they have more narrow optimal ranges for Mg^{++} levels and they prefer lower levels than does KlentaqLA. It is difficult to hit the optimum magnesium concentration if the chelating activity of the dNTPs is ignored. The level of each dNTP ranges from 100 to 250 µM among recipes. This can have a dramatic effect on the level of magnesium that is left over to bind to the magnesium-binding site on the enzyme at the active site. Excess dNTP over magnesium does not completely prevent DNA polymerase activity and PCR, but it does prevent long PCR.

- *Tip 4. Include betaine at a level of 1.3 M or even 2.0 M final.* Betaine will ensure that high-GC targets amplify easily (Baskaran et al. 1996). Some average GC targets (perhaps those with inverted repeats) nevertheless respond well to betaine, and it never seems to hurt. Even more betaine (2 M final) is recommended for the amplification of whole vectors containing the colE1 replicon. A 3 M level of betaine is being currently evaluated with promising results (W.M. Barnes, unpubl.). See Chapter 5 for more information regarding betaine.

  Use a melting temperature of 91–93°C instead of 94–95°C. Why lower the melting temperature by 2–3°C? Although Baskaran et al. (1996) found that KlentaqLA is more resistant to betaine than TaqLA, the PCR (presumably the enzyme) is a degree or two more sensitive to heat in betaine. Concomitantly, we find that the DNA melts at least two degrees lower. It may not seem like much, but every degree less melting temperature means somewhat less DNA damage by depurination. The relative sensitivity of *Taq* to betaine may become welcomely insignificant at the lower melting temperatures (per cycle) of 91–93°C.

  Although no cases of this are known to the author, it seems reasonable to suppose that the annealing temperature step of a PCR should also be lowered by 2–3°C for short primers that are at the edge of stability. Alternatively, at the same annealing temperature, some PCRs become more specific with (high) betaine, because alternate primer sites are prevented.

  How does betaine work? This is not known for sure, but perhaps it interferes with that third hydrogen bond that GC base pairs have over AT base pairs. At 5.2 M betaine, this makes all the DNA melt as easily as AT regions without affecting specificity (Reese et al. 1993). Preliminary evidence shows that betaine has no effect on fidelity (W.M. Barnes, unpubl.).

- *Tip 5. Do not unnecessarily depurinate your template by preheating at 95°C.* Many recipes, even at the step of preparation of the DNA template, subject the template to a preheat step of 95–100°C for several minutes. This is very bad for the template. The main danger is depurination, leading to abasic sites on the template. DNA polymerases cannot cross an abasic site (exceptions are repairasomes not used for PCR mixtures, at least not yet). Consider reducing any heat step before the PCR to 80°C or even lower, e.g., 68°C. In moderate salt, this will also prevent melting of the DNA. Double-stranded DNA template is much less sensitive to mispriming during the reaction setup and warmup, because how can the primers prime on double-stranded DNA?

  An exception to this rule occurs when high-quality genomic DNA is the template. The PCR cannot start until the DNA template is denatured to the single-stranded form. High-quality DNA is so long (1 million bp) that it cannot denature efficiently in 2 seconds. Therefore, it is helpful for the first, and only the first, heat step at 93–95°C to be 1 or 2 minutes. The resulting DNA damage limits the length of PCR product that is obtainable, but it allows the DNA to denature and become template.

- *Tip 6. Minimize the melt time of each PCR cycle.* For longer PCR, this is a variable worth optimizing for each brand of PCR machine, tubes, and the volume of the reactions. For the Stratagene Robocycler, set the heat step to 30 seconds, plus about 1/4 second per microliter of reaction volume. The manufacturer's recommendations should be considered a good starting point, but let the PCRs be their own thermometer. Using a fairly high GC target, experiment to find the minimum, then add 5–10 seconds for reliability.

- *Tip 7. Use tips with filters in them for setting up PCRs and separate pipettors for before and after PCR.* These tips are an essential part of good PCR and are covered in Chapter 1. With beginner's luck, some PCR targets amplify well for the first week or two in the lab. With beginner's (or expert's) carelessness, repeated amplification of the same target can lead to two kinds of failure: (1) target in the no-template control and (2) an apparently huge product that photographs as an ugly stain in the well, instead of a band down in the gel. These problems can travel from person to person with the mere use of the same pipettors. This is an indication that the problem can travel in the air of a pipette barrel (Barnes 1994b).

  To delay the appearance of these problems, start off right and follow the suggestions in Chapter 1. For pipetting template in the pre-PCR area, use tips with filters in them, and always use a dedicated set of pipettors for PCR setup. Separate post-PCR manipulations such as loading of gels as much as possible from the pre-PCR area. Occasionally, the problem can be cured, at least temporarily, by bleach treatment of pipette barrels (10% Clorox, 30 minutes followed by rinsing with autoclaved water), combined with fresh stock solutions for all reaction components.

- *Tip 8. Try less proteinase K.* Most proteinase K is provided without any calcium, even though it is a calcium-requiring enzyme. It is best to at least store it with some calcium, about 1 mM $CaCl_2$. Although famous for being a powerful protease, proteinase K cannot work in very high salt, either! When these two items are allowed for, some recipes need only 1/100 as much proteinase K. For instance, 1 μg of proteinase K is sufficient to prepare DNA template for a mouse cell suspension in 20 mM Tris, pH 7.9, 0.5% Tween 20, and 2 mM EDTA, at 65°C for 1 hour. At 65°C, proteinase K also eats up itself, so the DNA is PCR-ready.

- *Tip 9. Use longer extension steps.* Longer extension times often increase economy, reliability, and yield of product, even for short targets. For targets up to 2 kb, 8–10 times less DNA polymerase (for KlentaqLA; 2–3 times less for TaqLA) can be used if extension steps are 2–5 times longer—that is, 10–20 minutes instead of 2–4 minutes.

- *Tip 10. Extend at 68°C or lower.* The author has never seen a PCR that worked better at 72°C than at 68°C (although he is usually testing long targets). For long PCR, 68°C is definitely better. For any PCR, the recommendation of 72°C by some companies is based on the optimum for *Taq* DNA polymerase activity in a standard assay that is not PCR. This assay, which dates from the pioneering first paper about DNA polymerase (Lehman et al. 1958), is based on incorporation at nicks, gaps, and/or staggered ends of calf thymus DNA that has been partially degraded and heated to form poorly defined structures. To optimize PCR, it is more appropriate to use PCRs for the assays.

- *Tip 11. Add the enzyme from a full-strength stock to a master mix for many reactions.* This is just generally good advice for enzyme stability. Enzymes are usually more stable in higher salt and at higher concentration. If the enzyme must be diluted, for KlentaqLA the storage buffer is 50% glycerol (v/v; 63% w/v), 222 mM ammonium sulfate, 20 mM Tris-HCl, pH 8.55, mM EDTA, 10 mM mercaptoethanol, 0.5% NP-40, and 0.5% Tween 20.

- *Tip 12. For the RT step of RT-PCR, do not preheat your RNA template in the presence of magnesium.* Magnesium and high temperature cause RNA degradation. (The author is grateful to J. Perrault [pers. comm.] for pointing this out in explaining why a RT-PCR did not work when the more-standard advice to heat-denature the RNA template was followed.) It turns out that a pre-denature of the RNA template is just not necessary, and several cDNA targets up to 12 kb have been amplified by skipping that step (W.M. Barnes and L. Ivanova, unpubl.).

Conflict of interest notice: The author is an inventor and provider of Klentaq1 and improvements to PCR technology.

REFERENCES

Barnes W.M. 1994a. PCR amplification of up to 35 kb DNA with high fidelity and high yield from λ bacteriophage templates. *Proc. Natl. Acad. Sci.* **91:** 2216–2220.

Barnes W.M. 1994b. Tips and tricks for long and accurate PCR. *Trends Biochem. Sci.* **19:** 342.

Barnes W.M. 1995. Thermostable DNA polymerase with enhanced thermostability and enhanced length and efficiency of primer extension. U.S. Patent No. 5,436,149.

Baskaran N., Kandpal R.P., Bhargava A.K., Glynn M.W., Bale A., and Weissman S.M. 1996. Uniform amplification of a mixture of deoxyyribonucleic acids with varying GC content. *Genome Res.* **6:** 633–638.

Hogrefe H.H., Hansen C.J., Scott B.R., and Nielson K.B. 2002. Archaeal dUTPase enhances PCR amplifications with archaeal DNA polymerases by preventing dUTP incorporation. *Proc. Natl. Acad. Sci.* **99:** 596–601.

Lasken R.S., Schuster D.M., and Rashtchian A. 1996. Archaebacterial DNA polymerases tightly bind uracil-containing DNA. *J. Biol. Chem.* **271:** 17692–17696.

Lehman I.R., Bessman M.J., Simms E.S., and Kornberg A. 1958. Enzymatic synthesis of deoxyribonucleic acid. *J. Biol. Chem.* **233:** 163–173.

Reese W.A., Yager T.D., Korte J., and von Hippel P.H. 1993. Betaine can eliminate the base pair composition dependence of DNA melting. *Biochemistry* **32:** 137–144.

7   PCR Primer Design

Arun Apte and Saurabha Daniel

PREMIER Biosoft International, Palo Alto, California 94303-4504

The objective of PCR is to amplify a specific DNA segment without any nonspecific by-products. In principle, each physical and chemical component of PCR can be modified to produce a potential increase in yield, specificity, or sensitivity. Yet the most critical parameter for successful PCR is optimal primer design. A poorly designed primer can result in little or no product, due to nonspecific amplification and/or primer–dimer formation leading to reaction failure, even when all the other parameters are properly optimized. This chapter provides general guidelines for PCR primer design, tips for development of primer pairs for more complex applications, and advice on the development of probes for real-time PCR. We close with a discussion of computer programs available for PCR primer design.

BASICS OF PRIMER DESIGN

Several parameters must be taken into account when designing primers for PCR. Each parameter is discussed in detail below.

Primer Length

Primer length critically affects PCR success by influencing specificity, melting temperature, and time of annealing. A primer length of 18–30 bases is optimal for most PCR applications. Shorter primers could lead to nonspecific PCR amplification. Longer primers are more specific but have a higher probability of containing secondary structures such as hairpin loops.

Primer Melting Temperature

The specificity of PCR strongly depends on the melting temperature (T_m) of the primers. For most PCR applications, the optimum melting temperature of primers ranges from 55°C to 60°C. In the absence of destabilizing agents, the T_m of a primer depends on its length, sequence composition, and concentration. The impact of ionic strength is negligible because the salt concentration does not vary significantly under different PCR conditions.

It is also important that all of the primers used in a reaction have similar melting temperatures. For most PCR applications, the primer pair T_m mismatch should not be more than 2–3°C. If the difference is greater, amplification will be less efficient or may not work at all because the primer with the higher T_m will misprime at a lower than optimal anneal-ing temperature, and the primer with the lower T_m will only bind in low concentrations at a higher than optimal annealing temperature.

The most accurate T_m can be estimated by formulas based on the nearest-neighbor ther-modynamic theory, which takes into account thermodynamic analysis of the duplex melt-ing process.

$$T_m = [\Delta H/\Delta S + R \ln (C)] - 273.15 \tag{1}$$

The changes in enthalpy (ΔH) and entropy (ΔS) of duplex formation are calculated from nearest-neighbor thermodynamic parameters, R is the molar gas constant, and C is the molar concentration of the oligonucleotide. This analysis determines the specific enthalpy and entropy contribution to the free energy of the duplex, made by each "nearest neigh-bor" in the sequence. The nearest neighbor starts at the 5´ end, and the enthalpy and entropy contributions are additive. A second term is added to Equation 1 to account empir-ically for the stabilizing effect of salt on the duplex:

$$T_m = [\Delta H/\Delta S + R \ln (C)] - 273.15 + 12.0 \, \text{Log} \, [\text{Na}^+] \tag{2}$$

Most primer design software uses either the Breslauer or SantaLucia nearest-neighbor parameter set to estimate the T_m of oligonucleotide duplexes (Breslauer et al. 1986; SantaLucia 1998, respectively).

A first-order approximation for short sequences with 20 bases or less can be calculated by the Wallace rule. The equation assumes a salt concentration of 0.9 M, common for dot-blot and hybridization assays:

$$T_m = 2°C \, (A + T) + 4°C \, (G + C) \tag{3}$$

A general rule of thumb is to use a temperature ~5°C lower than the primer melting temperature. Often, the annealing temperature determined this way will not be optimal, and empirical experiments will have to be performed to determine the optimal tempera-ture. This is most easily accomplished using a gradient thermal cycler. Alternatively, more accurate equations can be used to calculate the T_a (optimum annealing temperature) (Rychlik et al. 1990).

$$T_a \, \text{Opt} = 0.3 \times (T_m \, \text{of primer}) + 0.7 \times (T_m \, \text{of product}) - 25 \tag{4}$$

where T_m of primer is the melting temperature of the less stable primer–template pair and T_m of product is the melting temperature of the PCR product, both calculated as shown above.

Amplicon Melting Temperature

In addition to calculating the melting temperatures of the primers, care must be taken to ensure that the melting temperature of the product is low enough to ensure complete melt-ing at 92°C. In general, products between 100 and 600 bp are efficiently amplified in PCR. This parameter will help ensure a more efficient PCR, but is not always necessary for suc-cessful PCR. The product T_m can be calculated using the formula

$$T_m = 81.5 + 16.6 \, (\log10[\text{K}^+]) + 0.41 \, (\%G+C) - 675/\text{length} \tag{5}$$

Secondary Structures

An important factor to consider when designing a primer is the presence of secondary structures. A primer sequence with regions of self-homology can snap back to form partially double-stranded hairpin structures. Similarly, interprimer homology can lead to the formation of primer–dimers. Because high concentrations of primers, relative to template, are used in the PCR, primers may anneal to each other much more readily than they anneal to the template. The presence of 3´ hairpins or 3´ dimers is especially detrimental to amplification, because these will lead to the nonspecific amplification of sharp background products. A simple way to avoid secondary structures is to select primers that are 50% G + C and deficient in one of the four bases.

Repeat and Runs

Primers with long runs of a single base or nucleotide repeats should generally be avoided. The effects of different types of repeats and runs are shown in Table 1.

GC Clamp

Including a G or C residue at the 3´ end of primers increases the priming efficiency. The so-called "GC clamp" helps to ensure correct binding at the 3´ end of the primer due to the stronger hydrogen bonding of GC residues, thus providing enhanced specificity. Primers that end with a thymidine residue tend to have reduced specificity.

Amplification of High GC Content Targets

For amplification of GC-rich DNA, primers with a higher T_m (preferably 75–80°C) are recommended. The strand separation temperature for GC-rich DNA is significantly higher

TABLE 1. Effects of primer repeats and runs

Repeat motifs	Description	Effect
Simple repeats	a repeated sequence of 4 or more nucleotides, which is repeated: (...AATCGA...AATCGA...)	Simple repeats can generate secondary binding sites for primers. Stable hybridization to secondary binding sites results in nonspecific amplification. Repeats >3–4.
Inverse repeats	a self-complementary sequence motif of 4 or more nucleotides, (stem loop or hairpin motifs): (...AATGGC....GCCATT...)	Inverse repeats can cause inefficient priming as they lead to formation of stable hairpins in the binding region, or within the amplicon.
Homopolymeric runs	a sequence of 4 or more identical nucleotides: (...AAAAA...)	Homopolymeric runs can be considered a special case of direct repeats. These can cause ambiguous binding of primers to their target site ("slippage effect"). Poly (A) and poly (T) stretches should also be avoided as these will "breathe" and open up stretches of the primer–template complex. Additionally, runs of 3 or more G residues can cause problems due to intermolecular stacking.

than for normal DNA. At lower temperatures, the two strands of PCR amplicons have a tendency to re-anneal faster and, as a result, compete with primer annealing. Higher annealing temperatures during PCR favor primer annealing and therefore increase amplification efficiency.

Unintended Homologies

Primer sequences should be searched using BLAST (http://www.ncbi.nlm.nih.gov/BLAST/) and checked for cross-homology with repetitive sequences or with other loci elsewhere in the genome. Such homologies could lead to false priming and the production of nonspecific amplicons.

MULTIPLEX PCR PRIMER DESIGN

Multiplex PCR (MPCR) uses one template and several sets of primers for the simultaneous amplification of multiple target sequences in a single PCR. It is an extremely useful technique that increases the throughput of PCR and allows more efficient use of each DNA sample. A variant of MPCR is combinatorial PCR. Combinatorial PCR uses several templates and several primer sets, all in the same reaction. The terms MPCR and combinatorial PCR are often used synonymously.

Design Strategy

MPCR requires that all the primer pairs in a reaction amplify their unique targets under a defined set of reaction conditions. Most multiplex reactions are restricted to amplification of five to ten targets. One reason for this is that a degree of flexibility is lost with each additional primer set included in the reaction. Increased numbers of primers also increase the probability of primer–dimer formation and nonspecific amplification. The development of an efficient MPCR requires strategic planning and often multiple attempts to optimize reaction conditions.

Ideally, all the primers in a multiplex reaction should amplify their individual target sequences with equal efficiency. Most often it is difficult to predict the efficiency of a primer pair, but oligonucleotides with near-identical annealing temperatures work well under similar conditions.

General Rules for Multiplex Primer Design and Optimization

When designing primers for use in MPCR, the general rules of primer design apply, but additional considerations must be taken into account. Generally, all the primers in a multiplex reaction should be matched for T_m. Care should be taken to avoid primers with complementary 3′ nucleotides. Each primer pair should be tested separately to determine optimal conditions. Once the panel of primer pairs is assembled, they need to be mixed sequentially and optimized:

1. The length of individual primers should be 18–24 bases. Longer primers are more likely to result in formation of primer–dimers.

2. Annealing temperature and cycle number are critical to the success of MPCR. The annealing temperature should be kept as high as possible. Identify the annealing tem-

peratures for each primer pair and use the lowest temperature in the multiplex reaction. Similarly, use the minimum number of cycles.

3. Since multiple templates are simultaneously amplified, the pool of enzyme and nucleotides in a MPCR can be a limiting factor, and more time is required for complete synthesis of all products. It is important to optimize reagent concentrations and extension times for each reaction. Longer extension times are needed than in single-target PCR.

NESTED PCR

Nested PCR is designed to increase the sensitivity of PCR using a second PCR to amplify the product directly from a primary PCR. The second reaction uses primers placed internal to the first primer pair. These internal primers are referred to as "inner" or "nested" primers. The amplification product from the first round acts as a template for the second round, significantly improving sensitivity without impairing specificity (Albert and Fenyo 1990). Because nested PCR uses two sets of primer pairs, a higher total number of cycles are possible with replenishment of reaction components such as *Taq* DNA polymerase. If a full nested PCR (using two internal primers) cannot be performed, sensitivity and specificity can be improved by designing a hemi-nested PCR using just one inner primer in conjunction with one of the outer primers from the first reaction.

The biggest problem with nested PCR is that it is prone to contamination. Tubes from the first PCR have to be opened so that the primary product can be transferred to a new tube for the second reaction. It also involves the additional cost of two rounds of PCR and additional primer synthesis.

To reduce the risk of contamination, a single-tube approach, referred to as drop-in/drop-out nested PCR, can be used. In this protocol, the inner primers are designed with a significantly lower melting temperature than the outer primer pair. Both primer pairs are included in the first reaction. During the first round of PCR, the annealing temperature is chosen so that only the outer pair will anneal and extend. During the second round of amplification, the annealing temperature is lowered so that the inner pair can function. Although this method is attractive from the contamination control perspective, it does not provide all the advantages of the original nested PCR protocol.

Design Strategy

The same general rules of PCR primer design can be applied to the design of primers of nested PCR. Because multiple primer pairs are used, the increased probability of primer–primer interaction should be carefully considered. When designing primers for single-tube nested PCR, it is important that the melting temperature of the inner (nested) primer pair is significantly lower than that of the outer (first-round) primer pair. The easiest way of achieving this is to reduce the length of the nested primers as compared to outer primers (e.g., 18–20 bases vs. 25–28 bases). The T_m of the outer primer pairs should be high enough to prevent the inner primers from annealing, ensuring that only the longer product is produced in the first-round PCR. Internal primers should be used in excess (typically 40 times more) compared to the outer primers. The use of shorter annealing and extension times in the second PCR favors the annealing of the shorter primers and production of the smaller amplicon.

cDNA/gDNA SELECTIVE PRIMER DESIGN

Design Strategy

The major problem in RT-PCR is the presence of contaminating genomic DNA (gDNA), which can lead to false-positive signals, reduced specificity, or overestimation of specific RNA. To prevent any interference by gDNA in RT-PCR applications, primers can be designed to anneal to splice junctions, unique to cDNA sequences, so that gDNA will not be amplified.

The cDNA-specific primers can be designed in three ways (as illustrated in Fig. 1):

1. **Primers spanning the exon–exon junction.** The primer is designed so that one-half of its sequence will hybridize to the 3′ end of one exon and the other half to the 5′ end of the adjacent exon. This primer will anneal to mRNA, or to cDNA synthesized from spliced mRNAs, but not to gDNA, thus eliminating amplification of contaminating gDNA. Either the forward or the reverse primers can be designed to cross the exon junction; the second primer can be designed to bind at a second exon junction, or at a site completely within the exon.

2. **Primers flanking the exon–exon junction.** RT-PCR primers can be designed to flank a region that contains at least one intron. Products amplified from cDNA will be smaller than those amplified from gDNA. This size difference can be used to detect the presence of contaminating DNA.

3. **Exon-primed intron-crossing primers.** For selective amplification of the gDNA from a mixture of cDNA and gDNA, primers should be designed to anneal across the exon–intron junctions. These primers are called exon-primed intron-crossing primers (EPIC primers) (see, e.g., Bierne et al. 2000). Because exon sequences are more highly conserved than intron sequences, EPIC primers can be used in phylogenetic studies to amplify homologous intron regions.

CROSS-SPECIES PRIMER DESIGN

In addition to the identification of genes and prediction of the encoded proteins, the current effort in genomics includes studies on genome organization to examine the interposition of genes with structural and regulatory elements. Cross-species analysis, which forms a major part of comparative genomics, is seen as an important method for the study of evolution, gene function, and human disease.

Cross-species analysis of microbial genomes, particularly bacteria, is beginning to identify genes conserved among bacteria, and virulence genes associated with subspecies (Fredricks and Relman 1996). These advances have facilitated the development of new, broad-based methods for the detection and discrimination of pathogenic microbes. Cross-species PCR can be used to study evolutionarily related species that share some common genomic properties; the conserved regions can be used identify species at the genomic level. The technique can also be used to amplify homologous genes in unsequenced genomes.

Design Strategy

Cross-species PCR primers are designed using multiple sequence alignment and basic primer design principles. The alignment of small numbers of large contiguous sequences identifies conserved regions, which constitute potential cross-species primer-binding sites. Cross-species comparisons require accurate alignment of a small number of large contigu-

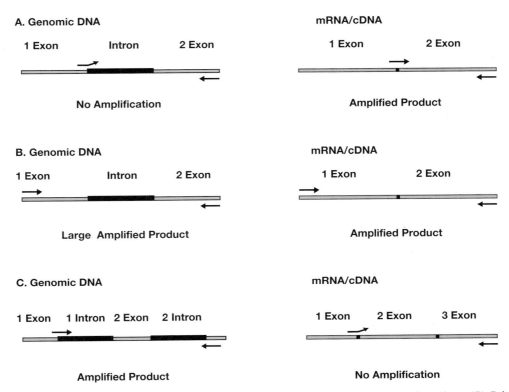

FIGURE 1. cDNA/gDNA selective primer design. (*A*) Primer across exon–exon junction. (*B*) Primer spanning exon–exon junction. (*C*) Exon-primed intron-crossing primers (EPIC primers).

ous sequences. Conserved regions are identified from the alignment as potential cross-species primer-designing sites. Primers annealing to the conserved regions can be used to amplify homologous loci in different species (see Fig. 2).

The following steps should be followed:

1. Align selected sequences using a multiple sequence alignment program such as Clustal (available from: http://www.ebi.ac.uk/clustalw).

2. Evaluate conserved regions of the alignment. Derive a consensus primer sequence by choosing the most commonly occurring nucleotide at each position of the conserved sequence.

3. The primer-binding site should lie entirely within the conserved region. For distantly related species, when sequences are not completely conserved, the minority consensus can also be used to design probes/primers. Degeneracy can be tolerated at the 5′ end of the primer, but mismatches at the 3′ end reduce both annealing specificity and PCR yield (Sommer and Tautz 1989).

4. Use general primer-design rules for PCR to avoid false priming and primer–dimer formation in cross-species PCR.

Real-Time PCR Probe Design

The introduction of real-time PCR has made it possible to quantify accurately the starting amounts of nucleic acid during the PCR, without the need for post-PCR analysis. In real-time PCR, a fluorescent reporter is used to monitor the PCR as it occurs (i.e., in real time).

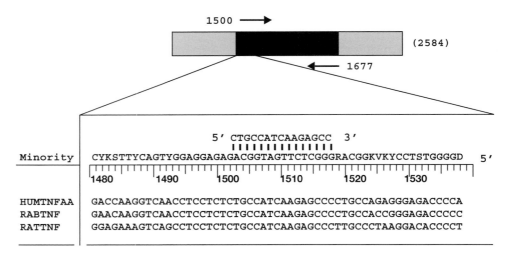

FIGURE 2. Cross-species primer design in conserved region of multiple sequence alignment for human tumor necrosis factor, rabbit tumor necrosis factor, and rat tumor necrosis factor gene sequence.

The fluorescence of the reporter molecule increases as products accumulate with each successive round of amplification. The reporter can be a nonspecific intercalating double-stranded DNA-binding dye or a sequence-specific fluorescent-labeled oligonucleotide probe. The oligonucleotide probe is labeled with both a reporter fluorescent dye and a quencher dye. While the probe is intact, the proximity of the quencher greatly reduces the fluorescence emitted by the reporter dye by Förster resonance energy transfer (FRET) through space. Adequate quenching is observed for probes with the reporter at the 5′ end and the quencher at the 3′ end. When a probe molecule is incorporated into an amplicon, quenching is disrupted and the probe fluoresces.

The ability to monitor the real-time progress of the PCR has completely revolutionized the approach to PCR-based quantification of DNA and RNA. It is now possible to quantify PCR products reliably by eliminating the variability associated with conventional quantitative techniques. Because reporter fluorescence is monitored externally, the reaction tubes do not need to be opened after the PCR is complete; this prevents aerosol contamination by PCR products and reduces the number of false-positive results.

TaqMan Assay

One popular real-time PCR probe strategy is the TaqMan assay (Applied Biosystems). This assay uses a hydrolysis probe, which exploits the 5′ exonuclease activity of *Taq* polymerase to cleave a labeled hybridization probe during the extension phase of PCR (Holland et al. 1991).

Strategy

The fluorogenic 5′ nuclease assay is a convenient, self-contained process. The probe is designed to anneal to the target sequence between the upstream and downstream primers. The probe is labeled with a reporter fluorochrome (usually 6-carboxyfluorescein [6-FAM]) at the 5′ end and a quencher fluorochrome (6-carboxy-tetramethyl-rhodamine [TAMRA]) at the 3′ end, or any T position.

During the PCR amplification, the probe anneals to its target on the template DNA. As the primer is extended, the newly synthesized DNA strand approaches the site of probe hybridization. On arrival, due to its intrinsic 5′ to 3′ nuclease activity, the polymerase cleaves the probe, releasing the reporter fluorochrome into the reaction buffer. When the reporter molecule parts company with the quencher, it starts to fluoresce, and synthesis of the DNA strand continues until the amplification cycle is complete. Because the reporter molecules are cleaved from the probes during every amplification cycle, the fluorescence intensity of the overall system increases proportionally to the amount of DNA amplified.

The following issues should be considered when designing TaqMan probes:

1. The 3′ end of the probe must be protected against chain elongation during PCR. To block the 3′-end phosphate, cordycepin, 2′,3′-dideoxynucleosides, inverse T, or the quencher dye itself can be used. Labeling the 3′ end with TAMRA or another quencher can be achieved by using a 3′ amino linker. It is also possible to label the 3′ end of oligonucleotides postsynthetically via an amino link. This method is available for all dyes, except Cy3TM, Cy5TM, Cy5.5TM, HEX, and TET.

2. The probe T_m should be 10°C higher than the amplification primer T_m and within 65–72°C for optimal *Taq* exonuclease activity.

3. Avoid incorporation of a G residue at the 5′ end of the probe; this will quench reporter fluorescence even after cleavage.

4. Probe must not have more Gs than Cs.

5. Avoid runs of single nucleotides, particularly G residues.

6. AT-rich sequences require longer primer and probe sequences to achieve the recommended T_m. Note that probes more than 40 bp long can exhibit inefficient quenching and produce lower yields.

7. Design the probe to anneal as close as possible to the primer without overlapping (at least one base away from primer 3′ end).

8. If a TaqMan probe is being designed for allelic discrimination, it is advisable to place the mismatching nucleotide (the polymorphic site) in the middle of the probe rather than at the ends. These probes should be as short as possible to provide the greatest discrimination by the mismatch.

Primer and Amplicon Properties

Amplicon size should range between 50 and 150 bp and should not exceed 400 bp. Smaller amplicons give more consistent results because PCR is more efficient and more tolerant of reaction conditions.

Commercially Available TaqMan Probes

Two types of TaqMan probes are commercially available, as shown in Table 2.

TABLE 2. TaqMan probes

TaqMan probe	5′ Label	3′ Label	Features
TaqMan	6-FAM, VIC, or TET	TAMRA	none
TaqMan MGB	6-FAM, VIC, or TET	nonfluorescent quencher	minor groove binder

Only three possible fluorophores—FAM, HEX, and TET—can be used as reporter dyes when TAMRA is used as a quencher. This limits the use of the probes for multiplex PCR due to a spectral overlap between these fluorescent reporters. A larger range of fluorescent reporters is available for use with dabcyl, a universal quencher for molecular beacons. Dabcyl can replace TAMRA at the 3′ end of the TaqMan probe, thus permitting the use of a large range of fluorescent reporters in the synthesis of probes, and simultaneous detection of different targets in a single reaction.

TaqMan Probes for Allelic Discrimination Assays

TaqMan MGB probes are most widely used for allelic discrimination assays. These probes have the following features:

1. A nonfluorescent quencher at the 3′ end. Because the quencher does not fluoresce, the fluorescence of the reporter dye can be measured more precisely.

2. A minor groove binder at the 3′ end. The minor groove binder increases the T_m of probes, allowing the use of shorter probes. Consequently, the TaqMan MGB probes exhibit greater differences in T_m values between matched and mismatched probes, which provides more accurate allelic discrimination.

Molecular Beacon Design

Molecular beacons (Tyagi and Kramer 1996) represent another example of specific fluorescent real-time probes (see Fig. 3). The molecular beacons enable dynamic, real-time detection of nucleic acid hybridization events both in vitro and in vivo (Tyagi and Kramer 1996; Kostrikis et al. 1998; Tyagi et al. 1998).

A molecular beacon is a dual-labeled oligonucleotide (25–40 nt) that forms a hairpin structure with a loop (probe) and self-complementary stem. There is a fluorescent reporter molecule at the 5′ end and a quencher at the 3′ end. At room temperature, the molecular beacon assumes the hairpin formation, and the fluorescent reporter and quencher molecule are brought together by the probe's self-complementary stem structure, thus suppressing the fluorescent signal. During the annealing step of the PCR, when the probe encounters a target DNA sequence, thermodynamics favor the binding of the beacon to its target, rather than reformation of the hairpin structure. Disruption of the stem structure separates the fluorescent reporter from its quencher, thus allowing the reporter to fluoresce (see Fig. 3).

Unlike conventional oligonucleotide hybridization, in which unhybridized probe molecules must be removed to eliminate background, hybridization using molecular beacons has an inherently low background, due to the stability of the beacon–quencher association in the absence of target sequences. The low background eliminates the need for washing and probe degradation steps. Moreover, with each successive cycle of amplification, the proportion of molecular beacons that bind target and emit light increases. Thus, the increase in fluorescence corresponds directly to the accumulation of product.

Due to the stability of the loop–stem structure, selectivity of molecular beacons is higher than that of a linear probe. Molecular beacons can discriminate between targets that have a single-nucleotide change, making them ideal tools for the investigation of single-nucleotide polymorphisms (SNPs). The perfectly matched probe–target hybrid is more stable than the single-stranded hairpin structure of the molecular beacon, whereas the mismatched probe–target hybrid is generally less stable, regardless of the base-pair combination of the mismatch. This thermodynamic feature is the key to the exquisite specificity of molecular beacons.

Molecular beacons are finding a multitude of uses in many different quantitative and qualitative target detection assays. They are used for real-time monitoring of DNA amplification during PCR (Tyagi and Kramer 1996); they can be multiplexed using multiple dye labels and used for real-time fluorescent genotyping (Kostrikis et al. 1998; Tyagi et al. 1998) and in the simultaneous detection of different pathogens in clinical samples (Vet et al. 1999). Molecular beacons have also been used to detect RNA transcripts in living cells (Sokol et al. 1998) and to detect DNA-binding proteins (Heyduk and Heyduk 2002). Molecular beacons have been adapted to real-time and end-point PCR, and to RT-PCR assays.

Design Guidelines

To monitor PCRs successfully, molecular beacons must hybridize to their targets at PCR annealing temperatures while the free molecular beacons remain closed and nonfluorescent. The annealing temperature and the buffer will influence probe specificity and must be carefully controlled. The loop or probe region of the beacon is target-specific and is flanked by the sequences that form the hairpin. The following general guidelines should be observed when designing beacons:

1. The probe region should be 15–33 nucleotides long. The loop sequence must be complementary to the target sequence of the assay.

2. The T_m of the probe region should be 7–10°C higher than the PCR annealing temperature (as predicted using the percent GC rule). The prediction should be made for the probe sequence alone, before adding the stem sequences.

3. To ensure preferential hybridization of the beacon to target sequences, the probe must be complementary to a region of minimal secondary structure. The template secondary structure can be evaluated by analyzing the sequence using folding software, such as Mfold (Michael Zuker, Rensselaer Polytechnic Institute, http://www.bioinfo.rpi.edu /applications/mfold/).

4. The molecular beacon should bind at or near the center of the amplicon. The distance between the 3′ end of the upstream primer and the 5′ end of the molecular beacon (stem) should be greater than 6 nucleotides.

5. The stem region of the molecular beacon should be 5–7 bp long, with a GC content of 70–80%. Longer stems will make the molecular beacon sluggish in binding to its target.

6. The length, sequence, and GC content of the stem should be chosen such that the melting temperature is 7–10°C higher than the annealing temperature of the PCR primers.

7. The melting temperature of the stem is usually calculated using the Mfold hairpin folding formula, which is part of the Mfold software. The formula is used to estimate the free energy of the formation of the stem hybrid from which its melting temperature can be predicted. Assuming a 100% GC content, a 5-bp stem will melt at 55–60°C, a 6-bp stem at 60–65°C, and a 7-bp stem at 65–70°C.

8. The more negative the free energy of a sequence, the more favorable and stable the structure will be. The stem-loop free-energy values obtained with Mfold should range between –3 and +0.5 kcal/mole.

9. Because G residues may act as quencher molecules, avoid designing molecular beacons with a G directly adjacent to the fluorescent dye (typically at the 5′ end of the stem sequence). A cytosine at the 5′ end of the stem is preferred.

10. The beacons should be checked for the presence of alternate secondary structure (other than the intended hairpin stem), because this can change the position of the fluorophore relative to the quencher, causing an increase in background signals.

11. Avoid complementarity between molecular beacon and PCR primers; this may cause the molecular beacon to bind to primers and increase background.

12. The amplification primers used in molecular beacon PCRs should be designed to produce a relatively short amplicon, preferably less than 150 bp. The molecular beacon is an internal probe, which must compete with the opposite strand of the amplicon for its complementary target. A shorter amplicon is more likely to fully complete DNA synthesis, assuring that the target sequence is present and, therefore, gives more reproducible results.

Choosing Fluorophore and Quencher Molecules

Another consideration in molecular beacon design is the choice of fluorophore and quencher. Dabcyl (4-[4′-dimethylaminophenylazo]benzoic acid) is the optimal choice of the quenching fluorophore molecules available for molecular beacons. Dabcyl is a neutral, hydrophobic molecule, making it an ideal partner for a variety of fluorophores. Dabcyl has a short operational range for quenching as compared to the total length of a beacon probe. This means that it must be in close contact with the fluorophore, for efficient energy-transfer quenching to be efficient. Consequently, a stem–loop beacon is quenched whereas a probe–target hybrid is not.

PRIMER DESIGN SOFTWARE

Beacon Designer 2.0

Beacon Designer 2.0 from PREMIER Biosoft International is a complete real-time PCR primer and probe design software tool for both molecular beacon and TaqMan probes. Probes can be designed for multiplex and allele discrimination assays. This software can be used to design beacons with stems of appropriate length, automatically adjusted for optimal T_m, and checked for cross-dimer formation with the amplifying primer pairs, thus preventing competition in multiplex reactions. The program checks sequences for secondary structures, and cross-hybridization to ensure high signal strength. The hairpin T_m is calculated using the highly accurate Mfold algorithm. Both wild-type and mutant beacons can

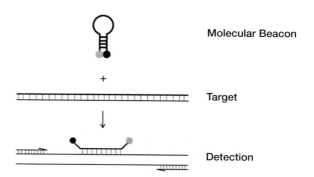

FIGURE 3. Schematic diagram of a molecular beacon and its principle for PCR detection.

be designed for use in genotyping. Beacon Designer is available for both Windows and Macintosh operating systems.

Primer Express 2.0

Primer Express 2.0 from Applied Biosystems is an NT-based oligonucleotide design program created to support Applied Biosystems PCR and Sequence Detection System instruments. Primer Express provides an oligonucleotide design facility for standard DNA PCR, RT-PCR, nested PCR, allele-specific PCR, multiplex PCR, and DNA PCR with TaqMan probes. It has a primer-testing document to evaluate predesigned primers on the basis of their T_m, secondary structure, and potential to form primer–dimers. The program also supports sequence annotation.

Primer Express supports the design of both TaqMan MGB probe assays and TaqMan Probe assays. Probes can be designed for single or multiple (up to 48) sequences. The program is available for both Windows and Macintosh.

Primer Premier 5

Primer Premier 5 from PREMIER Biosoft International integrates multiple-sequence alignment with primer design to facilitate the design of cross-species primers. The program uses a proprietary algorithm to calculate a minority consensus and design primers in highly conserved regions of the sequences. Allele-specific primers of the minority consensus sequence of a gene family can be designed graphically.

Primer Premier provides comprehensive primer-design features that allow primers to be designed automatically or manually, and multiplex/nested primer design ensures that there are no cross-homologies. Search results can be saved in the built-in database, and a synthesis order form is provided to process ordering. The program is available for both Windows and Macintosh.

Primer Designer 4.1

Primer Designer 4.1 assists the design of primers for sequencing, for PCR, or for use as oligonucleotide probes. Primer Designer reads files in GenBank, EMBL, FASTA, or Clone Manager file formats or sequence data in simple ASCII text files. It saves primer search information in a built-in database and uses a new unified set of nearest-neighbor parameters (SantaLucia 1998) for primer T_m calculations. This program is available for Windows.

REFERENCES

Albert J. and Fenyo E.M. 1990. Simple, sensitive and specific detection of human immunodeficiency virus type 1 in clinical specimens by polymerase chain reactions with nested primers. *J. Clin. Microbiol.* **28:** 1560–1564.

Bierne N., Lehnert S.A., Bedier E., Bonhomme F., and Moore S.S. 2000. Screening for intron-length polymorphisms in Penaeid shrimp using exon-primed intron-crossing (EPIC) PCR. *Mol. Ecol.* **9:** 233–235.

Breslauer K.J., Frank R., Blocker H., and Markey L.A. 1986. Predicting DNA duplex stability from the base sequence. *Proc. Natl. Acad. Sci.* **83:** 3746–3750.

Fredricks D.M. and Relman D.A. 1996. Sequence-based identification of microbial pathogens: A reconsideration of Koch's postulates. *Clin. Microbiol.* **9:** 18–33.

Heyduk T. and Heyduk E. 2002. Molecular beacons for detecting DNA binding proteins. *Nat. Biotechnol.* **20:** 171–176.

Holland P., Abramson R.D., Watson R., and Gelfand D.H. 1991. Detection of specific polymerase chain reaction product by utilizing the 5′-3′ exonuclease activity of *Thermus aquaticus* DNA polymerase. *Proc. Natl. Acad. Sci.* **88:** 7276–7280.

Kostrikis L.G., Tyagi S., Mhlanga M.M., Ho D.D., and Kramer F.R. 1998. Spectral genotyping of human alleles. *Science* **279:** 1228–1229.

Rychlik W., Spencer W.J., and Rhoads R.E. 1990. Optimization of the annealing temperature for DNA amplification in vitro. *Nucleic Acids Res.* **18:** 6409–6412.

SantaLucia J., Jr. 1998. A unified view of polymer, dumbbell, and oligonucleotide DNA nearest-neighbor thermodynamics. *Proc. Natl. Acad. Sci.* **95:** 1460–1465.

Sokol D.L., Zhang X., Lu P., and Gewirtz A.M. 1998. Real time detection of DNA.RNA hybridization in living cells. *Proc. Natl. Acad. Sci.* **95:** 11538–11543.

Sommer R. and Tautz D. 1989. Minimal homology requirements for PCR primers. *Nucleic Acid Res.* **17:** 6749.

Tyagi S. and Kramer F.R. 1996. Molecular beacons: Probes that fluoresce upon hybridization. *Nat. Biotechnol.* **14:** 303–308.

Tyagi S., Bratu D.P., and Kramer F.R. 1998. Multicolor molecular beacons for allele discrimination. *Nat. Biotechnol.* **16:** 49–53.

Vet J.A., Majithia A.R., Marras S.A.E., Tyagi S., Dube S., Poiesz B.J., and Kramer F.R. 1999. Multiplex detection of four pathogenic retroviruses using molecular beacons. *Proc. Natl. Acad. Sci.* **96:** 6394–6399.

WWW RESOURCES

http://www.bioinfo.rpi.edu/applications/mfold/ Bioinformatics Applications: RNA & DNA Folding, Michael Zuker, Center for Bioinformatics, Rensselaer Polytechnic Institute

http://www.ebi.ac.uk/clustalw European Bioinformatics Institute, EMBL-EBI, ClustalW sequence alignment tool

http://www.ncbi.nlm.nih.gov/BLAST/ NCBI BLAST home page

http://www.premierbiosoft.com PREMIER Biosoft International, molecular beacon, TaqMan, and multiple primer design software

http://www.scied.com Sci-ed software, primer design software

http://www-genome.wi.mit.edu/cig-bin/primer/primer3_www.cgi On-line PCR primer design software

8 Nonradioactive Cycle Sequencing of PCR-amplified DNA

Susan R. Haynes

Department of Biochemistry and Molecular Biology, F. Edward Hébert School of Medicine, Uniformed Services University of the Health Sciences, Bethesda, Maryland 20814

Many basic science and clinical studies depend on the ability to sequence a specific region of genomic DNA from multiple samples rapidly and efficiently. The combination of PCR amplification of target DNAs and cycle sequencing has dramatically reduced the time and effort required to obtain DNA sequence, particularly from genomic DNA templates, because it is no longer necessary to clone the DNA prior to sequencing. Cycle sequencing is a linear amplification protocol (Carothers et al. 1989; Gyllensten 1989; Murray 1989) that uses a thermostable DNA polymerase and temperature cycling (Fig. 1). This linear amplification dramatically increases the signal intensity as compared to single-round methods. As a result, only femtomole quantities of template are required, and a typical PCR provides sufficient material for multiple sequencing reactions. A further increase in signal intensity is provided by dideoxy terminators labeled with energy-transfer dyes, which are analyzed by highly sensitive fluorescence detectors (Prober et al. 1987). These dye terminators have distinct spectra, allowing all four dideoxy nucleotides to be combined in a single reaction, thus providing additional savings in time and materials. Further improvements in the dye terminators have increased the uniformity of peak heights and thus the accuracy of automated base-calling (Rosenblum et al. 1997). There is an additional benefit to direct sequencing of PCR products: The effect of any nucleotide substitutions introduced during PCR amplification is minimized, because each is present in only a small fraction of the PCR products. In contrast, if the PCR product is cloned prior to sequencing, it is necessary either to use a high-fidelity polymerase in the PCR or to sequence multiple clones, in order to ensure that the sequence obtained accurately reflects that of the original template.

This chapter provides methods for performing nonradioactive cycle sequencing of PCR fragments. The reactions are designed to be analyzed on automated sequencing equipment.

CHOOSING REAGENTS

DNA Template

The sequencing template is the most critical variable in the procedure. Therefore, it is worth devoting some time to optimizing the quality and quantity of the product. In the ideal situation, the PCR yields a single strong band of the correct size. Although PCRs using cloned templates usually produce a single product, it is not unusual to see multiple bands when amplifying genomic DNA. These unwanted bands can often be eliminated by

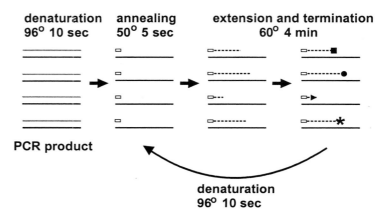

FIGURE 1. Schematic overview of the cycle sequencing reaction with dye terminators. The open rectangles represent the oligonucleotide primers, and the square, circle, arrowhead, and asterisk represent the four different dideoxy dye terminators. See text for details.

employing hot-start reaction conditions (D'Aquila et al. 1991; Chou et al. 1992), optimizing the annealing temperature, or selecting different primers for amplification. Alternatively, a small portion of the band of interest can be isolated and used as the template in a second PCR. A needle or pipette tip is used to transfer a small punch of agarose containing the desired fragment directly into a fresh PCR. A Pasteur pipette can also be used to take a larger piece of gel, but in this case the fragment should be eluted in ~20 µl of deionized H_2O for several hours, or overnight at 4°C, and only 1 µl of eluate used in the PCR. Agarose gel buffers typically contain EDTA, which chelates magnesium, and putting a large gel plug directly into the PCR could inhibit the reaction. To minimize DNA damage, prolonged exposure of the gel to UV light should be avoided.

Occasionally, the second PCR produces the same contaminants as the initial reaction. In these cases, the problem is often caused by mispriming of one, or both, primers at internal sites on the amplified fragment. The desired band can be excised from the gel and purified, but DNA losses may be high. There is also a risk of carryover of contaminants from the gel that could interfere with the sequencing reaction. If the unwanted band(s) is due to internal mispriming, it may be possible to obtain adequate sequence directly from the mixture of fragments by using an internal sequencing primer, rather than the amplification primers.

Cycle Sequencing Kit

This protocol was developed using the ABI PRISM BigDye Terminator Cycle Sequencing Ready Reaction Kit version 2 (Applied Biosystems). The sequencing reactions were analyzed on the ABI PRISM 377 DNA Sequencer (polyacrylamide gels) or the ABI PRISM 3100 Genetic Analyzer (capillary electrophoresis), both from Applied Biosystems. Other reagent kits and automated sequencing instruments can be used. However, it is important to ensure that the instrument is configured appropriately for the dye terminators that are being used in the sequencing reaction. This may involve specifying the correct filters, laser settings, etc., for the dyes; the operations manual for the instrument should be consulted for details.

Sequencing Primers

Standard oligonucleotide primers (18–25 nt) synthesized for PCR generally work well for sequencing. Primer design software is available in several commercial sequence analysis programs and via the Internet (e.g., http://www-genome.wi.mit.edu/cgi-bin/primer/-primer3_www.cgi). Chromatographic or gel purification of primers is usually not required. The sequencing primer can be either one of the primers used to generate the fragment or an internal primer if the sequence of the fragment is known. The choice depends on the region of interest in the PCR fragment, taking the following factors into consideration. (See also the Troubleshooting section, below.) A single sequencing reaction should yield at least 600 nucleotides of high-quality sequence, with reliable readings beginning ~20–40 nucleotides from the end of the primer. Longer sequences can be obtained, but it may become increasingly difficult to count the number of bases in homopolymeric regions. Remember that DNA polymerases extend the primer, and thus the sequence obtained will be the same strand as the primer.

PROTOCOLS

Once the desired template has been prepared by PCR, there are four steps required to obtain reliable sequence data:

> Protocol 1: PCR product purification
>
> Protocol 2: Quantification
>
> Protocol 3: Sequencing reaction
>
> Protocol 4: Sequencing reaction purification

PROTOCOL 1   PCR PRODUCT PURIFICATION

Carryover of primers or nucleotides with the template PCR product can give suboptimal results in the sequencing reaction. Excess nucleotides can alter the ratio between the dNTPs and dye-labeled terminators in the reaction mix, and the primers will initiate synthesis at unwanted sites. Any method that efficiently removes dNTPs and primers should be satisfactory. The following protocol is a rapid and simple method to accomplish this, using Amicon Microcon-PCR Filter Units (Millipore). According to the manufacturer, these filter units retain fragments smaller than ~130 bp and therefore should not be used to purify small templates. Recovery ranges from 60% to 70% for slightly larger fragments, to 98% for fragments larger than 300 bp.

MATERIALS

▼ CAUTION

See Appendix for appropriate handling of materials marked with <!>.

BUFFERS AND SOLUTIONS
  TE (10 mM Tris-HCl<!>, 1 mM EDTA, pH 7.5) or deionized H_2O

NUCLEIC ACIDS AND OLIGONUCLEOTIDES
  PCR product to be sequenced

CENTRIFUGES AND ROTORS
  Microcentrifuge with a fixed-angle rotor; e.g., Eppendorf 5415C

SPECIAL EQUIPMENT
  Amicon Microcon-PCR Filter Units (Millipore) or equivalent
  Microcentrifuge tubes or collection vials

METHOD

1. For each PCR to be purified, place a filter unit in a collection vial, with the reservoir (filter) side up. Add 400 µl of deionized H_2O or TE to the reservoir, being careful not to touch the membrane. Add 100 µl of PCR product, avoiding any carryover of mineral oil. If the volume of the PCR is less than 100 µl, make up the volume with H_2O or TE. Close the cap on the collection vial.

2. Place the assembly in a fixed-angle centrifuge rotor with the cap hinge facing the center of the rotor. Spin at 1000*g* for 15 minutes (e.g., 3300 rpm in an Eppendorf 5415C centrifuge).

3. Remove the filter unit and place it in a clean collection tube, filter side up. (The filtrate can be saved until the sample has been analyzed, if desired.) Carefully add 20 µl of deionized H_2O or TE to the filter unit without touching the membrane. Invert the filter unit and place it back in the same clean tube, so that the side with the fluid faces down.

4. Spin at 1000*g* for 2 minutes to recover the purified fragment.

PROTOCOL 2 QUANTIFICATION

Cycle sequencing reactions require very little template, in the range of 80–100 fmoles per reaction, and both too much and too little template give poor results. Therefore, it is important to quantitate the template prior to sequencing. Several different methods can be used, depending on the amount of DNA available. Standard absorption spectroscopy using a spectrophotometer is highly accurate, but may require as much as 500 ng of DNA, depending on the size of the cuvette. This is often an unacceptably high proportion of the purified sample, and more sensitive methods may be preferable. Two alternative methods are described here: fluorometry and comparative ethidium bromide staining.

METHOD A: FLUOROMETRY

Fluorometry can give accurate measurements in the nanogram range. Fluorometers and DNA quantitation kits are available from several manufacturers. It is important to be certain that the fluorometer is equipped with the appropriate excitation and emission filters for the specific dye being used. This protocol is designed for the Mini-Fluorometer with the Blue LED light source and filter set, minicell adapter, and cuvettes (Turner Designs), and the PicoGreen dsDNA Quantitation Kit (Molecular Probes).

MATERIALS

▼ CAUTION

See Appendix for appropriate handling of materials marked with <!>.

BUFFERS AND SOLUTIONS
    Lambda DNA standard (100 µg/ml)
    TE (10 mM Tris-HCl<!>, 1 mM EDTA, pH 7.5)

NUCLEIC ACIDS AND OLIGONUCLEOTIDES
    Purified PCR product

SPECIAL EQUIPMENT
Equipment for fluorometry
Foil
Microcentrifuge tubes

ADDITIONAL ITEMS
PicoGreen dsDNA Quantitation Kit<!> (Molecular Probes, P-7589)

METHOD

1. Prepare the concentration standards. Dilute 1 µl of the lambda DNA standard (100 µg/ml) into 49 µl of TE (final concentration is 2 ng/µl). Prepare tubes with 60, 20, and 2 ng of DNA standard, and bring the final volume to 50 µl with TE. Also prepare a blank sample containing only 50 µl of TE.

2. Prepare the samples. Usually 0.5 µl or less of the purified PCR fragment is sufficient. Bring the final volume to 50 µl with TE.

3. Dilute the stock PicoGreen reagent 1:200 in TE and cover with aluminum foil to protect from light. To each DNA sample and blank, add 50 µl of diluted PicoGreen reagent. Mix well and let sit in the dark for 5 minutes.

4. Blank the fluorometer and calibrate using the most concentrated standard. Read the other standards to check the linearity of the calibration, and then read the samples.

METHOD B: COMPARATIVE ETHIDIUM BROMIDE STAINING

If a fluorometer is not available, it is possible to obtain an adequate estimate of the concentration of the PCR fragment using agarose gel electrophoresis. The intensity of the fragment after staining with ethidium bromide<!> is compared to that of a known mass of DNA. The High DNA and Low DNA Mass Ladders (Invitrogen, 10496016 and 10068013, respectively) are useful mass standards, because each is an equimolar mixture of DNA fragments of different sizes and different known mass amounts of DNA. The High Mass Ladder covers the size range from 1 to 10 kb, and the Low Mass Ladder covers the range from 0.1 to 2 kb.

PROTOCOL 3   SEQUENCING REACTION

Only 80–100 fmoles of PCR fragment is required for reasonable signal intensity in the sequencing reaction. For PCR fragments in the range of 0.5–1 kb, use 30–50 ng of DNA. Adjust the amount of DNA proportionately for larger or smaller fragments. Each reaction requires 10–20 pmoles of primer, for a final concentration of 0.5–1.0 nM in the sequencing reaction. This protocol is designed to work with the ABI PRISM Big Dye Terminator Ready Reaction mix (Applied Biosystems) and a GeneAmp PCR System 9700 (Applied Biosystems) thermal cycler.

MATERIALS

BUFFERS AND SOLUTIONS
ABI PRISM Big Dye Terminator Ready Reaction mix (Applied Biosystems) or other cycle sequencing kit

NUCLEIC ACIDS AND OLIGONUCLEOTIDES
Primer and template

SPECIAL EQUIPMENT
Microcentrifuge tubes (thin-walled for DNA amplification)
Thermal cycler, GeneAmp PCR System 9700 (Applied Biosystems) or equivalent

METHOD

1. Mix the appropriate amounts (as described above) of primer and template for each reaction in a PCR tube.

2. Add 4 µl of Terminator Ready Reaction mix to each sample.

   This is half the amount recommended in the manufacturer's manual; this volume works well and saves reagents.

3. Adjust the volume to 20 µl with deionized H_2O. Mix well and centrifuge briefly in a microcentrifuge.

4. Perform the following thermal cycle series in a GeneAmp PCR System 9700 (Applied Biosystems) or equivalent:

Cycle number	Denaturation	Annealing	Polymerization/Extension
24	10 sec at 96°C	5 sec at 50°C	4 min at 60°C
1	10 sec at 96°C	5 sec at 50°C	4 min at 60°C, then hold at 4°C

5. At the end of the series, the sequencing reactions can be purified immediately or stored at –20°C for later purification.

PROTOCOL 4   SEQUENCING REACTION PURIFICATION

It is necessary to purify the sequencing reaction products before electrophoresis to remove unincorporated dye terminators. If the dye terminators are not completely removed, they will obscure the reading of sequence near the primer. Spin columns are a rapid and efficient method for removing unincorporated molecules. Other methods, such as ethanol precipitation, can also be used. This protocol is designed for Performa DTR Gel Filtration Cartridges (Edge BioSystems).

MATERIALS

BUFFERS AND SOLUTIONS
Sample loading solution

   The loading solution used will depend on the type of machine that will be used. For the ABI PRISM 3100 Genetic Analyzer, highly deionized formamide (such as Hi-Di* Formamide from ABI) is recommended.

NUCLEIC ACIDS AND OLIGONUCLEOTIDES
Sequencing reactions, obtained above in step 5

SPECIAL EQUIPMENT
ABI PRISM 3100 Genetic Analyzer
Microcentrifuge tubes

Performa DTR Gel Filtration Cartridges (Edge BioSystems) or equivalent
Vacuum concentrator

METHOD

1. Remove the required number of cartridge assemblies from the package and inspect them carefully to be certain that the matrix has not dried out. If there is no excess fluid in the tube or the matrix appears dried, add ~200 µl of deionized H_2O and allow the matrix to rehydrate for a few minutes.

2. Centrifuge the cartridge for 2 minutes at 1360g in a microcentrifuge (e.g., 4000 rpm in an Eppendorf 5415C centrifuge using a fixed-angle rotor).

   This step must be done whether or not extra water was added to the cartridge.

3. Remove the cartridge from the tube and place in a clean collection tube. Carefully load the sample in the middle of the matrix.

4. Place the cartridge assembly back in the microcentrifuge, maintaining the same orientation as originally used. Centrifuge at 1360g for 1–1.5 minutes.

   The shorter time gives a lower product recovery (75–85%) but removes essentially all the unincorporated dye terminators. The longer time gives up to 95% recovery, but yields some terminators in the eluate, which can obscure the sequence near the primer.

5. At this point, the eluate can be frozen at –20°C or processed immediately for analysis on the automated sequencer.

6. Dry the eluate in a vacuum concentrator, with no heat, for 20–30 minutes. Avoid excessive drying time, as the dye terminators are light-sensitive and may begin to degrade.

7. Resuspend the dried sample in 10 µl of the appropriate loading solution, as recommended by the manufacturer of the sequencer. The amount to load will depend on the machine used; for the ABI PRISM 3100 Genetic Analyzer, a 1.0- to 1.5-µl amount is loaded on each capillary.

8. Perform electrophoretic separation and analysis of the samples as recommended by the manufacturer of the sequencing instrument.

▶ TROUBLESHOOTING

The data from sequencing reactions are usually available in two forms, a nucleotide sequence file generated by the base-calling software and a tracing of the signal from the dye terminators. The tracing provides diagnostic information in the event of problems with the sequencing reaction, and it is important to look at the tracing to assess the quality of the data. Most problems with sequence quality arise from problems with template quality or quantity. There is a narrow range of optimal template quantity; because PCR fragments are generally short, a small difference in mass makes a large difference in molarity. Too little template gives weak signals (low peak heights), whereas too much template produces off-scale traces. In both cases, base-calling can be inaccurate. In particular, it can be difficult to determine the number of nucleotides in homopolymeric stretches if the traces are off-scale.

Primer–dimers generated during PCR should be removed by the presequencing purification step. Occasionally, the first ~100 nucleotides of sequence will not be interpretable, and the tracing will give the appearance of two sequences superimposed on each other.

This is often due to a small contaminating fragment that was present in the original PCR product. Its mass may be low but, because it is small, the molarity is high. Generally the sequence beyond this point will be fine, and can be used to design a primer to obtain sequence data from the opposite strand of the region in question.

False terminations, in which the polymerase stalls but does not incorporate a dideoxy nucleotide, are not a problem with dye terminators because these products are unlabeled. However, compression artifacts are sometimes seen. Compression refers to the sequence-dependent aberrant migration of some fragments with G and C residues at the 3′ end, and is seen on both capillary and gel systems as abnormal spacing of the dye-terminated fragments. When there is compression, fragments of length n and $n + 1$ have nearly identical mobilities. The traces will show two peaks close together, with an unusually large gap before the $n + 2$ fragment. The ABI PRISM BigDye Terminator Cycle Sequencing Ready Reaction Kit contains dITP, which eliminates most problems with compression of GC-rich regions. In addition, base-calling software is often sufficiently sophisticated to handle these artifacts. Usually visual inspection of the traces can resolve the sequence ambiguity, but the region should be sequenced on the opposite strand, as it is very rare to find compressions on both strands at identical locations.

Finally, sequence-independent compression is typically seen at one or more locations within 10–30 nucleotides from the beginning of the readable sequence. This is due to a salt front that distorts fragment migration but usually does not interfere significantly with base-calling in this region. However, if absolute accuracy is required and only one strand is being sequenced in this region, it is advisable to pick primers such that the critical sequence area is located well beyond the salt front.

ACKNOWLEDGMENTS

I thank members of my laboratory for comments on the manuscript, and Mike Flora of the U.S.U.H.S. Biomedical Instrumentation Center for helpful discussions on sequencing.

REFERENCES

Carothers A.M., Urlaub G., Mucha J., Grunberger D., and Chasin L.A. 1989. Point mutation analysis in a mammalian gene: Rapid preparation of total RNA, PCR amplification of cDNA, and Taq sequencing by a novel method. *BioTechniques* **7:** 494–499.

Chou Q., Russel M., Birch D.E., Raymond J., and Bloch W. 1992. Prevention of pre-PCR mis-priming and primer dimerization improves low-copy-number amplifications. *Nucleic Acids Res.* **20:** 1717–1723.

D'Aquila R.T., Bechtel L.J., Videler J.A., Eron J.J., Gorcyca P., and Kaplan J.C. 1991. Maximizing sensitivity and specificity of PCR by pre-amplification heating. *Nucleic Acids Res.* **19:** 3749.

Gyllensten U.B. 1989. PCR and DNA sequencing. *BioTechniques* **7:** 700–708.

Murray V. 1989. Improved double-stranded DNA sequencing using the linear polymerase chain reaction amplified DNA. *Nucleic Acids Res.* **17:** 8889.

Prober J.M., Trainor G.L., Dam R.J., Hobbs F.W., Robertson C.W., Zagursky R.J., Cocuzza A.J., Jensen M.A., and Baumeister K. 1987. A system for rapid DNA sequencing with fluorescent chain-terminating dideoxynucleotides. *Science* **238:** 335–341.

Rosenblum B.B., Lee L.G., Spurgeon S.L., Khan S.H., Menchen S.M., Heiner C.R., and Chen S.M. 1997. New dye-labeled terminators for improved DNA sequencing patterns. *Nucleic Acids Res.* **25:** 4500–4504.

WWW RESOURCE

http://www-genome.wi.mit.edu/cgi-bin/primer/primer3_www.cgi Primer3, web software, Whitehead Institute/MIT Center for Genome Research.

SAMPLE PREPARATION

Sample preparation—the purification and concentration of nucleic acids from specimens collected for analysis—is the critical first step for successful PCR. Because of the variety of PCR applications, the starting materials for nucleic acid extraction may include such diverse specimens as 17- to 20-million-year-old fossilized magnolia leaves (Golenberg et al. 1990), mitochondrial DNA from dogs (Leonard et al. 2002), and human remains. Clearly, each specific sample type places unique constraints and requirements on the method of nucleic acid extraction.

DNA ISOLATION

The chapter "DNA Purification" (p. 87), contains methods for isolation of DNA from blood; tissues, such as mouse tails, buccal smears, or plant tissues; and cultured cells. This chapter also provides methods for automating the DNA isolation process. All of the methods presented in this chapter are based on binding of nucleic acids to silica gel. This procedure allows the DNA to bind to the silica gel in the presence of strong protein denaturants and after washes to eliminate the presence of contaminants; the DNA is eluted in a low-salt buffer. This methodology reduces the number of steps in DNA purification by eliminating organic solvent extraction and alcohol precipitation. In addition, it reduces handling of the nucleic acid, which minimizes the risk of sample-to-sample contamination. The prevention of contamination of PCR by either sample-to-sample contamination or PCR product carry-over is discussed in Section 1.

Although the methods described here reproducibly liberate the DNA from the cells and tissues, prior handling and age of the specimen may affect the average length of the resulting DNA fragments. Therefore, consideration should be given to the size of the region to be amplified, based on the source of the DNA. It would be unreasonable to expect nucleic acid isolated from a formalin-fixed, paraffin-embedded tissue section to serve as template for products longer than 800 bp. However, with appropriate specimens, the isolated DNA can be used as template for PCR products in the 25- to 40-kb range.

This is possible because the method of nucleic acid isolation described in this chapter does not introduce single-strand nicks, but breaks the DNA to an average size of greater than 40 kb. As described in "Tips for Long and Accurate PCR" (p. 53) in Section 1, template DNA for this application requires additional special handling. Specifically, excess heating of the nucleic acid should be avoided. Temperatures of 95°C or higher for more than 5 minutes can result in the deamination of cytosine, giving rise to a number of deoxyuridine sites in the DNA. As described in "High-Fidelity PCR Enzymes" (p. 21), the archaeal thermophilic DNA polymerases irreversibly bind to these sites and block amplification. Because of this problem, the judicious use of heat to denature the DNA template is recommended.

LACK OF AMPLIFICATION

There are occasions when samples cannot be successfully amplified. This could be due to a variety of problems with the template, including high number of deoxyuridines, presence of contaminants, or degradation of the sample. Often the specimens for PCR analysis are not replaceable. The chapter "Strategies for Overcoming PCR Inhibition" (p. 149) provides a number of approaches for salvaging these specimens. This chapter covers a number of strategies to overcome lack of amplification, including changing the DNA polymerase to one that is less sensitive to the contaminants and the presence of deoxyuridine. When the DNA is too degraded, new primers designed to amplify shorter products, in the 130- to 150-bp range, are recommended.

A good method for handling and purification of PCR products is important for a number of applications, and there are several commercially available kits for this purpose. However, in some instances, it becomes important to minimize the chance of aerosol contamination of the laboratory with the product to be purified. The silica-based method is fast, simple, and effective, and contamination is less likely to occur than with other methods.

mRNA ISOLATION

Isolation of mRNA for reverse transciptase (RT)-PCR presents a new set of challenges. The RNA must be released from within the cell, protected from endogenous nucleases, and recovered as quantitatively as possible. Additionally, the RNA must be free of salts and detergents, which inhibit or disrupt the enzymatic steps need for RT-PCR. "RNA Purification" (p. 117) describes the techniques used for specimen processing and storage as well as the steps of RNA purification. For RNA isolation, the protocols presented in this chapter detail the evolution of the technology.

The methods developed by Chirgwin and coworkers (1979) use strong chaotropic agents to simultaneously liberate the RNA and denature all the cellular proteins. Then the RNA is pelleted using cesium chloride equilibrium gradient ultracentrifugation, which keeps the cellular DNA in solution. One advantage of this method is that no organic solvents are used during the RNA extraction. This remains the method of choice when processing large amounts of tissue per sample or when the specimen contains large quantities of endogenous RNase. One disadvantage of this method is that it is very time-consuming.

The second method developed by Chomczynski and Sacchi (1987) uses a phenol-based extraction system combined with strong protein denaturants. The RNA partitions into the aqueous phase, and the denatured proteins and the bulk of the genomic DNA stay at the aqueous:organic interface. A limitation with this method is the variable amount of genomic DNA that is isolated with the RNA. If the RNA population is used directly in RT-PCR, then under certain circumstances the contaminating DNA could be misinterpreted as higher levels of gene expression. Treatment of all samples with RNase-free DNase removes this problem, but this adds a significant number of steps to the procedure.

The third method uses the selective binding and elution of the RNA from a silica gel membrane or matrix. A major advantage of the silica-based systems is the incorporation of the DNase treatment into the protocol so that the isolated RNA is ready for RT-PCR or other uses.

Equally important to the isolation procedure is the handling and storage of specimens. The standard method of sample processing requires the rapid processing of the specimens—from chilling of the sample, followed by the isolation of the cells and tissues, then either immediate lysis and RNA extraction, or rapid freezing of the specimen in liquid nitrogen or a dry ice–ethanol bath. If samples are flash-frozen, the specimens must be processed rapidly at the time of RNA extraction to prevent any thawing and RNA degradation prior to complete lysis. Ambion has developed a solution to this problem. By storing the cells and tissues in RNAlater, samples can be collected and stored at 4°C for short periods of time or at –20°C until all samples are ready to be batch processed. This greatly improves the reproducibility of RNA expression studies.

A major limitation of RNA expression studies is that the analysis reflects the average expression from all the cells in the tissue. "In Situ Localization of Gene Expression Using Laser Capture Microdissection" (p. 135) details a method for the visualization of the tissue and the selection and isolation of cells or groups of cells from the tissue section. This methodology extends the power of in situ analysis by providing a means to dissect tissues and perform gene expression studies of numerous genes using real-time PCR. Standard in situ methods are limited by the limit of detection and the single-use nature of a specimen.

Collectively, these chapters provide isolation methods for RNA and DNA for nearly every possible PCR application.

REFERENCES

Chirgwin J.M., Przybyla A.E., MacDonald R.J., and Rutter W.J. 1979. Isolation of biologically active ribonucleic acid from sources enriched in ribonuclease. *Biochemistry* **18:** 5294–5299.

Chomczynski P. and Sacchi N. 1987. Single-step method of RNA isolation by acid guanidinium thiocyanate-phenol-chloroform extraction. *Anal. Biochem.* **162:** 156–159.

Golenberg E.M., Giannasi D.E., Clegg M.T., Smiley C.J., Durbin M., Henderson D., and Zurawski G. 1990. Chloroplast DNA sequence from a miocene Magnolia species. *Nature* **344:** 656–658.

Leonard J.A., Wayne R.K., Wheeler J., Valadez R., Guillen S., and Vila C. 2002. Ancient DNA evidence for Old World origin of New World dogs. *Science* **298:** 1613–1616.

9 DNA Purification

Craig Smith, Paul Otto, Rex Bitner, and Gary Shiels

Promega Corporation, Madison, Wisconsin 53711-5399

The amplification of nucleic acids from small amounts of starting material is a critical tool for basic research and diagnostics laboratories. Numerous methods have been developed, including self-sustaining sequence replication (3SR; Abravaya et al. 1995), ligase chain reaction (LSR; Shimer and Backman 1995), and rolling circle amplification (Lizardi et al. 1998). However, PCR (Kleppe et al. 1971; Saiki et al. 1985) has become the standard.

Methods to isolate nucleic acids are a necessary part of the overall PCR process. They provide the means both to isolate source materials for amplification and to clean up, or purify, amplification products before their use in downstream applications. Genomic DNA is the most commonly used template for PCR amplification. DNA may be isolated from a wide range of starting materials, including whole blood, plant or animal tissue, and cultured cells. For some diagnostics and forensic applications, specialized techniques for isolation of DNA from processed food, buccal swabs, bodily fluids, and paraffin-embedded tissues are necessary. Post-PCR cleanup of amplified DNA fragments requires the removal of excess primers, salts, and nucleotides, as well as efficient recovery of the amplified product. In some cases, limitations in primer design can result in the amplification of multiple products of varying size. These contaminating fragments must also be removed before use in downstream applications.

The successful isolation of DNA requires methods that prevent nuclease degradation and eliminate PCR inhibitors, which can be problematic in some samples (Akane et al. 1994; Wilson 1997; Al-Soud et al. 2000). Methods that provide template in a concentrated form and in buffers that are compatible with downstream applications are also desirable. Numerous methods have evolved to isolate genomic DNA for amplification, as well as to purify PCR products post-amplification. Which method is the best depends on a number of factors, including the nucleic acid type and size, yield, purity, the presence or absence of contaminating fragments, throughput and the need for automation, and price per isolation.

Binding and elution from silica has become the method of choice for both template isolation and purification of amplification products. Silica-based methods have been developed to isolate all classes of nucleic acids from a range of sample types. Different formats are available to handle various throughput needs, ranging from individual samples to high-throughput, fully automated methods capable of processing thousands of samples per day. The high concentrations of chaotropes used to bind nucleic acids to silica surfaces are potent denaturants that, when used in combination with various detergents, serve to disrupt tissue, lyse cells, and inactivate endogenous nucleases. A hallmark of silica methods is the elution of nucleic acids in nuclease-free water or other low-ionic-strength buffer such as TE (10 mM Tris HCl, pH 8.0; 0.1 mM EDTA), allowing use of isolated nucleic acids in downstream applications without further processing. Typically, DNA isolated by the following methods is stable for up to a year at 4°C, depending on the source material, and can be stored indefinitely at –20 to –70°C.

For large-scale isolation of milligram quantities of DNA, solution extraction and precipitation methods based on the differential solubility of DNA, RNA, and protein have been developed. These methods are best suited to the reagent volumes required for processing large amounts of starting material and can be readily scaled to isolate from 1 to 100 mg of nucleic acid.

OVERVIEW OF METHODS

Here we outline a series of sample protocols for (1) purification of genomic DNA from blood, tissues, and cultured cells and (2) purification of PCR products from agarose gels and from solution—including the cleanup of dye terminator sequencing reactions. Solution extraction, silica membrane, and magnetic particle-based methods are included. Where applicable, more than one method is presented to provide users with a choice of methods to meet their individual throughput requirements.

Genomic DNA Isolation

The purification of genomic DNA for downstream use in PCR can be achieved in a number of ways. With the advent of genomic-based DNA analyses, the requirements for easily processing large numbers of samples, particularly using automated or semiautomated methods, have led to greater use of solid-phase extraction methods. The common element with these methods is the reversible binding of DNA to a solid matrix, most commonly silica. The Wizard SV 96 technology, based on a high-density 96-well filter plate with a silica-impregnated membrane, and the Wizard MagneSil silica paramagnetic particle chemistry, are two technologies that satisfy these requirements.

Here we present a solution-based method for the isolation of genomic DNA from 10 ml of whole blood, a silica membrane-based method for isolation of genomic DNA from cultured cells and tissue, and magnetic silica paramagnetic particle-based methods for small-scale isolation of genomic DNA from whole blood, and for simultaneous isolation and quantitation of genomic DNA from buccal swabs.

PROTOCOL 1 SOLUTION-BASED METHOD FOR GENOMIC DNA PURIFICATION FROM WHOLE BLOOD

Isolation of genomic DNA from whole blood (10 ml) using the Wizard Genomic DNA Purification Kit is based on a four-step process. The first step in the purification procedure involves lysis of the red blood cells and removal of the soluble fraction, followed by lysis of the white blood cells and their nuclei. An RNase digestion step may be included at this time; it is optional for some applications (see Table 1). Cellular proteins are selectively removed by a salt precipitation step, which leaves the high-molecular-weight genomic DNA in solution. Finally, the genomic DNA is concentrated and desalted by isopropanol precipitation (Fig. 1).

The Wizard Genomic DNA Purification Kit protocol is scaleable and can be used to isolate genomic DNA from 20 μl to 10 ml of whole blood. A 96-well microwell plate format

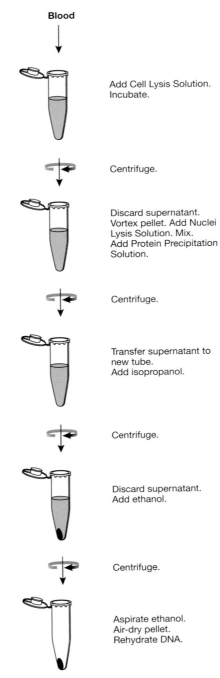

Blood

Add Cell Lysis Solution.
Incubate.

Centrifuge.

Discard supernatant.
Vortex pellet. Add Nuclei
Lysis Solution. Mix.
Add Protein Precipitation
Solution.

Centrifuge.

Transfer supernatant to
new tube.
Add isopropanol.

Centrifuge.

Discard supernatant.
Add ethanol.

Centrifuge.

Aspirate ethanol.
Air-dry pellet.
Rehydrate DNA.

FIGURE 1. An overview of genomic DNA purification using the Wizard Genomic DNA Purification Kit. (Reprinted, with permission, from Wizard Genomic DNA Purification Kit protocol card #FB022, fig. 2818MA11_9A [© 2003 Promega Corporation].)

has been developed for processing volumes less than 50 μl. Purifications of equal quality and yield are achieved from whole-blood samples collected in EDTA-, heparin-, or citrate-coated tubes (Beutler et al. 1990).

MATERIALS

▼ CAUTION

See Appendix for appropriate handling of materials marked with <!>.

BUFFERS, SOLUTIONS, AND REAGENTS

Ethanol, 70%<!>, room temperature

Isopropanol<!>, room temperature

RNase A

> Dissolve RNase A to 4 mg/ml in DNA Rehydration Solution, and boil for 10 minutes to remove contaminating DNase. Store in aliquots at –20°C.

SPECIAL EQUIPMENT

Agarose gel electrophoresis equipment

Centrifuge tubes, sterile 50-ml

ADDITIONAL ITEMS

Wizard Genomic DNA Purification Kit (Promega; includes Cell Lysis Solution, Nuclei Lysis Solution, Protein Precipitation Solution, and DNA Rehydration Solution)

Molecular-weight markers for use in agarose gel electrophoresis

CELLS AND TISSUES

Whole blood<!> (10 ml)

METHOD

RED BLOOD CELL LYSIS

1. Add 30 ml of Cell Lysis Solution to a sterile 50-ml centrifuge tube.

2. Gently rock the blood sample tube until thoroughly mixed; then transfer blood (10 ml) to the tube containing the Cell Lysis Solution. Invert the tube 5 or 6 times to mix.

3. Incubate the mixture for 10 minutes at room temperature (invert 2 or 3 times once during the incubation) to lyse the red blood cells. Centrifuge at 2000*g* for 10 minutes at room temperature.

4. Remove and discard as much supernatant as possible without disturbing the visible white pellet. Approximately 1.4 ml of residual liquid will remain. If the blood sample has been frozen, add an additional 30 ml of Cell Lysis Solution, invert 5 or 6 times to mix, and repeat steps 3 and 4 until the pellet is nearly white. There may be some loss of DNA from frozen samples.

   > At step 4, some red blood cells or cell debris may be visible along with the white blood cells. If the pellet appears to contain only red blood cells, add an additional aliquot of Cell Lysis Solution after removing the supernatant above the cell pellet, and then repeat steps 3 and 4.

5. Vortex the tube vigorously until the white blood cells are resuspended (10–15 seconds).

NUCLEI LYSIS AND PROTEIN PRECIPITATION

1. Add 10 ml of Nuclei Lysis Solution to the tube containing the resuspended cells. Pipette the solution 5 or 6 times to lyse the white blood cells. The solution should become very viscous. If clumps of cells are visible after mixing, incubate the solution at 37°C until the clumps are disrupted. If the clumps are still visible after 1 hour, add an additional 3 ml of Nuclei Lysis Solution and repeat the incubation.

2. *Optional:* Add RNase A, to a final concentration of 20 µg/ml, to the nuclear lysate and mix by inverting the tube 2–5 times. Incubate the mixture at 37°C for 15 minutes, then cool to room temperature.

3. Add 3.3 ml of Protein Precipitation Solution to the nuclear lysate and vortex vigorously for 10–20 seconds. Small protein clumps may be visible after vortexing.

4. Centrifuge at 2000*g* for 10 minutes at room temperature. A dark brown protein pellet should be visible.

DNA PRECIPITATION AND REHYDRATION

1. Transfer the supernatant to a clean 50-ml centrifuge tube containing 10 ml of room-temperature isopropanol.

2. Gently mix the solution by inversion until the white, thread-like strands of DNA form a visible mass.

3. Centrifuge at 2000*g* for 1 minute at room temperature. The DNA will be visible as a small white pellet.

4. Decant the supernatant and add 10 ml of room-temperature 70% ethanol to the DNA. Gently invert the tube several times to wash the DNA pellet and the sides of the tube. Centrifuge as in step 3.

5. Carefully aspirate the ethanol. The DNA pellet is very loose at this point, and care must be used to avoid aspirating the pellet into the pipette. Invert the tube on clean, absorbent paper and air-dry the pellet for 10–15 minutes.

6. Add 800 µl of DNA Rehydration Solution, and rehydrate the DNA by incubating at 65°C for 1 hour. Periodically mix the solution by gently tapping the tube. Alternatively, rehydrate the DNA by incubating the solution overnight at room temperature or at 4°C.

7. Store the DNA at 2–8°C.

CHARACTERIZATION OF ISOLATED DNA

DNA yield can be measured by absorbance at 260 nm or by agarose gel electrophoresis of purified samples relative to standards. This latter method is preferred over spectrophotometric determinations if precise concentrations are required, because absorbance readings can be artificially high due to the presence of low-molecular-weight, UV-absorbing material that may be present in the sample (Glasel 1995). A typical yield from 10 ml of whole blood is 250–500 µg (Table 1).

The quality of DNA isolated by this procedure can be assessed by length determination and functionality in PCR amplification (Saiki et al. 1985). The majority of the DNA isolated using this method is 50–200 kb, as determined by pulsed-field gel electrophoresis (Fig. 2). Compatibility of the isolated DNA with PCR amplification was demonstrated by successful amplification of three polymorphic human loci. Samples (25 µg) from five individuals amplified efficiently, producing an identical pattern of alleles for each locus (Fig. 3).

PROTOCOL 2   SILICA MEMBRANE-BASED METHOD FOR ISOLATION OF GENOMIC DNA FROM TISSUES AND CULTURED CELLS IN A 96-WELL FORMAT

Silica chemistry has become the method of choice for the purification of high-quality genomic DNA from samples such as mouse tail clippings, animal tissue, and tissue culture cells. The Wizard SV Genomic DNA Purification System uses a solid-phase silica-impregnated filter membrane for the purification of genomic DNA. The procedure consists of tissue disruption (if necessary), cell lysis, binding of DNA to the SV membrane, washing to remove residual contaminants, and elution of DNA in nuclease-free water. Sample and reagents can be driven through the membrane by either centrifigation or vacuum filtration.

TABLE 1. DNA yield from various starting materials using the Wizard Genomic DNA Purification Kit

Species and material	Amount of starting material	Typical DNA yield	RNase treatment
Human whole blood			
(Yield depends on the	300 µl	5–15 µg	optional
quantity of white	1.0 ml	25–50 µg	optional
blood cells	10.0 ml	250–500 µg	optional
present)			
96-well plate	50 µl/well	0.2–0.7 µg	optional
Mouse whole blood			
EDTA (4%) treated	300 µl	6 µg	optional
Heparin (4%) treated	300 µl	6–7 µg	optional
96-well plates	50 µl/well	0.2–0.7 µg	optional
Cell lines			
K562	3×10^6 cells	15–30 µg	required
COS	1.5×10^6 cells	10 µg	required
NIH-3T3	2.25×10^6 cells	9.5–12.5 µg	required
PC12	8.25×10^6 cells	6 µg	required
CHO	$1–2 \times 10^6$ cells	6–7 µg	required
Animal tissue			
Mouse liver	11 mg	15–20 µg	required
Mouse tail	0.5–1 cm of tail	10–30 µg	optional
Insect			
Sf9 cells	5×10^6 cells	16 µg	required
Plant tissue			
Tomato leaf	40 mg	7–12 µg	required
Gram-negative bacteria			
Escherichia coli JM109	1 ml	20 µg	required
overnight culture,	5 ml	75–100 µg	required
$\sim 2 \times 10^9$ cells/ml			
Enterobacter cloacae,	1 ml	20 µg	required
overnight culture,	5 ml	75–100 µg	required
$\sim 6 \times 10^9$ cells/ml			
Gram-positive bacteria			
Staphylococcus epidermidis,	1 ml	6–13 µg	required
overnight culture,			
$\sim 3.5 \times 10^8$ cells/ml			
Yeast			
Saccharomyces cerevisiae,	1 ml	4.5–6.5 µg	required
overnight culture,			
$\sim 1.9 \times 10^8$ cells/ml			

The procedure can be performed manually or on liquid-handling robots such as the Biomek FX and Biomek 2000 Laboratory Automation Workstations (Beckman Coulter). Individual minicolumns can be used for manual processing of low numbers of samples. Amplifiable DNA can be isolated from up to 5×10^6 cells, 20 mg of tissue, or from up to 1.2 cm of mouse tail without centrifugation of the lysates. Isolation from tissue or mouse tails requires an overnight digestion with proteinase K. The genomic DNA isolated with this system is of high quality and serves as an excellent template for PCR.

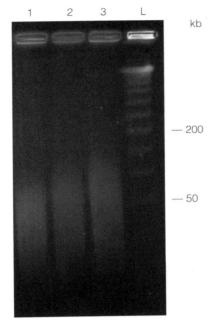

FIGURE 2. Pulsed-field gel electrophoresis of genomic DNA isolated from whole blood using the Wizard Genomic DNA Purification Kit. DNA samples were resolved by electrophoresis through a 1% agarose gel using a CHIEF Mapper apparatus (BioRad, Hercules, CA). The initial and final switch intervals were varied with a time ramp of 2 seconds (initial) to 30 seconds (final). Electrophoresis was performed in 0.5x TBE buffer at 14°C for 20 hours. (Lane *L*) Lambda Ladders Molecular Weight Markers. DNA was purified from whole blood collected in a tube coated with EDTA (lane *1*), heparin (lane *2*), or citrate (lane *3*). (Reprinted, with permission, from Micka et al. 1996 Promega Notes 56, p. 2, fig. 0896GA12 [©2003 Promega Corporation].)

MATERIALS

BUFFERS, SOLUTIONS, AND REAGENTS

1x PBS (for tissue culture cells)

Dissolve the following in 1 liter of sterile, deionized H_2O. Autoclave. The pH of 1x PBS should be 7.4.

Na_2HPO_4<!>	1.15 g
KH_2PO_4<!>	0.2 g
NaCl	8 g
KCl<!>	0.2 g

ENZYMES AND ENZYME BUFFERS

Digestion Solution Master Mix (for tissues and mouse tails):

For each tissue sample, combine the following:

Nuclei Lysis Solution	200 μl
EDTA (0.5 M)	50 μl
Proteinase K<!> (20 mg/ml)	20 μl
RNase A solution (4 mg/ml)	5 μl

Proteinase K<!> (20 mg/ml solution; for tissues and mouse tails)

Resuspend proteinase K with nuclease-free H_2O to a concentration of 20 mg/ml. Dispense into working volumes determined by the average number of preps done at a time. Store at –20°C and thaw on ice. Avoid multiple freeze–thaw cycles.

▼ CAUTION

See Appendix for appropriate handling of materials marked with <!>.

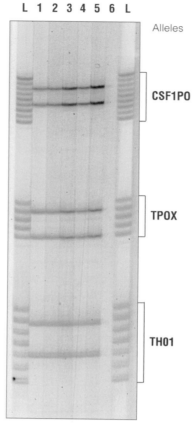

FIGURE 3. PCR amplification of polymorphic short tandem repeat (STR) sequences from genomic DNA isolated using the Wizard Genomic DNA Purification Kit. Amplification was performed using the *GenePrint* CSF1PO, TPOX, TH01 Multiplex (Promega, DC6001). Products were separated by electrophoresis in a 4% denaturing polyacrylamide gel and detected by silver staining. (Lane *L*) STR multiplex CSF1PO, TPOX, TH01 Ladder Mix. (Lanes *1–5*) Replicate amplification reactions from samples purified from the same blood sample. (Lane *6*) Negative (no template) control. (Reprinted, with permission, from Micka et al. 1996, Promega Notes 56, p. 2, fig. 0897GA12 [©2003 Promega Corporation].)

SPECIAL EQUIPMENT

3 single-position labware ALP (Beckman Coulter)
4 P250 tip rack assemblies (Beckman Coulter)
4 × 4 position labware ALP (Beckman Coulter)
96-well Tip Wash ALP (Beckman Coulter)
Adhesive plate sealers (foil)
Heating and Cooling ALP (Beckman Coulter)
Liquid-handling robot (for automated procedure) such as the Biomek FX or Biomek 2000 Laboratory Automation Workstation (Beckman Coulter)
Orbital Shaker ALP (Beckman Coulter)
Tip Loader ALP (Beckman Coulter)
Vac-Man 96 Vacuum Manifold (Promega)
Vacuum pump capable of 15–20 inches of Hg
Vacuum trap for waste collection (1-liter size)
Vacuum tubing
Water bath, 55°C

ADDITIONAL ITEMS

> Wizard SV 96 Genomic DNA Purification System (Promega; includes Binding Plate, 96-well deep-well plate, DNA elution plate, Nuclei Lysis Solution, 0.5 M EDTA, Wizard SV Lysis Buffer, Wizard SV Wash Solution, RNase A solution, and nuclease-free H$_2$O)

CELLS AND TISSUES

> Samples for genomic DNA purification (mouse tail clippings, animal tissue, or tissue culture cells)

METHOD

PREPARATION OF MOUSE TAIL AND TISSUE LYSATES FOR AUTOMATED PURIFICATION

1. Place a 0.5- to 1.2-cm mouse tail clipping or up to 20 mg of other tissue into each well of a 96-well, deep-well plate.

   > 20 mg of tissue is the maximum sample size. Exceeding this amount may result in clogging of the SV 96 Binding Plate during purification.

2. Add 275 µl of the Digestion Solution Master Mix per sample. Cover the plate with an adhesive seal.

3. Place the plate in a 55°C water bath and incubate overnight (16–18 hours).

   > If genomic DNA will not be purified directly after overnight proteinase K digestion, the sample lysate may be stored frozen at –70°C. Before proceeding with automated genomic DNA purification, the frozen lysate must be warmed to 55°C for 1 hour.

   > *Optional:* If undigested hair and cartilage remain after overnight proteinase K digestion, spin the plate in a centrifuge at 2000*g* to pellet the undigested sample. Transfer the supernatant to a new 96-well, deep-well plate.

AUTOMATED PURIFICATION OF GENOMIC DNA FROM PREPARED TISSUE LYSATES

After proteinase K digestion of sample tissue, all subsequent genomic DNA purification steps can be automated on a liquid-handling workstation without further user intervention. After the required tools, reagents, and the warm tissue lysate plate are placed on the deck of the automated liquid-handler, perform the following steps automatically.

1. Add 250 µl of Wizard SV Lysis Buffer to each sample well.

2. Mix the contents of each well by pipetting several times.

3. Transfer the tissue lysates to the wells of the Binding Plate sitting in the vacuum manifold. Bind genomic DNA to the plate by applying a vacuum until all of the lysate has passed through the Binding Plate.

4. Add 1 ml of Wizard SV Wash Solution to each well of the Binding Plate.

5. Apply the vacuum until the wash solution passes through the Binding Plate.

6. Repeat steps 4 and 5 for a total of 3 washes.

7. After the wells have emptied, continue to apply vacuum for an additional 6 minutes to allow the binding matrix to dry.

8. Prepare the vacuum manifold assembly for elution. Move the Binding Plate and position a deep-well elution plate in the vacuum manifold. Then position the Binding Plate on top of the elution plate in the same orientation so that the Binding Plate tips are centered on the deep-well elution plate.

9. Add 200 µl of room-temperature, nuclease-free H$_2$O to each well of the Binding Plate, and incubate for 2 minutes at room temperature.

10. Apply the vacuum for 1 minute to elute purified genomic DNA from the SV 96 Binding Plate into the 96-well deep-well plate.

11. Repeat steps 9–10 for a total elution volume of 400 µl.

12. Purification of genomic DNA is complete, and the purified sample is contained in the 96-well deep-well plate. Purified samples can be stored at –20°C or –70°C by covering the plate tightly with a plate seal.

AUTOMATED PURIFICATION OF GENOMIC DNA FROM TISSUE CULTURE CELLS

The following procedure is for automated purification of genomic DNA from tissue culture cells in 96-well culture plates. A maximum of 5 x 10^6 cells per well can be purified. Exceeding this number of cells per well may cause clogging of the SV 96 Binding Plate during purification. The number of cells may need to be adjusted depending on cell type and function.

1. Wash the cells once with sterile 1x PBS before automated purification. All subsequent steps can be automated on a liquid-handling workstation without further user intervention. After the required tools, reagents, and cells (without media or PBS) are placed on the deck of the automated liquid-handler, perform the following steps automatically.

2. Add 150 µl of Wizard SV Lysis Buffer to the washed cells. Mix by pipetting.

> If genomic DNA will not be purified immediately, the method may be stopped after the addition of the Wizard SV Lysis Buffer and the samples can be stored at –70°C. With thawed sample lysates, purification of genomic DNA may resume at step 3.

3. Transfer the cell lysates to the Binding Plate sitting in a vacuum manifold assembly. Bind the genomic DNA to the binding matrix by applying a vacuum until the lysate has passed through the Binding Plate.

4. Add 1 ml of Wash Solution to each well of the Binding Plate. Apply the vacuum until the Wash Solution passes through the Binding Plate.

5. Repeat step 4 two more times for a total of 3 washes with the Wash Solution.

6. After the wells have emptied, continue to apply the vacuum for an additional 6 minutes to dry the binding matrix.

7. Prepare the vacuum manifold assembly for elution. Move the Binding Plate, and position a deep-well elution plate in the vacuum manifold. Then position the Binding Plate on top of the elution plate in the same orientation so that the Binding Plate tips are centered on the deep-well elution plate.

8. Add 200 µl of room-temperature, nuclease-free H$_2$O to each well of the Binding Plate and incubate for 2 minutes at room temperature.

9. Apply a vacuum for 1 minute to elute purified genomic DNA from the SV 96 Binding Plate to the 96-well deep-well plate.

10. Purification of genomic DNA is complete, and the purified sample is contained in the 96-well, deep-well plate. Purified samples can be stored at –20°C or –70°C by covering the plate tightly with a plate seal.

> RNA may be copurified with genomic DNA. To remove copurified RNA, add 2 µl of RNase A Solution per 250 µl of nuclease-free H$_2$O before elution of genomic DNA from the column. After elution, incubate the purified genomic DNA at room temperature for 10 minutes. Alternatively, the RNase A Solution (2 µl) may be added after elution.

CHARACTERIZATION OF ISOLATED DNA

Yield and quality of Wizard SV-purified samples can be determined by spectrophotometric analysis at A_{260} and A_{280}. Typical yield of high-molecular-weight genomic DNA is 5–20 µg, depending on sample source (Table 2), with A_{260}/A_{280} ratios of 1.7 ± 0.08. Yield and quality of individual mouse tail samples processed by a Biomek 2000 robotic workstation are given in Figure 4. The utility of purified genomic DNA as a template in PCR amplification was also assessed. One µl of genomic DNA purified from either mouse tail or tissue culture samples was amplified by PCR. Mouse-tail samples were evaluated by amplifying with interleukin-1β (IL-1β)-specific primers. Genomic DNA isolated from HeLa cells was evaluated by amplifying with primers specific for Factor V DNA (Fig. 5).

PROTOCOL 3

MAGNETIC PARTICLE-BASED METHODS FOR HIGH-THROUGHPUT GENOMIC DNA ISOLATION FROM WHOLE BLOOD AND BUCCAL SWABS

The Wizard MagneSil Genomic DNA Purification System uses a solid-phase silica paramagnetic particle for the purification of genomic DNA. This technology was developed as an alternative to vacuum filtration and centrifugation-based purification formats to allow cost-efficient, high-throughput sample processing. MagneSil technology is readily adaptable to robotic platforms, allowing complete automation of the purification process in either 96- or 384-well microwell plate formats.

A significant advantage of using magnetic particles for bioseparation applications is their utility with variously sized starting samples. With most nucleic acid purification systems, the amount of starting material is limited by the fixed size of the product's purification matrix. With MagneSil particles, the amount of magnetic solid phase can be scaled to the amount of starting material. Thus, a single system provides increased flexibility with protocols covering small- to large-scale purifications. Magnetic particles can also be used with most off-the-shelf plasticware, making them a cost-effective alternative to membrane-based purification methods.

Magnetic DNA purification systems based on MagneSil particles can be used to isolate genomic DNA from sample types such as whole blood, cultured cells, a variety of foods and foodstuffs, and plant leaf or seed tissue. Two systems have been developed for the isolation of genomic DNA from human whole blood: The first processes 200 µl of whole blood, generating 4–9 µg of high-quality genomic DNA. This method is scaleable down to 50 µl, providing added flexibility for downstream DNA testing.

For users who prefer to avoid the need for post-purification DNA quantitation, the DNA IQ System provides a magnetic-based method for the purification of a predetermined

TABLE 2. Typical yields of genomic DNA isolated from a variety of sources using the Wizard SV Genomic DNA Purification System

Sample	Amount	Average yield
Mouse tail clipping	20 mg	20 µg
Mouse liver	20 mg	15 µg
Mouse heart	20 mg	10 µg
Mouse brain	20 mg	6 µg
CHO cells	1×10^6 cells	5 µg
NIH-3T3 cells	1×10^6 cells	9 µg
293 cells	1×10^6 cells	8 µg

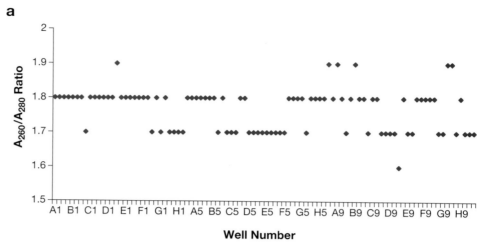

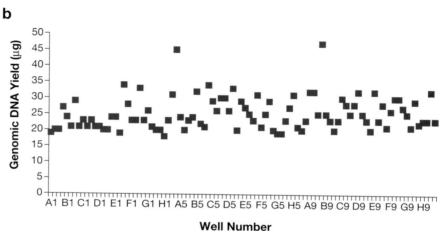

FIGURE 4. Purity (*a*) and yield (*b*) measurements of genomic DNA from 96 separate 20-mg mouse tail clippings purified with the Wizard SV 96 Genomic DNA Purification System. Purity was measured by A_{260}/A_{280} ratio and yield was measured by absorbance at 260 nm. (Reprinted, with permission, from Grunst and Worzella 2002, Promega Notes 81, p. 9, fig. 3694MA04_2A [© 2003 Promega Corporation].)

amount of DNA from a blood or tissue sample. Methods have been developed for the purification of DNA in the ranges of either 100 ng or 1.0 μg from human whole blood. These applications have been adapted to several laboratory robotic workstations, including the Biomek FX and 2000 (Beckman Coulter) and Genesis (Tecan) systems.

METHOD 1: HIGH-THROUGHPUT ISOLATION OF GENOMIC DNA FROM 200 μL OF WHOLE BLOOD USING A LIQUID-HANDLING ROBOTIC WORKSTATION

This method is designed to process 200-μl blood samples in a 96-well format on a liquid-handling robot such as the Biomek FX and Biomek 2000 Laboratory Automation Workstations (Beckman Coulter). Single-plate processing time is less than an hour with a single-head instrument.

Blood is mixed with a detergent-chaotrope lysis/binding buffer to disrupt cell and nuclear membranes and denature proteins. Released DNA adsorbs to the surface of the

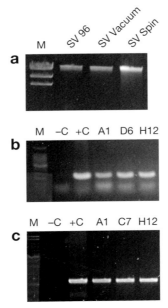

FIGURE 5. Analysis of genomic DNA from HeLa cells and mouse tail clippings isolated by using the Wizard SV and SV 96 Genomic DNA Purification Systems. (*a*) Agarose gel analysis of genomic DNA isolated from 5 x 10^6 HeLa cells. (*b*) PCR amplification of Factor V DNA from HeLa cells. (Lanes *A1*, *D6*, *H12*) Sample wells, 250-bp amplification products. (*c*) PCR amplification of IL-1β DNA from mouse tail samples. (Lanes *A1*, *C7*, *H12*) Sample wells, 1.2-kb amplification products. (Lane *M*) DNA markers, (–C) negative control, (+C) positive control. (Reprinted, with permission, from Grunst and Worzella 2002, Promega Notes 81, p. 9, fig. 3711TA02_2A [©2003 Promega Corporation].)

MagneSil particles contained in the lysis/binding buffer. The particle–adherent target complex is captured to the side of the well by an external magnetic field, and the lysate is removed as waste. Residual contaminants are removed with a series of washes, first with lysis/binding buffer, then with a salt wash solution, and finally with an alcohol/salt wash to remove the last traces of heme. After drying the particles, the DNA is eluted in TE buffer or water on a heat transfer plate.

MATERIALS

SPECIAL EQUIPMENT

3 single-position labware ALP (Beckman Coulter)
4 P250 tip rack assemblies (Beckman Coulter)
4 x 4 position labware ALP (Beckman Coulter)
96-well collection plate (Promega, A9161 or comparable)
96-well Tip Wash ALP (Beckman Coulter)
Biomek FX Robotic Workstation fitted with a single pod and a 96-well pipetting head (Beckman Coulter)
Deep-Well MagnaBot 96 Magnetic Separation Device (Promega)
Deep-well multiwell plates (Marsh, AB-0932 [2.2 ml] and AB-0787 [1.2 ml] or comparable)
Heat transfer block (Promega, Z3271 or comparable)
Heating and Cooling ALP (Beckman Coulter)

MagnaBot Spacer, 1/8 inch (Promega)
Orbital Shaker ALP (Beckman Coulter)
Pyramid-bottom, 96-well reservoir (Innovative Microplates, S30014 or comparable)
Tip Loader ALP (Beckman Coulter)

ADDITIONAL ITEMS

MagneSil Blood Genomic, Max Yield System (Promega; includes alcohol wash, antifoam reagent, Lysis Buffer, Salt Wash, Elution Buffer, and MagneSil Paramagnetic Particles)

CELLS AND TISSUES

Whole blood, buccal swabs, or sample stains<!>

METHOD

1. Add 400 µl of Lysis Buffer and 40 µl of MagneSil particles to 200 µl of blood. Use extensive robotic mixing to lyse the white blood cells. The released DNA binds the magnetic particles (PMPs).

2. Capture the particles with the MagnaBot magnetic array and discard the liquid.

3. Wash the PMPs extensively with 360 µl of Lysis Buffer.

4. Capture the particles with the MagnaBot magnetic array and discard the liquid.

5. Wash the PMPs extensively with 360 µl of Salt Wash.

6. Capture the particles with the MagnaBot magnetic array and discard the liquid.

7. Repeat steps 5 and 6 for a total of 2 washes with Salt Wash.

8. Wash the PMPs extensively with 360 µl of Alcohol Wash Buffer.

9. Capture the particles with the MagnaBot magnetic array and discard the liquid.

10. Repeat steps 8 and 9 for a total of 3 washes with Alcohol Wash Buffer.

11. Dry the PMPs on a heat block to eliminate any residue of the Alcohol Wash Buffer.

12. Elute the DNA in 210 µl of Elution Buffer, with extensive mixing, on a heat block set at 80°C. The temperature inside the wells will be ~65°C.

13. Store the eluted DNA at –20°C.

METHOD 2: DNA IQ ISOLATION OF GENOMIC DNA FROM STAINS AND BUCCAL SWABS

The DNA IQ System uses a novel approach for DNA isolation. The MagneSil chemistry is used to capture and release a consistent amount of DNA across a wide range of samples. The particles have a defined DNA capacity and, in the presence of excess DNA, will only bind a fixed amount of nucleic acid (Fig. 6). This property is used to isolate ~100 ng of DNA, independent of sample size. The DNA is eluted into 100 µl of Elution Buffer to give a final concentration of 1 ng/µl. As a result, post-purification DNA quantitation is not necessary when using this system. An overview of the protocol is given in Figure 7.

MATERIALS

BUFFERS, SOLUTIONS, AND REAGENTS

Dithiothreitol (DTT)<!>, 1 M

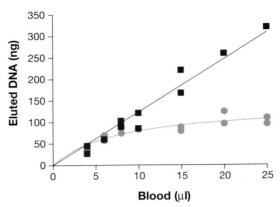

FIGURE 6. Effect of sample volume on the amount of DNA isolated using the DNA IQ System. A high-capacity resin shows a linear relationship between eluted DNA versus amount (μl) of blood extracted (represented by squares). In contrast, the DNA IQ System Resin is saturated using the blood volumes tested and thus gives approximately the same amount of DNA regardless of sample size (represented by circles). (Reprinted, with permission, from Promega DNA IQ System-Database Protocol Technical Bulletin #TB297, fig. 3236MA02_1A [©2003 Promega Corporation].)

Ethanol, 95–100%<!>
Isopropyl alcohol<!>
Lysis Buffer<!>
    Determine the total amount of Lysis Buffer to be used (Table 3). For each 100 μl of Lysis Buffer (supplied with the DNA IQ System), add 1 μl of 1 M DTT. Mix by inverting several times. Mark and date the label to indicate that DTT has been added. The Lysis Buffer can be stored at room temperature for up to one month.

SPECIAL EQUIPMENT
    Aerosol-resistant micropipette tips
    DNA IQ Spin Baskets (Promega, V1221)
    Heat block
    MagneSphere Technology Magnetic Separation Stand (Promega)
    Microcentrifuge tubes, 1.5-ml conical

ADDITIONAL ITEMS
    DNA IQ System (Promega; includes Resin, Lysis Buffer, 2x Wash Buffer, and Elution Buffer)
    Microtubes, 1.5 ml (Promega, V1231)

METHOD

1. Place sample (see Table 3) in a 1.5-ml microcentrifuge tube. Keep in mind that the recommended amount of resin can capture a maximum of ~100 ng of DNA.

2. Add the appropriate amount of prepared Lysis Buffer. Different samples require different volumes of Lysis Buffer; see Column 2 ("Lysis Buffer 1") of Table 3 for the appropriate volume to add at this point. Close the lid and place the tube in a heat block at 95°C for 30 minutes.

    For small stains, an alternative approach is to place the stained material in a DNA IQ Spin Basket seated in 1.5-ml Microtubes, and add 100–150 μl of Lysis Buffer to the basket.

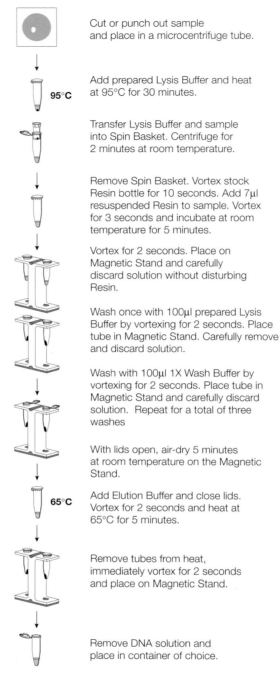

FIGURE 7. Schematic of DNA isolation from stains or swabs using the DNA IQ System. (Reprinted, with permission, from Promega DNA IQ System-Database Protocol Technical Bulletin #TB297, fig. 3237MA02_1A[©2003, Promega Corporation].)

Carefully close the lid and heat at 95°C for 30 minutes. Most of the Buffer should remain in the basket if the indicated Microtubes and Spin Baskets are used. Proceed to step 4. This does not work reliably with samples requiring more than 150 μl of Lysis Buffer.

3. Remove the tube from the heat block and transfer the Lysis Buffer and sample to a Spin Basket placed in a standard 1.5-ml conical microcentrifuge tube.

TABLE 3. Amount of DNA IQ Lysis Buffer required per sample

Material	Lysis Buffer 1	Lysis Buffer 2	Total Buffer
Liquid blood	100 μl	100 μl	200 μl
Cotton swab	250 μl	100 μl	350 μl
1/4th CEP swab	250 μl	100 μl	350 μl
15–50 mm² S&S 903 paper	150 μl	100 μl	150 μl
3–30 mm² FTA paper	150 μl	100 μl	150 μl

It is important to centrifuge Lysis Buffer with the stained matrix to obtain maximum recovery.

4. Centrifuge at room temperature for 2 minutes at maximum speed. Remove the Spin Basket.

5. Vortex the stock Resin bottle for 10 seconds at high speed or until thoroughly mixed. Add 7 μl of Resin to the DNA solution. Keep the stock Resin resuspended while dispensing to obtain uniform results.

6. Vortex the sample/Lysis Buffer/Resin mix for 3 seconds at high speed. Incubate at room temperature for 5 minutes.

7. Vortex the tube for 2 seconds at high speed and place it in the Magnetic Stand. Separation will occur instantly.

8. Carefully remove and discard all of the solution without disturbing the Resin on the side of the tube.

9. Add 100 μl of prepared Lysis Buffer. Remove the tube from the Magnetic Stand and vortex for 2 seconds at high speed.

10. Return the tube to the Magnetic Stand and discard all Lysis Buffer.

11. Add 100 μl of prepared 1x Wash Buffer. Remove tube from the Magnetic Stand and vortex for 2 seconds at high speed.

12. Return the tube to the Magnetic Stand and discard all Wash Buffer.

13. Repeat steps 11 and 12 two more times for a total of 3 washes. Make sure that all of the solution has been removed after the last wash.

14. Open the lid of the tube and air-dry the Resin in the Magnetic Stand for 5 minutes.

15. Add 100 μl of Elution Buffer.

16. Close the lid, vortex the tube for 2 seconds at high speed, and place the tube at 65°C for 5 minutes.

17. Remove the tube from the heat block and vortex for 2 seconds at high speed. Immediately place the tube on the Magnetic Stand.

18. Transfer the solution to a container of choice. The DNA concentration will be ~1 ng/μl, ready for use in an amplification reaction.

The DNA solution can be kept at 4°C for short-term storage, or at –20°C or –70°C for long-term storage.

POST-AMPLIFICATION CLEANUP OF PCR PRODUCTS

The purification of PCR products can be divided into two types: (1) direct purification from the PCR, sometimes referred to as solution purification, and (2) purification from agarose gels when separation from contaminating fragments is required.

A number of methods have been developed for solution purification. The most commonly used methods are based on ultrafiltration, precipitation, or binding and elution from silica.

Ultrafiltration uses small plastic devices containing an ultrafiltration membrane that retains larger fragments and allows nucleotides and small primers to pass through. Several rounds of dilution and filtration may be required for efficient removal of contaminating primers and nucleotides. Precipitation may also be required to concentrate the sample. The need for centrifugation during this method makes processing of large numbers of samples cumbersome. For the same reason, this type of method can be difficult to automate.

Ethanol, isopropanol, and polyethylene glycol can be used to precipitate nucleic acids in the presence of various salts. These methods are low in cost and can meet various throughput needs ranging from single samples to 384 samples at a time. Multiple centrifugation steps are required, first to precipitate the nucleic acid and second to remove contaminating salts that can be inhibitory to enzymes used in some downstream applications. The multiple centrifugation steps make hands-off automation difficult. In contrast, the use of silica provides a balance between ease of use, cost per isolation, and the ability to automate the process.

When a target DNA must be isolated from other contaminating fragments, the mixture can be fractionated by agarose gel electrophoresis followed by extraction of the fragment of interest from the gel. Standard methods for extracting DNA from agarose gel slices include electrophoresis onto DEAE-cellulose membranes, electro-elution, anion-exchange chromatography, organic extraction, and enzymatic digestion with agarase. These purification methods work well in certain situations but frequently are limited by low recovery rates and poor quality of the recovered DNA. These methods can also be time-consuming and labor-intensive, and they often require further purification or extraction steps. Silica-based methods solve most of these problems.

PROTOCOL 4   PURIFICATION OF PCR FRAGMENTS FROM AGAROSE GELS

The Wizard SV Gel and PCR Clean-Up System provides a silica-based method for the purification of PCR fragments from agarose gel slices or directly from PCR samples. In the gel-based method, the DNA band(s) of interest is excised from the gel and dissolved in a guanidine isothiocyanate solution. This mixture is passed through a binding membrane where DNA adsorbs to the silica surface. Nonadherent material is removed with a wash solution, and the purified DNA is eluted in water. This chemistry is designed to purify fragments of 100 bp to 10 kb from standard or low-melting-point agarose in either Tris-acetate (TAE) or Tris-borate (TBE) buffer with 70–95% recovery. Processing time is 15–20 minutes, depending on the number of samples. The purified samples can be used for a number of downstream applications, including fluorescent DNA sequencing, cloning, labeling, amplification, restriction digestion, and in vitro transcription/translation without further processing.

MATERIALS

. . . .

▼ CAUTION

See Appendix for appropriate handling of materials marked with <!>.

BUFFERS, SOLUTIONS, AND REAGENTS

    Ethanol (95%)<!>

    Intercalating dye such as ethidium bromide<!>

    TAE electrophoresis buffer

        40 mM Tris-acetate<!>, pH 8.2

        1 mM EDTA<!>

OR
> TBE electrophoresis buffer
> > 90 mм Tris-borate<!>, pH 8.3
> > 2 mм EDTA

NUCLEIC ACIDS AND OLIGONUCLEOTIDES
> PCR fragments from agarose gel slices

GELS
> Agarose gel (standard or low-melt)

SPECIAL EQUIPMENT
> Microcentrifuge tubes, 1.5-ml
> Scalpel or razor blade
> UV lamp, long-wavelength
> Vacuum manifold and vacuum adapters (Promega; for purification with vacuum)

ADDITIONAL ITEMS
> Wizard SV Gel and PCR Clean-Up System (Promega; includes SV Minicolumns, collection tubes, Membrane Binding Solution, Membrane Wash Solution, and nuclease-free H_2O)

CELLS AND TISSUES
> PCR samples for purification

METHOD

Standard safety apparel should be worn, especially when handling ethidium bromide-stained agarose gels. This includes gloves and a UV-blocking face shield to protect the eyes and face from UV light.

DISSOLVING THE GEL SLICE

1. Load the PCR samples on an agarose gel and run according to an established protocol.

2. Weigh a 1.5-ml microcentrifuge tube for each DNA fragment to be isolated; record the weight.

3. Visualize and photograph the DNA using a long-wavelength UV lamp and an intercalating dye such as ethidium bromide. To reduce nicking, irradiate the gel for the absolute minimum time possible (Grundemann and Schomig 1996; Zimmermann et al. 1998). Excise the DNA fragment of interest in a minimal volume of agarose using a clean scalpel or razor blade. Transfer the gel slice to the weighed microcentrifuge tube and record the weight. Subtract the weight of the empty tube from the total weight to obtain the weight of the gel slice.

4. Add Membrane Binding Solution at a ratio of 10 μl of solution per 10 mg of agarose gel slice.

5. Vortex the mixture and incubate at 50–65°C for 10 minutes or until the gel slice is completely dissolved. Vortex the tube every few minutes to increase the rate of agarose gel melting. Centrifuge the tube briefly at room temperature to ensure the contents are at the bottom of the tube. Once the agarose gel is melted, the gel will not resolidify at room temperature.

6. DNA may then be purified by centrifugation or vacuum filtration.

DNA PURIFICATION BY CENTRIFUGATION

1. Place one SV Minicolumn in a Collection Tube for each dissolved gel slice.

2. Transfer the dissolved gel mixture to the SV Minicolumn assembly and incubate for 1 minute at room temperature.

3. Centrifuge the SV Minicolumn assembly at 10,000g for 1 minute. Remove the Minicolumn from the Spin Column assembly and discard the liquid in the Collection Tube. Return the Minicolumn to the Collection Tube.

4. Wash the column by adding 700 µl of Membrane Wash Solution to the Minicolumn. Centrifuge the Minicolumn assembly for 1 minute at 10,000g. Empty the Collection Tube as before and place the Minicolumn back in the Collection Tube. Repeat the wash with 500 µl of Membrane Wash Solution and centrifuge the Minicolumn assembly for 5 minutes at 10,000g.

5. Remove the Minicolumn assembly from the centrifuge, being careful not to wet the bottom of the column with the flowthrough. If the column becomes wet, empty the Collection Tube and re-centrifuge the column assembly for 1 minute.

6. Transfer the Minicolumn to a clean 1.5-ml microcentrifuge tube. Apply 50 µl of nuclease-free H_2O directly to the center of the column without touching the membrane with the pipette tip. Incubate at room temperature for 1 minute. Centrifuge for 1 minute at 10,000g.

7. Discard the Minicolumn and store the microcentrifuge tube containing the eluted DNA at 4°C or –20°C.

DNA PURIFICATION BY VACUUM

1. Attach one Vacuum Adapter with a Luer-Lok fitting to one port of the manifold for each dissolved gel slice or PCR. Insert an SV Minicolumn into each Vacuum Adapter until it fits snugly in place.

2. Transfer the dissolved gel mixture or PCR to the Minicolumn and incubate for 1 minute at room temperature. Apply a vacuum to pull the liquid completely through the Minicolumn.

3. Wash the column by adding 700 µl of Membrane Wash Solution to the Minicolumn. Make sure that any droplets remaining on the sides of the Minicolumn from the last step are washed away. Apply a vacuum to pull the liquid through the Minicolumn. Repeat this wash a second time with 500 µl of Membrane Wash Solution.

4. Turn off the vacuum source and open an unused port to vent the manifold. Remove the minicolumn from the vacuum manifold and transfer the Minicolumn to a Collection Tube. Centrifuge the Minicolumn assembly for 5 minutes at 10,000g to remove any remaining Membrane Wash Solution from the column.

5. Transfer the Minicolumn to a clean 1.5-ml microcentrifuge tube. Apply 50 µl of nuclease-free H_2O directly to the center of the column, without contacting the membrane. Incubate at room temperature for 1 minute. Centrifuge for 1 minute at 10,000g.

6. Discard the Minicolumn and store the microcentrifuge tube containing the eluted DNA at 4°C or –20°C.

PROTOCOL 5 PURIFICATION OF PCR PRODUCTS FROM SOLUTION, SILICA MEMBRANE-BASED METHOD

The Wizard SV 96 PCR Clean-Up System provides a membrane-based system that can be used for high-throughput PCR cleanup in a 96-well multiwell plate format. In the Membrane Binding Solution, PCR products adsorb to the silica surface. Primers, low-molecular-weight double-stranded fragments such as primer–dimers, and other reaction components have no affinity for the solid phase and can be removed by vacuum filtration. Residual nonadherent material is removed by washing, and purified target DNA is eluted in water.

Fully automated methods have been developed for a number of automated liquid-handling robotic workstations, including the Biomek 2000 and FX. Recovery is independent of variable PCR conditions that may include different DNA polymerases, PCR additives, or the presence or absence of mineral oil. Fragments from 100 bp to 10 kb can be purified with up to 90% recovery. The purified PCR fragments are suitable for automated fluorescent DNA sequencing, cloning, labeling, restriction enzyme digestion, or DNA microarray analysis without further manipulation.

MATERIALS

▼ CAUTION

See Appendix for appropriate handling of materials marked with <!>.

BUFFERS, SOLUTIONS, AND REAGENTS
Ethanol<!>, 80% (75 ml/plate; freshly made before each use)

NUCLEIC ACIDS AND OLIGONUCLEOTIDES
PCR samples in 96-well format (20–100 μl per sample)

SPECIAL EQUIPMENT
Vac-Man 96 Vacuum Manifold (Promega) or suitable vacuum manifold
Vacuum pump capable of 15–20 inches of Hg
Vacuum trap for waste collection (1-liter size)
Vacuum tubing

ADDITIONAL ITEMS
Wizard SV 96 PCR Clean-Up System (Promega; includes Binding Plate, Collection Plate, DNA Elution Plate, Membrane Binding Solution, and nuclease-free H_2O)

METHOD

1. Prepare the vacuum manifold as shown in Figure 8. Place the Binding Plate in the vacuum manifold base. To ensure that the well numbers on the sample plate correspond to the numbers on the Binding Plate, orient the Binding Plate in the vacuum manifold with the numerical column headers toward the vacuum port. Attach the vacuum line to the vacuum port on the Manifold Base.

2. Add an equal volume of Membrane Binding Solution to each volume of PCR sample in a 96-well plate (e.g., add 100 μl of Membrane Binding Solution to each 100 μl of PCR sample).

3. Mix by pipetting, and transfer the entire sample volume to the wells of the Binding Plate sitting on the vacuum manifold. Incubate for 1 minute at room temperature.

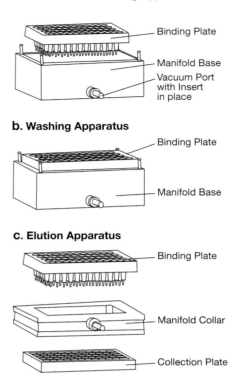

FIGURE 8. The Vac-Man 96 Vacuum Manifold with the Wizard SV 96 PCR Clean-Up System. *a, b,* and *c* show the manifold and plate combinations necessary to accomplish PCR product binding, washing, and elution, respectively, for manual PCR product cleanup. For automated cleanup on the Beckman Biomek 2000, the Beckman vacuum manifold (Beckman part no. 609670) and collar (Beckman part no. 609597) are required. (Reprinted, with permission, from Promega Wizard SV 96 PCR Clean-Up System Technical Bulletin #TB311, fig. 3744MA06_2A [© 2003 Promega Corporation].)

4. Apply a vacuum until the samples pass through the Binding Plate (~30 seconds). Release the vacuum.

5. Add 200 μl of 80% ethanol to each well of the Binding Plate. Incubate for 1 minute at room temperature. Apply the vacuum until the ethanol passes through the plate (~30 seconds). Release vacuum.

6. Repeat step 5 for a total of 3 x 200 μl, 80% ethanol washes.

7. After the wells of the Binding Plate have emptied from the final wash, continue to apply the vacuum for an additional 4 minutes to allow the binding matrix to dry.

8. Turn off the vacuum. Release the vacuum line from the Manifold Base and snap it into the vacuum port in the Vacuum Manifold Collar. Remove the Binding Plate from the Manifold Base and blot by gently tapping onto a clean paper towel to remove residual ethanol.

9. Place a 96-well, U-bottom Collection Plate in the Manifold Bed and position the Vacuum Manifold Collar on top. Make sure to orient the U-bottom Collection Plate with the numerical column headers toward the vacuum port.

10. Position the Binding Plate on top of the Manifold Collar and the Collection Plate as shown in Figure 8. The Binding Plate tips must be centered on the Collection Plate wells, and both plates must be in the same orientation. Add 100 µl of nuclease-free H_2O to each well of the Binding Plate and incubate for 1 minute at room temperature. Apply the vacuum until the solution passes through the plate (~1 minute).

11. Release the vacuum and remove the Binding Plate. Carefully remove the Manifold Collar, making sure that the Collection Plate remains positioned in the Manifold Bed. If droplets are present on the walls of the Collection Plate, briefly centrifuge the plate to collect the droplets on the bottom of the wells. Eluate volumes may vary, but are roughly 75 µl.

PROTOCOL 6   PURIFICATION OF PCR PRODUCTS FROM SOLUTION, MAGNETIC PARTICLE-BASED METHOD

MagneSil is a silica-paramagnetic particle that acts as a mobile solid phase to capture double-stranded DNA fragments. The particle–adherent target complex is washed to remove residual contaminants, and purified DNA is eluted in water, ready for downstream fluorescent DNA sequencing and microarray analyses.

The MagneSil system can be scaled up for higher yield or scaled down for use in 384-well formats. PCR products greater than 150 bp in size are selectively isolated from primers and primer–dimers with high yields and purity. This technology has been adapted for use with a number of robotic workstations, including Beckman Coulter's Biomek 2000 and FX Laboratory Automation Workstations. The plate-processing times are less than 20 minutes with a single POD Biomek FX. Typical recovery is greater than 80% for a 1000-base product with negligible carryover of primers or nucleotides. Two methods have been developed to purify either a single plate or multiple plates of PCR samples; multiple-plate processing is achieved by the incorporation of a high-capacity stacker carousel to deliver consumables.

MATERIALS

▼ CAUTION

See Appendix for appropriate handling of materials marked with <!>.

BUFFERS, SOLUTIONS, AND REAGENTS
Ethanol, 80%<!>

NUCLEIC ACIDS AND OLIGONUCLEOTIDES
PCR samples in 96-well format (up to 100 µl/well)

SPECIAL EQUIPMENT
3 single-position labware ALP (Beckman Coulter)
4 P250 tip rack assemblies (Beckman Coulter)
4 x 4 position labware ALP (Beckman Coulter)
96-well Tip Wash ALP (Beckman Coulter)
Heating and Cooling ALP (Beckman Coulter)
Liquid-handling robot such as the Biomek FX or Biomek 2000 Laboratory Automation workstation (Beckman Coulter)
MagnaBot 96 Magnetic Separation Device (Promega)
Nonpermeable plate sealer
Orbital Shaker ALP (Beckman Coulter)
Tip Loader ALP (Beckman Coulter)
Plate Clamp 96 (Promega)

ADDITIONAL ITEMS

Wizard MagneSil PCR Clean-Up System (Promega; includes MagneSil YELLOW [paramagnetic particles], Wash Solution, nuclease-free H_2O, and Collection Plates)

METHOD

The following protocol contains the steps and timing that are intended for use of the MagneSil PCR Clean-Up System in the development of an automated process using a robot. These same steps, including the specified reagent volumes, may also be used to perform the procedure manually. This system was designed to purify PCR products from 100-μl sample volumes.

DNA BINDING

1. Begin with completed PCR samples in a 96-well plate. Adjust the volume of each sample to 100 μl with nuclease-free H_2O.
2. Resuspend MagneSil particles thoroughly by shaking. Add 100 μl to each well.
3. Mix the contents of each well by pipetting up and down four times.
4. Incubate for 2 minutes at room temperature. Mix by pipetting up and down four times halfway through the incubation.
5. Transfer the mixed samples to a 96-well Collection Plate. Place the plate onto the MagnaBot 96 Magnetic Separation Device to magnetize the particles.
6. Remove and discard the supernatant.

WASHING

1. Remove the Collection Plate containing the Paramagnetic Particles from the MagnaBot 96 Magnetic Separation Device. Add 200 μl of Wash Solution to each well.
2. Mix the contents of each well by pipetting up and down four times.
3. Place the Collection Plate onto the MagnaBot 96 Magnetic Separation Device to magnetize the particles.
4. Remove and discard the supernatant.
5. Remove the Collection Plate from the MagnaBot 96 Magnetic Separation Device. Add 200 μl of 80% ethanol to each well.
6. Mix the contents of each well by pipetting up and down four times.
7. Place the Collection Plate onto the MagnaBot 96 Magnetic Separation Device to magnetize the particles.
8. Remove and discard the supernatant.
9. Repeat steps 5–8 with 200 μl of 80% ethanol, for a total of two ethanol washes.
10. Leave the plate to dry on the MagnaBot 96 Magnetic Separation Device for 5–10 minutes. Remove any residual ethanol from the wells by pipetting.

ELUTION OF DNA

1. Remove the plate from the MagnaBot 96 Magnetic Separation Device. Add 100 μl of nuclease-free H_2O to elute the PCR products from the Paramagnetic Particles. Mix each well by pipetting up and down four times. Incubate for 2 minutes at room temperature.
2. Place the Collection Plate onto the MagnaBot 96 Magnetic Separation Device to magnetize the particles.

3. Transfer the eluted PCR products to a clean Collection Plate.

4. If particles have been carried over with the eluate, place the Collection Plate onto the MagnaBot 96 Magnetic Separation Device and transfer the eluate to a clean Collection Plate. Alternatively, place the Collection Plate onto the MagnaBot 96 Magnetic Separation Device when withdrawing samples from the Collection Plate.

5. Seal the plate tightly with a nonpermeable plate sealer and store at 4°C for 1–2 days or at –20°C for long-term storage.

PROTOCOL 7 DYE TERMINATOR CLEANUP

Automated fluorescent sequencing using the ABI PRISM BigDye terminator chemistry is the method of choice for high-throughput DNA sequence determination. Sequence ladders are generated by the standard Sanger method (Sanger et al. 1977) using fluorescently labeled chain-terminator nucleotides (Prober et al. 1987). These BigDye terminators are dideoxynucleotide triphosphates linked to an energy transfer dye composed of a fluorescein donor and one of four d-Rhodamine-based acceptors (Rosenblum et al. 1997). Automated sequencers separate the labeled sequence extension products using a molecular sieve matrix and identify the terminal nucleotide based on its emission wavelength.

If not removed before analysis, certain reaction components will interfere with data collection. Buffer salts may also give rise to aberrant electrophoretic migration (Ruiz-Martinez et al. 1998) and unincorporated dye-labeled terminators, and related species interfere with the legitimate signal from sequencing extension products. Common purification methods based on gel filtration and precipitation by ethanol are problematic for high-throughput robotic applications because they can require multiple centrifugation and/or vacuum drying steps.

MagneSil particles can be applied to the high-throughput purification of BigDye terminator DNA sequencing reactions prior to automated sequence analysis. Procedures have been developed on several robotic workstations using the standard 96-well amplification plate format. Sequence quality is similar to or better than standard manual methods and offers the distinct advantage of a hands-off protocol. No user intervention is required from the time the amplification plate is put on the robot deck until the samples are loaded onto the DNA sequencer.

The MagneSil Sequencing Reaction Clean-Up System is designed to work with the ABI PRISM 310, 3100, 377, and 3700 DNA sequencers. Sequence quality has been assessed by accuracy (and/or PHRED score), read length, and start position. The system is qualified to provide sequence data that are >98% accurate over 600 bases with a start position of less than primer plus 25 bases when using the control template and primer included in the ABI PRISM BigDye terminator cycle sequencing reaction kit. An overview of the MagneSil Sequencing Reaction Clean-Up System protocol is given in Figure 9.

MATERIALS

. . . .

BUFFERS, SOLUTIONS, AND REAGENTS
Elution/loading solution for sequencing gel<!> (see Table 4)
90% Ethanol wash solution<!>

SPECIAL EQUIPMENT
MagnaBot II Magnetic Separation Device (Promega)

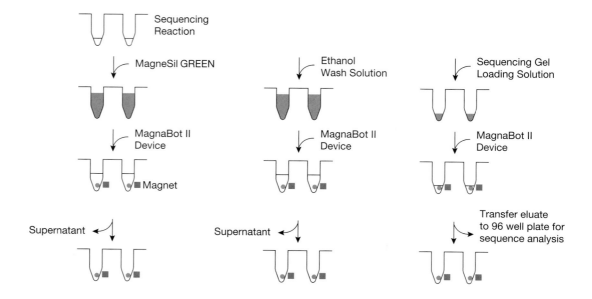

Binding

The Binding Solution has been optimized so that the PMPs selectively bind the sequencing extension products. A magnet is used to capture and hold the PMPs against the side of the well. Binding Solution and nonbound sequencing reaction components are removed to waste.

Washing

Samples are removed from the magnet, and nonspecifically bound material is removed by washing with ethanol. The PMPs are resuspended in ethanol wash solution, mixed and captured. The wash solution containing nonspecifically bound terminators and residual salts are removed to waste. The washes are repeated twice, then the samples are air-dried.

Elution

Sequencing extension products are eluted in gel loading solution. The particles are resuspended in the appropriate sequencing gel loading solution, then captured. Eluted sample is removed and ready to analyze.

FIGURE 9. Schematic diagram of the Wizard MagneSil Sequencing Reaction Clean-Up System protocol. (Reprinted, with permission, from Otto and Bjerke 2001, Promega Notes 78, p. 2, fig. 3269MA12_0A [©2003 Promega Corporation].)

Plate Clamp 96 (Promega)
Plate Stand (Promega)

ADDITIONAL ITEMS

Wizard MagneSil Sequencing Reaction Clean-Up System (Promega; includes MagneSil GREEN [paramagnetic particles])
96-well plate

METHOD

1. Assemble the 96-well plate into the Plate Clamp 96 (Fig. 10a,b).

2. Use the Plate Stand (Fig. 10c) to position the plate on the robotic deck.

3. Resuspend the MagneSil particles by vigorously shaking the bottle. Add 180 µl of the MagneSil particles to each 20-µl sequencing reaction.

4. Incubate at room temperature for 5 minutes. Mix by pipetting at 0, 2.5, and 5 minutes.

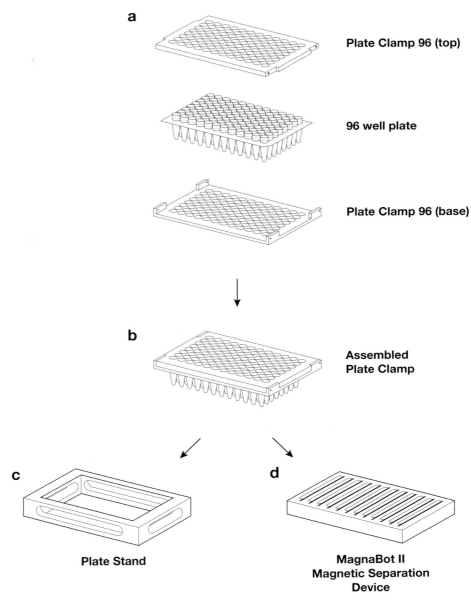

a

Plate Clamp 96 (top)

96 well plate

Plate Clamp 96 (base)

b

Assembled Plate Clamp

c

Plate Stand

d

MagnaBot II Magnetic Separation Device

FIGURE 10. Accessories for the Wizard MagneSil Sequencing Reaction Clean-Up System. To eliminate plate warping, the 96-well plate is placed into the Plate Clamp 96 (*a* and *b*). During the Clean-Up procedure, the Plate Clamp assembly can be held by the Plate Stand (*c*) or placed onto the MagnaBot II Device for magnetic separation of the MagneSil Paramagnetic Particles (*d*). (Reprinted, with permission, from Promega Wizard MagneSil Sequencing Reaction Clean-Up System Technical Bulletin #TB287, fig. 3222MA01_2A [©2003 Promega Corporation].)

5. Place the plate onto the MagnaBot II Magnetic Separation Device (Fig. 10c) to capture the particles.

6. Remove and discard the liquid. Avoid removing any particles.

7. Remove the plate from the MagnaBot II Device and place on Plate Stand.

8. Add 100 µl of 90% ethanol to each sample.

TABLE 4. Recommended elution/loading solution volumes for the Magnesil Sequencing Reaction Clean-Up System protocol

Instrument	Elution/Loading solution	Volume
ABI 377	formamide/blue dextran/EDTA	6 µl
ABI 310	template suppression reagent supplied with ABI POP-6 polymer	10–20 µl
ABI 3100	Hi-Di formamide	10–20 µl
ABI 3700	formamide or water	10 µl

9. Incubate at room temperature for 5 minutes. Mix at 0, 2.5, and 5 minutes.

10. Place the plate on the MagnaBot II Device to capture particles.

11. Remove and discard the liquid. Avoid removing any particles.

12. Repeat steps 7–11 for a total of two washes.

13. Allow the particles to air-dry for ~10 minutes at room temperature.

14. Add appropriate elution/loading solution (Table 4).

15. Incubate at room temperature for 1–2 minutes.

16. Place the plate on the MagnaBot II Device to capture particles.

17. Transfer purified sequencing reactions to a clean 96-well plate. Be careful to avoid removing any particles.

SUMMARY

The purification of nucleic acids is an integral part of the PCR process, providing the source nucleic acid to be amplified and the methods to clean up amplified product after PCR. The choice of purification method depends on the efficiency of the amplification reaction, PCR fragment size, the researcher's needs for template purity, purification scale, and number of samples to be handled, and the need for automation. Today's methods have eliminated the need for hazardous organic extractions, have reduced time requirements, and have provided increased convenience and reproducibility. Methods based on binding and elution from silica, as well as solution extraction methods, have been developed to handle the majority of researchers' current needs. Advances in technology such as capillary sequencing, continued assay miniaturization, and the push for more automation will place higher demands on the nucleic acid purification systems of the future.

REFERENCES

Abravaya K., Carrino J.J., Muldoon S., and Lee H.H. 1995. Detection of point mutations with a modified ligase chain reaction (Gap-LCR). *Nucleic Acids Res.* **23:** 675–682.

Akane A., Matsubara K., Nakamura H., Takahashi S., and Kimura K. 1994. Identification of the heme compound copurified with deoxyribonucleic acid (DNA) from bloodstains, a major inhibitor of polymerase chain reaction (PCR) amplification. *J. Forensic Sci.* **39:** 362–372.

Al-Soud W.A., Jonsson L.J., and Radstrom P. 2000. Identification and characterization of immunoglobulin G in blood as a major inhibitor of diagnostic PCR. *J. Clin. Microbiol.* **38:** 345–350.

Beutler E., Gelbart T., and Kuhl W. 1990. Interference of heparin with the polymerase chain reaction. *BioTechniques* **9:** 166.

Glasel J.A. 1995. Validity of nucleic acid purities monitored by 260nm/280nm absorbance ratios. *BioTechniques* **18:** 62–63.

Grundemann D. and Schomig E. 1996. Protection of DNA during preparative agarose gel electrophoresis against damage induced by ultraviolet light. *BioTechniques* **21:** 898–903.

Kleppe K., Ohtsuka E., Kleppe R., Molineux I., and Khorana H.G. 1971. Studies on polynucleotides. XCVI. Repair replications of short synthetic DNA's as catalyzed by DNA polymerases. *J. Mol. Biol.* **56:** 341–361.

Lizardi P.M., Huang X., Zhu Z., Bray-Ward P., Thomas D.C., and Ward D.C. 1998. Mutation detection and single-molecule counting using isothermal rolling-circle amplification. *Nat. Genet.* **19:** 225–232.

Prober J.M., Trainor G.L., Dam R.J., Hobbs F.W., Robertson C.W., Zagursky R.J., Cocuzza A.J., Jensen M.A., and Baumeister K. 1987. A system for rapid DNA sequencing with fluorescent chain-terminating dideoxynucleotides. *Science* **238:** 336–341.

Rosenblum B.B., Lee L.G., Spurgeon S.L., Khan S.H., Menchen S.M., Heiner C.R., and Chen S.M. 1997. New dye-labeled terminators for improved DNA sequencing patterns. *Nucleic Acids Res.* **25:** 4500–4504.

Ruiz-Martinez M.C., Salas-Solano O., Carrilho E., Kotler L., and Karger B.L. 1998. A sample purification method for rugged and high-performance DNA sequencing by capillary electrophoresis using replaceable polymer solutions. A. Development of the cleanup protocol. *Anal. Chem.* **70:** 1516–1527.

Saiki R.K., Scharf S., Faloona F., Mullis K.B., Horn G.T., Erlich H.A., and Arnheim N. 1985. Enzymatic amplification of beta-globin genomic sequences and restriction site analysis for diagnosis of sickle cell anemia. *Science* **230:** 1350–1354.

Sanger F., Nicklen S., and Coulson A.R. 1977. DNA sequencing with chain-terminating inhibitors. *Proc. Natl. Acad. Sci.* **74:** 5463–5467.

Shimer G.H., Jr. and Backman K.C. 1995. Ligase chain reaction. *Methods Mol. Biol.* **46:** 269–278.

Wilson I.G. 1997. Inhibition and facilitation of nucleic acid amplification. *Appl. Environ. Microbiol.* **63:** 3741–3751.

Zimmermann M., Veeck J., and Wolf K. 1998. Minimizing the exposure to UV light when extracting DNA from agarose gels. *BioTechniques* **25:** 586.

10 RNA Purification

Michael A. Connolly, Peter A. Clausen, and James G. Lazar

Marligen Bioscience, Inc., Ijamsville, Maryland 21754

Francis Crick and Leslie Orgel in 1968 proposed that RNA was the carrier of genetic information (Crick 1968; Orgel 1968). Since that time, more than 350,000 peer-reviewed studies have been published that rely, at least in part, on the analysis of RNA. The total RNA population in a cell or tissue comprises several different RNA species including ribosomal (rRNA), transfer RNA (tRNA), small nuclear RNA (snRNA), and messenger RNA (mRNA). Table 1 lists these types of RNA and their relative abundance in a typical cell.

Specific protocols have been developed to isolate the various species of RNA, and these generally begin with the isolation of total RNA. Relative to rRNA, snRNA, and tRNA, mRNA continues to hold the vast majority of interest on the basis of its role as the template for protein synthesis. Although total RNA is often the desired starting material for downstream applications designed to analyze mRNA populations, protocols have been developed to isolate mRNA directly from cells and tissues. This chapter provides protocols that are used for preparing total RNA from cell and tissue sources as well as protocols for the purification of mRNA.

YIELD AND QUALITY OF RNA

Independent of the protocol, the primary goal of RNA isolation is to obtain a yield and quality of RNA that will enable successful downstream applications. Yield is typically assessed by measuring the absorbance of the RNA solution at a wavelength of 260 nm. RNA concentration (and therefore yield) is calculated from the A_{260} value by assuming that one A_{260} unit represents an RNA concentration of 40 µg/ml. RNA quality is assessed by evaluating the A_{260}/A_{280} ratio and by visualization of RNA samples on agarose gels. The $A_{260/280}$ ratio is an indicator of RNA purity and should fall between 1.9 and 2.2. Lower ratios generally indicate contamination with protein, phenol, or guanidinium salts. Formaldehyde gels can provide an additional indication of yield and sample quality. Dyes such as ethidium bromide are used to detect RNA in gels, and yield can be estimated by comparison with a standard by direct visualization or with computer imaging software. The rate of migration of RNA molecules in formaldehyde gels is directly related to their size. Gel analysis can be used to estimate the average size distribution of RNA in a sample and can also be used to assess the level of degraded RNA present within a sample. Visualization of RNA on formaldehyde gels is sensitive to ~5 nanograms and makes gel analysis especially useful for quality control when there are not sufficient quantities of RNA for spectrophotometric analysis.

TABLE 1. RNA species and their relative abundance

RNA type	Relative abundance %	Picogram/ cell
rRNA	80	10–30
snRNA	15	2–5
tRNA		
mRNA	5	0.5–2

Modified from Stryer (1988).

The two variables that most often affect RNA yield and quality are the condition of the cells or tissue samples being analyzed and the presence of RNase activity. For example, quiescent cells tend to contain less mRNA relative to cells in log-phase growth, and cells that are overgrown or cultured to confluence tend to yield more degraded RNA. Some sources of cells and tissues, such as pancreas, contain high levels of endogenous RNase activity that can be highly problematic for the isolation of intact RNA. When using frozen samples, even small amounts of sample thawing prior to addition or contact with chaotropic agents may cause leakage of compartmentalized RNase activity and the subsequent degradation of the RNA. Extra care must be taken to avoid this possibility when dealing with larger cell pellets and tissue samples that must dissociate before complete lysis is possible.

EXOGENOUS RNase CONTAMINATION

Exogenous RNase contamination is another hazard to handling RNA in the laboratory. Exogenous RNase activity can be present on hands, bench tops, and laboratory equipment, but the risk of contamination can be easily managed by following basic RNA handling procedures.

The risk of introducing exogenous RNase activity is minimized by appropriate cleaning of bench tops, plasticware, and laboratory equipment, and also by the appropriate handling of samples. Skin is an excellent source of RNase activity, and gloves should be worn at all times. Mild detergent or chemicals such as RNaseAway (Molecular BioProducts) can be used to decontaminate laboratory surfaces and equipment prior to handling RNA samples. Although most unopened plasticware is rendered RNase-free by the molding process, the use of materials that are certified as RNase-free is advisable. Glassware to be used for RNA work should be cleaned and then baked at 180°C for at least 8 hours before use (Sambrook et al. 1989). Solutions should be prepared using chemicals that are free of RNase activity, and water should be treated with diethylpyrocarbonate (DEPC) to inactive RNase (Sambrook et al. 1989). However, DEPC readily reacts with amines and is not compatible with certain reagents such as Tris-based buffers. A wide variety of chemicals, reagents, and buffers certified as "RNase-free" are now commercially available and help make RNA handling much simpler than in the past.

SAMPLE PROCESSING

With the exception of paraffin-embedded or fixed tissues, the initial sample-processing steps for RNA purification are similar for most protocols. It is preferable to use fresh cells (60–80% confluency) or tissues as the starting material for RNA isolation. If samples must be stored for later processing, manageable-sized cell pellets or tissues should be snap-frozen using a dry ice bath or liquid nitrogen. When processing frozen material, extreme care must be taken to avoid even minimal amounts of thawing, which could detrimentally affect the integrity of an RNA preparation. All protocols for isolating RNA utilize lysis buffers containing denaturants such as phenol, sodium dodecyl sulfate (SDS), and guanidinium salts that effectively inhibit RNase. These lysis buffers can be used to lyse cells directly. However, fresh or frozen tissues must be ground to a fine powder prior to lysis. Grinding is performed under liquid nitrogen using a prechilled (liquid nitrogen) mortar and pestle. In most cases, a finer grind provides a better RNA yield from both plant and animal tissues due to the higher surface area and faster kinetics for inactivation of RNase. As an alternative to the conventional snap-freeze, Ambion Inc offers RNAlater, a storage solution for cells and tissues that will reportedly preserve RNA in tissues and cells prior to processing. Although the authors have not used this product, RNAlater has been reported to stabilize RNA in cells prior to freezing. Grotzer et al. (2000) were able to isolate total RNA from tissue samples stored for up to 7 days at room temperature. No significant differences were seen between the samples stored in RNAlater at room temperature and samples snap-frozen and stored at –70°C. RNAlater may provide the capability for long-term storage of samples at –20°C and short-term (up to 1 month) storage at 4°C and may also obviate the need for processing in liquid nitrogen prior to lysis.

For protocols other than those relying on phenol-based phase separation, homogenization of the sample using a rotor-stator tissue homogenizer or similar device is recommended. Homogenization facilitates contact of all cell surfaces with lysis buffer and facilitates rapid inactivation of RNases. Homogenization also shears genomic DNA, which results in reduced sample viscosity and is of particular importance in protocols that employ solid-phase RNA-binding matrices that are readily clogged by viscous samples. When processing larger samples ($>10^6$ cells or 10 mg of tissue) on solid-phase binding matrices, homogenization is critical to obtain a good yield and quality of RNA.

RNA ISOLATION STRATEGIES

Following the initial processing of samples, RNA is generally isolated using one or more of three strategies: phenol-based phase separation, silica binding, and oligo(dT) capture. Phenol-based strategies have evolved from the method developed by Chomczynski and Sacchi (1987). Using phenol-based phase separation, cellular proteins and lipids are extracted into an organic phase, and RNA is precipitated from the aqueous phase. These protocols involve gentle homogenization in buffered phenol solution, phase separation, precipitation and washing of RNA, and resuspension of the RNA pellet. Current silica-binding methods are derived from a combination of the Chirgwin et al. (1979) RNA isolation protocol and a modification to the Vogelstein and Gillespie (1979) technique for purifying DNA. These methods rely on homogenization in a chaotropic lysis buffer, selective binding of RNA to silica in the presence of high salt, and elution of RNA using water or low-salt buffer. These protocols eliminate the need for hazardous chemicals like phenol and chloroform that require special handling and enable rapid (<30 minutes) isolation of highly purified RNA.

A number of protocols for obtaining messenger RNA (mRNA) directly from cells and tissues are also available. These protocols are adaptations of the strategy described by Berger and Kimmel (1987) and Sambrook et al. (1989) that uses oligo(dT) to capture mRNA through binding of the polyadenylated tails that are present on most mRNA molecules. The direct isolation of mRNA typically requires homogenization in lysis buffer, shearing of genomic DNA, binding of polyadenylated RNA to oligo(dT), and washing and elution of the RNA sample. Using oligo(dT) capture, a single round of selection on oligo(dT) will provide a sample enriched in the poly(A) fraction of RNA, and additional rounds of oligo(dT) purification are recommended to obtain samples of pure mRNA.

Phenol-based, silica-based, and oligo(dT)-based protocols have been used successfully to isolate RNA from animal cells and tissues, but have met with varying degrees of success in obtaining RNA from plant cells and tissues. Frequently these reagents do not produce a high yield or quality of RNA because plants often contain high levels of starches, polysaccharides, resins, or polyphenolic compounds that interfere with the isolation of pure RNA or that may copurify with the RNA. One adaptation used to isolate RNA from difficult plant tissues is to add polyvinylpyrrolidone (PVP) to the chaotropic lysis solution. PVP is used to bind and precipitate contaminating polysaccharides and polyphenolics, but this approach also has met with limited success. More recently, specific reagents and methods have emerged that are effective for isolating RNA from difficult plant tissues such as spruce, potato tuber, grape, and seeds (Hughes and Galau 1988; Wan and Wilkins 1994; Geuna et al. 1998; Scott et al. 1998; Wang et al. 1999; Lönnenborg and Jensen 2000; Gao et al. 2001; Hosein 2001). Methods representative of the three basic strategies for the isolation of RNA and specific protocols for obtaining RNA from plants and tissues are detailed below.

In all RNA purification strategies, the final step in purification is to redissolve or elute RNA samples in a low-salt buffer or water. In the absence of chaotropic agents, the RNA samples are susceptible to the activity of RNase. For this reason, RNase inhibitors such as RNasin (Promega) or vanadyl-ribonucleoside complexes can be used to ensure long-term integrity of purified RNA (Berger and Kimmel 1987). If the use of RNase inhibitors is desirable, it is important to consider the compatibility of inhibitors with downstream applications. For example, vanadyl-ribonucleoside complexes inhibit in vitro translation of RNA and are not recommended for preserving samples intended for this purpose.

Purified RNA preparations invariably contain at least trace amounts of DNA, which may affect downstream applications such as reverse transcriptase (RT)-PCR or microarray analysis. DNase I is typically used to digest contaminating genomic DNA, while not affecting the RNA. Traditionally, digestion of DNA is done as a final step after the elution or resuspension of purified RNA. More recently, silica-based protocols have been developed that streamline the process by incorporating the DNase digestion on a column containing the bound RNA. However, the authors' experience, using purified RNA for real-time PCR, is that on-column digestion is not as effective as post-elution DNase treatment for removing 100% of contaminating DNA.

Several published protocols are available for the isolation of total RNA from cells and tissues (Chirgwin et al. 1979; Auffray and Rougeon 1980; Favaloro et al. 1980; Birnboim 1988), and many RNA purification kits and reagents are commercially available. Although there are a limited number of manufacturers of phenol-based RNA purification reagents, there are more than 25 manufacturers of silica- and oligo(dT)-based kits. Protocols that are representative of phenol-based, silica-based, and oligo(dT)-based strategies for RNA isolation are presented below, beginning with the guanidinium isothiocyanate/cesium chloride ultracentrifugation method. This protocol is a modification of the original method developed by Chirgwin et al. (1979). All other protocols presented in this chapter were selected on the basis of the authors' experience both in terms of "ease of use" and their relative superiority for consistently providing the highest yield and quality of RNA.

PROTOCOL 1   PREPARATION OF RNA USING GUANIDINIUM ISOTHIOCYANATE/CESIUM CHLORIDE ULTRACENTRIFUGATION

This is a modification of the protocol outlined by Chirgwin et al. (1979). Adaptations of the reagents defined by Chirgwin et al. are found in many kits, and reagents are currently available for the isolation and purification of RNA.

MATERIALS

BUFFERS, SOLUTIONS, AND REAGENTS

Chloroform<!>

CsCl solution, 5.7 M (5.7 M CsCl, 25 mM sodium acetate<!>, pH 6.0, 1 mM EDTA, pH 8.0)

Diethyl pyrocarbonate (DEPC)<!>-treated H_2O

Dithiothreitol (DTT) <!>, 0.2 mM

EDTA, 1 mM, pH 8.0

Ethanol 70%<!> (v/v)

Guanidine isothiocyanate lysis buffer (4 M guanidine isothiocyanate <!>, 25 mM sodium acetate, pH 6.0, 1 mM EDTA, pH 8.0)

Isopropanol<!>

N-Laurylsarcosine 20%<!> (w/v)

Lithium chloride (LiCl)<!>, 8 M

Liquid nitrogen<!>

β-Mercaptoethanol<!>, 10 M

Phenol<!>, H_2O-saturated, pH 4.0

CENTRIFUGES AND ROTORS

Beckman Ti50 rotor or equivalent

SPECIAL EQUIPMENT

Mortar and pestle, washed in DEPC-treated H_2O, prechilled

Polyallomer ultracentrifuge tubes or equivalent

Polytron or equivalent homogenizer

CELLS AND TISSUES

Source cells/tissues, frozen in liquid nitrogen<!>

METHOD

1. Grind frozen tissue samples under liquid nitrogen using a mortar and pestle to produce a fine powder. Homogenize the tissues or cells (up to 1 x 10^9) in 7 ml of guanidine lysis buffer using a Polytron or other suitable apparatus.

2. Place 4 ml of 5.7 M CsCl solution in a polyallomer ultracentrifuge tube.

3. Add 1 ml of the 5.7 M CsCl solution, 120 µl of 10 M 2-mercaptoethanol, and 240 µl of *N*-laurylsarcosine to the lysate.

4. Mix the cesium chloride/lysate mixture and carefully layer it over the cesium chloride cushion in the centrifuge tube.

5. Centrifuge at 180,000*g* for 21 hours at 20°C.

6. Remove the cesium chloride supernatant, being careful not to disrupt the RNA pellet in the bottom of the tube.

7. Gently redissolve the RNA pellet in 750 µl of guanidine isothiocyanate lysis buffer.

8. Add an equal volume of phenol and mix.

9. Centrifuge for 15 minutes at 10,000*g*.

10. Transfer the aqueous phase to a fresh tube. Repeat the phenol extraction of the aqueous phase until the interface is clear.

11. Add 1 volume of chloroform to the clear aqueous phase.

12. Mix and centrifuge at 10,000g for 10 minutes.

13. Transfer the aqueous phase to a clean tube. Add an equal volume of isopropanol and 1/10 volume of 8 M lithium chloride. Centrifuge at 10,000g for 10 minutes.

14. Remove the supernatant. Wash the RNA pellet with 750 μl of 70% ethanol and centrifuge at 10,000g.

15. Remove the ethanol, air-dry the pellet, and redissolve in 50–500 μl of 0.2 mM DTT.

16. Store the RNA at –70°C.

PROTOCOL 2   PREPARATION OF RNA FROM PARAFFIN-EMBEDDED FIXED TISSUE

This protocol is a modification of the Fisher protocol published by Godfrey et al. (Fisher 1988; Godfrey et al. 2000). Additional protocols for the purification of RNA from fixed tissues have been published previously (Stanta and Schneider 1991; Cairns et al. 1997; Krafft et al. 1997; Coombs et al. 1999; Masuda et. al. 1999). Commerical kits for the purification of RNA from paraffin-embedded tissues are available from Ambion and Roche.

MATERIALS

▼ CAUTION

See Appendix for appropriate handling of materials marked with <!>.

BUFFERS, SOLUTIONS, AND REAGENTS

Deionized formamide<!>
Diethyl pyrocarbonate (DEPC)<!>-treated H$_2$O
Digestion buffer (1 M guanidine isothiocyanate<!>, 25 mM 2-mercaptoethanol<!>, 0.5% *N*-lauroylsarcosine<!>, 20 mM Tris-HCl, pH 7.5)
Ethanol<!>, 100% and 70%
Glycogen
Isopropanol<!>
Phenol<!> (pH 4.3):chloroform<!>(70%:30%)
Proteinase K<!> (60 mg/ml at 20 units/mg)
Trizol (Invitrogen)<!>
Xylene<!>

SPECIAL EQUIPMENT

Microfuge tubes, 2 ml, RNase-free
Incubator, preset to 37°C
Water bath or incubator, preset to 55°C

CELLS AND TISSUES

Paraffin-embedded tissue samples, cut to 5–50 x 5-μm sections

METHOD

1. Place tissue sections in 2.0-ml microcentrifuge tubes and add 1.8 ml of xylene to the sample. Incubate at 37°C for 20 minutes.

2. Centrifuge the sample briefly and remove the supernatant. Add fresh xylene and repeat the incubation and centrifugation.

3. Wash the tissue with 0.5 ml of ethanol. Remove the ethanol and air-dry.

4. Add 80 μl of proteinase K (60 mg/ml at 20 units/mg) and 720 μl of digestion buffer. Vortex and incubate overnight at 55°C.

5. Add an additional 80 µl of proteinase K; vortex and incubate overnight at 55°C. On the following day, add a further 80 µl of proteinase K; vortex and incubate again overnight at 55°C.

6. Add an equal volume of phenol:chloroform, mix, and centrifuge at maximum speed in a microcentrifuge for 5 minutes. Transfer the aqueous phase to a fresh RNase-free tube.

7. Add an equal volume of isopropanol and 2 µg of glycogen. Incubate at –20°C for 30 minutes.

8. Centrifuge at maximum speed in a microcentrifuge for 30 minutes. Remove the supernatant and wash the pellet with 70% ethanol.

9. Centrifuge at maximum speed in a microcentrifuge for 30 minutes and remove the supernatant. Air-dry the pellet and redissolve in 20 µl of DEPC-treated H_2O. Add 500 µl of Trizol. Follow Trizol protocol (see below) with one modification: After the first Trizol treatment, transfer the aqueous phase to a fresh RNase-free tube and treat the sample a second time with 500 µl of Trizol.

10. Precipitate the RNA by adding 500 µl of isopropanol to the aqueous phase from the second Trizol treatment.

11. Redissolve the RNA pellet in 20 µl of deionized formamide. Store the RNA at –20°C.

PROTOCOL 3 · PURIFICATION OF RNA USING TRIZOL

There are a limited number of manufacturers of phenol-based reagents for RNA isolation. These reagents are based on the Chomczynski protocol (Chomczynski and Sacchi 1987) and make use of guanidine buffers as lysis reagents and phenol chloroform for extraction of proteins and lipids. Today, these protocols are most useful for obtaining RNA from tissues rich in RNase, such as pancreas. Presented here is a summary of the protocol that accompanies the Trizol reagent sold by Invitrogen. This protocol was adapted, with permission, from Invitrogen.

MATERIALS

▼ CAUTION

See Appendix for appropriate handling of materials marked with <!>.

BUFFERS, SOLUTIONS, AND REAGENTS

Trizol<!> (Invitrogen)
Chloroform<!>
Isopropanol<!>
Ethanol<!>, 75%, in diethyl pyrocarbonate (DEPC)<!>-treated H_2O
DEPC-treated H_2O

CELLS AND TISSUES

50- to 100-mg source tissue, finely ground

METHOD

1. Add 1 ml of Trizol to 50–100 mg of finely ground tissue in a microcentrifuge tube. Incubate for 5 minutes at room temperature.

   It is important that the tissue be of a fine grind to aid in the disruption of the tissue.

2. Add 200 µl of chloroform for each 1 ml of Trizol. Mix by inverting the tube several times.

> Vortexing is not recommended because it may result in shearing of genomic DNA and excessive DNA contamination of the RNA preparation.

3. Incubate for 3 minutes and centrifuge at 1500*g* for 15 minutes.

4. Transfer the aqueous phase to a fresh tube. Add 500 µl of isopropanol for each 1 ml of Trizol. Incubate for 10 minutes.

5. Centrifuge at 1500*g* for 10 minutes. RNA should be visible at the bottom of the tube.

6. Remove the supernatant, being careful not to disrupt the pellet.

7. Add 1 ml of 75% ethanol for each 1 ml of Trizol used. Resuspend the pellet. Centrifuge at 1500*g* at 4°C for 5 minutes.

8. Remove the alcohol and air-dry the pellet.

9. Redissolve in DEPC-treated H_2O.

> When using phenol-based reagents, it may be necessary to subject purified RNA to a second round of phenol-chloroform extraction to obtain a sufficient level of purity.

PROTOCOL 4   PREPARATION OF RNA FROM PLANT TISSUE USING TRIZOL

The following protocol is a modification of the Trizol protocol to be used for the isolation of RNA from plant tissues. A high-salt isopropanol precipitation step has been added to precipitate RNA selectively while maintaining polysaccharides and proteoglycans in solution (Sewall and McRae 1998).

MATERIALS

▼ CAUTION

See Appendix for appropriate handling of materials marked with <!>.

BUFFERS, SOLUTIONS, AND REAGENTS
Trizol<!> (Invitrogen)
Chloroform<!>
Isopropanol<!>
Ethanol <!>, 70% in diethyl pyrocarbonate (DEPC) <!>-treated H_2O
DEPC-treated H_2O
Liquid N_2<!>
NaCl, 1.2 M
Sodium citrate<!>, 0.8 M

SPECIAL EQUIPMENT
Rotor-stator homogenizer (or equivalent)
Mortar and pestle, washed in DEPC-treated H_2O, prechilled in liquid N_2
50-ml conical tube, e.g., Falcon

CELLS AND TISSUES
Appropriate plant tissue

METHOD

1. Grind ~1 g of plant tissue in a N_2-chilled mortar using enough liquid N_2 to cover the plant tissue.

> It is important to grind the tissue to obtain a very fine powder.

2. Transfer the powder to a 50-ml conical tube containing 10 ml of Trizol per gram of tissue.

3. Homogenize the tissue using a rotor-stator homogenizer for 30 seconds.

4. Centrifuge the lysate at 1500*g* for 15 minutes at 4°C to remove debris.

5. Transfer the supernatant to a fresh tube. Add 200 µl of chloroform per ml of Trizol used.

6. Mix by inverting the tube several times. Incubate for 5 minutes at room temperature.

7. Centrifuge at 1500*g* for 20 minutes at 4°C. Transfer the aqueous phase to a fresh 50-ml tube.

8. Add 0.5 volume of isopropanol and 0.5 volume of a 0.8 M sodium citrate, 1.2 M sodium chloride solution. For example, to a 500-µl volume of aqueous phase, add 250 µl of isopropanol and 250 µl of sodium citrate solution.

9. Mix and incubate at room temperature for 10 minutes. Overnight incubation at –20°C may improve the RNA yield.

10. Centrifuge at 4°C for 20 minutes at 1500*g*.

11. Remove the supernatant and add 1 ml of 70% ethanol for each 1 ml of Trizol used for the initial extraction.

12. Centrifuge at 1500*g* for 15 minutes. Carefully remove the supernatant being careful not to disrupt the pellet. Air-dry the pellet.

13. Redissolve the RNA in DEPC-treated H_2O.

> A typical yield from 1 g of mature leaf tissue is ~500 µg of RNA.
>
> This protocol may not work well with plant tissues that contain polyphenolic compounds or resins. The Concert Plant Reagent (Invitrogen) is recommended for difficult plant tissues that are rich in polyphenolics, resins, or starch.

PROTOCOL 5 PURIFICATION OF RNA FROM PLANT TISSUE USING THE CONCERT PLANT

Summarized below is the protocol included with the Concert Plant Reagent (Invitrogen). This protocol is not phenol based, but does require the addition of chloroform. This reagent is intended for the isolation of RNA from a wide variety of plant tissues including blue spruce needles, potato tuber, corn seeds, and cotton leaves. This protocol was adapted, with permission, from Invitrogen.

MATERIALS

▼ CAUTION

See Appendix for appropriate handling of materials marked with <!>.

BUFFERS, SOLUTIONS, AND REAGENTS
Concert Plant RNA Reagent<!> (Invitrogen), prechilled to 4°C
NaCl, 5 M (RNase-free)
Chloroform<!>
Isopropanol<!>
Ethanol<!>, 75% in DEPC-treated H_2O
DEPC-treated H_2O

CELLS AND TISSUES
Appropriate plant tissue (see step 1), frozen and finely ground

METHOD

1. Add 0.5 ml of cold (+4°C) Concert Plant RNA Reagent for up to 0.1 g of frozen, ground plant tissue. Mix by vortexing briefly until the sample is suspended in the reagent.

2. Incubate the tube on its side for 5 minutes at room temperature.

3. Centrifuge for 2 minutes at 12,000g to remove debris.

4. Transfer the supernatant to a new RNase-free tube and add 0.1 ml of 5 M NaCl. Mix the tube by tapping it on the bench.

5. Add 0.3 ml of chloroform and mix by inverting the tube several times.

6. Centrifuge the sample at 4°C for 10 minutes at 12,000g.

7. Transfer the aqueous phase to a fresh RNase-free tube. Add an equal volume of isopropanol to the aqueous phase. Mix by inverting the tube several times. Incubate at room temperature for 10 minutes.

8. Centrifuge the sample for 10 minutes at 12,000g at 4°C.

9. Carefully remove the supernatant, being careful not to disrupt the pellet.

10. Add 1 ml of 75% ethanol to the pellet. Centrifuge at room temperature for 1 minute at 12,000g.

11. Remove the supernatant, being careful not to disrupt the pellet. Centrifuge briefly and remove any remaining liquid with a pipette.

12. Redissolve the pellet in 10–30 μl of DEPC-treated H_2O. If any cloudiness is observed, centrifuge the solution at 12,000g for 1 minute at room temperature. Transfer the supernatant to a fresh RNase-free tube and store at –70°C.

COMMERCIALLY AVAILABLE KITS FOR ISOLATION OF TOTAL RNA

A number of silica-based kits are commercially available for the isolation of total RNA from animal cells and tissue. Table 2 provides a listing of some of these kits. Protocol 6 summarizes the protocol that is supplied with Rapid Total RNA System from Marligen.

Several companies (Ambion, Marligen, Promega, Qiagen, and Stratagene, to name a few), in addition to single-tube formats, offer 96-well plate formats for the high-throughput purification of total RNA. These high-throughput systems generally do not accommodate the large sample sizes that can be processed using single spun columns. The stated capacity for most 96-well plate systems is 1 x 10^6 cells, with an expected yield of between 10 and 30 μg. In general, 96-well plates can be processed using vacuum and centrifugation. Vacuum protocols are well suited for use with robotic platforms but require a vacuum manifold that can accommodate the plate format. Vacuum manifolds that can be used with 96-well plate formats are available from BioRad, Whatman, and Qiagen. Processing plates using centrifugation make the systems difficult for use on robotic platforms, but they work well for manual processing. The centrifugation method also requires the centrifuge to accommodate the plate "stack," consisting of the RNA purification plate and the sample collection plate. In most cases, special plate carriers are required to accommodate the plate stack. In the authors' experience, processing RNA purification plates using a centrifugation protocol will slightly improve the quality and quantity of the RNA obtained compared to the vacuum method.

TABLE 2. Comparison of selected commercially available total RNA purification kits

Vendor	Qiagen	Ambion	Promega	Marligen	Stratagene	Invitrogen
Product name	RNeasy Mini	RNAqueous	SV Total RNA isolation system	Rapid RNA Purification system	StrataPrep Total RNA Miniprep system	Micro-to-Midi RNA purification system
Lysis method	guanidine isothiocyanate	guanidine isothiocyanate	guanidine isothiocyanate	guanidine isothiocyanate	guanidine isothiocyanate	guanidine isothiocyanate
Binding matrix	silica-gel membrane	glass fiber filter	silica membrane	silica membrane	silica-based fiber	silica fiber
Starting material	cells or tissue	cells or tissue	cells or tissue	cells or tissue	cells or tissue	cells or tissue
Capacity	1×10^7 cells or 30 mg of tissue	1×10^7 cells or 75 mg of tissue	5×10^6 cells or 60 mg of tissue	1×10^8 cells or 200 mg tissue	1×10^7 cells or 40 mg of tissue	1×10^8 cells or 200 mg of tissue
Yield	100 µg	100 µg	100 µg	600 µg	120 µg	600 µg

PROTOCOL 6   PURIFICATION OF TOTAL RNA USING THE MARLIGEN RAPID TOTAL RNA SYSTEM

The Rapid Total RNA System offered by Marligen Biosciences provides reproducible yields and quality over a broad range of sample sizes.

MATERIALS

▼ CAUTION

See Appendix for appropriate handling of materials marked with <!>.

BUFFERS, SOLUTIONS, AND REAGENTS
Ethanol<!>, 70% and 100%
2-Mercaptoethanol<!>

SPECIAL EQUIPMENT
Rotor-stator, or equivalent homogenizer

ADDITIONAL ITEMS
Marligen Rapid Total RNA System<!> (Marligen Biosciences)

METHOD

1. Prepare Lysis Solution by addition of 2-mercaptoethanol.

2. Add Lysis Solution to cell or tissue samples according to manufacturer's protocol. (Lysis volume is dependent on the size of the sample being processed.) Homogenize with a rotor-stator or similar device.

3. Add one volume of 70% ethanol to the lysate. Homogenize lysate to disperse precipitate.

4. Apply lysate to spin column provided with the kit and centrifuge at 12,000–16,000g in a microfuge.

5. Wash once with 700 µl of Wash Buffer 1 and centrifuge at 12,000–16,000g for 1 minute.

6. Wash twice with 500 µl of Wash Buffer 2 and centrifuge at 12,000–16,000g for 1 minute.

7. Centrifuge the spun column at 12,000–16,000g for 2 minutes to remove residual wash buffer.

8. Elute RNA from the column with H_2O according to manufacturer's recommendations. Store the RNA at –70°C.

   Smaller samples can be eluted in 45 µl of H_2O, whereas larger samples should be eluted in 200 µl.

COMMERCIALLY AVAILABLE KITS FOR THE ISOLATION OF mRNA

There are numerous kits available for the isolation of mRNA directly from cells and tissues, some of which are compared in Table 3. The majority of these kits rely on oligo(dT), coupled to a solid phase, to bind the polyadenylic acid (poly[A]) residues that reside at the 3´ terminus of the mRNA molecule (Berger and Kimmel 1987; Sambrook et al. 1989). Ambion's MicroPoly(A)Pure can be used to obtain reproducible yields of high-quality Poly(A) RNA, and this kit protocol is summarized in Protocol 7.

PROTOCOL 7 ISOLATION OF mRNA USING THE MICRO POLY(A) PURE KIT FROM AMBION

MATERIALS

▼ CAUTION

See Appendix for appropriate handling of materials marked with <!>.

BUFFERS, SOLUTIONS, AND REAGENTS
   Ethanol<!>, 100%

SPECIAL EQUIPMENT
   Tissue homogenizer; *optional*, see step 4
   Water bath, preset to 70°C

ADDITIONAL ITEMS
   Micro Poly(A) Pure Kit (Ambion)

CELLS AND TISSUES
   Source cells

METHOD

1. Prewarm the Elution Buffer to 70°C.

2. Pellet cells at 250g and then remove the supernatant.

3. Wash the cell pellet once in PBS by gentle resuspension and centrifugation. Discard the supernatant.

4. Add 250 µl of Lysis Solution. Homogenize with a tissue homogenizer or by vigorous vortexing or pipetting. Measure the volume of the lysate. This is the "Starting Volume."

5. Add 2 Starting Volumes of dilution buffer to the lysate. Mix by inversion or shaking for 10 seconds.

6. Centrifuge at 12,000g at 4°C for 15 minutes. Transfer the supernatant to a fresh tube.

7. Add one vial of oligo(dT) cellulose to the sample. Mix by inversion. Incubate for 30–60 minutes with gentle agitation at room temperature.

TABLE 3. Comparison of commercially available mRNA purification kits

Vendor	Active Motif	Ambion	Invitrogen	Promega	Qiagen	Roche	Stratagene
Product name	mTRAP Midi, mTRAP Maxi	Poly(A)Pure, Micro Poly(A)Pure	FastTrack 2.0, Micro-FastTrack 2.0	PolyAT tract mRNA Isolation System, PolyATract System 1000	Oligotex mRNA Purification System	mRNA Isolation kit	Poly(A) Quick mRNA Isolation Kit
Isolation method	lysis in guanidine isothiocyanate poly(T) peptide nucleic acid binding	lysis in guanidine isothiocyanate, d(T) binding, spun column purification	detergent lysis, batch d(T) binding	magnetic separation lysis in guanidine isothiocyanate, magnetic separation	lysis in guanidine isothiocyanate, oligotex binding, spun column	detergent lysis, binding to oligo d(T), magnetic separation	batch binding to d(T), push column
Starting material	2×10^7 cells or 200 mg of tissue (Midi), 2×10^8 cells or 1 g of tissue (Maxi)	2×10^8 cells or 1 g of tissue [Poly(A)Pure], 5×10^6 cells or 200 mg of tissue (Micro)	3×10^8 cells or 1 g or tissue (FastTrack 2.0), 5×10^6 cells or 200 mg of tissue (Micro)	5 mg of total RNA, 4×10^8 cells or 2 g of tissue	1×10^8 cells or 1 g of tissue	2.5 mg of total RNA, 1×10^8 cells or 1 g of tissue	500 µg of total RNA
Yield	10–20 µg, 100–200 µg	60 µg per gram of tissue, not reported	40 µg for 1×10^8 Sf9 cells, 1.8 µg for 5×10^6 HeLa cells	1–2% of total RNA, 70–200 ng of mRNA/mg tissue	30–200 µg	25 µg of mRNA from 2.5 mg of total RNA. 14 µg of 1×10^8 cells	1.5% of total RNA

8. Centrifuge at 4000g for 3 minutes at room temperature.

9. Remove the supernatant. Save the supernatant until mRNA isolation is verified.

10. Add 1 ml of Binding Buffer to the oligo(dT) cellulose pellet. Mix by inversion. Pellet by centrifugation at 4000g for 3 minutes. Remove the Binding Buffer and repeat this step two times.

11. Add 1 ml of Wash Buffer to the pellet. Mix by inversion. Pellet by centrifugation at 4000g for 3 minutes. Remove the Wash Buffer and repeat this step two times.

12. Place a spun column in a wash tube. Resuspend the oligo(dT) cellulose in 400 µl of Wash Buffer. Transfer the solution to a spun column and centrifuge briefly at 4000g. Discard the flowthrough.

13. Add 500 µl of Wash Buffer to the spun column. Mix the resin with a pipette tip, being careful not to damage the membrane in the column. Centrifuge briefly at room temperature. Discard the flowthrough and repeat this step. Save the flowthrough from the last wash.

14. Determine the A_{260} of the flowthrough. If the A_{260} is less than 0.05, continue with the protocol. If the A_{260} is greater than 0.05, repeat washes with 500 µl of Wash Buffer until the A_{260} is less than 0.05.

15. Place the spun column into a fresh microfuge tube. Add 100 µl of 70°C Elution Buffer to the spun column. Centrifuge immediately at 5000g. Add an additional 100 µl of warm Elution Buffer to the spun column and centrifuge briefly at 5000g. Discard the spun column.

16. Add 20 µl of 5 M ammonium acetate, 1 µl of glycogen, and 550 µl of 100% ethanol to the eluted mRNA. Incubate overnight at –20°C.

17. Recover the mRNA by centrifugation at 12,000g or more for 20 minutes at 4°C.

18. Remove the supernatant carefully. Centrifuge again briefly and remove any remaining fluid using a fine-tipped pipette.

19. Dissolve the pellet in 1–20 µl of DEPC-treated H_2O/EDTA provided with the kit. Vortex vigorously to completely redissolve the pellet. Store the RNA at –70°C.

SUMMARY AND TROUBLESHOOTING

Commercial reagents and kits are available for the isolation of high-quality RNA. Due to the ease and reproducibility associated with the silica-based strategies, many laboratories have adopted these systems as the method of choice to isolate RNA from animal cells and tissues. Although these systems frequently are not suitable for the isolation of RNA from plant tissues, novel reagents such as the Concert Plant Reagent fill this void. The reagents and protocols presented in this chapter can be used to provide quality RNA contingent on suitable condition of the starting materials and adherence to the recommended protocols. As a primary concern for any RNA isolation procedure, careful sample handling and adequate preparation of work surfaces minimize the potential for the introduction of RNases into samples. RNA obtained using these protocols is of a quality suitable for downstream applications such as northern blotting, RNase protection assays, in vitro translation, cDNA synthesis, microarray analysis, and quantitative RT-PCR.

Depending on the particular downstream application, it is important to consider that contamination of purified RNA with small to modest amounts of genomic DNA is inherent to all strategies used for the purification of RNA. In cases where downstream applications will be affected by contaminating DNA, DNase I treatment of samples is recommended. Typically, DNase I treatment is performed on the final RNA sample that has been redissolved or eluted in H_2O. In the case of the silica-based strategies, protocols including on-column solid-phase DNase I treatment have been developed in an effort to streamline the DNA removal. Nonetheless, in the authors' experience, on-column DNase I treatment is not effective for the complete removal of contaminating DNA. Therefore, samples to be used for applications that demand RNA completely devoid of contaminating DNA should be treated with DNase I after elution of the RNA. Additional protocols for "on-column" and post-elution treatment of RNA with DNase I are presented here.

ADDITIONAL PROTOCOL 1 POST-ELUTION REMOVAL OF CONTAMINATING DNA

The following protocol is a modified version of the Dilworth and the Huang protocols (Dilworth and McCarey 1992; Huang et. al. 2000). It can be used to remove contaminating DNA from RNA prepared using most commercial kits and reagents.

MATERIALS

▼ CAUTION

See Appendix for appropriate handling of materials marked with <!>.

BUFFERS, SOLUTIONS, AND REAGENTS
  Diethyl pyrocarbonate (DEPC) <!>-treated H_2O
  EDTA, 25 mM

ENZYMES AND ENZYME BUFFERS
  DNase I , amplification grade (RNase-free)
  DNase I buffer, 10x

NUCLEIC ACIDS AND OLIGONUCLEOTIDES
  RNA sample

SPECIAL EQUIPMENT
  Water bath, preheated to 65°C

METHOD

1. To an RNase-free tube add the following:

   RNA   up to 80 µg
   DNase I buffer, 10x   1 µl
   DNase I (1unit/µl)   1 µl
   Bring total volume to 10 µl with DEPC-treated H_2O.

2. Incubate at room temperature for 15 minutes.

3. Inactivate DNase I by adding 1 µl of 25 mM EDTA solution and heating for 10 minutes at 65°C.

Samples treated in this manner are suitable for direct use in RT-PCR.

ADDITIONAL PROTOCOL 2 ON-COLUMN REMOVAL OF CONTAMINATING DNA

This protocol can be used with RNA isolation strategies that utilize silica-based RNA-binding columns, e.g., Protocol 6. However, 100% DNA removal may not be achieved.

MATERIALS

ENZYMES AND ENZYME BUFFERS
DNase I, amplification grade (RNase-free)
DNase I buffer, 10x

CELLS AND TISSUES
Cell lysate obtained using a Total RNA Purification Kit (such as the Marligen Rapid Total RNA Purification System)

METHOD

1. Load the lysate onto a spin column included in a total RNA Purification kit (such as the Marligen Rapid Total RNA Purification System).

2. Following the loading of the RNA purification membranes with lysates, add 50 µl of DNase I solution containing 5 units of DNase I in 1x DNase I buffer to each spin column.

3. Incubate at room temperature for 15 minutes.

4. Continue with the Marligen purification protocol at step 6. The subsequent washing steps in the protocol will remove the DNase I.

REFERENCES

Auffray C. and Rougeon F. 1980. Purification of mouse immunoglobulin heavy-chain messenger RNAs from total myeloma tumor RNA. *Eur. J. Biochem.* **107:** 303–314.

Berger S.L. and Kimmel A.R. 1987. *Guide to molecular cloning techniques.* Academic Press, Orlando, Florida.

Birnboim H.C. 1988. Rapid extraction of high molecular weight RNA from cultured cells and granulocytes for Northern analysis. *Nucleic Acids Res.* **16:** 1487–1497.

Cairns M.T., Church S., Johnston P.G., Phenix K.V., and Marley J.J. 1997. Paraffin-embedded tissue as a source of RNA for gene expression analysis in oral malignancy. *Oral Dis.* **3:** 157–161.

Chirgwin J.M., Przybyla A.E., MacDonald R.J., and Rutter W.J. 1979. Isolation of biologically active ribonucleic acid from sources enriched in riboculease. *Biochemistry* **18:** 5294–5299.

Chomczynski P. and Sacchi N. 1987. Single-step method of RNA isolation by guanidinium thiocyanate-phenol-chloroform extraction. *Anal. Biochem.* **162:** 156–159.

Coombs N.J., Gough A.C., and Primrose J.N. 1999. Optimisation of DNA and RNA extraction from archival formalin-fixed tissue. *Nucleic Acids Res.* **27:** e12.

Crick F.H.C. 1968. The origin of the genetic code. *J. Mol. Biol.* **38:** 367–379.

Dilworth D.D. and McCarey J.R. 1992. Single-step elimination of contaminating DNA prior to reverse transcriptase PCR. *PCR Methods Appl.* **1:** 279–282.

Favaloro J., Treisman R., and Kamen R. 1980. Transcription maps of polyoma virus-specific RNA: Analysis by two-dimensional nuclease S1 gel mapping. *Methods Enzymol.* **65:** 718–749.

Fisher J.A. 1988. Activity of proteinase K and RNase in guanidium thiocyanate. *FASEB J.* **2:** A1126 (Abstr.).

Gao J., Liu J., Li B., and Li Z. 2001. Isolation and purification of functional total RNA from blue-grained wheat endosperm tissues containing high levels of starches and flavonoids. *Plant Mol. Biol. Rep.* **19:** 185a–185i.

Geuna F., Hartings H., and Scienza A. 1998. A new method for rapid extraction of high quality RNA from recalcitrant tissues of grapevine. *Plant Mol. Biol. Rep.* **16:** 61–67.

Godfrey T.E., Kim S.-H., Chavira M., Ruff D.W., Warren R.S., Gray J.W., and Jensen R.H. 2000. Quantitative mRNA expression analysis from formalin-fixed, paraffin-embedded tissue using 5′

nuclease quantitative reverse transcription-polymerase chain reaction. *J. Mol. Diagn.* **2:** 84–91.

Grotzer M.A., Patti R., Geoerger B., Eggert A., Chou T.T., and Phillips P.C. 2000. Biological stability of RNA isolated from RNA*later*-treated brain tumor and neuroblastoma xenografts. *Med. Pediatr. Oncol.* **34:** 438–442.

Hosein F. 2001. Isolation of high quality RNA from seeds and tubers of the Mexican yam bean (*Pachyrhizus erosus*). *Plant Mol. Biol. Rep.* **19:** 65a–65e.

Huang L., Lee J., Sitaraman K., Gallego A., and Rashtchian A. 2000. A new highly sensitive two-step RT-PCR system. *Focus* **22:** 6–7.

Hughes D.W. and Galau G.A. 1988. Preparation of RNA from cotton leaves and pollen. *Plant Mol. Biol. Rep.* **6:** 253–275.

Krafft A.E., Duncan D.W., Bijwaard K.E., Taubenberger J.K., and Lichy J.H. 1997. Optimization of the isolation and amplification of RNA from formalin-fixed, paraffin-embedded tissue: The Armed Forces Institute of Pathology Experience and Literature review. *Mol. Diagn.* **2:** 217–230.

Lönneborg A. and Jensen M. 2000. Reliable and reproducible method to extract high-quality RNA from plant tissues rich in secondary metabolites. *BioTechniques* **29:** 714–718.

Masuda N., Ohnishi T., Kawamoto S., Monden M., and Okubo K. 1999. Analysis of chemical modifications of RNA from formalin-fixed samples and optimization of molecular biology applications for such samples. *Nucleic Acids Res.* **27:** 4436–4443.

Orgel L.E. 1968. Evolution of the genetic apparatus. *J. Mol. Biol.* **38:** 381–393.

Sambrook J., Fritsch E.F., and Maniatis T. 1989. *Molecular cloning: A laboratory manual,* 2nd edition. Cold Spring Harbor Laboratory Press, Cold Spring Harbor, New York, pp. 7.1–7.83.

Scott D.L., Jr., Clark C.W., Deahl K.L., and Prakash C.S. 1998. Isolation of functional RNA from periderm tissue of potato tubers and sweet potato storage root. *Plant Mol. Biol. Rep.* **16:** 3–8.

Sewall A. and McRae S. 1998. RNA isolation with TRIzol reagent. *Focus* **20:** 36–37.

Stanta G. and Schneider C. 1991. RNA extracted from paraffin-embedded human tissues is amenable to analysis by PCR amplification. *BioTechniques* **11:** 304–308.

Stryer L. 1988. *Biochemistry,* 3rd edition. W.H. Freeman, New York, p. 92.

Vogelstein B. and Gillespie D. 1979. Preparative and analytical purification of DNA from agarose. *Proc. Natl. Acad. Sci.* **76:** 615–619.

Wan C.Y. and Wilkins T.A. 1994. A modified hot borate method significantly enhances the yield of high-quality RNA from cotton (*Gossypium hirsutum L.*). *Anal. Biochem.* **223:** 7–12.

Wang S.X., Hunter W., and Plant A. 1999. Isolation and purification of functional total RNA from woody branches and needles of sitka and white spruce. *BioTechniques* **28:** 292–296.

In Situ Localization of Gene Expression Using Laser Capture Microdissection

Motoko Morimoto,[1*] Masahiro Morimoto,[2,3*] Jeannette Whitmire,[1] Robert A. Star,[4] Joseph F. Urban, Jr.,[3] and William C. Gause[1]

[1]Department of Microbiology and Immunology, Uniformed Services University of the Health Sciences, Bethesda, Maryland 20814-4799; [2]Department of Veterinary Pathology, Yamaguchi University, Yamaguchi 753-8515, Japan; [3]Nutrient Requirements and Functions Laboratory, Beltsville Human Nutrition Research Center, U.S. Department of Agriculture, Beltsville, Maryland 20705-2350; [4]Renal Diagnostics and Therapeutics Unit, and Laboratory of Pathology, National Institutes of Health, Bethesda, Maryland 20892-1268

The lymph nodes of the immune system are complex organs with specific microenvironments in which lymphoid lineages differentiate and become committed to specific effector functions. Elucidation of the proteins and genes expressed by lymphocytes in situ can potentially provide important information regarding their function in their natural milieu. The technique of laser capture microdissection (LCM), which was originally developed to study gene expression in tumor cells (Emmert-Buck et al. 1996; Bonner et al. 1997; Simone et al. 1998), is well-suited for examination of cell populations in lymphoid organs or other heterogeneous tissue microenvironments. Other techniques that examine cell function in situ include immunohistology and in situ hybridization. These techniques are frequently not sufficiently sensitive to detect physiological up-regulation of proteins or specific mRNA expression (Fend and Raffeld 2000), and they usually cannot be used to detect more than one or two expressed molecules at a time.

Various microdissection techniques have been used to remove single cells from tissue sections for further analysis (Isenberg et al. 1976; Meier-Ruge et al. 1976; Whetsell et al. 1992). For example, micropipettes have been successfully used to remove single cells from germinal centers of B-cell follicles in the lymph node to study B-cell differentiation (Kuppers et al. 1993). However, these techniques are laborious and difficult to perform and thus are not practical in many laboratories. The development of LCM provides a rapid and readily learned technique for removal of specific cells or tissue regions from a tissue section on a microscope slide. RNA or protein can be purified and used in a variety of assays in the laboratory, including real-time reverse transcriptase (RT)-PCR, allowing the analysis of the expression of a number of different genes from the same captured sample.

---

*These authors contributed equally to this work.

BASIC TECHNIQUE

The basic technique involves the use of a microscope and a laser to select single cells or regions of tissue from a thin section on a microscope slide (Fig. 1). A thermoplastic membrane, attached to a cap, is placed over the target area and a laser pulse is delivered to the thermoplastic membrane, which then melts onto the selected area. When the cap is removed, the targeted cell or tissue remains attached to the membrane, allowing it to be removed selectively from the rest of the tissue section on the slide. The cap can then be transferred to a solution for isolation of DNA, RNA, or protein. Our laboratory and many others currently perform microdissection on a PixCell II (Arcturus Engineering, Mountain View, CA). Here we present a detailed protocol that yields high-quality mRNA isolated from captured cells or regions in tissue sections.

METHODS STRATEGY

There are five basic steps involved in the analysis of gene expression of cells and tissues using the LCM technique, as shown in Figure 2:

1. Preparation of frozen tissue blocks and sectioning

2. Staining of the slides (if necessary, immunohistology)

3. Performing LCM

4. RNA extraction from small numbers of cells and cDNA synthesis

5. Real-time PCR

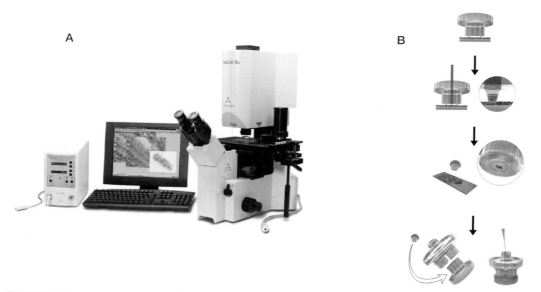

FIGURE 1. Laser-capture microdissection (LCM) system. The Arcturus LCM system allows rapid isolation of targeted cell populations from tissue sections. (A) The capture process is visualized with a specially adapted microscope. (B) Using this microscope, a plastic cap coated with a thermoplastic film is brought in close proximity to the tissue region of interest. A laser passes through the cap and melts the film, which adheres to the tissue. When the cap is lifted, the selected cells remain attached and captured for analysis. (Reprinted, with permission, from Arcturus, www.arctur.com.)

PROTOCOL 1 PREPARATION OF FROZEN TISSUE BLOCKS AND SECTIONING

For optimal results, freshly obtained tissue should be used, because storage of tissue blocks at –80°C for more than 24 hours reduces the RNA integrity and yield. Although several groups have reported good recovery of undegraded RNA from paraffin-embedded tissue by LCM (Schutze and Lahr 1998; Shibutani et al. 2000; Specht et al. 2001), we recommend frozen blocks.

MATERIALS

BUFFERS, SOLUTIONS, AND REAGENTS

Acetone <!>
Distilled H_2O
Dry ice<!>
Phosphate-buffered saline (PBS)
Tissue-Tek O.C.T Compound (Sakura Finetek U.S.A., Torrance, CA)

SPECIAL EQUIPMENT

Air-tight container
Beaker
Cryostat
Dissecting tools
Flat plate
Peel-A-Way Disposable Histology Molds (Pelco International, Redding, CA)
Plain uncoated slides
Sakura Low Profile blades (IMEB, San Marcos, CA)

CELLS AND TISSUES

Tissue, freshly obtained (see step 2)

METHOD

Preparation of Frozen Tissue Blocks

1. Cover the cryomold with embedding medium (Tissue-Tek O.C.T. Compound is recommended). Prepare the dry ice–acetone bath (a simple mixture of acetone and dry ice in beaker).

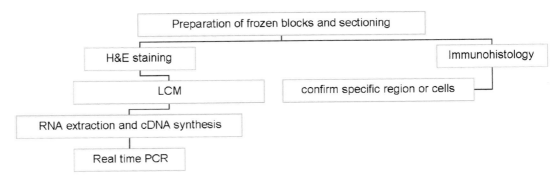

FIGURE 2. Flow chart of the combination of LCM and quantitative real-time PCR for analysis of gene expression in situ. This approach uses navigational LCM, as described in the text, to locate the cells or tissue region of interest.

2. Sacrifice the animals and remove the desired tissues. Mix the tissues with embedding medium on a flat plate.

3. Place the tissue against the bottom of the cryomold. To facilitate cutting with the cryostat, place the desired cutting face flush against the bottom of the cryomold.

4. Touch the bottom of the cryomold to the surface of the dry ice–acetone bath, which will gradually freeze the sample in 3–4 minutes. The sample will turn white when frozen. Do not drop cryomold into acetone during the freezing procedure.

5. Remove the block from the cryomold and place it at –80°C until ready for sectioning. If the blocks are stored for a prolonged period, place them in an airtight container with frozen distilled H_2O to maintain humidity.

Sectioning and Slide Storage

1. Attach a frozen block of tissue to the chuck in the cryostat with O.C.T. Compound, using standard frozen tissue-sectioning methods.

2. Allow the block to equilibrate to the cryostat temperature (–20 to –25°C) for about 15–20 minutes.

   If the block is too cold during cutting, this time may need to be extended.

3. Cut thin sections (1–5 µm) for LCM sampling and place them on a plain uncoated glass slide.

   Tissue adheres too strongly to coated slides, making LCM sampling difficult.

4. After placing the section on the slide, *immediately* place the slides on dry ice. If LCM is to be performed that day, the slides should be kept on dry ice. Alternatively, they may be stored in paper slide boxes at –80°C until needed.

PROTOCOL 2   H&E STAINING OF THE SLIDES FOR LCM

Before performing LCM, the sectioned tissue sample is fixed, stained if necessary, and dehydrated. We typically stain with H&E for gut and lymph node tissue. In addition, we prepare serial slides for immunohistochemistry to distinguish specific regions or cells. There is also a commercially available staining kit, HistoGene LCM Frozen Section Staining Kit (KIT0401, Arcturus, Mountain View, CA) that can process 32 slides. Once the slide is ready, LCM should be performed within 45 minutes. All reagents, tubes, and other materials for the RNA extraction step should be prepared before staining.

MATERIALS

▼ CAUTION

See Appendix for appropriate handling of materials marked with <!>.

BUFFERS, SOLUTIONS, AND REAGENTS
   100% ethanol<!>
   70% ethanol<!>
   EosinY (Sigma)
   Mayer's hematoxylin (Sigma)
   DEPC-treated Ultraspec Water<!> (Biotecx Laboratories, store at 4°C)
   Xylene <!>, mixed (Sigma)

SPECIAL EQUIPMENT
   Conical tubes, 50-ml
   Forceps
   Glass jars (for xylene)

CELLS AND TISSUES
  Slides from Protocol 1

METHOD

Staining should be performed as closely as possible to the scheduled LCM transfer time using solution baths that are replaced regularly. In particular, H_2O and the last ethanol and xylene baths should be fresh every time. Keep in mind that the extent of RNA degradation is in proportion to the time the tissue is in aqueous phase after fixation (5 minutes is the limit for obtaining good RNA quality). Use DEPC-treated H_2O for all steps of this protocol. The last ethanol and xylene steps can be modified depending on the humidity. If the humidity is high, we recommend dipping the slide in 100% ethanol three times and keeping it in the second xylene wash for 2 minutes. If the humidity is low, skip the third ethanol wash and shorten the xylene washes to 30 seconds.

Proceed rapidly through each of the following steps, wiping off the edges and back side of the slide between solutions. Sometimes the slides repel the solution, making it necessary to dip the slide repeatedly. Use forceps to hold the slides.

1. Prepare the staining solution set (Fig. 3).

2. Fix in 70% ethanol (30 seconds).

3. Purified H_2O wash (don't substitute PBS for H_2O; if PBS is used, tissue will not adhere well to LCM membrane). Dip up and down (5–30 seconds).

4. Mayer's hematoxylin (30 seconds).

5. Purified water wash (don't substitute PBS for water; if PBS is used, tissue will not adhere well to LCM membrane). Dip up and down (5–30 seconds).

6. 70% ethanol wash (30 seconds).

7. 100% ethanol wash (30 seconds).

8. Eosin Y (30 seconds).

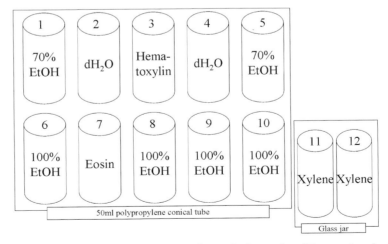

FIGURE 3. Rapid Hematoxylin & Eosin staining. In this technique, the slides are incubated for a minimal period to reduce exposure to aqueous solutions, which is associated with RNA degradation. After the slides are removed from the freezer, they are maintained at room temperature until ice crystals at the slide edge start melting, and then immersed in 70% ethanol (the first 50-ml conical tube). (See Protocol 2 for further details.)

9. 100% ethanol wash (30 seconds), two or three times.

10. Xylene wash for 0.5–2 minutes twice.

The sample is now ready for LCM.

PROTOCOL 3 PERFORMING LCM

The workspace for LCM methods must be clean, dry, and RNase-free. Dispenser, forceps, and other equipment should be wiped with RNaseAway. If humidity is high, a dehumidifier may be required to maintain dry conditions. The laser beam of the PixCell II must be properly focused to obtain a good specimen. Because differences in tissue thickness affect laser focus, tissue sections must be of uniform thickness.

Coupling immunohistological staining with LCM is an obvious advantage, permitting capture of specific cell populations recognized by antibodies in situ. Its use for analysis of cytokine gene expression is limited by the loss of undegraded mRNA during the staining procedures (Jin et al. 1999; Kohda et al. 2000). Immunofluorescence staining is preferred because it avoids the use of enzymes and multiple incubations in aqueous solutions required for many immunohistochemical procedures. Recently, a rapid immunofluorescence LCM that minimizes mRNA loss was developed (Murakami et al. 2000). Another approach is to use serial sections on different slides, with one slide dedicated to LCM and the other to immunohistological staining (Wong et al. 2000).

The LCM slide is usually stained with H & E so that the cell populations which have been specifically identified through immunohistological staining can be generally localized. This approach, called intervening or navigated LCM, becomes cumbersome and inaccurate because smaller numbers of cells are selected or the cells are in heterogeneous tissue microenvironments. However, this LCM method can be very effectively used when only one cell type in the region selected expresses a particular gene or when the region selected is predominantly of one cell type.

MATERIALS

BUFFERS, SOLUTIONS, AND REAGENTS
RNaseAway (7000, Molecular BioProducts, San Diego, CA)

SPECIAL EQUIPMENT
Adhesive tape
CapSure LCM Caps (Arcturus Engineering)
Dehumidifier
Dispenser, P200, P2
Forceps
LCM system
PixCell II (Arcturus Engineering)
RNase-free tube, 0.5 ml

METHOD

Perform LCM following the manufacturer's instruction. A brief description follows.

1. Place sample slide on the translation stage of the PixCell II apparatus.

2. Manually position the slide to locate the transfer area.

3. Turn on the vacuum chuck to hold the slide in place.

4. Place the LCM cap on the slide by rotating the placement arm.

5. Rotate the laser interlock keyswitch and wait 10 seconds.

6. Push the laser enable button.

7. Proceed to a higher magnification and adjust the focus.

> The laser beam should be located in the center of the field of view. Note that the laser beam is invisible through the microscope. When the laser is enabled, you should carry out microdissection watching the monitor. Select the smallest spot size (7.5 μm) and fire a test pulse outside tissue area to adjust focus of the laser beam.

8. Use the joystick to move to other areas on the tissue and capture other cells of interest.

After LCM, the LCM cap should be removed immediately and inserted onto an RNase-free 0.5-ml tube containing RNA isolation solution.

◗ TROUBLESHOOTING

- If it is difficult to focus the laser beam, dust may be present on the LCM cap. Dust can be removed from LCM caps with adhesive tape. Otherwise, you can try to shoot another area. If the thickness of the section varies, refocusing may be necessary.

PROTOCOL 4   RNA EXTRACTION FROM LIMITED CELLS

RNA samples can be kept at –80°C for 6 months. Total cellular RNA is extracted using the Stratagene RNA isolation kit following the manufacturer's instructions. Alternatively, the PicoPure RNA Isolation Kit is available and is particularly useful for RNA extraction from a few cells. Quantitation of isolated RNA through UV spectrometry is often not practical, because reagents in the RNA isolation kits affect absorbance values and because the total amount of RNA from LCM samples is frequently too little to detect with conventional spectrometers. Instead, one can use ribosomal RNA values as a standard to normalize for differences in undegraded RNA between samples. Collecting the same number of laser spots per sample can also help to maintain similar RNA quantities between samples.

Reverse transcription is performed immediately following the RNA extraction, and first-strand cDNA is synthesized with a random hexamer and SuperScript II RNase H⁻ reverse transcriptase (see Protocol 5). Once the RNA is collected and purified, real-time PCR commercial kits (PE Applied Biosystems, Foster City, CA) specific for different cytokines or ribosomal RNA are used, and all data are normalized to constitutive ribosomal RNA values (see Protocol 6). The Applied Biosystems 7700 sequence detector (PE Applied Biosystems) is used for amplification of target mRNA, and quantitation of differences between treatment groups is calculated according to manufacturer's instructions.

MATERIALS

....

BUFFERS, REAGENTS, AND SOLUTIONS

75% Ethanol <!> with Ultraspec Water

Glycogen, 5 mg/ml (Ambion, Austin, TX)

Stratagene RNA isolation kit (200345, Stratagene Cloning Systems, La Jolla, CA)

> Includes Denaturing solution, phenol <!> saturated with H_2O (store at 4°C), β-mercaptoethanol <!> (store at 4°C), chloroform<!>:isoamyl alcohol <!>, 2 M sodium acetate, isopropanol<!> OR use PicoPure RNA Isolation Kit (KIT0202, Arcturus, Mountain View, CA)

DEPC-treated Ultraspec Water <!> (Biotecx Laboratories, Houston, TX) (store at 4°C)

SPECIAL EQUIPMENT
Pipette
RNase-free tube, 1.5 ml

CELLS AND TISSUES
LCM caps from Protocol 3

METHOD

1. Add 200 μl of Denaturing solution and 1.6 μl of β-mercaptoethanol to a 0.5-ml tube.

2. Insert the LCM cap into the tube and vortex for 5 seconds.

3. Invert the tube and incubate for 5 minutes.

4. Remove the LCM cap.

    Samples processed to this point can be stored at –80°C up to 6 months.

5. Transfer the solution to a 1.5-ml tube.

6. Add 20 μl of 2 M sodium acetate. Mix gently.

7. Add 220 μl of H$_2$O-saturated phenol (bottom layer). Mix gently.

8. Add 60 μl of chloroform:isoamy alcohol. Mix vigorously.

9. Put on wet ice for 15 minutes.

10. Centrifuge at 14,000 rpm for 30 minutes at 4°C.

11. Transfer upper layer to a new tube.

12. Add 2 μl of glycogen (final 10–150 μg/ml).

13. Add 200 μl of cold isopropanol.

14. Place the tube in a –80°C freezer overnight.

15. If the sample is frozen solid, allow it to melt a little until it is a slurry. This will facilitate formation of a more compact pellet. Centrifuge at 14,000 rpm for 30 minutes at 4°C.

16. Remove the supernatant and wash the pellet with 400 μl of 75% ethanol.

17. Centrifuge at 14,000 rpm for 5 minutes at 4°C.

18. Remove the supernatant. Use smaller pipette to remove the supernatant completely.

19. Air-dry for 5–10 minutes.

20. Resuspend the RNA with 25 μl of Ultraspec Water.

PROTOCOL 5 cDNA SYNTHESIS

METHOD

BUFFERS, SOLUTIONS, AND REAGENTS
0.1 M Dithiothreitol <!>
dNTPs (2.5 mM each dNTP)
Random primer (20–40 units/ml) (1034731, Roche Applied Science, Indianapolis, IN)
RNA sample from Protocol 4
RNasin (Promega)

▼ CAUTION

See Appendix for appropriate handling of materials marked with <!>.

5x RT buffer
RT master mix (x sample number)

5x RT buffer	10 µl
dNTP	5 µl
0.1 M DTT	4 µl
random primer	4 µl
RNasin	1 µl

SuperScript II (18064-014, Invitrogen, Carlsbad, CA)
> Store reagents at –20°C.

SPECIAL EQUIPMENT

PCR tube, 0.2 ml

Thermal cycler (e.g., Applied Biosystems 7700 sequence detector; PE Applied Biosystems, Foster City, CA)

METHOD

1. Dispense 24 µl of RT reaction mixture to each tube.

2. Add 25 µl of RNA sample from Protocol 4.

3. Denature at 70°C for 5 minutes and then quench on ice for 5 minutes.

4. Add 1 µl of SuperScript II (final reaction volume, 50 µl).

5. Incubate the mixture at 37°C for 60 minutes, denature at 90°C for 5 minutes, and quench on ice for 5 minutes.

6. The cDNA samples should be stored at –20°C until use.

PROTOCOL 6   REAL-TIME PCR

Once the RNA is collected and purified, real-time PCR commercial kits specific for different cytokines or ribosomal RNA are used, and all data are normalized to constitutive ribosomal RNA values. The Applied Biosystems 7700 sequence detector is used for amplification of target mRNA, and quantitation of differences between treatment groups is calculated according to manufacturer's instructions.

MATERIALS

▼ CAUTION

See Appendix for appropriate handling of materials marked with <!>.

BUFFERS, SOLUTIONS, AND REAGENTS

cDNA sample from Protocol 5

20x 18S rRNA primers and probe (Applied Biosystems)

20x Target primers and probe (Applied Biosystems)

2x TaqManUniversal PCR Master Mix (4304437, Applied Biosystems, Foster City, CA)

DEPC-treated Ultraspec Water <!> (Biotecx Laboratories) (Store at 4°C.)
> Store reagents at –20°C.

SPECIAL EQUIPMENT

Applied Biosystems 7700 sequence detector (PE Applied Biosystems, Foster City, CA)

METHOD

1. Dilute the cDNA sample from Protocol 5 1/100.

2. Use 1 µl of diluted solution for the control reaction and 10 µl for the target reaction.

Solutions	Control	Target
Water	21.5 µl	12.5 µl
20x Master Mix	25 µl	25 µl
2x primers and probe	2.5 µl	2.5 µl
cDNA sample	1 µl	10 µl
Total	50 µl	50 µl

3. PCR cycle conditions:

    1. 50°C for 2 minutes

    2. 95°C for 10 minutes

    3. PCR cycle conditions – n cycles

        a. 95°C for 15 seconds

        b. 60°C for 1 minute

ANALYSIS OF RESULTS

The results from a real-time PCR analysis of ribosomal RNA isolated from LCM-captured mouse lymph node samples are shown in Figure 4. This procedure used a 30-µm cap diameter. A linear correlation between cycle number and number of spots was obtained, indi-

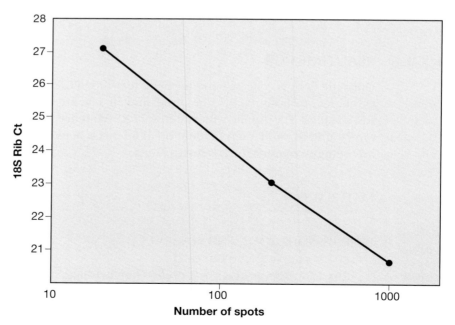

FIGURE 4. The relationship between number of spots (captured samples) and real-time PCR amplification of ribosomal RNA. Cells were captured from the lymph node using 30-µm-diameter laser spots. Total RNA was extracted and reverse-transcribed with random primer and SuperScript II reverse transcriptase. The numbers of spots pooled into each amplification reaction are listed on the x axis, and the threshold level of detection of the ribosomal RNA signal is graphed on the y axis.

cating that one can equilibrate RNA by taking an equal number of spots per sampling area. To show the practicality of the method, results of the cytokine gene expression in localized microenvironments in the gut after infection of mice with the intestinal nematode parasite *Heligmosomoides polygyrus* are shown in Figure 5. Figure 5A is a micrograph of a cyst containing developing larvae at day 4 after a primary inoculation. Figure 5B shows the region after tissue samples (the area immediately around the cyst) have been collected by LCM. To examine the gene expression of individual microenvironments in the small intestine of nematode-infected mice, cells were captured from (1) the lamina propria; (2) the Peyer's patch; (3) the area immediately surrounding the cyst; and (4) the area inside the cyst. RNA was purified from pooled LCM samples, reverse-transcribed, and interleukin (IL)-4, IL-13, and ribosomal RNA were assessed by real-time RT-PCR. All data were normalized to constitutive ribosomal RNA values. Nematode-infected mice that received anti-CD4 antibodies at day 7 after inoculation showed markedly reduced elevations in both IL-4 and IL-13 in the region surrounding the cyst (Fig. 6).

DISCUSSION

Various assays have been used to measure protein and mRNA expression in situ, and each has distinct advantages and disadvantages. Immunohistological techniques (immunohisto-chemistry and immunofluorescence) can measure protein directly, but, because many proteins are rapidly made, secreted, and utilized, sufficient quantities may not be present at a given time point to be detected. Furthermore, the technique is dependent on the availability of specific antibodies effective in staining tissue sections. In situ hybridization can localize mRNA expression and, coupled with IH or IF, can detect specific cell populations expressing the target mRNA. However, in situ hybridization is often not sufficiently sensitive to detect tightly regulated mRNA species. In addition, expression of only one or two genes can be analyzed in a single assay, making analysis of multiple mRNA species cumbersome.

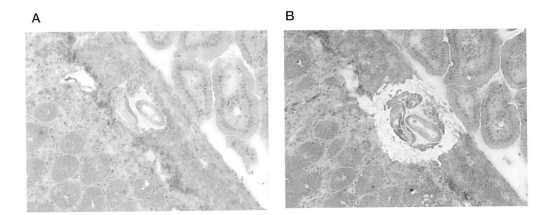

A **B**

FIGURE 5. Laser capture microdissection at the host–pathogen interface. Mice were inoculated orally with the intestinal nematode parasite, *H. polygyrus*. At day 8 after inoculation, tissues were embedded in O.C.T., frozen, and sectioned with a cryostat. LCM was performed to examine the localized immune response in the microenvironment surrounding the invading parasite. Intestinal cyst containing the nematode is shown in serial sections: (*A*) before LCM, (*B*) after LCM. The cells were taken from the area immediately around the cyst in this picture. The cells were also captured from the lamina propria, Peyer's patch, and inside the cyst.

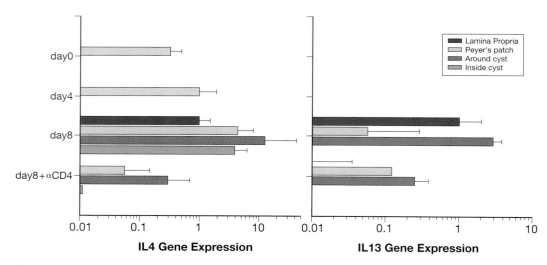

FIGURE 6. RT-PCR analysis of LCM-captured samples at the host–pathogen interface during primary immune response to an intestinal nematode parasite. IL-4 and IL-13 mRNA were detected in the lamina propria, and in the localized region around and inside the cyst by day 8 after inoculation. Anti-CD4 antibody treatment at day 7 after inoculation inhibited elevations in both IL-4 and IL-13. Mice were inoculated orally with the infectious parasite, as described in Fig. 5. All data were normalized to ribosomal RNA and expressed relative to day-8 lamina propria. The results are expressed as the mean and standard error for each treatment group. Similar results were obtained in two experiments.

The recently developed LCM technique adds an important methodology that greatly extends in situ analyses of tissue sections. The ability to remove targeted cell populations selectively from their tissue microenvironment for nucleotide and protein analysis allows the use of sensitive laboratory techniques available for identifying and quantitating cellular molecules in solution (Nagle 2001; Wittliff and Erlander 2002). Real-time PCR is well-suited for gene expression of samples captured by LCM (Specht et al. 2000; Trogan et al. 2001; Vincent et al. 2002). After cDNA synthesis, multiple targets can be analyzed from a single sample, and the sensitivity of the assay allows successful amplification, even with small amounts of target material. When used as described in the accompanying protocols, the LCM technique can collect significant quantities of undegraded mRNA. However, slight changes in these protocols, particularly involving the sectioning and storage of tissue sections, can result in a considerable loss in RNA integrity. Combining LCM with immunohistology staining can be done under certain conditions, particularly with IF, but it is still not widely used due to significant loss of mRNA. The ability to capture IF-stained cells directly, rather than being limited to using navigational LCM, will greatly extend the capability of this technique.

Another area of refinement will be the diameter of the focused laser beam. The current minimum diameter is 7.5 μm, but most studies have suggested that the actual area taken is about twice that diameter (L. Barisoni and R.A. Star, in prep.), which in many cases can result in several cells, perhaps heterogeneous, being captured at a single site. There are adjustments that can be made to the current LCM microscope (PixCell II) which may further tighten the beam, including additional focusing lenses and specially designed caps, all of which are less than optimal. A new LCM microscope has recently been developed by Arcturus (PixCell IIe) that has a narrower laser beam and improved image quality. More studies are required to determine whether this apparatus can be used routinely to collect single cells, but if it does have this capability, it would be a marked improvement.

The LCM technique has developed in recent years into a versatile and robust tool that can be routinely used in the laboratory after the initial fairly expensive purchase of the equipment. It greatly extends our capability for analyzing both protein and gene expression in the tissue microenvironment (Liotta and Kohn 2001).

REFERENCES

Bonner R.F., Emmert-Buck M., Cole K., Pohida T., Chuaqui R., Goldstein S., and Liotta L.A. 1997. Laser capture microdissection: Molecular analysis of tissue. *Science* **278**: 1481–1483.

Emmert-Buck M.R., Bonner R.F., Smith P.D., Chuaqui R.F., Zhuang Z., Goldstein S.R., Weiss R.A., and Liotta L.A. 1996. Laser capture microdissection. *Science* **274**: 998–1001.

Fend F. and Raffeld M. 2000. Laser capture microdissection in pathology. *J. Clin. Pathol.* **53**: 666–672.

Isenberg G., Bielser W., Meier-Ruge W., and Remy E. 1976. Cell surgery by laser microdissection: A preparative method. *J. Microsc.* **107**: 19–24.

Jin L., Thompson C.A., Qian X., Kuecker S.J., Kulig E., and Lloyd R.V. 1999. Analysis of anterior pituitary hormone mRNA expression in immunophenotypically characterized single cells after laser capture microdissection. *Lab. Invest.* **79**: 511–512.

Kohda Y., Murakami H., Moe O.W., and Star R.A. 2000. Analysis of segmental renal gene expression by laser capture microdissection. *Kidney Int.* **57**: 321–331.

Kuppers R., Zhao M., Hansmann M.L., and Rajewsky K. 1993. Tracing B cell development in human germinal centres by molecular analysis of single cells picked from histological sections. *EMBO J.* **12**: 4955–4967.

Liotta L.A. and Kohn E.C. 2001. The microenvironment of the tumour-host interface. *Nature* **411**: 375–379.

Meier-Ruge W., Bielser W., Remy E., Hillenkamp F., Nitsche R., and Unsold R. 1976. The laser in the Lowry technique for microdissection of freeze-dried tissue slices. *Histochem. J.* **8**: 387–401.

Murakami H., Liotta L., and Star R.A. 2000. IF-LCM: Laser capture microdissection of immunofluorescently defined cells for mRNA analysis. *Kidney Int.* **58**: 1346–1353.

Nagle R.B. 2001. New molecular approaches to tissue analysis. *J. Histochem. Cytochem.* **49**: 1063–1064.

Schutze K. and Lahr G. 1998. Identification of expressed genes by laser-mediated manipulation of single cells. *Nat. Biotechnol.* **16**: 737–742.

Shibutani M., Uneyama C., Miyazaki K., Toyoda K., and Hirose M. 2000. Methacarn fixation: A novel tool for analysis of gene expressions in paraffin-embedded tissue specimens. *Lab. Invest.* **80**: 199–208.

Simone N.L., Bonner R.F., Gillespie J.W., Emmert-Buck M.R., and Liotta L.A. 1998. Laser-capture microdissection: Opening the microscopic frontier to molecular analysis. *Trends Genet.* **14**: 272–276.

Specht K., Richter T., Muller U., Walch A., and Hofler M.W. 2000. Quantitative gene expression analysis in microdissected archival tissue by real-time RT-PCR. *J. Mol. Med.* **78**: B27.

Specht K., Richter T., Muller U., Walch A., Werner M., and Hofler H. 2001. Quantitative gene expression analysis in microdissected archival formalin-fixed and paraffin-embedded tumor tissue. *Am. J. Pathol.* **158**: 419–429.

Trogan E., Choudhury R.P., Dansky H.M., Rong J.X., Breslow J.L., and Fisher E.A. 2001. Laser capture microdissection analysis of gene expression in macrophages from atherosclerotic lesions of apolipoprotein E-deficient mice. *Proc. Natl. Acad. Sci.* **99**: 2234–2239.

Vincent V.A., DeVoss J.J., Ryan H.S., and Murphy Jr., G.M. 2002. Analysis of neuronal gene expression with laser capture microdissection. *J. Neurosci. Res.* **69**: 578–586.

Whetsell L., Maw G., Nadon N., Ringer D.P., and Schaefer F.V. 1992. Polymerase chain reaction microanalysis of tumors from stained histological slides. *Oncogene* **7**: 2355–2361.

Wittliff J.L. and Erlander M.G. 2002. Laser capture microdissection and its applications in genomics and proteomics. *Methods Enzymol.* **356**: 12–25.

Wong M.H., Saam J.R., Stappenbeck T.S., Rexer C.H., and Gordon J.I. 2000. Genetic mosaic analysis based on Cre recombinase and navigated laser capture microdissection. *Proc. Natl. Acad. Sci.* **97**:12601–12606.

WWW RESOURCE

www.arctur.com Arcturus home page.

12 Strategies for Overcoming PCR Inhibition

Peter Rådström,[1] Charlotta Löfström,[1,2] Maria Lövenklev,[1] Rickard Knutsson,[1] and Petra Wolffs[1]

[1]Applied Microbiology, Center for Chemistry and Chemical Engineering, Lund Institute of Technology, Lund University, SE-221 00 Lund, Sweden; [2]AnalyCen Nordic AB, SE-531 19 Lidköping, Sweden

The use of conventional and real-time PCR is to some extent restricted by the presence of PCR inhibitors. This is particularly so when the techniques are applied directly to complex biological samples such as clinical, environmental, or food samples for the detection of microorganisms. PCR inhibitors may originate from the sample itself, or as a result of the method used to collect or otherwise prepare the sample (Rossen et al. 1992). Either way, inhibitors can dramatically reduce the sensitivity and amplification efficiency of PCR. Altered amplification efficiency will change the kinetics of an amplification reaction, generating ambiguous data in quantitative PCR. Some inhibitors, originating from clinical, environmental, and food samples, are shown in Table 1 (for a comprehensive list, see Wilson 1997). Most of these inhibitors are known to interact with the structure and function of proteins, and the thermostable DNA polymerase is doubtless the most important target site of the inhibitors. The polymerase can be degraded by proteinases, denatured by phenol or detergents, and inhibited by the blocking of the active site by the inhibitor, as is the case with heme (Akane et al. 1994). Furthermore, it is well known that the polymerization activity of DNA polymerase varies with, for example, ionic strength, pH, sulfhydryl, and the presence of other chemical agents. However, there is a need to specifically and systematically identify and study the mechanisms of PCR inhibitory compounds.

PCR inhibition can be categorized into (1) inhibition caused by inhibitory molecules and particles in the sample and (2) sample-independent inhibition, including random amplification failure, probably caused by human factors (Cone et al. 1992). The latter category is rather uncommon, and it is difficult to discriminate between a failure in sample and reagent handling and an actual failure of amplification. The inhibitory mechanisms at work in conventional PCR may include (1) inactivation of the thermostable DNA polymerase, (2) degradation or capture of the nucleic acids, and/or (3) interference with the release of target nucleic acids from cells (Wilson 1997). However, the list of inhibitors and the mechanisms by which they may act has grown since the introduction of real-time PCR. New inhibitors may interfere with fluorescent double-stranded DNA-intercalating dyes, such as SYBR Green I, or with different types of fluorogenic sequence-specific probes used in these complex reactions. Inhibitory compounds may also generate background fluorescence or scatter the emission from the fluorogenic molecules.

TABLE 1. PCR inhibitors

Inhibitor	Proposed mechanism	Reference
Proteinases	degradation of DNA polymerase	Powell et al. 1994
IgG	binding to nucleic acids	Abu Al-Soud et al. 2000
Exopolysaccharides	binding to DNA polymerase	Monteiro et al. 1997
Lactoferrin	release of iron ions	Abu Al-Soud and Rådström 2001
Calcium ions	competing with Mg^{++} as polymerase cofactor	Bickley et al. 1996
Phenol	denaturation of DNA polymerase	Katcher and Schwartz 1994
EDTA	chelation of Mg^{++} ions	Rossen et al. 1992
Heparin	binding to nucleic acids	Satsangi et al. 1994

SAMPLE PREPARATION

A great deal of effort is being focused on the pretreatment of samples to reduce PCR inhibition, with little understanding of the mechanisms by which the inhibitors act (for review, see Lantz et al. 1994a). The treatments aim either to reduce the effect of the inhibitors, to remove them, or to provide a combination of both. For example, aqueous two-phase systems (Lantz et al. 1994b), boiling (Magistrado et al. 2001), density gradient centrifugation (Lindqvist et al. 1997), dilution (Abu Al-Soud et al. 1998), DNA extraction (Dahlenborg et al. 2001), filtration (Lantz et al. 1999), immunological techniques (Nogva et al. 2000), and PCR-compatible enrichment media (Payne and Kroll 1991; Knutsson et al. 2002b) are some of the techniques that are presently used. Particular attention should be paid to the final concentration of target molecules and inhibitory substances. The optimal method of sample preparation will (1) concentrate the target molecules to suit the practical operating range of the PCR, (2) remove or reduce the effects of PCR-inhibitory substances, (3) minimize sample variation, and (4) be simple to perform.

Many of the currently used sample preparation methods are technically challenging and time-consuming. Rapid and robust sample preparation methods are needed before PCR can be used for routine analysis to facilitate detection of targets in recalcitrant biological samples.

CHOICE OF DNA POLYMERASE AND AMPLIFICATION FACILITATORS

As an alternative to optimizing sample preparation, PCR inhibition can sometimes be overcome by using inhibitor-tolerant thermostable polymerases or including amplification facilitators in the reaction (Abu Al-Soud and Rådström 1998, 2000). Thermostable DNA polymerases vary in their tolerance of inhibitors. For example, the DNA polymerases from *Thermus aquaticus*, Ampli*Taq* Gold and *Taq*, are completely inactive in the presence of 0.004% (v/v) human blood, whereas the Hot*Tub*, *Pwo*, r*Tth*, and *Tfl* DNA polymerases are unaffected by the presence of at least 20% (v/v) blood (Abu Al-Soud and Rådström 1998). A number of thermostable DNA polymerases with various properties are now commercially available (Table 2). Table 3 presents a survey of PCR applications that have benefited from a careful choice of DNA polymerase and/or inclusion of facilitators in the reaction mix.

Amplification facilitators have been found to improve the specificity of PCR and to allow the amplification of GC-rich regions (Sarkar et al. 1990; Varadaraj and Skinner 1994). Facilitators can increase the fidelity of DNA synthesis (Wu and Yeh 1973), as well as enhance amplification efficiency in the presence of PCR-inhibitory substances (Abu Al-

TABLE 2. The thermostable DNA polymerases

Polymerase	Source	Molecular weight (kD)	Thermal stability[a]	Exonuclease activity	References
AmpliTaq[c]	Thermus aquaticus	94	95°C/40 min	5′–3′	Barnes 1994
Deep Vent[R,b]	Pyrococcus sp. GB-D	91	100°C/3.4 h	3′–5′	Barnes 1994; Takagi et al. 1997
DyNAzyme[d]	Thermus brockianus			5′–3′	
Pfu[e]	Pyrococcus furiosus	90	100°C/2.9 h	3′–5′	Klimezak et al. 1986
Pwo[f]	Pyrococcus woesei	90	100°C/2 h	3′–5′	Perler et al. 1996
Stoffel fragment[c]	Thermus aquaticus	61	95°C/80 min	–	Lawyer et al. 1989; Vainshtein et al. 1996
Tfl/HotTub[g,h]	Thermus flavus	94	70°C/2 h	5′–3′	Kaledin et al. 1981
rTth[c]	Thermus thermophilus	94	95°C/20 min	5′–3′	Rüttimann et al. 1985; Wnendt et al. 1990
Ultma[c]	Thermotoga maritima	70	97.5°C/50 min	3′–5′	Perler et al. 1996
Vent[R](Exo–)[b]	Thermococcus litoralis	90	100°C/2 h	–	Mattila et al. 1991; Barnes 1994; Perler et al. 1996
Tth[i]	Thermus thermophilus	95		5′–3′	Rüttimann et al. 1985
Hot start					
AmpliTaq Gold[c]	Thermus aquaticus	94	95°C/40 min	5′–3′	Kebelmann-Betzing et al. 1998
PlatinumTaq[h]	Thermus aquaticus	94	95°C/40 min	5′–3′	Westfall et al. 1995, 1998
FastStart	Thermus aquaticus				
Enzyme mixture					
Expand PCR system[f]	Thermus aquaticus	95	100°C/5 min	5′–3′	Barnes 1994
	Pyrococcus woesei	90	100°C/2 h	3′–5′	
ELONGASE[h]	Thermus aquaticus	94	95°C/40 min	5′–3′	Westfall et al. 1995
	Pyrococcus sp. GB-D	91	100°C/3.4 h	3′–5′	
TaqPlus PCR System[e]	Thermus aquaticus	94	95°C/40 min	5′–3′	
	Pyrococcus furiosus	90	100°C/2.9 h	3′–5′	

These data were obtained from published data and from commercial sources.

[a]Half-life of the enzyme at the specified temperature.

[b]Deep Vent[R] is the registered trademark of New England Biolabs, Beverly, Maryland.

[c]AmpliTaq, AmpliTaq Gold, Stoffel fragment, and rTth are registered trademarks of Applied Biosystems, Foster City, California. rTth can be used as reverse transcriptase in the presence of Mn[++] and as DNA polymerase after chelation of Mn[++] by EGTA and addition of Mg[++].

[d]DyNAzyme is a registered trademark of Finnzymes Oy, Riihitontuntie, Finland.

[e]Pfu and TaqPlus PCR System are registered trademarks of Stratagene, La Jolla, California.

[f]Pwo, Tth, and Expand PCR system are registered trademarks of Roche Diagnostics, Mannheim, Germany. Expand PCR system includes a group of products such as Expand High fidelity, Expand Long template, and Expand 20 kb[PLUS].

[g]Tfl is a registered trademark of Promega, Madison, Wisconsin.

[h]ELONGASE and PLATINUMTaq are registered trademarks of Life Technologies, Inc., Rockville, Maryland.

[i]HotTub is a registered trademark of Amersham Pharmacia Biotech, Cleveland, Ohio.

TABLE 3. Examples of DNA polymerases and/or PCR facilitators used in different PCR-based applications

Application	Objective	DNA polymerase	PCR facilitator	Reference
Diagnostic PCR	overcome inhibition from meat, blood, and feces	r*Tth* (Applied Biosystems)	BSA, gp32, Tween 20, Betaine, PEG 400	Abu Al-Soud and Rådström 1998; 2000
	overcome inhibition from vitreous fluids	*Tth* (Promega Corporation)	TritonX-100	Wiedbrauk et al. 1995
	increase sensitivity	*Tth* (Roche Diagnostics)	—	Shames et al. 1995; Kim et al. 2001
	increase sensitivity	r*Tth* (Applied Biosystems)	—	Dahlenborg et al. 2001
	increase specificity and sensitivity	*Tth* (MBI Fermentas)	W1 detergent (Gibco-BRL), BSA	Laigret et al. 1996
	overcome inhibition from phenol	*Tth* (Amersham)	—	Katcher and Schwartz 1994
Reverse Transcription-PCR	increase sensitivity	r*Tth* (Applied Biosystems)	—	Schwab et al. 2001
	reduce intra-PCR variability and increase sensitivity	*Pwo* (Roche Diagnostics)	—	Mullan et al. 2001
	perform PCR of GC-rich sequences	3 different DNA polymerases	Betaine	Henke et al. 1997
Cloning and sequencing	reduce amplification errors	*Pfu* (Stratagene)	—	Cline et al. 1996, Shafikhani 2002
	reduce amplification errors	*Pfu* (Stratagene)	—	
Genotyping	improve amplification capacity	DyNazyme (Finnzymes)	—	Zsolnai and Fesus 1997
	increase yield and specificity	AmpliTaqGold (Applied Biosystems)	—	Moretti et al. 1998
Real-time PCR	reveal polymorphism	Dynazyme EXT (Finnzymes)	Betaine	Diakou and Dovas 2001
	overcome inhibition from sludge	AmpliTaqGold (Applied Biosystems)	Polyvinylpyrrolidone (PVP), gp32	Monpoeho et al. 2000
	overcome inhibition from buffered peptone water (BPW)	r*Tth* (Applied Biosystems)	Glycerol	Knutsson et al. 2002c
	enable detection	15 different DNA polymerases	—	Kreuzer et al. 2000
	increase sensitivity	Platinum *Taq* (Gibco BRL)	—	Hein et al. 2001

Soud and Rådström 2000). Bovine serum albumin (BSA), for example, has proved to be a particularly potent facilitator by relieving the inhibitory effects of hemoglobin and lactoferrin (Abu Al-Soud and Rådström 2001). The ability of BSA to relieve inhibition may be related, at least in part, to its ability to bind inhibitors (Akane et al. 1994). Thus, the PCR-inhibitory effects of biological samples can be reduced, or even eliminated, by the use of an appropriate combination of DNA polymerase and amplification facilitator(s). The following sections discuss optimization strategies for overcoming PCR inhibition in conventional and real-time PCR.

CONVENTIONAL PCR

PCR-inhibition Studies

PCR inhibitors originating from the sample itself or from sample preparation, e.g., residual phenol from the DNA purification step, have been shown to affect both the yield and specificity of PCR (Katcher and Schwartz 1994). The effect of PCR inhibitors can be studied by increasing the concentration of the purified template DNA, and/or by adding different concentrations of the PCR-inhibitory substance. Increasing the concentration of target DNA may overcome the effect of inhibitors that interfere with DNA and/or bind reversibly to the DNA-binding domain of the DNA polymerase. Adding different concentrations of inhibitory substances such as blood, feces, or heme to reactions allows their inhibitory effect to be evaluated. Characterization of PCR-inhibitory effects is necessary for the development of more efficient sample preparation methods, which will eliminate the need for extensive processing of biological samples prior to diagnostic PCR.

Detection Limit and Detection Probability

In diagnostic PCR applications, "sensitivity" and "detection limit" are loosely defined and often used interchangeably. The terms are related and refer to the lower concentration range of amplification of target DNA/cells. Both sensitivity and detection limits are strongly affected by reaction conditions, the quality of the target DNA, and the presence of extraneous matter (Schmidt 1997). When studying the sensitivity and detection limit of a reaction, it should be borne in mind that PCR employs very small sample volumes. When analyzing a sample for the presence of a microorganism, 5 µl might be a typical sample volume. If the detection limit of the assay is stated to be 1 CFU/ml, the probability of the target being present in a 5-µl PCR sample is very low; i.e., one positive reaction out of 200 (assuming no concentration step is included in the pre-PCR processing of the sample). To improve the reliability of the assay, the detection limit should be viewed in association with the probability of detecting the target DNA/cell at a certain concentration. For instance, in Figure 1, the vertical line illustrates the detection level at 95% probability.

Mathematical and statistical models can be used to define the method detection limit (MDL) of an assay, which provides an idea of how well an assay works and how well it can be reproduced. It is determined from replicate analyses and is based on statistical analyses, which incorporate an acceptable level of risk of false positive or false negative results. The advantage of using models is that the results can be interpreted objectively. This is especially important for the identification of false negatives in microbiological assays. A model was used to determine the MDL of a PCR detection assay for *Cryptosporidium parvum* (Walker et al. 1998). However, other models have also been used to describe the detection probability of a PCR assay. Since conventional PCR, based on agarose gel electrophoresis, is a qual-

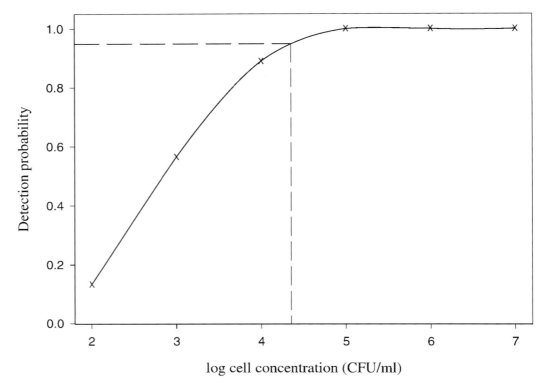

FIGURE 1. Detection probability of target cells (log CFU/ml) using a PCR assay. The PCR detection is described by a logistic regression model, estimated from a number of replicate experiments, and represented by the solid line. By plotting the concentration of target against the observed relative frequencies (X) of positive PCR detection calculated as the number of positive samples divided by the total number of replicates, it can be verified that the model applied fits the observed data. The model can then be used to describe the detection probability of the PCR studied and to calculate the detection probability at various target concentrations or to determine the probability of detecting a certain amount of target, i.e., the dashed line indicates the concentration of cells (log CFU/ml) at a detection probability of 0.95.

itative method, the response is binary, being 1 for detection and 0 for no detection. By assuming that the response is a binary random variable and that it follows a binomial distribution, a logistic regression model has been applied to describe the detection probability of a *Yersinia enterocolitica* PCR assay (Knutsson et al. 2002a).

PCR Optimization

In addition to characterizing the general effects of inhibitors and the detection probability of PCR assays, the individual demands of each PCR-based application must be met. For instance, in applications that require repeated amplification of the same target sequence, the prevention of carryover contamination is of great importance. Furthermore, studying the nucleic sequence of a genome in, for example, sequencing applications requires highly accurate PCR. Other examples are the need for high sensitivity or specificity. To ensure a reliable method, all the steps in the PCR procedure should be optimized for the intended use.

The choice of a suitable DNA polymerase and buffer has been shown to affect many PCR-based applications, such as diagnostic PCR, genotyping using restriction fragment length polymorphism (RFLP) (Zsolnai and Fesus 1997), short tandem repeats (STR)

(Moretti et al. 1998), and randomly amplified polymorphic DNA (RAPD) (Diakou and Dovas 2001), as well as cloning, sequencing, RT-PCR (Haag and Raman 1994), quantitative PCR, and real-time PCR (Table 2) (Kreuzer et al. 2000). The various DNA polymerases differ in many features that are important for PCR, such as susceptibility to PCR inhibitors (Abu Al-Soud and Rådström 1998), fidelity (Cline et al. 1996), the ability to incorporate modified bases (Sakaguchi et al. 1995), and termination of primer extension (Modin et al. 2000). As discussed earlier, thermostable DNA polymerases differ in their tolerance of inhibitors (Abu Al-Soud and Rådström 1998), and the polymerase of choice will depend on the application being used. It is also important to optimize the buffer conditions of a reaction. Using the most suitable polymerase under the correct buffer conditions is critical to the success of RT-PCR (Brooks et al. 1995). For cloning and sequencing applications, where high fidelity is required, *Pfu* DNA polymerase has been found to reduce amplification errors (Shafikhani 2002).

REAL-TIME PCR

Amplification Efficiency and Linear Range

The development of real-time PCR has provided new opportunities to study the PCR process, because it is now possible to follow the amplification of the product by monitoring accumulation of a fluorescent marker, which is incorporated into the amplicon during the reaction. In contrast to end-point measurements in conventional PCR, real-time PCR measures the first significant increase in fluorescence, or the point at which the fluorescence crosses a certain threshold level. This level is known as the Cp (Wittwer et al. 1997; Kyger et al. 1998) or Ct value (Heid et al. 1996). By measuring the threshold values for samples of known concentration, standard curves can be produced (Fig. 2). From these standard curves, new information can be derived. These data illustrate the linear range of amplification; i.e., the range over which the logarithm of the target concentration versus the threshold value forms a linear relationship. Furthermore, the slope of the standard curve over the linear range can be used to determine the amplification efficiency (AE), using the following equation:

$$AE = 10^{-1/\text{slope}} - 1$$

Quantitative PCR

For absolute quantitative real-time PCR, comparison with a standard curve is required, whereas for relative quantification, comparison with a reference target/gene is sufficient (Orlando et al. 1998). Small differences in amplification efficiency can have a significant impact on the productivity of a PCR. For example, a 5% difference in amplification efficiency between two targets with identical starting concentrations can lead to a twofold difference in amplicon concentration after just 26 cycles (Freeman et al. 1999). Differences in amplification efficiency between the target and reference will lead to mistakes in relative quantification unless corrections are made according to newly developed methods (Pfaffl 2001). The length of the standard curve, i.e., the linear range of amplification, influences mainly absolute quantification. As with any method that employs a standard curve, it is important that the linear range be as wide as possible, because for correct quantification, the unknown samples must fall within this window.

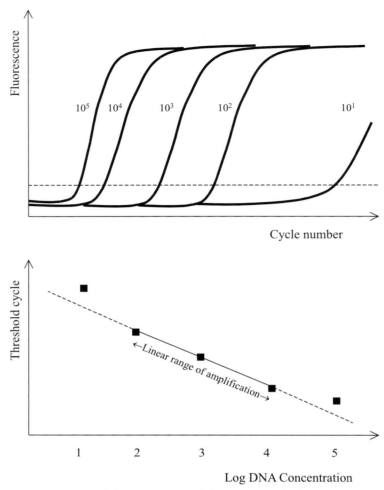

FIGURE 2. Schematic overview of the generation of data and standard curves by real-time quantitative PCR.

PCR Optimization

Real-time PCR offers several novel ways to study PCR inhibition and the effects of sample preparation. By studying the amplification efficiency and the linear range of amplification, possible PCR inhibition can be easily spotted. For example, by comparing these two parameters for a DNA dilution series in water and another in carcass rinse from poultry, it is clear that the reaction is less efficient and also the linear range of amplification is narrower in carcass rinse (Fig. 3A,B). In addition to studying the pre-PCR processing effect of removing inhibitors or reducing the inhibition (Fig. 3C), the effects of target concentration and homogenization of the sample should also be studied.

The effects of PCR inhibitors on DNA synthesis may be investigated using a reaction system based on a single-stranded poly(dA) template with an oligo(dT) primer annealed to the 3′ end (Abu Al-Soud and Rådström 2001). For objective characterization of PCR inhibitors, the use of mathematical models is recommended. Models enable (1) comparison of the effects of different inhibitors, (2) kinetic analysis of the DNA polymerase in the presence and absence of inhibitors, and (3) evaluation of the effect of amplification facilitators. The model developed by Knutsson et al. (2002c) was used to study the effect of buffered

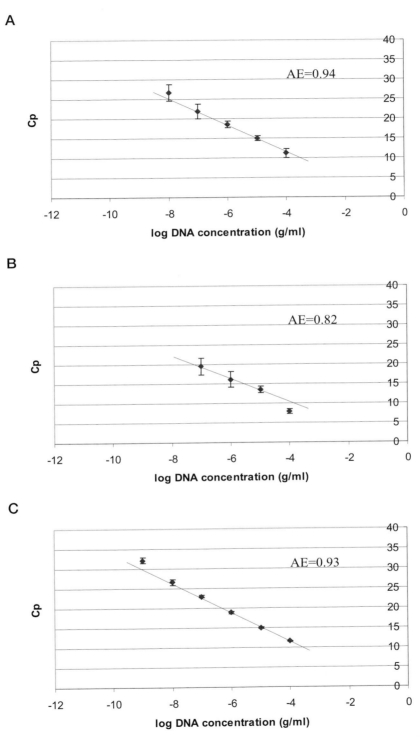

FIGURE 3. Standard curves made testing 12 samples with DNA concentrations between 1 mg/ml and 1 fg/ml. All the results that gave positive product peaks between 88°C and 92°C in the melting curve analysis are shown. Results show the linear range of amplification and the amplification efficiency. (*A*) *Taq* DNA polymerase with DNA samples in water. (*B*) *Taq* DNA polymerase with DNA samples in tenfold diluted carcass-rinse of chicken. (*C*) *Tth* DNA polymerase with DNA samples in tenfold diluted carcass-rinse of chicken.

peptone water (BPW) on real-time PCR performance. Results showed that BPW partially inhibited Ampli*Taq* Gold mixture, whereas r*Tth* maintained the same PCR performance as the positive control.

Due to the complexity of the real-time PCR mix, and the associated high sensitivity to PCR inhibition, nearly all sample preparation methods for this technique involve DNA purification, perhaps the most rigorous of sample treatments. However, even PCR using purified DNA templates can be enhanced by using an alternative DNA polymerase and/or changing the buffer composition (P. Wolffs and P. Rådström, unpubl.). The performance of real-time PCR using *Taq*Man probes relies completely on the exonuclease activity of the DNA polymerase. This means that only certain DNA polymerases are suitable for use in this technique (Kreuzer et al. 2000).

SUMMARY

To design a reliable and sensitive conventional or real-time PCR method, it is crucial to select the optimal DNA polymerase(s) and/or amplification facilitator(s) for the chemicals present in the samples during amplification. PCR optimization currently tends to focus on primer design, buffer, and thermocycling conditions to obtain specific PCR products. The amplification mixture can be modified by the addition of PCR facilitators, or by the choice of the appropriate polymerase, to improve proofreading activity, PCR yield, length of amplicon, etc., or to suit specific applications, such as cloning of PCR products, in vitro mutagenesis, in situ PCR, multiplex PCR, PCR ELISA, reverse-transcription PCR. However, less effort is being devoted to overcoming the effects of PCR-inhibitory compounds. Ideally, the number of steps required to generate PCR samples should be minimized. The use of appropriate DNA polymerases and amplification facilitators, in combination with optimized sampling techniques, reduces the amount of sample handling involved in the analysis. In the future development of PCR technology, research in pre-PCR processing is likely to expand in response to the growing demand for rapid, robust, and simple PCR protocols. A future challenge in pre-PCR processing strategies is to design PCR protocols that integrate sampling and DNA amplification in an automated manner. Furthermore, to monitor PCR performance in the presence of biological compounds, mathematical models are required for the objective interpretation of PCR results. Once the validation parameters of the PCR assay of interest have been studied, the specific sample matrix can be characterized with respect to PCR interference. After obtaining this information, pre-PCR processing strategies can be designed to optimize the protocol for a specific sample.

ACKNOWLEDGMENTS

This work was supported by grants from the Swedish Agency for Innovation Systems (VIN-NOVA), the Foundation of the Swedish Farmers Supply and Crop Marketing Cooperation (SL-stiftelsen), the Swedish Foundation for Strategic Research, and the Commission of the European Community within the program, "Quality of Life and Management of Living Resources," QLRT-1999.00226.

REFERENCES

Abu Al-Soud W. and Rådström P. 1998. Capacity of nine thermostable DNA polymerases to mediate DNA amplification in the presence of PCR-inhibiting samples. *Appl. Environ. Microbiol.* **64:** 3748–3753.

———. 2000. Effects of amplification facilitators on diagnostic PCR in the presence of blood, feces, and meat. *J. Clin. Microbiol.* **38:** 4463–4470.

————. 2001. Purification and characterization of PCR-inhibitory components in blood cells. *J. Clin. Microbiol.* **39:** 485–493.

Abu Al-Soud W., Jönsson L.J., and Rådström P. 2000. Identification and characterization of immunoglobulin G in blood as a major inhibitor of diagnostic PCR. *J. Clin. Microbiol.* **38:** 345–350.

Abu Al-Soud W., Lantz P.-G., Bäckman A., Olcén P., and Rådström P. 1998. A sample preparation method which facilitates detection of bacteria in blood cultures by the polymerase chain reaction. *J. Microbiol. Methods* **32:** 217–224.

Akane A., Matsubara K., Nakamura H., Takahashi S., and Kimura K. 1994. Identification of the heme compound copurified with deoxyribonucleic acid (DNA) from bloodstains, a major inhibitor of polymerase chain reaction (PCR) amplification. *J. Forensic Sci.* **39:** 362–372.

Barnes W.M. 1994. PCR amplification of up to 35-kb DNA with high fidelity and high yield from lambda bacteriophage templates. *Proc. Natl. Acad. Sci.* **91:** 2216–2220.

Bickley J., Short J.K., McDowel D.G., and Parkes H.C. 1996. Polymerase chain reaction (PCR) detection of *Listeria monocytogenes* in diluted milk and reversal of PCR inhibition caused by calcium ions. *Lett. Appl. Microbiol.* **22:** 153–158.

Brooks E.M., Sheflin L.G., and Spaulding S.W. 1995. Secondary structure in the 3′ UTR of EGF and the choice of reverse transcriptases affect the detection of message diversity by RT-PCR. *BioTechniques* **19:** 806–812, 814–815.

Cline J., Braman J.C., and Hogrefe H.H. 1996. PCR fidelity of *Pfu* DNA polymerase and other thermostable DNA polymerases. *Nucleic Acids Res.* **24:** 3546–3551.

Cone R.W., Hobson A.C., and Huang M.-L.W. 1992. Coamplified positive control detects inhibition of polymerase chain reactions. *J. Clin. Microbiol.* **30:** 3185–3189.

Dahlenborg M., Borch E., and Rådström P. 2001. Development of a combined selection and enrichment PCR procedure for *Clostridium botulinum* types B, E, and F and its use to determine prevalence in fecal samples from slaughtered pigs. *Appl. Environ. Microbiol.* **67:** 4781–4788.

Diakou A. and Dovas C.I. 2001. Optimization of random-amplified polymorphic DNA producing amplicons up to 8500 bp and revealing intraspecies polymorphism in *Leishmania infantum* isolates. *Anal. Biochem.* **288:** 195–200.

Freeman W.M., Walker S.J., and Vrana K.E. 1999. Quantitative RT-PCR: Pitfalls and potential. *BioTechniques* **26:** 112–122, 124–125.

Haag E. and Raman V. 1994. Effects of primer choice and source of *Taq* DNA polymerase on the banding patterns of differential display RT-PCR. *BioTechniques* **17:** 226–228.

Heid C.A., Stevens J., Livak K.J., and Williams P.M. 1996. Real time quantitative PCR. *Genome Res.* **6:** 986–994.

Hein I., Lehner A., Rieck P., Klein K., Brandl E., and Wagner M. 2001. Comparison of different approaches to quantify *Staphylococcus aureus* cells by real-time quantitative PCR and application of this technique for examination of cheese. *Appl. Environ. Microbiol.* **67:** 3122–3126.

Henke W., Herdel K., Jung K., Schnorr D., and Loening S.A. 1997. Betaine improves the PCR amplification of GC-rich DNA sequences. *Nucleic Acids Res.* **25:** 3957–3958.

Kaledin A.S., Sliusarenko A.G., and Gorodetskii S.I. 1981. Isolation and properties of DNA-polymerase from the extreme thermophilic bacterium *Thermus flavus*. *Biokhimiya* **46:** 1576–1584.

Katcher H.L. and Schwartz I. 1994. A distinctive property of *Tth* DNA polymerase: Enzymatic amplification in the presence of phenol. *BioTechniques* **16:** 84–92.

Kebelmann-Betzing C., Seeger K., Dragon S., Schmitt G., Moricke A., Schild T.A., Henze G., and Beyermann B. 1998. Advantages of a new *Taq* DNA polymerase in multiplex PCR and time-release PCR. *BioTechniques* **24:** 154–158.

Kim C.H., Khan M., Morin D.E., Hurley W.L., Tripathy D.N., Kehrli Jr., M., Oluoch A.O., and Kakoma I. 2001. Optimization of the PCR for detection of *Staphylococcus aureus nuc* gene in bovine milk. *J. Dairy Sci.* **84:** 74–83.

Klimezak L.J., Grummt F., and Burger K.J. 1986. Purification and characterization of DNA polymerase from the Archaebacterium *Thermoautotrophicum thermoautotrophicum*. *Biochemistry* **25:** 4850–4855.

Knutsson R., Blixt Y., Grage H., Borch E., and Rådström P. 2002a. Evaluation of selective enrichment PCR procedures for *Yersinia enterocolitica*. *Int. J. Food Microbiol.* **73:** 35–46.

Knutsson R., Fontanesi M., Grage H., and Rådström P. 2002b. Development of a PCR-compatible enrichment medium for *Yersinia enterocolitica*: Amplification precision and dynamic detection range during cultivation. *Int. J. Food Microbiol.* **72:** 185–201.

Knutsson R., Löfström C., Grage H., Hoorfar J., and Rådström P. 2002c. Modeling of 5′ nuclease real-time responses for optimization of a high-throughput enrichment PCR procedure for *Salmonella enterica. J. Clin. Microbiol.* **40:** 52–60.

Kreuzer K.A., Bohn A., Lass U., Peters U.R., and Schmidt C.A. 2000. Influence of DNA polymerases on quantitative PCR results using TaqMan probe format in the LightCycler instrument. *Mol. Cell. Probes* **14:** 57–60.

Kyger E.M., Krevolin M.D., and Powell M.J. 1998. Detection of the hereditary hemochromatosis gene mutation by real-time fluorescence polymerase chain reaction and peptide nucleic acid clamping. *Anal. Biochem.* **260:** 142–148.

Laigret F., Deaville J., Bove J.M., and Bradbury J.M. 1996. Specific detection of *Mycoplasma iowae* using polymerase chain reaction. *Mol. Cell. Probes* **10:** 23–29.

Lantz P.-G., Hahn-Hägerdal B., and Rådström P. 1994a. Sample preparation methods in PCR-based detection of food pathogens. *Trends Food Sci. Technol.* **5:** 384–389.

Lantz P.-G., Stålhandske F.I., Lundahl K., and Rådström P. 1999. Detection of yeast by PCR in sucrose solutions using a sample preparation method based on filtration. *World J. Microbiol. Biotechnol.* **15:** 345–348.

Lantz P.G., Tjerneld F., Borch E., Hahn-Hägerdal B., and Rådström P. 1994b. Enhanced sensitivity in PCR detection of *Listeria monocytogenes* in soft cheese through use of an aqueous two-phase system as a sample preparation method. *Appl. Environ. Microbiol.* **60:** 3416–3418.

Lawyer F.C., Stoffel S., Saiki R.K., Myambo K., Drummond R., and Gelfand D.H. 1989. Isolation, characterization and expression in *Escherichia coli* of the DNA polymerase gene from *Thermus aquaticus. J. Biol. Chem.* **264:** 6427–6437.

Lindqvist R., Norling B., and Thisted Lambertz S. 1997. A rapid sample preparation method for PCR detection of food pathogens based on buoyant density centrifugation. *Lett. Appl. Microbiol.* **24:** 306–310.

Magistrado P.A., Garcia M.M., and Raymundo A.K. 2001. Isolation and polymerase chain reaction-based detection of *Campylobacter jejuni* and *Campylobacter coli* from poultry in the Philippines. *Int. J. Food Microbiol.* **70:** 197–206.

Mattila P., Korpela J., Tenkanen T., and Pitkänen K. 1991. Fidelity of DNA synthesis by the *Thermococcus litoralis* DNA polymerase an extremely heat stable enzyme with proofreading activity. *Nucleic Acids Res.* **19:** 4967–4973.

Modin C., Pedersen F.S., and Duch M. 2000. Comparison of DNA polymerases for quantification of single nucleotide differences by primer extension assays. *BioTechniques* **28:** 48–51.

Monpoeho S., Dehee A., Mignotte B., Schwartzbrod L., Marechal V., Nicolas J.C., Billaudel S., and Ferre V. 2000. Quantification of enterovirus RNA in sludge samples using single tube real-time RT-PCR. *BioTechniques* **29:** 88–93.

Monteiro L., Bonnemaison D., Vekris A., Petry K.G., Bonnet J., Vidal R., Cabrita J., and Megraud F. 1997. Complex polysaccharides as PCR inhibitors in feces: *Helicobacter pylori* model. *J. Clin. Microbiol.* **35:** 995–998.

Moretti T., Koons B., and Budowle B. 1998. Enhancement of PCR amplification yield and specificity using Ampli*Taq* Gold DNA polymerase. *BioTechniques* **25:** 716–722.

Mullan B., Kenny-Walsh E., Collins J.K., Shanahan F., and Fanning L.J. 2001. Inferred hepatitis C virus quasispecies diversity is influenced by choice of DNA polymerase in reverse transcriptase-polymerase chain reactions. *Anal. Biochem.* **289:** 281–288.

Nogva H.K., Rudi K., Naterstad K., Holck A., and Lillehaug D. 2000. Application of 5′-nuclease PCR for quantitative detection of *Listeria monocytogenes* in pure cultures, water, skim milk, and unpasteurized whole milk. *Appl. Environ. Microbiol.* **66:** 4266–4271.

Orlando C., Pinzani P., and Pazzagli M. 1998. Developments in quantitative PCR. *Clin. Chem. Lab. Med.* **36:** 255–269.

Payne M.J. and Kroll R.G. 1991. Methods for the separation and concentration of bacteria from foods. *Trends Food Sci. Technol.* **2:** 315–319.

Perler F.B., Kumar S., and Kong H. 1996. Thermostable DNA polymerases. *Adv. Protein Chem.* **48:** 377–435.

Pfaffl M.W. 2001. A new mathematical model for relative quantification in real-time RT-PCR. *Nucleic Acids Res.* **29:** E45.

Powell H.A., Gooding C.M., Garret S.D., Lund B.M., and McKee R.A. 1994. Proteinase inhibition of the detection of *Listeria monocytogenes* in milk using the polymerase chain reaction. *Lett. Appl. Microbiol.* **18:** 59–61.

Rossen L., Nørskov P., Holmstrøm K., and Rasmussen O.F. 1992. Inhibition of PCR by components of food samples, microbial diagnostic assays and DNA-extraction solutions. *Int. J. Food Microbiol.* **17:** 37–45.

Rüttimann C., Cotoras M., Zaldivar J., and Vicuna R. 1985. DNA polymerases from the extremely thermophilic bacterium *Thermus thermophilus* HB-8. *Eur. J. Biochem.* **149:** 41–46.

Sakaguchi A.Y., Sedlak M., Harris J.M., and Sarosdy M.F. 1995. Cautionary note on the use of dUMP-containing PCR primers with *Pfu* and Vent$_R$ DNA polymerases. *BioTechniques* **21:** 368–370.

Sarkar G., Kapelner S., and Sommer S.S. 1990. Formamide can dramatically improve the specificity of PCR. *Nucleic Acids Res.* **18:** 7465.

Satsangi J., Jewell D.P., Welsh K., Bunce M., and Bell J.I. 1994. Effect of heparin on polymerase chain reaction. *Lancet* **343:** 1509–1510.

Schmidt B.L. 1997. PCR in laboratory diagnosis of human *Borrelia burgdorferi* infections. *Clin. Microbiol. Rev.* **10:** 185–201.

Schwab K.J., Neill F.H., Le Guyader F., Estes M.K., and Atmar R.L. 2001. Development of a reverse transcription-PCR-DNA enzyme immunoassay for detection of "Norwalk-like" viruses and hepatitis A virus in stool and shellfish. *Appl. Environ. Microbiol.* **67:** 742–749.

Shafikhani S. 2002. Factors affecting PCR-mediated recombination. *Environ. Microbiol.* **4:** 482–486.

Shames B., Fox J.G., Dewhirst F., Yan L., Shen Z., and Taylor N.S. 1995. Identification of widespread *Helicobacter hepaticus* infection in feces in commercial mouse colonies by culture and PCR assay. *J. Clin. Microbiol.* **33:** 2968–2972.

Takagi M., Nishioka M., Kakihara H., Kitabayshi M., Inouke H., Kawakami B., Oka M., and Imanaka T. 1997. Characterization of DNA polymerase from *Pyrococcus* sp. strain KOD1 and its application to PCR. *Appl. Environ. Microbiol.* **63:** 4504–4510.

Vainshtein I., Atrazhev A., Eom S.H., Elliott J.F., Wishart D.S., and Malcolm B.A. 1996. Peptide rescue of an N-terminal trunction of the Stoffel fragment of *Taq* DNA polymerase. *Protein Sci.* **5:** 1785–1792.

Varadaraj K. and Skinner D.M. 1994. Denaturants or cosolvents improve the specificity of PCR amplification of a G + C-rich DNA using genetically engineered DNA polymerases. *Gene* **140:** 1–5.

Walker M.J., Montemagno C., Bryant J.C., and Ghiorse W.C. 1998. Method detection limits of PCR and immunofluorescence assay for *Cryptosporidium parvum* in soil. *Appl. Environ. Microbiol.* **64:** 2281–2283.

Westfall B., Darfler M., Solus J., and Xu R.H. 1998. Biochemical characterization of Platinum™ *Taq* DNA polymerase. *BioTechniques* **20:** 17–18.

Westfall B., Sitaraman K., Berninger M., and Mertz L.M. 1995. Elongase™ reagents for amplification of long DNA templates. *BioTechniques* **17:** 62-65.

Wiedbrauk D.L., Werner J.C., and Drevon A.M. 1995. Inhibition of PCR by aqueous and vitreous fluids. *J. Clin. Microbiol.* **33:** 2643–2646.

Wilson I.G. 1997. Inhibition and facilitation of nucleic acid amplification. *Appl. Environ. Microbiol.* **63:** 3741–3751.

Wittwer C.T., Herrmann M.G., Moss A.A., and Rasmussen R.P. 1997. Continuous fluorescence monitoring of rapid cycle DNA amplification. *BioTechniques* **22:** 130–131, 134–138.

Wnendt S., Hartmann R.K., Ulbrich N., and Erdmann V.A. 1990. Isolation and physical properties of the DNA-directed RNA polymerase from *Thermus thermophilus* HB8. *Eur. J. Biochem.* **191:** 467–472.

Wu J.R. and Yeh Y.C. 1973. Requirement of a functional gene 32 product of bacteriophage T4 in UV, repair. *J. Virol.* **12:** 758–765.

Zsolnai A. and Fesus L. 1997. Enhancement of PCR-RFLP typing of bovine leukocyte adhesion deficiency. *BioTechniques* **23:** 380–382.

RT-PCR METHODS

Reverse transcriptase (RT)-PCR has become the method of choice for analysis of gene expression due to its sensitivity, speed, and minimal sample size. In addition, many improvements have made RT-PCR quantitative, as described in the real-time PCR chapters of this section. This methodology greatly expands the applications of this technique. Significant advantages of RT-PCR over other methods to study gene expression are the high-throughput capacity of the method and the ability to measure multiple target sequences in each sample. Applications of RT-PCR related to the analysis of differentially expressed genes and as a tool to clone new cDNAs are discussed in other chapters of this book.

Other methods have been developed for the purpose of quantifying RNA, including NASBA (van Gemen et al. 1994) and bDNA (Urdea 1993). These methodologies are used to measure a defined RNA species, usually a viral genomic RNA. Because NASBA and bDNA are generally used in clinical settings rather than in research laboratories, they are not discussed further in this book.

IS RT-PCR THE RIGHT TOOL?

Before beginning a study of analysis of gene expression, it is worth determining whether RT-PCR is the best technique to answer the research question. First, is the focus of the experiment the regulation of a specific gene or genes, or is it changes in the expression pattern following a researcher-defined event? If the answer to this question is the former, an RT-PCR is the best option. When focusing on known genes, RT-PCR provides greater accuracy and throughput than more conventional methods, such as northern blot analysis. If the answer is the latter, we recommend using one of the differential display methods in Section 5, or an expression array method, which can be prepared by the method described in "Applying PCR for Microarray-based Gene Expression Analysis" (p. 281).

Second, RT-PCR is the method of choice if the mRNA under study is expressed at low levels. Low-level expression of an mRNA species may arise if a few cells in the tissue express the gene of interest or if all cells express the gene at low levels. In either case, RT-PCR is the most sensitive technique available to detect the mRNA of interest. When studying gene expression in this low-level range, it is important to determine whether this expression is biologically relevant. One approach to this question is to identify the cells expressing the mRNA under study. Here, in situ methods, such as those described in "In Situ Localization of Gene Expression Using Laser Capture Microdissection" (p. 135) and "Reverse Transcriptase In Situ PCR" (p. 251), should be considered.

RT-PCR METHODOLOGY

All RT-PCR methods begin with the copying of RNA into cDNA. Methods for the preparation of quality RNA templates from any source are given in "RNA Purification" (p. 117) in Section 2. A variety of reverse transcriptases are available for the synthesis of cDNA prior to RT-PCR. These enzymes include Moloney murine leukemia virus (MMLV) RT, avian myeloblastosis virus (AMV) RT, and two RNase H-mutants of MMLV RT, which are marketed as SuperScript and SuperScript II. All of these enzymes have been used successfully in RT-PCR. Because RT-PCR depends on the reverse transcription of RNA into cDNA, maximum conversion of RNA into cDNA is of critical importance to the success of the experiment. Depending on the purpose and design of the RT-PCR, the requirement for synthesis of full-length cDNA varies. When using RT-PCR to clone new cDNAs, obtaining a full-length cDNA from reverse transcription of the RNA is of critical importance. This is also true if the target region selected for the amplification is near the 5´ end of the mRNA. The RNase H⁻ derivatives of MMLV RT can convert a greater proportion of the RNA into cDNA and can synthesize longer cDNAs than other enzymes (Kotewicz et al. 1988; Gerard et al. 1992). These enzymes also operate at a higher temperature (50°C) than their wild-type counterparts and AMV RT. This property allows the synthesis of longer cDNAs from RNA templates with secondary structure that are difficult to copy at lower temperatures.

Most of the RT-PCR assays are performed in a two-tube format: first reverse transcription of the RNA into cDNA is done, then an aliquot of the RT reaction is used for PCR. This has made it difficult to use UNG-based contamination control procedures. Additionally, the extra handling and pipetting add to the variability of the results. Recently, an improved one-tube RT-PCR enzyme system was developed. As described in "Amplification of RNA: High-temperature Reverse Transcription and DNA Amplification with a Magnesium-activated Thermostable DNA Polymerase" (p. 211), this method has the advantage of a single-tube format, is real-time PCR-compatible, and can tolerate the inclusion of contamination control systems.

The choice of a primer for cDNA synthesis is largely dictated by the specific application of the RT-PCR. A first-strand cDNA synthesis reaction may be primed using three different methods. The relative specificity of each approach influences the amount and variety of cDNA synthesized. For nearly all RT-PCR-based gene expression methods, random hexamer oligonucleotide primers are used. cDNA produced by this method is equally distributed over all the RNA present in the reaction, including ribosomal RNA. For differential display and cloning techniques such as "Fluorescent mRNA Differential Display in Gene Expression Analysis" (p. 327) or "Identification of Differential Genes by Suppression Subtractive Hybridization" (p. 297) a method specific for mRNA is required. Here oligo(dT), which hybridizes to 3´ poly(A) tails, is the method of choice. In some instances, an antisense gene-specific primer, which ideally converts into cDNA exclusively the RNA of interest, is utilized for the RT portion of the reaction. Examples of this can be found in "Beyond Classic RACE (Rapid Amplification of cDNA ends)" (p. 359).

GENE EXPRESSION

When measuring gene expression by RT-PCR, it is critical that the minute amounts of contaminating DNA do not interfere and add to the PCR product derived from the mRNA. Two methods have been developed to circumvent this problem. First is treatment of the RNA with RNase-free DNase, and a procedure for DNase treatment of RNA is provided in "RNA Purification (p. 117)." The second approach, when possible, is to place the PCR primers on different exons within the gene, eliminating the colinearity between the gene and the mRNA. This approach is discussed in detail in "PCR Primer Design" (p. 61).

The methods presented in this section demonstrate the real versatility and power of PCR. No other method yet developed can be applied to both the measurement of gene expression and the identification of yet-undefined sequences that are differentially regulated.

REFERENCES

Gerard G.F., Schmidt B.J., Kotewicz M.L., and Campbell J.H. 1992. cDNA synthesis by Moloney murine leukemia virus RNase H-minus reverse transcriptase possessing full DNA polymerase activity. *Focus* **14:** 91.

Kotewicz M.L., Sampson C.M., D'Alessio J.M., and Gerard G.F. 1988. Isolation of a cloned Moloney murine leukemia virus reverse transcriptase lacking ribonuclease H activity. *Nucleic Acids Res.* **16:** 265–277.

Urdea M. 1993. Synthesis and characterization of branched DNA (bDNA) for direct and quantitative detection of CMV, HBV, HCV and HIV. *Clin. Chem.* **39:** 725–736.

van Gemen B., van Beuningen R., Nabbe A., van Strijp D., Jurriaans S., Lens P., and Kievits T. 1994. A one-tube quantitative HIV-1 RNA NASBA nucleic acid amplification assay using electrochemiluminescent (ECL) labelled probes. *J. Virol. Methods* **49:** 157–168.

The Use of Internal Controls for Relative and Competitive Quantitative RT-PCR

Renée M. Horner

Ambion, Inc., Austin, Texas 78744

Reverse transcription coupled with the polymerase chain reaction (RT-PCR) amplifies cellular RNA for use in analyzing gene expression (Raemaekers 1996; Reidy et al. 1995). In this procedure, an RNA transcript is copied by using a retroviral reverse transcriptase to produce complementary DNA (cDNA). The cDNA can be amplified by a thermal stable DNA polymerase in a subsequent PCR. In this way, gene expression can be measured using small amounts of starting material, typically as low as 10–100 copies of the RNA transcript (Innis et al. 1990). To make the RT-PCR quantitative, the amount of product of the transcript of interest is compared to an internal control (relative RT-PCR) or a competing synthetic transcript (competitive RT-PCR).

Results for relative RT-PCR are expressed as x-fold difference or a percent difference relative to a control sample, giving semi-quantitative results. For relative quantitative RT-PCR to be meaningful, the PCR products must be measured during the exponential phase of amplification (Dukas et al. 1993). To normalize for sample-to-sample variation, internal controls such as β-actin, glyceraldehyde-3-phosphate dehydrogenase (GAPDH), or 18S rRNA are used (Suzuki et al. 2000).

In competitive RT-PCR, the results are expressed as copy number or mass of the transcript, providing absolute quantitation. Competitive RT-PCR requires the use of a synthetic RNA transcript, or competitor, as an exogenous control (Dillon and Rosen 1990; Freeman et al. 1999). The competitor is designed to have the same reaction kinetics as the endogenous transcript and is titrated into replicate RNA samples at increasing amounts to create a standard curve. By this method, the amount of PCR product for the sample and the competitor can be directly compared.

RELATIVE RT-PCR

The process of designing and optimizing a relative quantitative RT-PCR experiment is shown schematically in Figure 1. First, a detection method should be chosen. For relative quantitation, a large dynamic range of detectable PCR products is desirable. The inclusion of a ^{32}P-labeled nucleotide in the reactions produces a linear dynamic range of 4 orders of magnitude. Although the direct incorporation of $[\alpha\text{-}^{32}P]dCTP$ into the PCR product is slightly less sensitive than end labeling of the primers, the procedure is more straightforward and is the protocol described in this chapter. The resulting PCR fragments are run on a denaturing polyacrylamide gel. The use of a denaturing, rather than native, gel avoids the appearance of single- and double-stranded products as a doublet. The gel is transferred to filter paper and dried, and the product is quantified using a PhosphorImager or other densitometer. The bands of product can also be quantified by excising them from the gel and using a scintillation counter.

An alternative method for quantifying the PCR product involves the use of 2% agarose gels stained with ethidium bromide, accompanied by a gel documentation system and associated software. This method, however, is relatively insensitive, with a dynamic range of 1–2 orders of magnitude. Staining by SYBR Green or silver staining will increase the sensitivity of detection and the dynamic range.

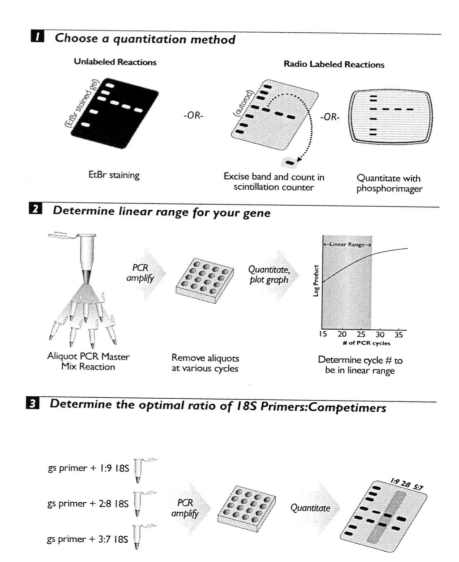

FIGURE 1. Experimental design for relative quantitative RT-PCR. (gs primers) Gene-specific primers. (Modified, with permission, from Ambion Gene Specific Relative RT-PCR Kit Instruction Manual version 0301 [©Ambion, Inc.].)

The next step of relative quantitation is to determine the linear range of amplification for the gene of interest. For the results of relative quantitation to be meaningful, the PCR products must be measured during the exponential phase of amplification, which must be determined for each target and primer pair experimentally. PCR products accumulate at a rate that is dependent on the amplification efficiency, where efficiency reflects the proportion of template molecules that are copied in a given amplification cycle. Factors that influence efficiency include length of the amplicon, GC content of the template, prime performance, and template quality. The exponential phase occurs when the amplification efficiency is at its maximum and is sustained over a period of cycles. As reagents become limiting, the amplification efficiency decreases, and the reaction plateaus. When targets vary in abundance, the cycle in which they enter exponential amplification will differ, and the cycle in which they plateau may differ. A reaction with an abundant template concentration, for example, will enter into exponential amplification earlier and may plateau earlier than one with a lower template concentration. Pilot experiments to determine the linear range of template amplification are performed using a sample with a high abundance of the transcript of interest. Replicate PCRs are assembled and removed from the thermal cycler every other cycle for analysis. By graphing the amount of product (on a log scale) versus the number of cycles, the linear range produces a straight line as indicated in Figure 2.

To normalize for sample-to-sample variation because of pipetting errors or errors in the concentration of the initial template, internal controls are used. Typically, this internal control is a housekeeping gene such as β-actin, GAPDH, or 18S rRNA. The choice of internal control is typically based on two considerations. First, the expression of the internal con-

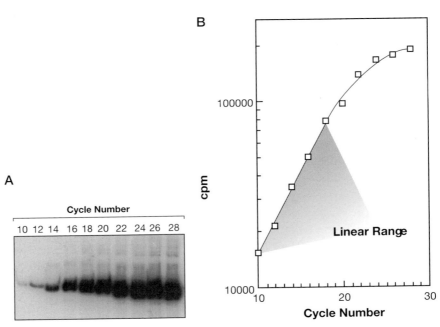

FIGURE 2. Determination of the linear range of amplification. (A) PCR master mix was prepared with [α-^{32}P]dCTP, according to Protocol 1. The master mix was split into 10 aliquots and subjected to PCR. Aliquots were removed from the thermal cycler at the indicated cycle numbers and run on a 5% denaturing polyacrylamide gel. (B) The products of the reactions in A were quantitated on a BioRad Molecular Imager. The log of the signal is plotted against the cycle number (note the log scale of the y axis). The linear range of amplification is indicated by the shaded area. (cpm) Counts per minute. (Reprinted, with permission, from Ambion Gene Specific Relative RT-PCR Kit Instruction Manual version 0301 [©Ambion, Inc.].)

trol cannot vary among the sample treatments or cell types. Second, the expression level of the internal control needs to be similar to that of the transcript of interest. 18S rRNA is commonly used because of its invariant expression across different treatments and different types of eukaryotic cell and tissue samples. The high abundance of housekeeping genes such as 18S rRNA, however, can interfere with the amplification of the transcript of interest because of competition for the reagents in the tube (Wong et al. 1994).

To avoid this competition for reagents, Competimers are added to the amplification reaction to attenuate the signal from the housekeeping gene. Competimers are modified primers that cannot be extended. Use of Competimers allows the overall PCR amplification efficiency of the 18S cDNA, for example, to be reduced without primer limitation and without the loss of relative quantitation. The primers:Competimers ratio needs to be determined experimentally to mimic the yield of the transcript of interest. Once this ratio has been determined, it can be used in conjunction with the cycling parameters of the linear range to quantitate the amount of transcript in RNA samples.

COMPETITIVE RT-PCR

The process of designing and optimizing a competitive, quantitative RT-PCR experiment is shown schematically in Figure 3. A competitor that is nearly identical to the endogenous transcript of interest is designed with a small deletion for identification purposes. Known quantities of the competitor RNA template are reverse-transcribed and PCR-amplified in the experimental RNA sample. The competitor and the endogenous transcript of interest compete equally for PCR reagents, so equivalent amounts of amplification product reflect equivalent amounts of starting template.

Competitive RT-PCR requires less optimization than relative RT-PCR. Because the reaction kinetics are the same, product will accumulate at the same rate for both the transcript of interest and the competitor, even after the PCR reagents become limiting. It is not necessary, therefore, to be in the linear range of amplification (Hirano et al. 2002). In pilot experiments, the competitor is titrated using 100-fold serial dilutions to get a standard curve of initial copy number. Because the dynamic range needed for this experiment is less than that of relative RT-PCR, ethidium bromide–stained agarose gels can be used to visualize the PCR products. A second pilot experiment is then performed using twofold serial dilutions of the competitor to narrow the quantitation range, thereby obtaining better resolution of quantitation.

The final competitive RT-PCR experiment entails reverse-transcribing and amplifying the diluted competitor and the endogenous gene of interest in the same tube to control for variation in the reverse transcription of the competitor and the endogenous transcript. As in relative RT-PCR, controls such as β-actin, GAPDH, or 18S rRNA can be used to normalize for sample-to-sample variations. These controls, however, are run in a separate tube from the transcript of interest.

PROTOCOL 1   RELATIVE RT-PCR: DETERMINATION OF THE LINEAR RANGE OF AMPLIFICATION

PART 1: SINGLE-STRAND cDNA SYNTHESIS

This protocol makes use of total RNA, because the use of 18S rRNA as an internal control in the subsequent protocols prohibits the use of poly(A) RNA. It is strongly suggested that the RNA be treated with DNase to remove any contaminating genomic DNA.

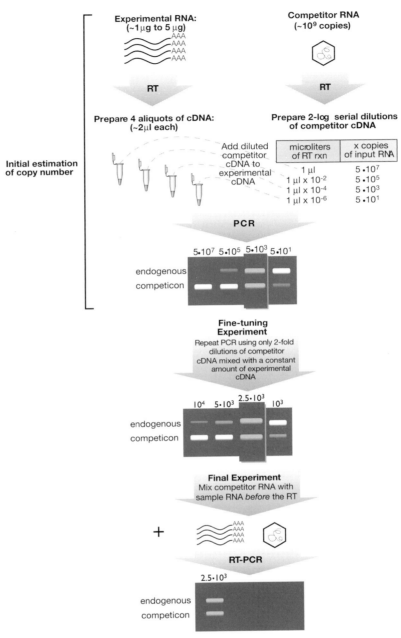

FIGURE 3. Experimental design for competitive RT-PCR. (Modified, with permission, from Ambion Competitive Quantitative RT-PCR Kit Instruction Manual [©Ambion, Inc.].)

MATERIALS

BUFFERS, SOLUTIONS, AND REAGENTS

RNase-free ddH$_2$O

10x RT-PCR buffer (RNase-free H$_2$O, 100 mM Tris-HCl<!>, pH 8.3, 500 mM KCl<!>, 15 mM MgCl$_2$<!>)

ENZYMES AND ENZYME BUFFERS
MMLV reverse transcriptase (100–200 units/μl)
SUPERaseIN RNase inhibitor (20 units/μl; Ambion)

NUCLEIC ACIDS AND OLIGONUCLEOTIDES
dNTP mix (10 mM, contains all four dNTPs)
Random primers (50 μM)
Total RNA (up to 2.5 μg)

ADDITIONAL ITEMS
Thin-walled PCR tubes

METHOD

1. Add 2 μl of 50-μM random primers to thin-walled PCR tubes on ice (1 tube per reaction).

2. Add up to 2.5 μg of total RNA.

3. Add 4 μl of 10 mM dNTP mix.

4. Bring the volume to 16 μl with RNase-free H_2O.

5. Heat for 10 minutes at 70°C to denature secondary structure, then immediately place on ice.

6. Add 2 μl of 10x RT-PCR buffer.

7. Add 1 μl of 20 units/μl SUPERaseIN.

8. Add 1 μl of 100–200 units/μl MMLV reverse transcriptase.

9. Incubate at 42°C for 1 hour.

10. Inactivate the reverse transcriptase by heating at 95°C for 10 minutes.

11. Proceed to the amplification step or store at –20°C.

PART 2: PCR AMPLIFICATION

When designing the gene-specific primers, choose primers that amplify a fragment between 200 bp and 500 bp. Designing the primers to span an intron–exon junction can reduce the potential for amplifying genomic DNA and facilitate the differentiation during gel electrophoresis of PCR products derived from RNA and genomic DNA. The primers should have melting temperatures within 2°C of each other and a GC content of 40–60%. Check the primers for secondary structure, for their ability to form primer–dimers, and for their specificity to the gene of interest. Set the annealing temperatures in the PCR program to be 2–5°C below the melting temperature of the primers. The hold times in these protocols are 30 seconds, but be aware that thermal cyclers with short ramp rates may need longer hold times (e.g., one minute for Stratagene's Robocycler).

This protocol is designed for 10 reactions with 10% excess, corresponding to the removal of samples from the thermal cyler after each odd-numbered cycle, starting at cycle 15 and ending at cycle 35.

MATERIALS

BUFFERS, SOLUTIONS, AND REAGENTS
RNase-free ddH$_2$O
10x RT-PCR buffer (100 mM Tris-HCl<!>, pH 8.3, 500 mM KCl<!>, 15 mM MgCl$_2$<!>)

ENZYMES AND ENZYME BUFFERS
Taq DNA polymerase (5 units/µl)

NUCLEIC ACIDS AND OLIGONUCLEOTIDES
dNTP mix (10 mM, contains all four dNTPs)
Gene-specific forward primer (5 µM)
Gene-specific reverse primer (5 µM)
Single-stranded cDNA (see previous Method)

RADIOACTIVE COMPOUNDS
[α-^{32}P]dCTP (10 µCi/µl) <!>

ADDITIONAL ITEMS
Thin-walled PCR tubes

SPECIAL EQUIPMENT
Programmable thermal cycler

METHOD

1. Prepare the PCR Master Mix:

cDNA reaction from previous method	11 µl
10x RT-PCR buffer	55 µl
10 mM dNTP mix	44 µl
5 µM gene-specific forward primer	22 µl
5 µM gene-specific reverse primer	22 µl
5 units/µl *Taq* polymerase	2.75 µl
RNase-free ddH$_2$O	387.75 µl
[α-^{32}P]dCTP (10 µCi/µl)	5.5 µl

2. Aliquot 50 µl of the PCR Master Mix into each of 10 thin-walled PCR tubes.

3. Run PCR in a suitable thermal cyler (e.g., GeneAmp PCR System 9700, Applied BioSystems) equipped with a heated cover.

   Use the following cycling parameters for PCR amplification:
   94°C for 30 seconds
   Annealing temperature for 30 seconds
   72°C for 30 seconds
   35 cycles

4. Remove samples and place on ice after each odd-numbered cycle starting with cycle 15 and ending with cycle 35.

5. Analyze the samples on a 6% acrylamide/8 M urea, denaturing polyacrylamide gel. Load 10 µl of each sample. Dry the gel on filter paper, and quantify the products with a PhosphorImager or X-ray film and a densitometer. Alternatively, excise the bands of product from the gel and quantify with a scintillation counter.

6. Plot the log of the signal on the *y* axis versus the cycle number on the *x* axis. Choose the cycle number that is in the center of the linear range. This cycle number will be used in subsequent reactions.

PROTOCOL 2   RELATIVE RT-PCR: DETERMINATION OF THE PRIMERS: COMPETIMERS RATIO

This protocol is used to determine the ratio of 18S rRNA primers to Competimers that mimics the yield of the transcript of interest. For most transcripts, a 3:7 ratio is appropriate. For rare targets, a ratio of 2:8 or 1:9 may be more appropriate. The PCR Master Mix is prepared for five reactions plus 10% excess.

MATERIALS

▼ CAUTION

See Appendix for appropriate handling of materials marked with <!>.

BUFFERS, SOLUTIONS, AND REAGENTS
RNase-free ddH$_2$O
10x RT-PCR buffer (100 mM Tris-HCl<!>, pH 8.3, 500 mM KCl<!>, 15 mM MgCl$_2$<!>)

ENZYMES AND ENZYME BUFFERS
Taq DNA polymerase (5 units/µl)

NUCLEIC ACIDS AND OLIGONUCLEOTIDES
cDNA from Protocol 1
dNTP mix (10 mM, contains all four dNTPs)
Gene-specific forward primer (5 µM)
Gene-specific reverse primer (5 µM)

RADIOACTIVE COMPOUNDS
[α-^{32}P]dCTP (10 µCi/µl) <!>

ADDITIONAL ITEMS
QuantumRNA Universal 18S standards (Ambion; includes 18S PCR Primer Pair and 18S PCR Competimers)
Thin-walled PCR tubes

SPECIAL EQUIPMENT
Programmable thermal cycler

METHOD

1. Prepare the primers:Competimers ratios as follows:

	1:9	2:8	3:7
18S rRNA primers	1 µl	2 µl	3 µl
18S rRNA Competimers	9 µl	8 µl	7 µl

2. Prepare the PCR Master Mix on ice:

cDNA	5.5 µl
10x RT-PCR buffer	27.5 µl
10 mM dNTP mix	22 µl
5 units/µl *Taq* polymerase	1.4 µl
RNase-free ddH$_2$O	177 µl
10 µCi/µl [α-^{32}P]dCTP	2.5 µl

3. Aliquot 42 μl of the PCR Master Mix into each of five thin-walled PCR tubes labeled 1–5.

4. Add 2 μl of the 5 μM gene-specific forward primer and 2 μl of 5 μM gene-specific reverse primer to tubes 1–4.

5. Add the 18S rRNA primer:Competimer mixtures as follows:

    4 μl of 1:9 to tube 2
    4 μl of 2:8 to tube 3
    4 μl of 3:7 to tubes 4 and 5

6. Perform PCR as described in Protocol 1, PCR Amplification. Use the optimal cycle number determined for the linear range of amplification.

7. Assess the results by denaturing gel electrophoresis, as described in Protocol 1. Identify the lane in which the level of 18S rRNA product is most similar to the level of the gene-specific product. The primers:Competimers ratio used in this sample should be used for the subsequent relative RT-PCR experiment.

| PROTOCOL 3 | RELATIVE RT-PCR: QUANTITATION OF SAMPLES |

RELATIVE RT-PCR: QUANTITATION OF SAMPLES

At this point, cycling parameters of the linear range have been determined and the 18S rRNA primer:Competimer ratio has been determined. This information can now be used to perform relative quantitative RT-PCR on your RNA samples. When preparing the master mixes, be sure to include 10% excess to account for pipetting error. The PCR Master Mix below is enough for nine reactions (four samples, four no-reverse transcription controls, and one no-template control) plus 10% excess.

MATERIALS

BUFFERS, SOLUTIONS, AND REAGENTS
RNase-free ddH$_2$O
10x RT-PCR buffer (100 mM Tris-HCl<!>, pH 8.3, 500 mM KCl<!>, 15 mM MgCl$_2$<!>)

ENZYMES AND ENZYME BUFFERS
MMLV reverse transcriptase (100–200 units/μl)
SUPERaseIN RNase inhibitor (20 units/μl; Ambion)
Taq DNA polymerase (5 units/μl)

NUCLEIC ACIDS AND OLIGONUCLEOTIDES
Total RNA (up to 2.5 μg per sample)
Random primers (50 μM)
dNTP mix (10 mM, contains all four dNTPs)
Gene-specific forward primer (5 μM)
Gene-specific reverse primer (5 μM)

RADIOACTIVE COMPOUNDS
[α-^{32}P]dCTP (10 μCi/μl)<!>

ADDITIONAL ITEMS
QuantumRNA Universal 18S standards (Ambion; includes 18S PCR Primer Pair and 18S PCR Competimers)
Thin-walled PCR tubes

SPECIAL EQUIPMENT
Programmable thermal cycler

METHOD

1. Assemble and run the single-strand cDNA synthesis reactions as described in Protocol 1. Use the same input amount of RNA that was used in the determination of the linear range for all of the samples.

2. On ice, prepare the PCR Master Mix:

10x RT-PCR buffer	50 µl
10 mM dNTP mix	40 µl
5 µM gene-specific forward primer	20 µl
5 µM gene-specific reverse primer	20 µl
18S rRNA primers + Competimers, at the ratio determined in Protocol 2	40 µl
5 units/µl *Taq* polymerase	2.5 µl
[α-^{32}P]dCTP (10 µCi/µl)	5 µl
RNase-free ddH$_2$O	312.5 µl

3. Aliquot 49 µl of the PCR Master Mix into each of 9 thin-walled PCR tubes.

4. Add 1 µl of ddH$_2$O to the no-template control tube.

5. Add 1 µl of each sample RNA to the corresponding no-reverse-transcription control tube.

6. Add 1 µl of each sample cDNA synthesis reaction from step 1 to its corresponding tube.

7. Perform PCR as described in Protocol 1, PCR Amplification. Use the optimal cycle number determined for the linear range of amplification.

8. Assess the results by denaturing gel electrophoresis, as described in Protocol 1. The no-template and no-reverse-transcription controls should produce no PCR product (see Troubleshooting). When quantifying the relative amounts of PCR products from the sample reactions, divide the signal obtained for the gene-specific product by the signal for the 18S rRNA. This will give a corrected relative value for the gene-specific product for each sample. These values can be compared among samples for determination of the relative expression of target RNA. By dividing the normalized signal for one sample by the normalized signal for a second sample, a comparison expressed in terms of *x*-fold difference or percent difference can be obtained. An example of relative quantitative RT-PCR is seen in Figure 4.

PROTOCOL 4   COMPETITIVE RT-PCR: COMPETITOR RNA CONSTRUCTION

PART 1: COMPETITOR DESIGN

A critical step in competitive RT-PCR is the synthesis and purification of the synthetic competitor. A competitor that is nearly identical to the endogenous transcript of interest is designed with a small deletion for identification purposes. The competitor is an RNA molecule designed to be reverse-transcribed and PCR-amplified with the same efficiency as the endogenous transcript of interest. The competitor RNA is generated by transcription of a DNA template, which is produced from PCR amplification of the endogenous target. For

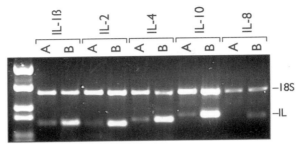

FIGURE 4. Example of relative quantitative RT-PCR. In this figure, mouse liver RNA was spiked with 1x (sample A) or 10x (sample B) sense-strand RNA for five human cytokines (IL-1β, IL-2, IL-4, IL-10, and IL-8). The reactions were reverse-transcribed using random primers. PCR was performed using a primer set specific for each cytokine and 18S rRNA primers and Competimers (Ambion). The resulting products were run on an ethidium bromide-stained agarose gel. (Reprinted, with permission, from Ambion 2003 Catalog [©Ambion, Inc.].)

this amplification, the upstream primer (5′ forward, P3 in Fig. 5) should include elements for efficient transcription by T7 RNA polymerase as well as regions homologous to the endogenous target. The addition of a GCG clamp (or a restriction site for cloning purposes) on the 5′ end of the forward primer may stabilize the promoter domain and provides increased transcription efficiency. The ten nucleotides following the T7 promoter are required so that transcription is initiated at the G nucleotide denoted at +1. These ten nucleotides must include only G and A residues. Ambion's Competitor Construction Kit contains modified C and U residues to make the competitor RNase-resistant; these modifications preclude the use of C or U in this region. The regions homologous to the endogenous target (B and C in Fig. 5) should be 18–20 nucleotides long. To facilitate separation of the competitor and target by gel electrophoresis in later steps, these two regions should flank a 10% deletion of the target. The total length of this primer is about 70 nucleotides.

The downstream primer (3′ reverse; P4 in Fig. 5) should be placed 50 nucleotides or more downstream of the reverse primer-binding site used in the competitive RT-PCR experiment. These extra sequences are necessary to ensure equivalent transcription of the competitive construct and the endogenous transcript.

The competitor construct then should be amplified using PCR to create enough template to use directly in a T7 RNA polymerase transcription reaction or, if restriction sites were added in the design of the P3 forward primer and P4 reverse primer, ligated into an appropriate vector.

PART 2: COMPETITOR RNA SYNTHESIS, PURIFICATION, AND QUANTITATION

In this protocol, the T7 RNA polymerase is used to prepare the competitor RNA template for use in competitive RT-PCR experiments (Milligan et al. 1987). If a plasmid vector is used as the template for transcription, it should first be linearized. Trace amounts of [α-^{32}P]ATP are added to the transcription reaction to aid in the purification and quantitation of the competitor. After the enzymatic reaction, the full-length competitor RNA is purified by denaturing gel electrophoresis. The competitor RNA concentration is determined in either micromolar amounts or copies/μl using a scintillation counter. The ratio of counts/minute (cpm) of the purified competitor versus the unpurified competitor is determined, and this ratio is multiplied by either the concentration of ATP/#A's in the competitor (for μM of competitor RNA) or by the number of molecules of ATP/#A's in competitor RNA (for copies/μl).

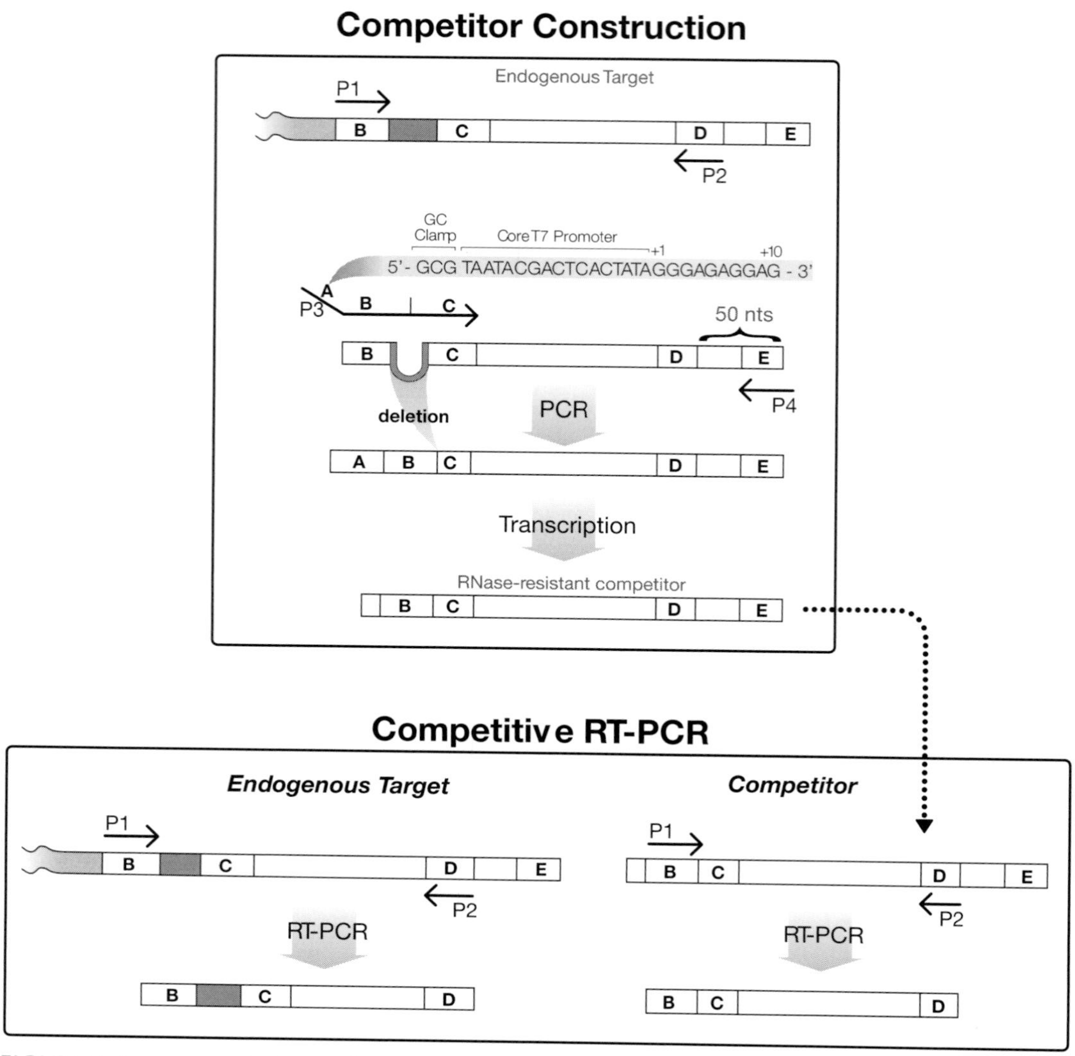

FIGURE 5. Competitor design and construction. In this figure, primers P1 and P2 are used to amplify the endogenous target and the competitor in RT-PCR. Primers P3 and P4 are used to produce the template for the competitor RNA. (Reprinted, with permission, from Ambion RT-PCR Competitor Construction Kit Instruction Manual Version 0106a [©Ambion, Inc.].)

MATERIALS

▼ CAUTION

See Appendix for appropriate handling of materials marked with <!>.

BUFFERS, SOLUTIONS, AND REAGENTS

Ethanol <!>

RNase-free ddH_2O

NUCLEIC ACIDS AND OLIGONUCLEOTIDES

DNA template (0.5–1.0 µg of linearized plasmid template or 0.1–0.5 µg of PCR template)

RADIOACTIVE COMPOUNDS

[α-^{32}P]ATP<!>, 10 mCi/ml

ADDITIONAL ITEMS

RT-PCR Competitor Construction Kit (Ambion; includes T7 RNA polymerase, 10x transcription buffer, 5x NTP solution, MnCl$_2$<!>, DNase I, gel loading buffer, and probe elution buffer)

METHOD

1. Prepare the transcription reaction at room temperature. Assemble the reagents in the order listed.

   RNase-free ddH$_2$O to bring the final reaction volume to 20 μl

10x Transcription buffer	2 μl
5x NTP solution	4 μl
DNA template	1–5 μl
[α-^{32}P]ATP (10 μCi/ml)	0.4 μl
T7 RNA polymerase	2 μl

2. Flick the tube to mix the reagents, and briefly centrifuge to collect the reaction mixture at the bottom of the tube.

3. Incubate the transcription reaction for 2–4 hours at 37°C.

4. Remove a 1-μl aliquot for later use in the determination of the competitor concentration.

5. Add 40 μl of probe elution buffer and 200 μl of ethanol to the transcription reaction.

6. Incubate at –80°C for 20 minutes.

7. Centrifuge at maximum rpm at 4°C to precipitate the transcription reaction products.

8. Aspirate the supernatant, wash the pellet with 70% ethanol, and air-dry the pellet.

9. Dissolve the pellet in 16 μl of RNase-free ddH$_2$O.

10. Heat-denature the resuspended pellet (containing the DNA template and the transcribed RNA) at 95°C for 2 minutes.

11. Add 2 μl of 10x transcription buffer and 2 μl of RNase-free DNase I (5 units/μl).

12. Flick the tube to mix the reagents, and briefly centrifuge to collect the reaction mixture at the bottom of the tube.

13. Incubate at 37°C for 1 hour.

Competitor Purification

14. Add 22 μl of gel loading buffer II to the DNase I reaction.

15. Denature the samples at 95°C for 3 minutes.

16. Run the sample on a 6% acrylamide/8 M urea, denaturing polyacrylamide gel until the bromophenol blue approaches the bottom of the gel.

17. Remove one glass plate from the gel, cover the gel with plastic wrap, and expose to X-ray film for 30 minutes to 2 hours.

18. Develop the film and align it under the gel. Cut the full-length RNA transcript out of the gel with a scalpel or razor blade.

19. Transfer the gel slice to a microcentrifuge tube. Add 300 μl of probe elution buffer and submerge the slice by briefly vortexing and centrifuging.

20. Incubate the tube overnight at 37°C.

21. Remove and discard the acrylamide gel slice from the tube. Add 600 μl of ethanol to the probe elution buffer containing the RNA transcript.

22. Incubate at –80°C for 20 minutes.

23. Centrifuge at maximum rpm for 15–20 minutes at 4°C.

24. Wash the pellet with 70% ethanol and allow to air-dry.

25. Dissolve the pellet in 15 μl of RNase-free ddH$_2$O.

Competitor RNA Quantitation

26. Transfer 1 μl of the transcription reaction and 1 μl of the purified RNA competitor to scintillation vials. Add scintillant fluid and count with a scintillation counter.

27. Use the resulting counts in the following equations to determine the competitor concentration in micromolar or in copies/μl.

μM competitor RNA = [cpm/μl of purified competitor RNA/cpm/μl of transcription reaction] × [500 μM ATP/#A's in the competitor RNA]

Copies of competitor RNA/μl = [cpm/μl of purified competitor/cpm/μl of transcription reaction] × [3 × 10^{14} molecules of ATP/μl/#A's in the competitor RNA]

28. The competitor should be stored at –20°C at concentrations greater than 6 × 10^{10} copies/μl or 0.1 μM, and should be stable for at least one year.

PROTOCOL 5   COMPETITIVE RT-PCR: ESTIMATION OF COPY NUMBER

In this pilot experiment, reverse transcription (RT) reactions for the sample RNA and competitor are performed separately. Then a constant amount of sample RT product is combined with a 2-log serial dilution of competitor RT product for the PCR. This procedure provides an approximate copy number for the sample. This value can be fine-tuned by repeating the experiment with a series of twofold dilutions of competitor. To control for sample-to-sample variations in RT efficiency, a final experiment should be performed in which competitor RNA is added to the sample RNA before reverse transcription.

PART 1: SINGLE-STRAND cDNA SYNTHESIS

MATERIALS

BUFFERS, SOLUTIONS, AND REAGENTS
RNase free ddH$_2$O
10x RT-PCR Buffer (100 mM Tris-HCl<!>, pH 8.3, 500 mM KCl<!>, 15 mM MgCl$_2$<!>)

ENZYMES AND ENZYME BUFFERS
MMLV reverse transcriptase (100–200 units/μl)
SUPERaseIN (20 units/μl) (Ambion)

NUCLEIC ACIDS AND OLIGONUCLEOTIDES
dNTP mix (10 mM, contains all four dNTPs)

Random primers (50 μM)
Total RNA, 1–5 μg

RADIOACTIVE COMPOUNDS
Competitive RNA transcript<!>, 10^9 copies (see Protocol 4)

ADDITIONAL ITEMS
Thin-walled PCR tubes

METHOD

1. Add 2 μl of 50 μM random primers to thin-walled PCR tubes on ice. Prepare one tube for the RNA sample and one tube for the competitive RNA transcript.

2. Add 1–5 μg of total RNA to the sample tube (maximum volume = 10 μl).

3. Add 10^9 copies of competitive RNA transcript to the appropriate tube (maximum volume = 10 μl).

4. Add 4 μl of 10 mM dNTP mix to each tube.

5. Bring the volume to 16 μl with RNase-free ddH$_2$O.

6. Heat for 10 minutes at 70°C to denature secondary structure, then immediately place on ice.

7. Add 2 μl of 10x RT-PCR buffer.

8. Add 1 μl of 20 units/μl SUPERaseIN.

9. Add 1 μl of 100–200 units/μl MMLV reverse transcriptase.

10. Incubate at 42°C for one hour.

11. Inactivate the reverse transcriptase by heating at 95°C for 10 minutes.

12. Proceed to the amplification step or store at –20°C.

PART 2: PCR AMPLIFICATION

The guidelines of this PCR amplification are similar to those described for relative RT-PCR (see Protocol 1). This protocol is designed for six reactions (with 10% excess). This will be adequate for one sample using four competitor dilutions, one no-competitor control, and one no-template control.

MATERIALS

▼ CAUTION

See Appendix for appropriate handling of materials marked with <!>.

BUFFERS, SOLUTIONS, AND REAGENTS
RNase-free ddH$_2$O
10x RT-PCR buffer (100 mM Tris-HCl<!>, pH 8.3, 500 mM KCl<!>, 15 mM MgCl$_2$<!>)

ENZYMES AND ENZYME BUFFERS
Taq DNA polymerase (5 units/μl)

NUCLEIC ACIDS AND OLIGONUCLEOTIDES
dNTP mix (10 mM, contains all four dNTPs)
Gene-specific forward primer (5 μM)
Gene-specific reverse primer (5 μM)
cDNA from above reverse transcription reaction

RADIOACTIVE COMPOUNDS

Competitive RNA reverse transcription reaction (undiluted = 5×10^7 copies/µl) <!>

ADDITIONAL ITEMS

Thin-walled PCR tubes

SPECIAL EQUIPMENT

Programmable thermal cycler

METHOD

1. Prepare the following PCR Master Mix:

10x RT-PCR buffer	35 µl
10 mM dNTP mix	28 µl
5 µM gene-specific forward primer	14 µl
5 µM gene specific reverse primer	14 µl
5 units/µl *Taq* polymerase	1.75 µl
RNase-free ddH$_2$O	236.25 µl

2. Aliquot 47 µl of the PCR Master Mix into each of 6 thin-walled PCR tubes labeled 1–6. Keep the tubes on ice.

3. Add 2 µl of the sample cDNA prepared above to tubes 1–5 (see Single-Strand cDNA Synthesis).

4. Serial-dilute the RT reaction containing the competitive RNA transcript as follows:

Undiluted:	1 µl of RT = 5×10^7 copies/µl
10^{-2} dilution:	dilute 1 µl of RT into 99 µl of TE = 5×10^5 copies/µl
10^{-4} dilution:	dilute 1 µl of 10^{-2} dilution into 99 µl of TE = 5×10^3 copies/µl
10^{-6} dilution:	dilute 1 µl of 10^{-4} dilution into 99 µl of TE = 5×10^1 copies/µl

5. Add 1 µl of the undiluted competitor to tube 1, corresponding to 5×10^7 copies.

6. Add 1 µl of the 10^{-2} dilution of competitor to tube 2, corresponding to 5×10^5 copies.

7. Add 1 µl of the 10^{-4} dilution of competitor to tube 3, corresponding to 5×10^3 copies.

8. Add 1 µl of the 10^{-6} dilution of competitor to tube 4, corresponding to 5×10^1 copies.

9. Add 1 µl of ddH$_2$O to tube 5 for a no-competitor control.

10. Add 3 µl of ddH$_2$O to tube 6 for a no-template control.

11. Run the PCR in a suitable thermal cyler equipped with a heated cover (e.g., GeneAmp PCR System 9700, Applied BioSystems).
    For PCR amplification use:
    94°C for 30 seconds
    Annealing temperature for 30 seconds
    72°C for 30 seconds
    30 cycles

12. Assess the results on a 2–2.5% agarose gel containing 1 µg/ml ethidium bromide. Load 5 µl of each reaction. Results are expressed in copy numbers or mass of transcript. Because the competitor is designed to have the same reaction kinetics as the gene of interest and is co-amplified with the gene of interest, the amount of PCR product for the sample and competitor in each reaction can be directly compared. The intensity of the PCR product obtained from the competitor is compared to replicate samples of the

endogenous RNA transcript. Equivalent intensities represent the amount of endogenous message present in the RNA sample. An example of a competitive RT-PCR experiment is shown in Figure 6.

13. To control for variations in the efficiency of reverse-transcribing the competitor and the endogenous transcript, perform a final RT-PCR experiment in which the sample RNA and the competitor RNA are reverse-transcribed in the same tube. Use several amounts of competitor RNA that are close to the amount indicated by the initial RT-PCR experiments, and perform the reverse transcription and amplification steps as described above. As before, the amount of competitor RNA that produces a signal of equal intensity to that produced by the endogenous message represents the amount of endogenous message in the sample.

◗ TROUBLESHOOTING

General RT-PCR

Low yield in RT-PCR can be caused by degraded or impure RNA. Inhibitors of reverse transcriptase include guanidinium, proteinase K, and alcohol, which can be carried over from the RNA isolation procedure. Typically, the contaminant can be removed by phenol:chloroform extraction, reprecipitation, and washing of the pellet.

Poor primer design or nonoptimized PCR conditions can lead to poor yield and nonspecific product formation. The magnesium concentration can be varied from 1.5 mM to 4 mM to optimize the PCR chemistry. The annealing temperature of the PCR can be raised to alleviate nonspecific priming. Additionally, the hold times used in the PCR cycling may also have to be adjusted to obtain the highest possible yields. Thermal cyclers with short ramp times may need longer hold times at each temperature.

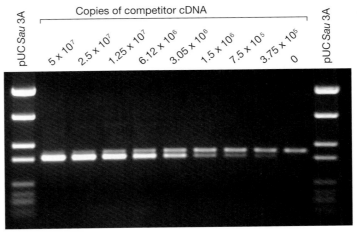

FIGURE 6. Example of competitive quantitative RT-PCR. A total of 2 μg of experimental RNA was reverse-transcribed with random primers. One μl of the resulting cDNA was combined with twofold serial dilutions of competitor cDNA and subjected to 30 cycles of PCR. Then 5 μl of the resulting PCR products were run on a 2% ethidium bromide-stained agarose gel. In this example, the transcript of interest is estimated to be 3.05×10^6 copies. (Modified, with permission, from Ambion RT-PCR Competitor Construction Kit Instruction Manual Version 0106a [©Ambion, Inc.].)

A PCR signal from the no-reverse-transcriptase controls could signify genomic DNA contamination. In this case, the RNA samples should be digested with RNase-free DNase (Ambion) for 15 minutes at 37°C, followed by the addition of EDTA to 2.5 mM and incubation at 70°C to deactivate the enzyme (Kwok and Higuchi 1989).

False positives in the no-template controls can signify contamination of one of the reagents or laboratory equipment by the target. Careful laboratory practices are essential; for example, PCRs should not be stored or used in the same location where PCR setup occurs. DNA*Zap* (Ambion) can be used to eliminate nucleic acid contamination on laboratory surfaces such as PCR machines, micropipettors, and racks. See Chapter 1 for information about contamination control.

Competitive RT-PCR

If low quantities of RNA competitor template are obtained, first increase the amount of template in the T7 RNA polymerase reaction (up to 2.5 μg). Also, confirm that there are no C or U bases in the first 10 nucleotides after the T7 promoter. The presence of these bases in this position will decrease the efficiency and yield of the competitor RNA transcription reaction. The addition of 1–3 mM final concentration of $MnCl_2$ to the transcription reaction, and increasing the incubation time to 8–16 hours, may aid in increasing the yield. To prevent precipitation of DNA, $MnCl_2$ should not be added to a cold reaction tube. Instead, first heat the transcription reaction to 37°C for 2 minutes. Poor yields may also be obtained from GC-rich transcription templates. In this case, the addition of single-stranded binding protein (2.6 μg/μg of template) may increase transcription efficiency (Ben Aziz and Soreq 1990).

If the transcription product of the competitor is only slightly longer than expected, it is most likely because of an undetected cloning problem or salt effect. In the case of a salt effect, the transcription product should be re-precipitated and washed in 70% ethanol. If the transcription product of the competitor is much larger than expected, incomplete digestion of plasmid template may have occurred. Small amounts of supercoiled uncut plasmid can result in large amounts of RNA because of continuous transcription of template. In this case, the plasmid template should be re-digested.

SUMMARY

RT-PCR semiquantitative and quantitative gene expression has revolutionized the field of gene expression (Reidy et al. 1995; Orlando et al. 1998). Relative RT-PCR is semiquantitative in that comparisons between samples are expressed as *x*-fold difference. Competitive RT-PCR is absolute quantitation in that copy numbers or micromolar amounts are determined.

RT-PCR has advantages over more traditional techniques such as northern blot analysis and RNase protection assays, because greater sensitivity allows the use of small samples. Relative RT-PCR and competitive RT-PCR have advantages over new methods such as real-time quantitative RT-PCR because there is no need for specialized equipment (Ferre 1992). These advantages make the techniques described in this chapter ideal for many laboratories that wish to quantitate gene expression.

The disadvantage of using any RT-PCR is the introduction of enzymatic steps to the quantitation process. These steps lead to an additional level of uncertainty because of the potential for differing efficiencies of the reactions (Peccoud and Jacob 1996). With proper experimental design and the introduction of appropriate controls, however, many of these concerns can be alleviated.

ACKNOWLEDGMENTS

I thank Ellen Prediger, Lori Martin, and Heath Thomas for supplying materials used in this chapter. I also thank James W. Horner II for critical reading of the manuscript.

REFERENCES

Ben Aziz R.B. and Soreq H. 1990. Improving poor in vitro transcription form GC-rich genes. *Nucleic Acids Res.* **18:** 3418.

Dillon P.J. and Rosen C.A. 1990. A rapid method for the construction of synthetic genes using the polymerase chain reaction. *BioTechniques* **9:** 298–300.

Dukas K., Sarfati P., Vaysse N., and Pradayrol L. 1993. Quantitation of changes in the expression of multiple genes by simultaneous polymerase chain reaction. *Anal. Biochem.* **215:** 66–72.

Ferre F. 1992. Quantitative or semi-quantitative PCR: Reality versus myth. *PCR Methods Appl.* **2:** 1–9.

Freeman W.M., Walker S.J., and Vrana K.E. 1999. Quantitative RT-PCR: Pitfalls and potential. *BioTechniques* **26:** 112–125.

Hirano T., Haque M., and Utiyama H. 2002. Theoretical and experimental dissection of competitive PCR for accurate quantification of DNA. *Anal. Biochem.* **303:** 57–65.

Innis M., Gelfand D., Sninsky J., and White T., eds. 1990. *PCR protocols: A guide to methods and applications.* Academic Press, San Diego, California.

Kwok S. and Higuchi R. 1989. Avoiding false positives with PCR. *Nature* **339:** 237–238.

Milligan J.F., Groebe D.R., Witherell G.W., and Uhlenbect O.C. 1987. Oligonucleotide synthesis using T7 RNA polymerase and synthetic DNA template. *Nucleic Acids Res.* **15:** 8783–8798.

Orlando C., Pinzani P., and Mazzagli M. 1998. Developments in quantitative PCR. *Clin. Chem. Lab. Med.* **36:** 255–269.

Peccoud J. and Jacob C. 1996. Theoretical uncertainty measurements using quantitative polymerase chain reaction. *Biophys. J.* **71:** 101–108.

Raemaekers L. 1996. A commentary on the practical applications of competitive PCR. *Genome Res.* **5:** 91–94.

Reidy M.C., Timm E.A., Jr., and Stewart C.C. 1995. Quantitative RT-PCR for measuring gene expression. *BioTechniques* **18:** 70–76.

Suzuki T. Higgins P.J., and Crawford D.R. 2000. Control selection for RNA quantitation. *BioTechniques* **29:** 332–337.

Wong H., Anderson W.D., Cheng T., and Riabowol K.T. 1994. Monitoring mRNA expression by polymerase chain reaction: The "primer dropping" method. *Anal. Biochem.* **223:** 251–258.

Comparison of Four Real-time PCR Detection Systems: Bio-Rad I-Cycler, ABI 7700, Roche LightCycler, and the Cepheid Smartcycler

Katherine E. Templeton and Eric C.J. Claas

Department of Medical Microbiology, Leiden University Medical Centre, 2300 RC Leiden, The Netherlands

Most diagnostic PCR assays are quickly being transformed into real-time PCR assays, which analyze PCR products during amplification. Such an approach has many advantages over conventional PCR. When real-time PCR is used, post-PCR product analysis by gel electrophoresis, dot-blot hybridization, or enzyme immunoassay (EIA) detection of PCR products is no longer required, and the results are generated more quickly and with greater accuracy and reliability. In addition, the risk of contamination is reduced because there is no need to open the PCR tubes after amplification.

The most significant advantage of real-time PCR is that it is quantitative. Conventional PCR is not quantitative; quantification is possible only by using an internal standard and setting up a complicated competitive PCR. By comparison, real-time PCR is intrinsically quantitative (Schutten and Niesters 2001). Real-time PCR machines are fitted with spectrofluorometers to measure the amount of product at every cycle of a reaction in which an intercalating, fluorescent dye such as SYBR Green, which only emits light upon excitation when bound to double-stranded (ds) DNA, is used. Another approach is using fluorescently labeled oligonucleotide probes; e.g., TaqMan probes, molecular beacons, or Scorpion primers. These probes bind to their target sequence and fluoresce only when the target–probe interaction has been achieved. In all systems, the amount of emitted fluorescence is directly proportional to the amount of PCR product. The different methods for detecting fluorescence and measuring different fluorescent signals in a single assay enable the real-time PCR methodology to be used for quantification, multiplex reactions, and single-nucleotide polymorphism (SNP) analysis (Wittwer 2001).

INSTRUMENTATION

Several instruments are available for the detection of real-time PCR fluorescence. The first available, and currently the most popular, machines are the ABI 7700 (Applied Biosystems) and the Roche LightCycler. However, other systems have since become (and will become) available. Presented here is a comparison of the above-mentioned systems with the Bio-Rad I-Cycler IQ detection system and the Cepheid Smartcycler. Protocols for design and optimization of real-time PCR and application to different systems are also provided.

Comparison of the Different Systems

A summary of different systems capable of real-time PCR is shown in Table 1. Essentially, each real-time PCR instrument consists of a computer-controlled thermocycler integrated with a fluorescence detection system and dedicated software to analyze the results. Some systems can detect four different wavelengths (I-Cycler, Mx4000 [Stratagene], and Smartcycler) whereas others can only detect two different wavelengths (LightCycler). The LightCycler and the Smartcycler are capable of performing rapid-cycle real-time PCR because the reactions are set up in capillaries or specifically designed tubes. Both have optimized heating–cooling characteristics. A complete amplification protocol can be performed in 15–30 minutes.

The Smartcycler is a combination of 16 individual, one-tube real-time PCR units. It is capable of performing a different PCR program on each of 16 reaction tubes. This is very useful for a rapid optimization of the assay, as many variables can be tested at the same time. The Bio-Rad I-Cycler IQ instrument can perform real-time amplification with a temperature gradient for specific PCR steps, allowing the optimizing of real-time PCR assays.

The spectrofluorometers in the thermocyclers have a number of differences, as detailed in Table 1. Laser-based systems are tuned to excite each fluorophore at a specific wavelength and provide maximum efficiency. Lamp-based systems provide a broad excitation range that can be filtered to work with a number of fluorophores. The laser source not only gives brighter illumination to the fluorophore signal, but also produces less background noise.

All real-time PCR machines detect fluorescence at each cycle of PCR, giving a real-time output of fluorescence against cycle number. However, the ABI 7700 units do not give a real-time output of fluorescence data; instead, the data are given at the end of a typical run once the data have been normalized. The software of each apparatus allows data to be displayed in a number of forms, including amplification plots, threshold cycle values, and standard curves, and has the ability to analyze melt curve and standard curve results (information from Web sites as stated in Table 1).

Quantification of Epstein-Barr Viral Load Using Different Systems

Real-time PCR assays for quantification of Epstein-Barr virus (EBV) viral loads were performed on each of the PCR systems. The EBV assay used here was developed for the ABI 7700 and has been published previously (Niesters 2001). The results for the ABI 7700, I-Cycler IQ, and Smartcycler were compared for differences in viral load levels. An attempt to run the assay on the LightCycler was unsuccessful. However, when using the specific reagents and probes as advised by the manufacturer, this assay could be adapted to the LightCycler as well. For the other real-time PCR systems, limited adaptation of the published protocol was required to run the assay. A dilution series of an EBV standard (Applied Biotechnologies) was prepared and used to generate an EBV standard curve for each assay. In addition, the viral load for 50 samples was determined using the three systems. Overall, the results obtained with the different machines were very similar (Templeton et al. 2001). Essentially, each of these machines had the flexibility to adopt a published, optimized assay and to generate comparable results with minor modifications.

TABLE 1. Real-time PCR systems

Company	PCR system	Number of fluorophores detected	Excitation and emission (nm)	Weight (kg)	Sample format (Maximum number of samples)	Excitation source	Emission detection
Applied Biosystems www.appliedbiosystems.com	ABI PRISM 7700 sequence detection system	2	500–660	145	microplate (96)	laser	CCD camera
Bio-Rad www.bio-rad.com	I-Cycler IQ real-time PCR detection system	4	460–650	17.5	microplate (96)	broad-wavelength lamp with filters to excite fluorophores	CCD camera
Cepheid www.cepheid.com	Smartcycler system	4	490–593	10	tubes (16)	laser	photodetector
Roche Molecular Biochemicals Biochem.roche.com	LightCycler	2	530–710	19	capillaries (32)	laser	photodiode
Stratagene www.stratagene.com	Mx4000 multiplex quantitative PCR system	4	350–780	50	microplate (96)	broad-wavelength lamp with filters to excite fluorophores	CCD camera

| PROTOCOL | ## SETTING UP A REAL-TIME PCR ASSAY |

The process of establishing a real-time PCR assay for a new target includes: (1) the design step—the selection of PCR primers and probes and (2) choice of a detection method (SYBR Green or a fluorescent nucleic acid probe) for the real-time PCR, depending on the requirement of the particular assay. Generally, real-time PCR with SYBR Green is used to optimize primers and is principally used to detect PCR products in a research setting. TaqMan probes and molecular beacons provide further specificity, necessary for diagnostic PCR and also to enable multiplexing of reactions.

Design of Primers and Probes

The design of a real-time PCR assay has more parameters than conventional PCR. Computer software is available to aid the design of real-time PCR; e.g., PE/ABI Primer Express software (Applied Biosystems) and Beacon Designer (Premier Biosoft). These programs include all of the design parameters required.

1. The first step in the design of a PCR is to choose a target region of the gene. If the target sequence contains regions of variability, then an alignment of sequences is necessary to find conserved areas, such as the 5′NCR of enteroviruses.

2. When selecting primers, they should define a 100- to 200-bp amplicon. The melting temperature (T_m) of the primers should be 7–10°C less than that of the probe in order to achieve probe binding prior to primer binding, which is crucial for the production of fluorescence. The best T_m for the primers is 55–60°C. There are many criteria in the computer programs for optimal primer and probe design. In general, the primers should be 20–25 bp long and fulfill general PCR requirements of having minimal anti-complementarity sequences and a maximum of two GCs at the 3′ end (Mitsuhashi 1996).

3. Once the primers have been selected, the target sequences need to be checked for secondary structures at the annealing temperature. This is best performed by putting the whole target sequence into the Mfold/Zuker program (http://www.bioinfo.math.rpi.edu/applications/mfold/old/dna) and entering the proposed annealing temperature and reaction conditions (e.g., 100 μM Na$^+$ and 3 mM Mg^{++}). The resulting structure should have little secondary structure and, more particularly, no folding where either primers or probe will bind.

4. The probe is designed in a similar way as the primers to fulfil general PCR requirements and to have a T_m 7–10°C higher than that of the primers. Molecular beacons have 5- to 7-bp stem sequences attached to the probe. These are GC-rich and, again, should form a stable hairpin structure in the Mfold program using the same parameters for the target as described above (www.molecular-beacons.org/). In some cases, a quenching effect of a 5′ G in a fluorescent probe has been described. However, initially a 5′ G was in the standard stem of molecular beacons, as deduced from one of the first reports of application of molecular beacons (Vet et al. 1999), and no problems were encountered.

5. Finally, the primers and probe should have their specificity checked for homology with other sequences using the BLAST program (http://www.ncbi.nlm.nih.gov/BLAST/). **N.B.: If significant homology is detected, it may be necessary to redesign the primer pair.**

Sample Preparation

For quantitative PCR to be meaningful, nucleic acids must be extracted in a reproducible manner for all sample types. Chapters 9 and 10 describe in detail the methods for isolation of amplification-ready DNA and RNA using solution- or silica-based nucleic acid extraction kits and automated procedures.

PCR Master Mixes

For all systems, the use of PCR master mixes, containing *Taq* DNA polymerase, dNTPs, and PCR buffer, is recommended. For the Roche LightCycler, LightCycler Master Hybridization Probes (Roche) is recommended. The other systems can use PCR mix with separate HotStart AmpliTaq, PCR buffer, and dNTPs. Importantly, HotStart polymerases are required for all these assays. The 2x master mixes, prepared by various manufacturers, provide standardized reagents with which to perform real-time PCR. Commonly used mixes are TaqMan Universal PCR master mix (Applied Biosystems), Hot StarTaq (Qiagen), Platinum Quantitative PCR superMix-UDG (Invitrogen), Smartcycler 2x mix (Eurogentec), LightCycler DNA FastStart mastermix (Roche), and the IQ Supermix (Bio-Rad).

MATERIALS

V CAUTION

See Appendix for appropriate handling of materials marked with <!>.

BUFFERS AND SOLUTIONS
$MgCl_2$ <!> (25 mM)
Probe (e.g., molecular beacons) or SYBR Green<!> (Molecular Probes)

ENZYMES AND ENZYME BUFFERS
HotStart Mix, 2x (containing *Taq* DNA polymerase, PCR buffer, and dNTPs)

NUCLEIC ACIDS AND OLIGONUCLEOTIDES
Forward and reverse amplification primers
Template DNA

SPECIAL EQUIPMENT
Real-time PCR machine
Appropriate reaction tubes, capillaries, or 96-well microtiter plates (see individual methods)

METHOD A: THE ABI 7700 OR I-CYCLER

1. In a 96-well microtiter plate, assemble the following PCR components for each reaction. Combine the HotStart mix with primer and probe/SYBR Green and then add the required amount of template. Include five 10-fold dilutions of the template DNA.

2x HotStart Mix	25 µl
Primer A	0.1–0.5 µM
Primer B	0.1–0.5 µM
Probe or SYBR Green	0.1–0.5 µM
$MgCl_2$ (25 mM)	3–6 mM
Template DNA (500 ng–50 pg)	10 µl
H_2O	to 50 µl

2. Use either two-step (all AB I7700 TaqMan assays) or three-step PCR.

For three-step PCR, carry out 40–50 cycles of amplification as follows:

Cycle number	Denaturation	Annealing	Polymerization/Extension
1	2–15* min at 95°C		
40–50	30 sec at 95°C	30 sec at 55°C	30 sec at 72°C

For two-step PCR, leave out the extension at 72°C and increase the annealing temperature to 60°C.

*Depending on HotStart protocol.

3. Continue with optimization protocol in step 4 below.

METHOD B: THE SMARTCYCLER

1. In an appropriate tube, assemble the following PCR components for each reaction. Combine the HotStart mix with primer and probe/SYBR Green and then add the required amount of template. Include five 10-fold dilutions of the template DNA.

2x HotStart Mix	2.5 µl
Primer A	0.3–0.8 µM
Primer B	0.3–0.8 µM
Probe or SYBR Green	0.1–0.5 µM
$MgCl_2$ (25 mM)	3–6 mM
Template DNA (500 ng–50 pg)	5 µl
H_2O	to 25 µl

The Smartcycler requires use of specific tubes.

2. Use either two-step or three-step PCR.

For three-step PCR, carry out 40 cycles of amplification as follows:

Cycle number	Denaturation	Annealing	Polymerization/Extension
1	2–15* min at 95°C		
40	5 sec at 95°C	10 sec at 55°C	10 sec at 65°C

For two-step PCR, leave out the extension at 72°C and increase the annealing temperature to 60°C.

*Depending on HotStart protocol.

3. Continue with optimization protocol in step 4 below.

METHOD C: THE LIGHTCYCLER PCR

1. In an appropriate tube, assemble the following PCR components for each reaction. Combine the HotStart mix with primer and probe/SYBR Green and then add the required amount of template. Include five 10-fold dilutions of the template DNA.

2x FastStart Mix	10 µl
Primer A	0.1–0.5 µM
Primer B	0.1–0.5 µM
Probe or SYBR Green*	0.1–0.5 µM
$MgCl_2$ (25 mM)	3–6 mM
Template DNA(500 ng–50 pg)	5 µl
H_2O	to 20 µl

*The LightCycler requires use of capillaries.

2. Using a three-step PCR, carry out 45 cycles of amplification as follows:

Cycle number	Denaturation	Annealing	Polymerization/Extension
1	10 min at 95°C		
40	10 sec at 95°C	10–15 sec at 55°C	10–20 sec at 72°C

3. Continue with optimization protocol in step 4 below.

Practical Optimization of the Assay (see Innis and Gelfand 1990; Harris and Jones 1997)

4. To optimize the amplicon and primers, analyze the real-time PCR products using the melt curve function, which is not directly available on ABI 7700.

> Results should show a specific melting peak for a single product with no secondary peaks that would indicate primer–dimer formation. When the melt curve data are presented as the first derivative, there should be a single peak. Multiple peaks indicate the presence of other amplified products. If multiple peaks are detected, the $MgCl_2$ concentration can be further optimized, but generally it is an indication that primers should be redesigned.

5. To optimize the probe, use the 10-fold target dilutions to generate a standard curve using arbitrary units; e.g., give the highest concentration of DNA the value 1^7. The 10-fold dilution series is therefore termed 1^7, 1^6, 1^5, 1^4, 1^3. Enter these values as starting quantity and use the software to generate a standard curve at the end of the assay. The threshold cycle (Ct) values and DNA starting quantity will allow reaction efficiency to be calculated.

6. Similar to conventional PCR, perform further runs with a set of 10-fold dilution standards to optimize for $MgCl_2$ concentration, primer concentration, probe concentration, annealing temperature, as well as adjustment of PCR protocol.

7. For systems that do not have an intrinsic rapid cycling protocol, shorter cycling protocols can be tested to reduce the length of the assay.

ANALYSIS OF RESULTS AND TROUBLESHOOTING

A real-time PCR assay produces a relative fluorescence intensity reading, which is analyzed by the individual software of the system. The fluorescence data can be collected either for real-time analysis or for post-run analysis. Data are simultaneously collected from all the wells or tubes during the entire dwell time of the steps defined for analysis. A number of different readings can be taken during a dwell time. Each apparatus takes several readings in order to give a fluorescent value. To obtain the Ct value for a sample (see below), the cycle at which the fluorescence intensity enters the exponential portion of the curve must be defined. To achieve this, several parameters need to be assessed:

- Range of baseline cycles for background fluorescence correction
- Threshold value
- Number of data points used for each cycle in the analysis
- Wells included in the analysis

These will all have an effect on the number assigned to the Ct value.

Background Fluorescence

Baseline cycles are used to characterize and correct for the background fluorescence over the course of the run. Background fluorescence is all fluorescence measured in the cycles before the traces begin to rise above background values. Improving the average of background fluorescence by including as many cycles as possible will result in better correction for background and therefore better data. Some of the software packages auto-calculate the baseline values.

Threshold Value

Each software program has a default value assigned to the threshold cycle. The value usually is calculated as being 10 times the average standard deviation of the background fluorescence for all wells over their baseline cycles. In addition, some systems allow a user-defined value of the threshold. Some systems use the fluorophore ROX in their mix as an internal standard to automatically correct for pipetting errors.

Reading of Results and Efficiency of PCRs

Positive samples will be identified by generation of a fluorescent signal that crosses the threshold at a certain cycle, which is defined as the Ct value, or threshold cycle. It is often necessary to check individual results of a real-time PCR run. For some samples, there is nonspecific crossing of the threshold, resulting in a reported Ct value. However, no "true positive" curve is seen when looking at the individual result in the graph.

Once the Ct values have been assigned to the standards of a PCR experiment, the data from that assay can be analyzed further. In an optimal reaction, the Ct values for each of the 10-fold dilution series should be 3.3 cycles apart (Fig. 1). When a 10-fold dilution series of a reference target is subjected to real-time PCR, the Ct values can be plotted as a standard curve (Fig. 2). The slope obtained for the standard curve can be used to assess the efficiency of the assay. The equation for the line is Y = MX + C, where M is the slope of the line and C is a constant where the standard crosses the y axis. Using the M value, the effi-

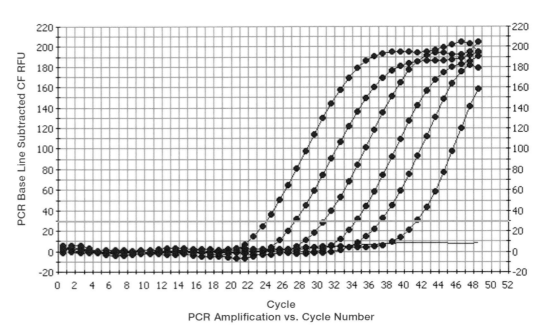

FIGURE 1. A 10-fold dilution series with 3.5 cycles between each dilution.

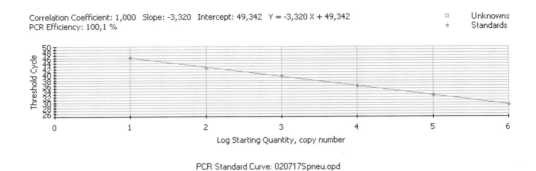

Correlation Coefficient: 1,000 Slope: -3,320 Intercept: 49,342 Y = -3,320 X + 49,342   □ Unknowns
PCR Efficiency: 100,1 %   ◇ Standards

PCR Standard Curve: 020717Spneu.opd

FIGURE 2. A standard curve obtained from 10-fold dilution series, with an efficiency of 100.1% and a correlation coefficient of 1.00.

ciency of the assay can be calculated via the equation $E = e^{\ln 10/-m} -1$, where E = the efficiency of assay, e = 2.718 (approximately), ln10 = 2.303, and M is the slope of the standard curve. Therefore, if the slope is −3.320 (as in Fig. 2), the corresponding efficiency is 100.1%.

An optimized assay should have an efficiency between 95% and 105%, and the correlation coefficient should be between 0.90 and 1.00. This is particularly important when the eventual aim is to develop a multiplex PCR assay. In theory, it should be easy to multiplex two 100% efficient assays provided the primers and probes do not interact. If the PCR efficiency is not close to 100%, with a correlation coefficient of 1, further optimization is required, as described above. The parameters that need optimization are $MgCl_2$ concentration, primer and probe concentration, dNTP and enzyme concentration, annealing temperature, and the cycling profile (e.g., two-step vs. three-step).

The use of a standard curve to optimize the correct probe concentration is shown in Figure 3. This figure shows the effect of probe concentration on efficiency. The efficiencies of the different assay runs show that 0.34 μM (Fig. 3b) gives the best efficiency at 100.1% as opposed to 119.3% (0.2 μM) and 104.9% (0.5 μM), as shown in Figure 3, a and c, respectively.

If there is poor quality of signal, or no signal, in real-time PCR, run the products on a 2% agarose gel to visualize products. If there is no product, the problem was obviously related to the amplification and, thus, either the PCR needs to be repeated or the nucleic acid was of poor quality. In the worst case, the PCR primer pair may need redesigning. If there are weak bands present, the PCR may still require redesigning, or more optimization may be required. There are housekeeping genes (e.g., human PBGB molecular beacon set from Biolegio) that are optimized for real-time PCR and can be purchased from real-time PCR manufacturers. These sequences provide a baseline upon which the apparatus and the assays can be standardized.

CONCLUSION

Real-time PCR is a powerful advancement of the basic PCR technique. The important steps in deciding which particular assay format to use are related to the types of data required. The requirements for a research laboratory are quite distinct from those of a diagnostic laboratory. For the latter, probe confirmation of the PCR product is an essential part of the assay, whereas SYBR Green detection may be sufficient for many other applications, such as quantifying expression of a gene. All of the real-time PCR machines analyzed were capable of detecting PCR products in real time, and a specific assay can be made operationally

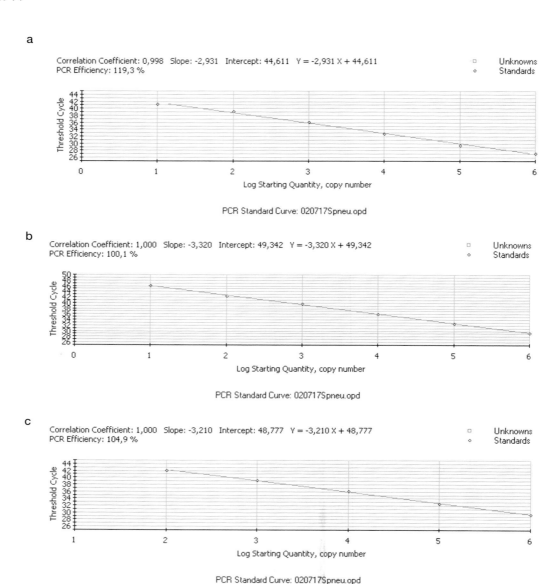

FIGURE 3. Use of a standard curve to optimize the correct probe concentration. (*a*) Standard curve obtained with a 0.2-μM probe. (*b*) Standard curve obtained with a 0.34-μM probe. (*c*) Standard curve obtained with a 5-μM probe.

on every system. However, there are some decisions to be made when selecting between the different formats. The choice of system is dependent on individual laboratory needs. Considering diagnostic applications, the LightCycler or Smartcycler may obtain faster results for urgent assays. This could reduce the time of analysis to result from 3–4 hours to 1.5 hours.

On the other hand, if sensitivity is the most important issue, these machines, with their smaller reaction volume and consequently lower sensitivity, would not be the first choice. The ABI 7700 and Bio-Rad I-Cycler IQ have a 96-well format, enabling higher throughput than other systems. The 384-well plates, as designed by ABI for use in the 7900HT system, can further enhance throughput.

For diagnostic applications, internal control of nucleic acid isolation and inhibition is essential to obtain valid results. This can be achieved using systems that enable multicolor detection, such as the I-Cycler IQ and the Smartcycler. Multiplex real-time PCRs can be developed for three different targets and an internal control by using the four detection wavelengths possible in multicolor detection. ABI recently launched the 7000 system, which also provides multicolor detection. Several real-time PCR machines are available in addition to those used in the study presented here; e.g., Stratagene Mx4000, MJ Research DNA Engine Opticon, and the Corbett/Westburg Rotor-gene. The conclusions reached for the systems analyzed above may also be applied to these systems.

In summary, the choice of which real-time system to use depends on the range of applications required. To achieve meaningful results, each assay must be validated and optimized for the particular system chosen.

REFERENCES

Harris S. and Jones D.B. 1997. Optimisation of the polymerase chain reaction. *Br. J. Biomed. Sci.* **54:** 166–173.

Innis M. and Gelfand D. 1990. Optimisation of PCRs. In *PCR protocols: A guide to methods and applications* (ed. M. Innis et al.), pp. 3–13. Academic Press, San Diego, California.

Mitsuhashi M. 1996. Technical report: Part 2. Basic requirements for designing optimal PCR primers. *J. Clin. Lab. Anal.* **10:** 285–293.

Niesters H.G. 2001. Quantitation of viral load using real-time amplification techniques. *Methods* **25:** 419–429.

Schutten M. and Niesters H.G. 2001. Clinical utility of viral quantification as a tool for disease monitoring. *Expert Rev. Mol. Diagn.* **1:** 153–162.

Templeton K., van Tol H., Scheltinga S., Thijssen J., van Soest R., and Claas E. 2001. Comparison of 4 real-time detection systems. In *Abstracts of the European Meeting of Molecular Diagnostics*, Scheveningham, The Netherlands, p. 83.

Vet J.A., Majithia A.R., Marras S.A., Tyagi S., Dube S., Poiesz B.J., and Kramer F.R. 1999. Multiplex detection of four pathogenic retroviruses using molecular beacons. *Proc. Natl. Acad. Sci.* **96:** 6394–6399.

Wittwer C. 2001. Rapid cycle real-time PCR: Methods and applications. In *Rapid cycle real-time PCR* (ed. S. Meuer et al.), pp. 1–11. Springer, Heidelberg, Germany.

WWW RESOURCES

For additional WWW resources, see Table 1.

http://www.molecular-beacons.org   Molecular beacons homepage. Hybridization probes for the detection of nucleic acids in homogeneous solutions.

http://www.bioinfo.math.rpi.edu/applications/mfold/old/dna Center for Bioinformatics, DNA folding.

http://www.ncbi.nlm.nih.gov/BLAST/ National Center for Biotechnology Information BLAST information page.

15   New Technology for Quantitative PCR

Wolfgang Kusser, Sandrine Javorschi, and Martin A. Gleeson

Invitrogen Corporation, Research and Development, Carlsbad, California 92008

Quantitative PCR has evolved from a low-throughput, gel-based analysis to higher-throughput fluorescence techniques, called real-time PCR, that do not require separation of the reaction products (Holland et al. 1991; Higuchi et al. 1992, 1993; Tyagi and Kramer 1996; Nazarenko et al. 1997). These new detection methods are based on fluorescent reporters and are characterized by their excellent sensitivity and broad dynamic range. Today, automated DNA synthesis by phosphoramidite chemistry provides easy access to oligonucleotides linked with fluorophores that are commonly used as detection agents, and a number of different instruments with built-in software are on the market to support this application.

The amount of DNA amplified by a PCR correlates with an increase in a fluorescent signal that results from an interaction between a fluorescent material and the other PCR reactants. Graphing the fluorescence output of a reaction at each cycle generates an amplification curve. The amount of starting template DNA is then estimated by determination of the cycle for which the fluorescence crosses a given threshold. These fluorescent techniques are faster because they do not require post-PCR handling and they may be facilitated by high-throughput robotics. They also reduce the risk of PCR contamination because there is no need to open the reaction tubes.

Several detection technologies are currently available for real-time PCR. The earliest techniques and practical applications used ethidium bromide (Higuchi et al. 1992). Subsequently, ethidium bromide was replaced by SYBR Green I, a double-stranded (ds) DNA-specific dye with a better fluorescent signal (Wittwer at al. 1997). These dyes bind to any dsDNA in the reaction vessel and also detect primer–dimers and unspecific amplification products. Furthermore, real-time PCR using dsDNA-specific dyes does not allow the detection of several DNA targets in one reaction vessel. This method, called multiplexing, employs more than one PCR primer pair in one tube. dsDNA-specific dyes like SYBR Green I are not able to discriminate between the different amplicons. The fluorogenic LUX (Light Upon Extension) primers (Invitrogen) allow multiplexing because they can be labeled with different fluorescent dyes that will allow the identification of a specific amplicon by the distinct emission spectrum of the dye attached to each primer. Some detection systems, among them SYBR Green and LUX fluorogenic primers, also allow the post-PCR analysis of data by melting curves (Ririe at al. 1997), a valuable tool to examine the specificity of the PCR and to distinguish specific product from artifact.

Several methods incorporate the use of an oligonucleotide probe labeled with a fluorophore and a quencher moiety. The quencher reduces the fluorescence of the fluorophore by fluorescence resonance energy transfer (FRET) when the two moieties are separated by less than 100 Å. During PCR, the fluorophore and quencher are separated, causing an increase in fluorescence. The separation occurs either by cleavage of the annealed probe, as in the TaqMan assay (Holland et al. 1991), or by a change in secondary structure of the oligonucleotide probe when it anneals to target DNA, as occurs with molecular beacons (Tyagi and Kramer 1996).

Here we describe a novel technique called LUX primers. It was observed that the fluorescence of some conjugated dyes was sensitive to the environment around the point of attachment (Cardullo et al. 1988; Murchie et al. 1989). Typically the dyes were quenched by guanosine, due to the electron-donating properties of this nucleotide (Steenken and Jovanovic 1997). Base-pairing of single-stranded oligonucleotides to the complementary sequence also quenches the conjugated fluorophores. On the basis of these data, a systematic study of the factors that can influence the fluorescence intensity of dyes linked to DNA was undertaken and published by Invitrogen/Life Technologies (Nazarenko et al. 2002a).

We have examined the importance of the location of the fluorophores within the sequence and the significance of secondary structure on fluorescence. The data on fluorescence intensity, polarization, and lifetime agree with the important role that guanosine plays as a quencher of fluorescence. However, the picture that emerges from our work is more complex than the standard model, and the actual fluorescence depends on secondary structure, local sequence, and the proximity of the 3′ or the 5′ end of the oligonucleotide. Depending on these factors, the fluorescence of a labeled oligonucleotide can increase or decrease upon hybridization to the complementary sequence. These factors have permitted us to modulate the fluorescence intensity of many conjugated dyes in the order of tenfold, providing the basis for many interesting applications including real-time PCR (Nazarenko et al. 2002b).

This method is unique, because a gain in fluorescent signal arises during PCR by the incorporation of a fluorogenic primer that is labeled with only one single fluorescent dye and no added quenching dye. The primer is designed in a hairpin conformation by extending the 5′ end of the sequence with bases complementary to the 3′ end. When the primer is incorporated into the double-stranded PCR product, the fluorescence increases and a signal is generated (Fig. 1). The label rules have been incorporated into a Web-based software for primer design (LUX Designer) that allows researchers to select and order primers on line (https://pf.invitrogen.com/primerf/pages/Default.cfm?cc=1). Most LUX applications have used FAM and JOE as labels, but other fluorophores, including HEX, TET, ROX, and TAMRA, also change their fluorescence depending on the sequence and structure of oligonucleotide.

Real-time quantitative PCR with LUX primers can be performed in multiplex and single dye format. Various primers, templates, and dye combinations were tested which demonstrate that fluorogenic primers can be used successfully to quantify at least two genes in a sample with high sensitivity and broad dynamic range ("multiplexing"). The method is an efficient, reliable, and cost-effective alternative to present methods for high-throughput, quantitative real-time PCR and can also be applied to allele-specific PCR.

PROTOCOL 1   PARAMETERS FOR REAL-TIME PCR

LUX primers are currently available with either a FAM label (6-carboxy-fluorescein) or a JOE label (6-carboxy-4′,5′-dichloro-2′,7′-dimethoxy-fluorescein). Fluorogenic PCR with LUX primers can be routinely used to quantify 100 or fewer copies of target in a back-

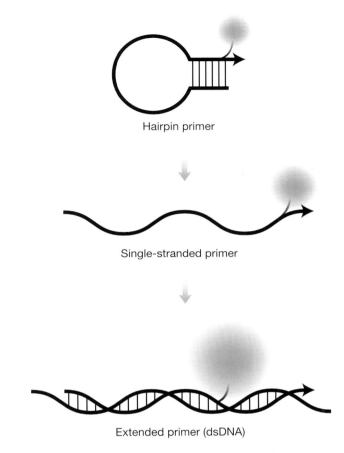

Hairpin primer

Single-stranded primer

Extended primer (dsDNA)

FIGURE 1. LUX principle. The fluorescent intensity is modulated due to the proximity of the fluorophore to specific primary and secondary structures of the DNA. The design rules have been incorporated into the proprietary software (LUX Designer) that generates LUX primers and unlabeled counterparts.

ground of nonspecific templates over a broad dynamic range of less than 100 to 10^7 copies. Using JOE- and FAM-labeled primers, the detection platform enables simultaneous detection of two targets. Quantitative PCR with LUX primers can be performed with real-time PCR instruments that detect FAM (all instruments) and JOE labels (most instruments).

METHOD

1. *Primers*. Primers for real-time PCR are gene-specific and typically amplify a short amplicon.
   a. The default parameters in the LUX primer design software (LUX Designer) have been set to result in amplicons of 75–200 bp in length. The default parameters for primer annealing temperatures are set at 60–68°C, and PCR programs with annealing temperatures between 55°C and 64°C are appropriate. The software contains algorithms to minimize self-complementarity and interactions of primers to each other. The primer design is flexible, and regions that span exon–exon boundaries or alternate splice forms can be targeted.

b. For PCR applications, primers have to be supplied in excess molar concentration. We recommend using LUX primers at a concentration of 200 nM for each primer. Optimal results, however, may require titrations of the primers between a 50 nM and 500 nM concentration for each primer. To optimize multiplex applications, we recommend limiting the primer concentration of the gene with the highest abundance (typically the housekeeping gene).

2. *Templates.* Target templates for real-time PCR are linear single-stranded or dsDNA, cDNA, or circular DNA (for example, plasmids). A total of 10 to 10^7 copies or 1 pg to 10 µg of template are routinely used.

a. The target template for real-time RT-PCR is RNA, usually total cellular RNA or mRNA; 1 pg to 100 ng of total RNA per assay is routinely used.

b. Purity and integrity of the RNA molecules have a direct impact on results.

c. RNase and genomic DNA contamination are the most common problems, and purification methods should be designed to avoid these.

d. The presence of RNase will degrade RNA template and lead to an underestimate of gene expression; the presence of genomic DNA will lead to a higher apparent expression.

e. The amplification of contaminating DNA can also be avoided by placing the real-time PCR primers over exon–exon junctions or in adjoining exons separated by large inserts. As this design strategy is not always feasible, we recommend the inclusion of a DNase I treatment in the standard RNA purification protocol.

3. *Magnesium and dNTP concentration.*

a. The optimum Mg^{++} concentration for a given target/primer combination can vary between 1 mM and 10 mM, but we usually recommend the use of $MgCl_2$ at a concentration of 3 mM. For multiplex applications, the $MgCl_2$ should be increased to 5 mM.

b. Optimum dNTP concentration is 200 µM each, for dATP/dCTP/dGTP/dTTP. When dUTP is used in place of dTTP, its concentration should be increased to 400 µM. For multiplex PCR, we recommend increasing the dNTP concentrations to 400 µM each.

4. *Enzymes.* We strongly recommend the use of hot-start enzymes.

a. Reaction mixes that contain all ingredients except primers and template provide ease of use and rapid assay setup protocols. For real-time PCR, we recommend the use of Platinum Quantitative PCR SuperMix UDG (Invitrogen, 11720-025, 500 Reactions) and for one-step RT-PCR, the use of ThermoScript One-Step-System (Invitrogen, 11731-023, 500 Reactions).

b. To avoid limiting the PCR efficiency in multiplex applications, the amount of polymerase enzyme can be doubled to 3 units in a 50-µl assay.

5. *Instruments and settings.* A variety of instruments are available for the real-time PCR. Refer to the specific instrument's user manual for operating instructions. The guidelines outlined in the following protocols have been optimized for the ABI PRISM 7700 sequence detection system.

PROTOCOL 2   REAL-TIME PCR

This protocol uses Platinum Quantitative PCR SuperMix-UDG. The real-time, fluorogenic PCR should be performed on instruments that incorporate fluorescence detection of the

PCR at each PCR cycle. The instrument must be able to excite the FAM- or JOE-labeled primers near their excitation wavelength maximum of 490 nm for FAM and 520 nm for JOE and must detect the emission of FAM- and JOE-labeled primers near their emission maximum of 520 nm for FAM and 550 nm for JOE. Either the forward or reverse primer can be labeled with the FAM or JOE fluorophore. Templates should be free of nuclease and protease activity.

MATERIALS

BUFFERS, SOLUTIONS, AND REAGENTS

Components for a 50-μl real-time PCR assay:

Platinum Quantitative PCR SuperMix-UDG (Invitrogen 11730-025)	25 μl
ROX reference dye (Invitrogen 12223012)	1 μl
Sterile distilled H_2O (Gibco 15230-162)	12 μl
10 μM Reverse primer	1 μl
10 μM Forward primer	1 μl
Subtotal 40 μl	
Template diluted in 0.1x TE or sterile dH_2O	10 μl
Total 50 μl	

SPECIAL EQUIPMENT

ABI PRISM 7700 sequence detection system

METHOD

1. The thermal cycling protocol is as follows:
   a. Three-step cycling protocol

      50°C, 2 minutes hold (UDG carryover digest)

      95°C, 2 minutes hold (denaturation of template, inactivation of uracil DNA glycosylase and activation of the hot-start *Taq* polymerase)

      45 cycles of 95°C for 15 seconds; 55°C for 30 seconds, 72°C for 30 seconds.
   b. Two-step cycling protocol

      50°C, 2 minutes hold (UDG carryover digest)

      95°C, 2 minutes hold (denaturation of template, inactivation of uracil DNA glycosylase and activation of the hot-start *Taq* polymerase)

      45 cycles of 95°C for 15 seconds; 60°C for 30 seconds.
2. The program for melting curve analysis is:
   a. Cycle the reactions at 40°C for 1 minute and then ramp to 95°C over a period of 19 minutes followed by incubation at 25°C for 2 minutes. Most real-time PCR instrumentation supplies software that generates the derivate

$$-\frac{d \text{ fluorescence}}{d \text{ temperature}}$$

   of the melting curve with a peak corresponding to the melting temperature.
   b. A successful result should only show one peak at the temperature of the intended PCR product, taking into account the ionic conditions used in the assay. If more then one amplicon and/or primer–dimers have been generated during the PCR, several peaks will appear, and primer–dimers usually have a substantially lower

melting temperature than the PCR amplicons. Thus, melting curve analysis provides rapid and efficient quality control of a PCR.

c. Results for a real-time PCR with c-*myc*-specific LUX primers (Table 1) are shown in Figure 2. Quantitative real-time PCR was performed with tenfold serial dilutions of

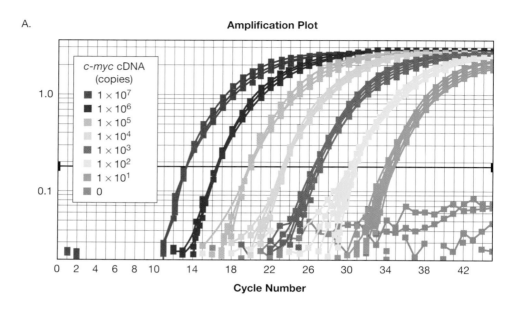

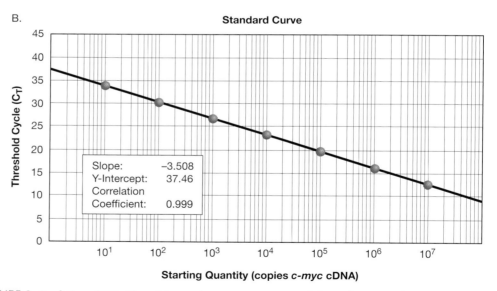

FIGURE 2. Real-time PCR. Plasmid templates were generated from first-strand cDNA by PCR with gene-specific primers and cloning into pCR4-TOPO using a TOPO-TA cloning kit (Invitrogen). The copy number of plasmids was quantified by OD_{260} measurements. As a conversion number, we used 9.1×10^{11} copies for 1 μg of plasmid, divided by the plasmid size (plus insert) in kilobases. (*A*) Real-time PCR of serial dilutions of a c-*myc* cDNA clone were performed using 200 nM FAM-labeled LUX primer, 200 nM unlabeled primer, Platinum Quantitative PCR SuperMix-UDG, and ROX Reference Dye. Reactions (12 replicates per dilution) were incubated for 2 minutes at 50°C and 2 minutes at 95°C, followed by 45 cycles of 95°C for 15 seconds; 60°C for 30 seconds using an ABI Prism 7700. (*B*) Standard curve showing the initial copy number of template versus the cycle threshold (C_T).

TABLE 1. Primer sequences used in real-time PCR

Target	Dye label	Forward primer	Reverse primer
c-*myc*	FAM	gacgcggggaggctattctg	<u>gactcg</u>tagaaatacggctgcaccgagtc*
β-Actin	FAM	<u>gatctt</u>cggcacccagcacaatgaagatc*	aagtcatagtccgcctagaagcat
Interleukin 4	JOE	gagttgaccgtaacagacatctt	<u>ctacag</u>tccttctcatggtggctgtag*

*The dye label is conjugated at the 3′ region of the primer sequence; the underlined sequence indicates the 5′ extension that forms a hairpin with the 3′end.

10 to 10^7 copies of cloned c-*myc* plasmid DNA. The assays were assembled as described above and SuperMix-UDG was used as a 2x reaction mix. The results show that the LUX platform can achieve real-time quantification results of 100 or fewer copies of target genes and achieve a dynamic range of 7 orders of magnitude.

PROTOCOL 3 MULTIPLEX REAL-TIME PCR

In multiplex real-time PCR, different sets of primers with different labels are used to amplify separate genes from the template DNA in one tube. LUX primers have been tested in multiplex reactions using FAM to label the gene of interest and JOE to label a housekeeping gene used as an internal control to normalize between different reactions. In a standard multiplex reaction, the additional primers can be added at the same volume and concentration as the primers in a singleplex reaction, as shown in the example mixture below.

MATERIALS

BUFFERS, SOLUTIONS, AND REAGENTS

Components for a 50-ml multiplex assay:

Platinum Quantitative PCR SuperMix-UDG (Invitrogen 11730-025)	25 μl
ROX reference dye (Invitrogen 12223012)	1 μl
Sterile distilled H_2O (Gibco 15230-162)	10 μl
10 μM Reverse primer target 1	1 μl
10 μM Forward primer target 1	1 μl
10 μM Reverse primer target 2	1 μl
10 μM Forward primer target 2	1 μl
Subtotal 40 μl	
Template diluted in 0.1 x TE or sterile dH_2O	10 μl
Total 50 μl	

SPECIAL EQUIPMENT

ABI PRISM 7700 sequence detection system

METHOD

1. Thermal cycling conditions are the same as for a single amplicon reaction given in Protocol 2.

   A typical result of a multiplex reaction is shown in Figure 3. The experiment was performed with interleukin-4 as the target gene and β-actin as the reference gene (housekeeping gene). The results show quantification of interleukin from 22 to 3×10^5 copies in the presence of a 10^6 copies of β-actin. The PCR was not further optimized, and the reagent concentrations were identical to a single amplicon PCR. In case of a decline in real-time PCR efficiency, the reactions can be optimized by sequentially performing the steps listed below.

2. To optimize multiplex conditions:

a. Reduce the primer concentration of the gene with the highest abundance (typically the housekeeping gene) to 1/4 the concentration of the other gene.

b. Increase the $MgCl_2$ in the reaction from 3 mM to 6 mM.

c. Increase the dNTP concentrations to 400 μm each.

d. Double the amount of polymerase enzyme (to 0.06 units per μl of reaction volume). Platinum *Taq* DNA polymerase stand-alone enzyme (Invitrogen 10966-018) can be used for this purpose.

PROTOCOL 4   REAL-TIME RT-PCR

The protocol uses the Superscript II First-Strand Synthesis system for the generation of cDNA. For the subsequent real-time PCR, either the forward or reverse primer can be labeled with the FAM or JOE fluorophore. The isolation of total RNA may be performed using either TRIzol reagent (Invitrogen 10296010) or methods described in Chapter 10. The amount of genomic DNA in the sample should be decreased by performing a DNase I digest (see chapter 10).

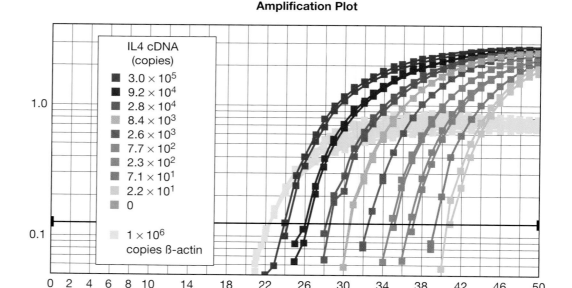

Amplification Plot

IL4 cDNA (copies)

- 3.0×10^5
- 9.2×10^4
- 2.8×10^4
- 8.4×10^3
- 2.6×10^3
- 7.7×10^2
- 2.3×10^2
- 7.1×10^1
- 2.2×10^1
- 0

1×10^6 copies ß-actin

FIGURE 3. Muliplex real-time PCR. Plasmids were prepared and quantified as described in Fig. 2. Serial dilutions of a human IL-4 cDNA clone and 1 x 10⁶ copies of a β-actin cDNA clone were co-amplified in the same tube using Platinum Quantitative PCR SuperMix-UDG in 50-μl volumes. IL-4 was targeted with 200 nM FAM-labeled LUX primer and 200 nM unlabeled primer, and the β-actin cDNA clone was targeted with 200 nM JOE-labeled LUX primer and 200 nM unlabeled primer. PCR assays were amplified in the presence of ROX reference dye for 2 minutes at 50°C and 2 minutes at 95°C, followed by 50 cycles of 95°C for 15 seconds; 55°C for 30 seconds; 72°C for 30 seconds on an ABI Prism 7700 sequence detection system.

MATERIALS

BUFFERS, SOLUTIONS, AND REAGENTS

Components for a 10-μl digest

RNA template (up to 1 μg/10 μl)	x μl
10x DNase buffer	1 μl
DNase I, 1 unit/μl (Invitrogen 18068-015)	1 μl
DEPC-treated H$_2$O <!>	8-x μl

Components for a 20-μl RT reaction

oligo(dT)$_{12-18}$ 0.5 μg/μl	0.5 μl
Random Hexamer 50 ng/μl	0.5 μl
RNA sample (up to 1 μg)	x μl
10x Buffer	2 μl
(20 mM Tris-HCl<!>, pH 8.4, 500 mM KCl<!>, 25 mM MgCl$_2$<!>)	
25 mM MgCl$_2$	4 μl
10 mM dNTP	1 μl
0.1 M dithiothreitol (DTT) <!>	2 μl
RNaseOUT (40 units/μl)	1 μl
SuperScript II (50 units/μl) system (Invitrogen 11904-018)	1 μl
DEPC-treated ddH$_2$O to 20 μl final volume <!>	
EDTA 1 μl of 25 mM	

METHOD

1. Assemble the components for the 10-μl digest listed above. Incubate at 25°C (room temperature) for 15 minutes, and then add 1 μl of 25 mM EDTA. Incubate the mixture at 65°C for 15 minutes to inactivate the DNase I.

2. For each assay, add the components listed for the 20-μl RT reaction to a tube on ice. For multiple reactions, a master mix without RNA may be prepared. Incubate the mixture for 10 minutes at 25°C, followed by 30–50 minutes at 42°C and 15 minutes at 70°C. Chill on ice.

3. For optional RNase H treatment, add 1 μl of RNase H and incubate at 37°C for 20 minutes. The cDNA can be used as template for real-time PCR with Platinum Quantitative PCR SuperMix-UDG as described above or stored at –20°C for future reference.

DISCUSSION

The LUX platform represents primers that are labeled with a single reporter dye but no quencher is required. The system can be used to detect nucleic acids efficiently by PCR in real time, or at the end point, without opening the reaction vessel. We report severalfold increases in signal of the PCR plateau over baseline at typical annealing temperatures, but an even higher increase in signal between the initial and end points of PCR can be reached at room temperature.

Fluorogenic PCR with LUX primers may be routinely used to quantify 100 or fewer copies of target in a background of nonspecific templates over a broad dynamic range of 10 to 10^7 copies. The method is highly specific, as demonstrated by the usual lack of signal when no template is added to the PCR. The relative quantification of cDNA using fluorogenic primers is comparable in sensitivity and dynamic range to other published methods of quantification, such as 5′ nuclease assay (Holland et al. 1991) or SYBR Green I (Wittwer et al. 1997).

Direct comparisons between methods of quantitation are difficult because these methods are functionally different and may be more or less sensitive to various factors. All probe-based technologies have inherent complexities related to the kinetics of the hybridization and amplification. The fluorogenic primer method is not susceptible to this problem. The detection methods involving DNA-binding dyes, such as SYBR Green I, are limited in their ability to detect multiple targets in a single reaction, whereas the LUX platform can be used in multiplex applications. DNA-binding dyes may also increase the stability of the double-stranded structures and, therefore, may facilitate the annealing of primers to nonspecific target.

Typically, a primer pair for a target of choice is designed using LUX Designer software that is accessible through the Invitrogen home page (www.invitrogen.com). The flexibility for primer design may be useful when creating quantitative PCR assays for highly mutated targets or for assays that target genes with multiple isoforms, which are problematic to detect with hybridization probes because of sequence alterations in the region of probe hybridization. A possible drawback of using LUX primers for real-time PCR is the potential for the amplification and detection of nonspecific products or primer–dimers, which may result in false positives or overestimates of input DNA. In this case, the nonspecific products can be identified through a melting curve experiment subsequent to the PCR, assuring quality control over the amplification specificity. The use of antibody-based hot-start PCR and LUX primer design algorithms avoid these artifacts, which go largely undetected in probe-based technologies but can affect PCR efficiencies.

The LUX platform enables simultaneous detection of multiple targets (multiplexing) and is compatible with real-time PCR instruments that detect FAM (all instruments) and JOE labels (most instruments).

The PCR using fluorogenic primers has great value as a tool in DNA and RNA detection, including quantitative, real-time PCR and single-nucleotide polymorphism (SNP) detection because of several important features. The synthesis of mono-labeled oligonucleotides is cost-effective, and the purification procedure, although effective, is less rigorous compared to dual-labeled probes and primers. For future developments, we plan to extend the LUX fluorogenic primer platform to SNP detections and to a universal format of detection in which a general, labeled LUX primer can be incorporated into different amplicons through the use of unlabeled primer pairs, where one unlabeled primer has an adapter tail (Nuovo et al. 1999; Myakishev et al. 2001).

ACKNOWLEDGMENTS

We thank Joseph Hayes for synthesis of LUX primers and Irina Nazarenko and David Schuster for their valuable contributions to the development of the LUX detection system.

REFERENCES

Cardullo R.A., Agrawal S., Flores C., Zamecnik P.C., and Wolf D.E. 1988. Detection of nucleic acid hybridization by nonradiative fluorescence resonance energy transfer. *Proc. Natl. Acad. Sci.* **91:** 8790–8794.

Higuchi R., Fockler C., Dollinger G., and Watson R. 1993. Kinetic PCR analysis: Real-time monitoring of DNA amplification reactions. *Bio/Technology* **11:** 1026–1030.

Higuchi R., Fockler C., Walsh P.S., Griffith R. 1992. Simultaneous amplification and detection of specific DNA sequences. *Bio/Technology* **10**: 413–417.

Holland P.M., Abramson R.D., Watson R., and Gelfand D.H. 1991. Detection of specific polymerase chain reaction products by utilizing the 5´-3´ exonuclease activity of DNA polymerase. *Proc. Natl. Acad. Sci.* **88**: 7276–7280.

Murchie A.I.H., Clegg R.M., von Kitzing E., Duckkett D.R., Diekmann S., and Lilley D.M.J. 1989. Fluorescence energy transfer shows that the four-way DNA junction is a right-handed cross of antiparallel molecules. *Nature* **341**: 763–766.

Myakishev M.V., Khripin Y., Hu S., and Hamer D.H. 2001. High-throughput SNP genotyping by allele-specific PCR with universal energy-transfer-labeled primers. *Genome Res.* **11**: 163–169.

Nazarenko I.A., Bhatnagar S.K., and Hohman R.J. 1997. A closed tube format for amplification and detection of DNA based on energy transfer. *Nucleic Acids Res.* **25**: 2516–2521.

Nazarenko I., Pires R., Lowe B., Obaidy M., and Rashtchian A. 2002a. Effect of primary and secondary structure of oligodeoxyribonucleotides on the fluorescent properties of conjugated dyes. *Nucleic Acids Res.* **30**: 2089–2195.

Nazarenko I., Lowe B., Darfler M., Ikonomi P., Schuster D., and Rashtchian A. 2002b. Multiplex quantitative PCR using self-quenched primers labeled with a single fluorophore. *Nucleic Acids Res.* **30**: e37.

Nuovo G.J., Hohman R.J., Nardone G.A., and Nazarenko I. 1999. In situ amplification using universal energy transfer-labeled primers. *J. Histochem. Cytochem.* **47**: 273–279.

Ririe K.M., Rasmussen R.P., and Wittwer C.T. 1997. Product differentiation by analysis of DNA melting curves during the polymerase chain reaction. *Anal. Biochem.* **245**: 154–160.

Steenken S. and Jovanovic V. 1997. How easily oxidizable is DNA? One-electron reduction potential of adenosine and guanosine radicals in aqueous solution. *J. Am. Chem. Soc.* **119**: 617–618.

Tyagi S. and Kramer F.R. 1996. Molecular beacons: Probes that fluoresce upon hybridization. *Nat. Biotechnol.* **14**: 303–308.

Wittwer C.T., Herrmann M.G., Moss A.A., and Rasmussen R.P. 1997. Continuous fluorescence monitoring of rapid cycle DNA amplification. *BioTechniques* **22**: 130–138.

WWW RESOURCES

https://pf.invitrogen.com/primerf/pages/Default.cfm?cc=1 Invitrogen, LUX™ Designer page

www.invitrogen.com Invitrogen home page or www.invitrogen.com/LUX

16

Amplification of RNA: High-temperature Reverse Transcription and DNA Amplification with a Magnesium-activated Thermostable DNA Polymerase

Edward S. Smith, Alvin K. Li, Alice M. Wang,[1] David H. Gelfand, and Thomas W. Myers

Program in Core Research, Roche Molecular Systems, Alameda, California 94501

Analysis of gene expression has historically relied on reverse transcription to produce cDNA copies of mRNA for cloning and sequencing. The two enzymes that typically have been used are either avian myeloblastosis virus (AMV) or Moloney murine leukemia virus (MuLV) reverse transcriptase. A major advance in the study of gene expression and virus detection was the coupling of the reverse transcription (RT) reaction to PCR amplification, and these techniques have been described in numerous laboratory manuals and texts (Innis et al. 1990; Mullis et al. 1994; Dieffenbach and Dveksler 1995; Innis et al. 1995, 1999). A common problem encountered when using these mesophilic reverse transcriptases, however, is the inability of the enzyme to synthesize cDNA through stable RNA secondary structures. Furthermore, the intrinsic RNase H activity of these native retroviral reverse transcriptases has been suggested to be detrimental to the production of full-length cDNA during reverse transcription (Kotewicz et al. 1988), and enzymes devoid of this activity have often been used.

In addressing the problems encountered when using mesophilic reverse transcriptases, we have sought to produce enzyme systems that could efficiently perform RT-PCR in a single tube. A recombinant DNA polymerase from the thermophilic eubacterium *Thermus thermophilus* (rTth pol) was found to possess efficient reverse transcriptase activity in the presence of Mn++ (Myers and Gelfand 1991). Because the reverse transcription step is typically carried out between 60°C and 70°C, it is likely that many of the potential RNA secondary structures are unstable and long cDNA synthesis can be achieved. An additional advantage gained by performing the RT step at elevated temperatures is an increase in specificity of primer hybridization and subsequent extension by the rTth pol. The rTth pol

---

[1] Current address: Celera Diagnostics, 1401 Harbor Bay Parkway, Alameda, California 94502.

does possess an inherent 5′→3′ exonuclease/RNase H activity similar to *E. coli* DNA polymerase I (Auer et al. 1995, 1996). However, r*Tth* pol does not cleave RNA•DNA hybrids endonucleolytically, as is the case with retroviral reverse transcriptases. The 5′-nuclease activity of the r*Tth* pol does enable the enzyme to be fully compatible with the TaqMan technology.

The high-temperature RT reaction can be followed by PCR amplification, resulting in the amplification of the newly synthesized cDNA in a two-step coupled process requiring only the addition of a single (thermostable) enzyme. The first protocols were based on performing the RT step in the presence of Mn^{++}, followed by chelation of the Mn^{++} with ethylenebis(oxyethylenenitrilo)tetraacetic acid (EGTA), and then performing a standard PCR amplification activated by Mg^{++}. Subsequently, reaction conditions were determined that allowed r*Tth* pol to efficiently perform both reverse transcription and DNA amplification in a single buffer containing Mn^{++} (Myers and Sigua 1995). This protocol has formed the basis of many virus detection assays, including those for hepatitis C virus (HCV) (Young et al. 1993) and human immunodeficiency virus (HIV) (Mulder et al. 1994).

The continual evolution of RT-PCR has led to the development of new and improved enzymes for high-temperature RT. A second-generation thermostable reverse transcriptase is a recombinant DNA polymerase of the thermophilic organism *Thermus* species Z05 (*T. Z05*) (Meng et al. 2001). This DNA polymerase is very similar both genotypically and phenotypically to the r*Tth* pol, but it is more thermostable (T. Myers, unpubl.). However, the r*Tth* pol and *T. Z05* DNA polymerases only exhibit efficient RT activity in the presence of Mn^{++}, and there are drawbacks associated with Mn^{++}-activated RT activity. For example, the divalent metal ion-catalyzed hydrolysis of RNA is increased with Mn^{++} (Brown 1974), and it is well known that the fidelity of DNA polymerases is lower in the presence of Mn^{++} when compared to reactions using Mg^{++} as the divalent metal ion activator (Beckman et al. 1985; Leung et al. 1989).

A Klenow fragment of the DNA polymerase isolated from the thermophilic organism *Carboxydothermus hydrogenoformans* (*C. therm*) has been discovered that is capable of performing efficient high-temperature RT using Mg^{++} as the divalent metal ion activator (Markau et al. 2002). This enzyme, however, is not thermostable and thus cannot be used as a single enzyme for both RT and PCR amplification. Although certain applications may benefit from dissociating the RT and PCR steps with the use of multiple enzymes such that both reactions can be independently optimized (while maintaining the advantage of high-temperature RT), a layer of complexity is added if one desires to maintain a single buffer for both RT and PCR amplification.

We have found point mutations in the polymerase domain of r*Tth* pol that alter the RT-associated divalent metal ion activator requirements (Smith et al. 2000). These mutant DNA polymerases have significantly improved RT efficiency compared to wild-type enzyme using either Mn^{++} or Mg^{++} ion activation. Certain mutations result in a 1000-fold increase in Mg^{++}-activated RT-PCR efficiency when compared to wild-type enzyme activated with Mg^{++}, while they remain equal to or better than wild-type enzyme with Mn^{++} activation. Similar results were achieved when these point mutations were made in the *T. Z05* DNA polymerase (Smith et al. 2000). Because both the RT step and the PCR amplification can be performed in a single Mg^{++}-activated buffer system, a significantly higher fidelity is achieved (E. Smith, unpubl.). This new DNA polymerase is called the GeneAmp AccuRT RNA PCR Enzyme.

Similar to the r*Tth* pol, the GeneAmp AccuRT RNA PCR Enzyme is completely compatible with the use of the dUTP/uracil N-glycosylase (UNG) carryover prevention system (Longo et al. 1990), which is critical for diagnostic laboratory settings. It should be noted that for optimal results when the GeneAmp AccuRT RNA PCR Enzyme and dUTP are used, the dUTP concentration needs to be increased to 400 μM (vs. 200 μM when using dTTP, the dATP, dCTP, and dGTP remain at 200 μM in all reactions), and the Mg^{++} concentration needs to be increased to 2.5 mM (vs. 1.5 mM when using dTTP). In addition to providing carry-over contamination control, the use of dUTP and UNG provides a "hot start" to reduce nonspecific amplification (Innis and Gelfand 1999). When dUTP and UNG are used, non-specific extension products containing dUMP that are formed during nonstringent setup conditions are degraded by UNG and cannot be used either as primers or as templates.

The use of dUTP/UNG for carryover contamination is essential for the diagnostic laboratory setting. However, many researchers find the use of dUTP cumbersome, especially for cloning applications, and therefore do not take advantage of this useful strategy for contamination control and for the added benefit of achieving a hot start. Although a manual hot start or the use of a wax-mediated hot start would be feasible, the use of either methodology for large numbers of samples would be impractical. The use of a reversibly chemically modified enzyme that is activated at elevated temperatures (typically 95°C for 10 minutes) provides an excellent hot-start technique for PCR (Birch et al. 1996, 1997; Zangenberg et al. 1999). Unfortunately, this method is contraindicated for single-enzyme RT-PCR, primarily because of the increased divalent metal ion-catalyzed hydrolysis of the RNA template at the high temperature (Innis and Gelfand 1999).

We have taken a different approach for providing a hot-start alternative for RT-PCR by using aptamers (aptamers are oligonucleotides that have an affinity to a specific molecule; in this case, they have a specific and strong affinity to the DNA polymerase that results in inhibition of enzymatic activity). The design of aptamers by the systematic evolution of ligands by exponential enrichment (SELEX) process (Tuerk and Gold 1990) directed against thermostable DNA polymerases has been described previously (Dang and Jayasena 1996; Lin and Jayasena 1997). We have engineered aptamers for optimal performance with *T.* Z05 DNA polymerase (T. Myers, unpubl.), and this enzyme–aptamer combination forms the basis of the GeneAmp AccuRT Hot Start RNA PCR Enzyme. A unique feature of this hot-start technique is that it is reversible. At elevated temperatures (>55°C), the aptamer begins to lose affinity for the DNA polymerase, and by 65°C the enzyme is fully active. Once the PCR amplification has been completed and the reaction temperature is reduced, the aptamer again binds to the DNA polymerase and enzymatic activity is inhibited. Because of the rapid kinetics involved, aptamer-based hot-start technology functions well in rapid thermal cycling instruments (T. Myers, unpubl.). Efficient high-temperature reverse transcription with the thermophilic and thermostable *T.* Z05 DNA polymerase, especially in conjunction with aptamer hot-start technology, is proving useful in a wide range of diagnostic and research applications.

PROTOCOL

HIGH-TEMPERATURE REVERSE TRANSCRIPTION AND DNA AMPLIFICATION WITH A MAGNESIUM-ACTIVATED THERMOSTABLE DNA POLYMERASE

MATERIALS

▼ CAUTION

See Appendix for appropriate handling of materials marked with <!>.

BUFFERS, SOLUTIONS, AND REAGENTS

Agarose

Dimethyl sulfoxide (see Note 1) (DMSO) <!>

Ethidium bromide <!>

Magnesium acetate (see Note 2) (Mg[OAc]$_2$) 25 mM <!>, (Applied Biosystems 4339582 or 4339585)

RNase-free H_2O

SYBR Green I Dye (see Note 1) <!> (Molecular Probes)

ENZYMES AND ENZYME BUFFERS

5x AccuRT Buffer (see Note 2) (Applied Biosystems 4339582 or 4339585)

GeneAmp AccuRT RNA PCR Enzyme (see Note 2): 3.75 units/µl (Applied Biosystems 4339582)

GeneAmp AccuRT Hot Start RNA PCR Enzyme (see Note 2): 3.75 units/µl (Applied Biosystems 4339585)

AmpErase uracil-N-glycosylase (UNG): 1 unit/µl (Applied Biosystems)

NUCLEIC ACIDS AND OLIGONUCLEOTIDES

Deoxynucleoside triphosphates: neutralized 10 mM solutions (Applied Biosystems, N808-0007 or N808-0095 containing 20 mM dUTP in place of dTTP)

Oligonucleotide primers: 15 µM in 10 mM Tris-HCl<!>, pH 8.3

RNA: generally stored in carrier RNA such as 30 µg/ml *E. coli* rRNA or poly(rA) in either RNase-free H_2O or 10 mM Tris-HCl<!>, pH 7.0, 1 mM EDTA, 0.1 mM NaCl

SPECIAL EQUIPMENT

Electrophoresis apparatus for agarose gels

GeneAmp PCR System 9700 or 2700 (Applied Biosystems), or equivalent

Real-Time quantitative PCR detection system (see Note 1): such as Applied Biosystem's GeneAmp 5700 Sequence Detection System or ABI PRISM 7000, 7700, or 7900HT Sequence Detection Systems or the Roche Applied Science LightCycler

ADDITIONAL ITEMS

MicroAmp reaction tubes (Applied Biosystems), or equivalent

NOTES

1. Required if performing dye-binding, real-time quantitative PCR.

2. Sold as a kit, which includes either the GeneAmp AccuRT RNA PCR Enzyme or GeneAmp AccuRT Hot Start RNA PCR Enzyme, 5x AccuRT Buffer, and 25 mM $Mg(OAc)_2$.

METHOD

RNA ISOLATION

The preparation of high-quality RNA for study is of paramount importance. There are numerous commercial products on the market, as well as detailed protocols available (Dieffenbach and Dveksler 1995; Farrell 1998; Sambrook and Russell 2001; Chapter 10).

COMBINED RT AND PCR AMPLIFICATION
Reactions Performed with dTTP and without UNG

1. Thaw all reagents and gently vortex. Avoid the generation of bubbles.

2. Spin down the enzyme and reagents in a microfuge and place on ice.

3. Combine the following reagents to make a master mix: (*Note:* Reagents in bold are required in different concentrations for reactions performed with dUTP and UNG.)

Reagent	Amount for one reaction (μl)	Final concentration
RNase-free H$_2$O	28.66	–
5x AccuRT Buffer	10.00	1x
10 mM dATP	1.00	200 μM
10 mM dCTP	1.00	200 μM
10 mM dGTP	1.00	200 μM
10 mM dTTP	**1.00**	**200 μM**
AccuRT or AccuRT Hot Start Enzyme (3.75 units/μl)	2.00	7.5 units /50 ml
15 μM "upstream" primer	0.67	0.20 μM
15 μM "downstream" primer	0.67	0.20 μM
25 mM Mg(OAc)$_2$	**3.00**	**1.5 mM**

4. Gently vortex the master mix, add 1.00 μl of RNA (up to 1 μg; see Troubleshooting and Tips), and gently vortex the entire reaction mix. Any combination of RNase-free H$_2$O and RNA can be used as long as the total volume of the master mix with RNA equals 50 μl.

5. Perform the RT reaction and PCR amplification consecutively in a GeneAmp PCR System 9700 or 2700 as follows:
   30 minutes at 65°C (RT step)
   1 minute at 95°C
   15 seconds at 95°C and 30 seconds at 65°C for 40 cycles (PCR amplification)
   7 minutes at 65°C

6. The reactions can be analyzed directly or stored at –20°C.

Reactions Performed with dUTP and UNG (for Carryover Contamination Control and UNG-mediated Hot Start)

1. Thaw all reagents and gently vortex. Avoid the generation of bubbles.

2. Spin down the enzyme and reagents in a microfuge and place on ice.

3. Combine the following reagents to make a master mix: (*Note:* Reagents in bold are required in different concentrations for reactions performed with dTTP and without UNG.)

Reagent	Amount for one reaction (μl)	Final concentration
RNase-free H$_2$O	25.66	–
5x AccuRT Buffer	10.00	1x
10 mM dATP	1.00	200 μM
10 mM dCTP	1.00	200 μM
10 mM dGTP	1.00	200 μM
20 mM dUTP	**1.00**	**400 μM**
AccuRT or AccuRT Hot Start Enzyme (3.75 unit/μl)	2.00	7.5 units/50 μl
AmpErase uracil *N*-glycosylase (UNG) (1.0 unit/μl)	1.00	1.0 unit/50 μl
15 μM "upstream" primer	0.67	0.20 μM
15 μM "downstream" primer	0.67	0.20 μM
25 mM Mg(OAc)$_2$	**5.00**	**2.5 mM**

4. Gently vortex the master mix, add 1.00 µl of RNA (up to 1 µg; see Troubleshooting and Tips), and gently vortex the entire reaction mix. Any combination of RNase-free H₂O and RNA can be used as long as the total volume of the master mix with RNA equals 50 µl.

5. Perform the RT reaction and PCR amplification consecutively in a GeneAmp PCR System 9700 or 2700 as follows:

> 2 minutes at 50°C (UNG step)
> 30 minutes at 65°C (RT step)
> 1 minute at 95°C
> 15 seconds at 95°C and 30 seconds at 65°C for 40 cycles (PCR amplification)
> 7 minutes at 65°C

6. Analyze the samples immediately after PCR amplification, or inactivate any remaining UNG by storing the reactions at –20°C or treating chemically (for example, by extracting with an equal volume of chloroform). The reactions can be stored long term at –20°C.

> The UNG does not need to be deactivated before analysis; however, it will destroy amplicons in reactions that are allowed to sit at room temperature or at elevated temperatures <55°C.

ANALYSIS OF REACTION PRODUCTS

To detect the amplification products, samples can be separated electrophoretically on a 2% agarose gel and stained with 0.5 µg/ml ethidium bromide. Alternatively, the protocols provided are completely compatible with real-time quantitative PCR by TaqMan technology (Meng et al. 2001), ethidium bromide (Higuchi and Watson 1999; Kang and Holland 1999; Kang et al. 2000; Holland 2002), or SYBR Green I Dye (Baranzini et al. 2000). Direct detection of PCR product can be monitored by measuring the increase in fluorescence caused by the intercalation of ethidium bromide (2.5 µM) or the binding of SYBR Green I Dye to double-stranded DNA. When using SYBR Green I Dye, the concentration of the dye used in the experiment is important. When adding SYBR Green I Dye to reactions, add it to the master mix at a final concentration of "0.2x" from a "20x" working stock in 100% DMSO (the "20x" working stock is a 500-fold dilution in DMSO of the SYBR Green I Dye obtained from Molecular Probes). Higher concentrations have been observed to inhibit the RT activity of many enzymes.

▶ TROUBLESHOOTING AND TIPS

• *Handling of reagents.* Before use, thaw frozen reagents, mix well by careful vortexing, spin down the reagents in a microfuge, and store on ice until used. Pipette reagents slowly to avoid pipetting errors and to reduce the generation of bubbles, especially with protein solutions. Add reagents in the order specified in the protocols. It is especially important to add the Mg(OAc)₂ either just before adding template or immediately afterward to minimize generating nonspecific extension products when the GeneAmp AccuRT Hot Start RNA PCR Enzyme is not used. Any combination of RNase-free water and template can be used as long as the total volume of the RT-PCR Master Mix, including buffer, dNTPs, enzyme, Mg(OAc)₂, primers, and template equals 50 µl. Care should be taken, however, to avoid adding high concentrations of EDTA or other metal ion chelators with the primers or template.

• *Poor results because of problems with the RNA template.* Although not addressed in this protocol, the preparation of "full-length" starting template RNA is also critically important. The isolation of intact RNA becomes especially critical as the target of choice increases in length. The experimental RNA sample should be diluted in RNase-free water. If the starting concentration of the RNA is low, the addition of carrier nucleic acid (e.g., *E. coli* rRNA or poly[rA]) will prevent nonspecific binding of RNA to sample tubes, etc. Low concentrations (copy number or molecules) of target RNA may require up to 35 or more cycles of PCR amplification to produce sufficient product for analysis. High levels of total RNA (>1 μg) may cause inhibition and should be avoided.

- *Tailor the reaction conditions to the annealing temperature of the primer pair.* The selection of 65°C for the RT temperature and the anneal–extend thermal cycling temperature was made because of optimal performance with multiple targets. However, it may be necessary to decrease or raise these temperatures for other primer–template pairs. It is important to remember that this temperature cannot be significantly decreased when using the GeneAmp AccuRT Hot Start RNA PCR Enzyme because of the presence of the aptamer, which will decrease enzymatic activity at lower temperatures. Additionally, maintaining these elevated temperatures will generally result in less nonspecific product formation. The optimum anneal–extend thermal cycling temperature can be determined empirically by testing small temperature increments until the maximum in specificity and product yield is achieved. The optimum temperature for the GeneAmp RNA PCR Enzymes is 65–80°C, depending on template and reaction buffer composition. High-temperature reverse transcription using r*Tth* pol and *T.* Z05 DNA polymerase helps to alleviate many of the problems typically encountered when amplifying RNA targets. However, increased metal ion-catalyzed hydrolysis of template RNA is observed at the elevated temperatures (Brown 1974) used with these enzymes. Therefore, factors such as increased enzyme activity and reaction specificity, as well as the reduction of RNA secondary structure, must be balanced with target degradation.

- *Target sequence length considerations.* The length of the target sequence being amplified will define the RT reaction time and the PCR amplification extension time. Longer targets and amplicons will require longer reaction times and sometimes can be more difficult to analyze using real-time PCR methods. Generally, RT reaction times greater than 60 minutes have been counterproductive. The thermal cycling parameters provided in the protocol were chosen for optimal performance with amplicons 200–500 nucleotides in length.

- *Optimization of RT-PCR amplifications.* The reagent concentrations provided in the protocol were chosen for optimal performance with multiple targets. Adjustment of reagent concentrations may be required to fully optimize any given RT-PCR. If the dNTP concentration in the reaction is modified, the concentration should remain equimolar (the exception being dUTP) so that misincorporation of the incorrect nucleotide by the enzyme is minimized. When using dUTP, a final concentration of 400 μM dUTP should be used, and the dATP, dCTP, and dGTP should remain at 200 μM in all reactions. The optimal $Mg(OAc)_2$ concentration should be determined empirically by titration between 1 and 4 mM, keeping in mind that the Mg^{++} requirement may vary depending on the total dNTP concentration and on the primers and template used. In general, 1.5 mM $Mg(OAc)_2$ is a good starting place when using dTTP, and 2.5 mM $Mg(OAc)_2$ is the preferred starting point when using dUTP.

  Reaction conditions are extremely important in the overall fidelity achieved in an RT-PCR. A key feature of the GeneAmp AccuRT RNA PCR Enzymes is that Mg^{++} activation provides increased fidelity (E. Smith, unpubl.). However, enhanced fidelity also may be achieved by using the minimal extension time and highest extension temperature possible, using low dNTP and divalent metal ion concentrations, and amplifying for as few cycles as necessary (Eckert and Kunkel 1990; Goodman 1995).

- *Use of cosolvents and primer design.* The addition of a final, total 1% DMSO concentration to the reaction often reduces nonspecific product formation and improves the results with the GeneAmp AccuRT RNA PCR Enzymes. However, the DMSO, glycerol, and/or other cosolvent concentrations present in the reaction must be taken into account when designing primers, because they effectively lower the T_m of the oligonucleotides (Landre et al. 1995). The use of sequence-specific oligonucleotide primers is preferred, especially when reaction specificity is critical. Primers are typically 15–30 nucleotides in length and should be purified by gel electrophoresis or high-performance liquid chromatography/ion-exchange chromatography. Primer sequences should not be complementary within themselves or to each other, particularly at the 3′ terminus, to reduce primer artifact formation. Another common technique to reduce terminal overlap of primers and primer artifacts, especially in complex multiplex PCR amplifications, is to create primers that have the dinucleotide "AA" at their 3′ termini (Zangenberg et al. 1999).

- *Importance of contamination control.* Finally, it is critical to realize the enormous amplification of products produced during PCR. Low levels of DNA contamination, especially from previous PCR amplifications, samples with high DNA levels, or positive control templates can result in product formation, even in the absence of intentionally added template DNA. In addition to following protocols to minimize carryover of amplified DNA (Higuchi and Kwok 1989; Kwok 1990), the use of the dUTP/UNG carryover prevention system (Longo et al. 1990 and Chapter 2) is strongly advised, especially for clinical diagnostic applications.

REFERENCES

Auer T., Landre P.A., and Myers T.W. 1995. Properties of the 5′→3′ exonuclease/ribonuclease H activity of *Thermus thermophilus* DNA polymerase. *Biochemistry* **34:** 4994–5002.

Auer T., Sninsky J.J., Gelfand D.H., and Myers T.W. 1996. Selective amplification of RNA utilizing the nucleotide analog dITP and *Thermus thermophilus* DNA polymerase. *Nucleic Acids Res.* **24:** 5021–5025.

Baranzini S.E., Elfstrom C., Chang S.Y., Butunoi C., Murray R., Higuchi R., and Oksenberg J.R. 2000. Transcriptional analysis of multiple sclerosis brain lesions reveals a complex pattern of cytokine expression. *J. Immunol.* **165:** 6576–6582.

Beckman R.A., Mildvan A.S., and Loeb L.A. 1985. On the fidelity of DNA replication: Manganese mutagenesis *in vitro*. *Biochemistry* **24:** 5810–5817.

Birch D.E., Laird W.J., and Zoccoli M.A. 1997. Nucleic acid amplification using a reversibly inactivated thermostable enzyme. U.S. Patent No. 5,677,152.

Birch D.E., Kolmodin L., Wong J., Zangenberg G.A., Zoccoli M.A., McKinney N., Young K.K.Y., and Laird W.J. 1996. Simplified hot start PCR. *Nature* **381:** 445–446.

Brown D.M. 1974. Chemical reactions of polynucleotides and nucleic acids. In *Basic principles in nucleic acid chemistry* (ed. P.O.P. Ts'o), pp. 43–44. Academic Press, New York.

Dang C. and Jayasena S.D. 1996. Oligonucleotide inhibitors of *Taq* DNA polymerase facilitate detection of low copy number targets by PCR. *J. Mol. Biol.* **264:** 268–278.

Dieffenbach C.W. and Dveksler G.S., eds. 1995. *PCR primer: A laboratory manual.* Cold Spring Harbor Laboratory Press, Cold Spring Harbor, New York.

Eckert K.A. and Kunkel T.A. 1990. High fidelity DNA synthesis by the *Thermus aquaticus* DNA polymerase. *Nucleic Acids Res.* **18:** 3739–3744.

Farrell R., ed. 1998. *RNA methodologies: A laboratory guide for isolation and characterization,* 2nd edition. Academic Press, San Diego, California.

Goodman M.F. 1995. DNA polymerase fidelity: Misinsertions and mismatched extensions. In *PCR strategies* (ed. M.A. Innis et al.), ch. 2, pp. 17–31. Academic Press, San Diego.

Higuchi R. and Kwok. S. 1989. Avoiding false positives with PCR. *Nature* **339:** 237–238.

Higuchi R. and Watson R. 1999. Kinetic PCR analysis using a CCD camera and without using oligonucleotide probes. In *PCR applications: Protocols for functional genomics* (ed. M.A. Innis et al.), ch. 16, pp. 263–284. Academic Press, San Diego, California.

Holland M.J. 2002. Transcript abundance in yeast varies over six orders of magnitude. *J. Biol. Chem.* **277:** 14363–14366.

Innis M.A. and Gelfand D.H. 1999. Optimization of PCR: Conversations between Michael and David. In *PCR applications: Protocols for functional genomics* (ed. M.A. Innis et al.), ch. 1, pp. 3–22. Academic Press, San Diego, California.

Innis M.A., Gelfand D.H., and Sninsky J.J. 1995. *PCR strategies.* Academic Press, San Diego, California.

Innis M.A., Gelfand D.H., and Sninsky J.J., eds. 1999. *PCR applications: Protocols for functional genomics.* Academic Press, San Diego, California.

Innis M.A., Gelfand D.H., Sninsky J.J., and White T.J., eds. 1990. *PCR protocols: A guide to methods and applications.* Academic Press, San Diego, California.

Kang J.J. and Holland M.J. 1999. Cellular transcriptome analysis using a kinetic PCR assay. In *PCR applications: Protocols for functional genomics* (ed. M.A. Innis et al.), ch. 27, pp. 429–444. Academic Press, San Diego, California.

Kang J.J., Watson R.M., Fisher M.E., Higuchi R., Gelfand D.H., and Holland M.J. 2000. Transcript quantitation in total yeast cellular RNA using kinetic PCR. *Nucleic Acids Res.* **28:** i–viii.

Kotewicz M.L., Sampson C.M., D'Alessio J.M., and Gerard G.F. 1988. Isolation of cloned Moloney murine leukemia virus reverse transcriptase lacking ribonuclease H activity. *Nucleic Acids Res.* **16:** 265–277.

Kwok S. 1990. Procedures to minimize PCR-product carry-over. In *PCR protocols: A guide to methods and applications* (ed. M.A. Innis et al.), ch. 17, pp. 142–145. Academic Press, San Diego, California.

Landre P.A., Gelfand D.H., and Watson R.M. 1995. The use of cosolvents to enhance amplification by the polymerase chain reaction. In *PCR strategies* (ed. M.A. Innis et al.), ch. 1, pp. 3–16. Academic Press, San Diego, California.

Leung D.W., Chen E., and Goeddel D.V. 1989. A method for random mutagenesis of a defined DNA segment using a modified polymerase chain reaction. *Technique* **1:** 11–15.

Lin Y. and Jayasena S.D. 1997. Inhibition of multiple thermostable DNA polymerases by a heterodimeric aptamer. *J. Mol. Biol.* **271:** 100–111.

Longo M.C., Berninger M.S., and Hartley J.L. 1990. Use of uracil DNA glycosylase to control carry-over contamination in polymerase chain reactions. *Gene* **93:** 125–128.

Markau U., Ebenbichler C., Achhammer G., and Ankenbauer W. 2002. Modified DNA-polymerase from *Carboxydothermus hydrogenoformans* and its use for coupled reverse transcription and polymerase chain reaction. U.S. Patent No. 6,399,320.

Meng Q., Wong C., Rangachari A., Tamatsukuri S., Sasaki M., Fiss E., Cheng L., Ramankutty T., Clarke D., Yawata H., Sakakura Y., Hirose T., and Impraim C. 2001. Automated multiplex assay system for simultaneous detection of hepatitis B virus DNA, hepatitis C virus RNA, and human immunodeficiency virus type 1 RNA. *J. Clin. Microbiol.* **39:** 2937–2945.

Mulder J., McKinney N., Christopherson C., Sninsky J., Greenfield L., and Kwok S. 1994. Rapid and simple PCR assay for quantitation of human immunodeficiency virus type 1 RNA in plasma: Application to acute retroviral infection. *J. Clin. Microbiol.* **32:** 292–300.

Mullis K.B., Ferré F., Gibbs R.A., eds. 1994. *The polymerase chain reaction.* Birkhäuser, Boston, Massachusetts.

Myers T.W. and Gelfand D.H. 1991. Reverse transcription and DNA amplification by a *Thermus thermophilus* DNA polymerase. *Biochemistry* **30:** 7661–7666.

Myers T.W. and Sigua C.L. 1995. Amplification of RNA: High temperature reverse transcription and DNA amplification with *Thermus thermophilus* DNA polymerase. In *PCR strategies* (ed. M.A. Innis et al.), ch. 5, pp. 58–68. Academic Press, San Diego, California.

Sambrook J. and Russell D., eds. 2001. *Molecular cloning: A laboratory manual,* 3rd edition. Cold Spring Harbor Laboratory Press, Cold Spring Harbor, New York.

Smith E., Wang A., Sigua C.L., Schönbrunner N., Myers T.W., and Gelfand D.H. 2000. High temperature magnesium (Mg^{2+})-activated reverse transcription by thermostable DNA polymerases. *Clin. Chem.* **46:** 1880.

Tuerk C. and Gold L. 1990. Systematic evolution of ligands by exponential enrichment: RNA ligands to bacteriophage T4 DNA polymerase. *Science* **249:** 505–510.

Young K.K.Y., Resnick R.M., and Myers T.W. 1993. Detection of hepatitis C virus RNA by a combined reverse transcription-polymerase chain reaction assay. *J. Clin. Microbiol.* **31:** 882–886.

Zangenberg G., Saiki R., and Reynolds R. 1999. Multiplex PCR: Optimization guidelines. In *PCR applications: Protocols for functional genomics* (ed. M.A. Innis et al.), ch. 6, pp. 73–94. Academic Press, San Diego, California.

DETECTION OF PCR PRODUCTS: SPECIALIZED APPLICATIONS

PCR is a multistep process—sample preparation, amplification, and the detection of the resulting DNA fragment. A detection system must accurately and reproducibly reflect the nature and quantity of the starting template. This section presents specialized methods of detection that are tailored to specific applications. The simplest detection method uses gel electrophoresis to define the size of the DNA product. The appearance of a discrete band of the correct size may be indicative of a successful reaction, but probing with an internal oligonucleotide is usually necessary to be certain of the identity of the product. This is particularly important when setting up a reaction with a new primer set, in which case the possible background application is unknown. In addition, the presence of a band of the predicted size does not provide any quantitative information.

CHOOSING THE BEST DETECTION METHODS

If the experiment requires quantitative measurement of the product, the best methods are real-time PCR techniques. These methods, described in Section 2, combine the amplification and detection steps and are sufficiently robust to allow the simultaneous measurement of up to four PCR products per reaction. For detection of multiple products, each reaction must contain a specific oligonucleotide probe for each amplified sequence. In the TaqMan system, the probe has a defined structure—the target sequence is flanked by GC-rich regions that form a stem-loop structure at temperatures below the T_{m} of the oligonucleotide probe, and the two ends of the molecule are labeled, respectively, with a fluorescent reporter and a quencher. The GC-rich stem-loop brings the fluorescent reporter and a quencher molecule into close proximity, preventing the molecule from fluorescing. Signal is generated during each cycle of the PCR when the probe hybridizes to its target sequence, followed by the processive degradation of this oligonucleotide probe by the 5′-exonuclease activity of *Taq* DNA polymerase. This releases the reporter molecule from the physical proximity of the quencher and produces a signal, and the amount of signal reflects the amount of template that has accumulated.

Other approaches to the detection of PCR products include systems using nonradioactive enzyme-linked immunofluorescent assay (ELISA)-based formats. These methods are useful in clinical research and diagnosis. The detection system described in "Applying PCR for Microarray-based Gene Expression Analysis" (p. 281) is an example of such an approach. ELISA-based formats are adaptable to numerous targets and can provide quantitative data.

Where specific PCR protocols end and unique detection methods begin is sometimes difficult to determine. The chapters "Single-strand Conformational Polymorphism Analysis"

(p. 237), "Sensitive and Fast Mutation Detection by Solid-phase Chemical Cleavage of Mismatches" (p. 265), and "Loss of Heterozygosity, A Multiplex PCR Method to Define Narrow Deleted Chromosomal Regions of a Tumor Genome" (p. 223) provide methods for identifying point mutations and the existence or loss of heterogeneity. The SSCP method relies on the resolving power of acrylamide gels to detect the sequence-dependent variation in the folding of single-stranded DNA. This method is easier to perform when compared with related methods such as denaturing gradient gel electrophoresis (Myers et al. 1987) and the GC clamp (Sheffield et al. 1989). However, SSCP misses a fraction of the sequence variants. An alternative approach that detects nearly all sequence changes in a defined region is provided in the chapter "Sensitive and Fast Mutation Detection by Solid-phase Chemical Cleavage of Mismatches (p. 265)." This technique is adaptable to a high-throughput format and can detect the mutation even when it is present in 5% of the analyzed material. Once a specific mutation or polymorphism has been defined, allele-specific detection systems can be developed. The most direct way to detect defined single-nucleotide sequence differences is to use sequence-specific oliogonucleotide probes. This technology was applied in some of the very first publications using PCR (Saiki et al. 1986).

To exploit the full power of PCR to detect the loss of heterozygosity, the sequence information from the Human Genome Project is used to select primers that flank simple tandem repeats—highly variable DNA segments that vary within alleles. Since each segment has a different length, the loss of one of the alleles is easy to detect using the resolving power of the Applied Biosystems GeneScan and GenoTyper systems.

Every PCR requires a method of analysis, and the methods described here are highly focused on the detection of previously unknown variations in DNA sequence at both the individual nucleotide level and when larger alterations, such as deletions, are suspected. The identification of these changes is the first step in the discovery of gene alterations that may be responsible for human disease.

REFERENCES

Myers R.M., Maniatis T., and Lerman L.S. 1987. Detection and localization of single base changes by denaturing gradient gel electrophoresis. *Methods Enzymol.* **155:** 501–527.

Saiki R.K., Bugawan T.L., Horn G.T., Mullis K.B., and Erlich H.A. 1986. Analysis of enzymatically amplified β-globin and HLA-DQα DNA with allele-specific oligonucleotide probes. *Nature* **324:** 163–166.

Sheffield V.C., Cox D.R., Lerman L.S., and Myers R.M. 1989. Attachment of a 40-base-pair G + C-rich sequence (GC-clamp) to genomic DNA fragments by the polymerase chain reaction results in improved detection of single-base changes. *Proc. Natl. Acad. Sci.* **86:** 232–236.

Loss of Heterozygosity, a Multiplex PCR Method to Define Narrow Deleted Chromosomal Regions of a Tumor Genome

Lise Lotte Hansen[1] and Just Justesen[2]

[1]Department of Human Genetics, [2]Department of Molecular Biology, University of Aarhus,
DK-8000 Aarhus C, Denmark

In the extensive search for new tumor suppressor genes, the effort to identify small well-defined lesions of the tumor genomes has intensified over the last decade. Comparative genome hybridization (CGH) can be used to scan an entire tumor genome for either large deletions or amplified regions. To narrow a particular (deleted) chromosomal region or a region flanking a candidate tumor suppressor gene, characterization of loss of heterozygosity (LOH) or allelic loss (AL) is very useful. The essence of the current LOH method is the comparison of allele intensities between PCR-amplified simple tandem repeats (STRs) from individual matched wild-type and tumor DNAs from a cohort of cancer patients with the same tumor type.

Access to the results from the Human Genome Project has improved the accuracy with which we can establish the borders of an allelic loss in a specific tumor. The locations of STRs are well documented and, importantly, it is possible to calculate the distance between the different STRs and the gene of interest, thereby defining the deleted region fairly accurately. An intragenic STR marker is very valuable, because it reveals whether a part of, or maybe an entire, gene is deleted from one chromosome in the tumor genome.

Initially, LOH analysis was carried out by labeling one primer with [γ-^{33}P]dCTP prior to PCR amplification of the STR. The alleles of two different STRs could, with careful planning, be separated by gel electrophoresis in a single run. Interpretation of the results was performed manually by careful examination of each sample from the autoradiography. Establishment of the order of the STR markers on the chromosome was tedious, being achieved either by comparison of maps from different sources or by assigning the marker by PCR amplification of the human/hamster hybrid systems. These methods are now obsolete, but the markers produced served as a skeleton for the sequencing of the human genome.

Most of the procedures have now been automated, resulting in a dramatic increase in throughput. Primers flanking each STR are labeled with fluorophores, which are available in increasing numbers of different colors. Several different fluorophores can be detected simultaneously during a single run. PCR amplifications are carried out in multiplex with four to five STR markers in a single reaction. These are pooled and, using five different colors, it is possible to analyze more than 20 STR markers in one run. One bottleneck has been avoided by the introduction of a multicapillary system in place of gels. Using four capillaries, more than 4000 STR markers can be analyzed in 24 hours. This increased throughput demands efficient software capable of handling large amounts of information. Some software is supplied with the automated electrophoresis instrument, but it is necessary to extend this with a spreadsheet as suggested below.

SELECTION OF STR MARKERS

STR markers can be selected from large chromosomal regions that have been identified by cytogenetic analysis, genome screening by comparative genome hybridization (CGH), or from either intragenic areas or regions flanking a putative tumor suppressor gene. Using the database of human genome sequences, it is relatively simple to locate uncharacterized STR sequences in introns or intergenic areas. Repeated sequences in translated exons are seldom polymorphic, but because the mutation rate is higher for repeated than nonrepeated sequences, it is worth analyzing the allele frequencies in tumor cells. Many tumor cells escape strict DNA repair, and analysis of allele sizes of exonic STRs is a fast way to locate tumor-specific mutations inhibiting normal gene expression.

Databases such as the National Center for Biotechnology Information (NCBI), The Human Genome Browser, Ensembl, and The Genome Database (GDB) provide detailed information on the genome structure and the precise location of well-characterized STR markers. Information on STR marker amplification is also available. This includes primer sequences, an indication of amplicon size, the degree of polymorphism, and the number of alleles. GDB is a particularly valuable source of information, and a specified PCR amplification protocol is often available for a particular STR.

STR Markers Are Primarily Selected According to:

1. Location in the genome.

    It is important to choose at least two STR markers flanking each side of the gene of interest. If one marker is noninformative (homozygous), the second marker may be informative.

2. Degree of polymorphism within the STR marker.

    Dinucleotide repeats tend to have a higher rate of polymorphism than tri- or tetra-nucleotide repeats.

Four or five STR markers may be amplified (PCR) in the same tube using 10–20 ng of genomic DNA. Each set of markers is selected according to the following criteria:

1. Annealing temperature.

    The primer pairs for the STR markers must be matched for T_m. There are several methods by which to determine the T_m of the primer pairs, including base content and computer programs. We find that using the very simple rule of thumb, that an AT bond adds 2°C and a CG bond 4°C to the melting temperature, starting from 0°C, is fast and reliable in the vast majority of PCR amplifications.

2. Size of the PCR product.

    When using ABI equipment, PCR products between 76 and 489 bp are measured using two size marker peaks that flank the DNA fragment. Fragments shorter than 75 bp should be avoided because they may disappear (be swallowed) in the large primer peak, which is detected initially during the run. The size of fragments larger than 489 bp cannot be established with the standard size marker. However, it is possible to use a custom-designed size marker. When setting up multiplex reactions, amplicons must differ by at least 20 bp to be sure of detection. This is straightforward if the markers and their alleles are well defined. If only the minimum length of the STR marker is provided by the database, add a spacing of 40–50 bp to ensure detection of larger alleles.

3. **PCR amplification efficiency.**

The primer concentrations may vary based upon conditions of amplification for each STR marker. It is often necessary to add more primer (1.5–2 pmoles) to long-range PCR amplifications to ensure an even representation of the different PCR products.

4. **Selection of STR markers.**

Where possible, avoid mononucleotide repeated sequences. These PCR products often contain many stuttered bands, making it difficult to interpret the correct alleles. The problem may be overcome by adding 1.5–2.0 mole/liter betaine to the PCR amplification. The selection criteria are illustrated in Figure 1.

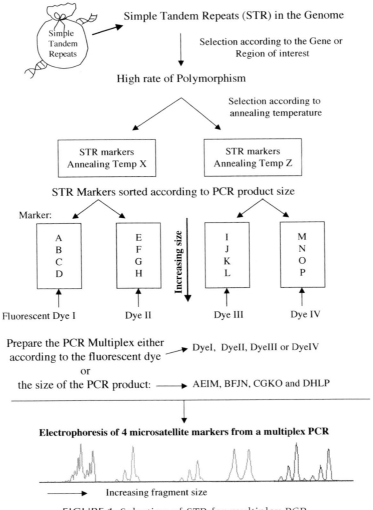

FIGURE 1. Selection of STR for multiplex PCR.

This chapter focuses on the characterization of LOH, from the isolation of high-quality DNA and the preparation of multiplex PCRs through to general data handling. We discuss the initial analysis of the electrophoresis products and the setting up of algorithms for automated LOH calculations and individual/specific tumor tissue LOH cutoff values using a "home-made" spreadsheet in Excel.

PROTOCOL 1   PURIFICATION OF HIGH-QUALITY DNA

DNA is purified using a modified version of the salt precipitation method published by Miller et al. (1988). The method is gentle, limits the breakage of the long chromosomal strands, and avoids the use of phenol and chloroform. Alternative DNA purification methods are described in Chapter 9. This method has been used successfully for blood samples, cultured cells, and breast tumor tissue, which has been subjected to hormone receptor analysis, leaving a pellet containing the nuclei (Hansen et al. 1998).

MATERIALS

▼ CAUTION

See Appendix for appropriate handling of materials marked with <!>.

BUFFERS, SOLUTIONS, AND REAGENTS

EDTA, 100 mM

Isopropanol<!>

Lysis buffer

    10 mM Tris-HCl<!>

    1 mM EDTA

    150 mM NaCl

    0.5% SDS<!>, pH 10.5

NaCl, 0.15 M and 6 M

Nuclei lysis buffer:

    75 mM NaCl

    24 mM EDTA, pH 8.0

Proteinase K<!> or Pronase, 20 mg/ml

SDS<!>, 20% (w/v)

Sucrose, 1 M

Tris-HCl<!>, 2 M, pH 9.2

Triton lysis buffer

    10 mM Tris-HCl<!>

    5 mM $MgCl_2$<!>

    1% Triton X-100

    TE buffer (10 mM Tris-HCl, 1 mM EDTA, pH 7.5)

Sucrose, pH 9.2

Triton X-100

SPECIAL EQUIPMENT

Water bath, preset to 55°C

Glass rod

CELLS AND TISSUES

Processed breast tumor tissue (per 100 mg tissue measured before the hormone receptor analysis) or 5×10^5 cultured cells washed twice in PBS

METHOD

1. To the chosen cells in a 10-ml tube, add 750 μl of lysis buffer and 200 μg of proteinase K (or pronase); incubate at 55°C for at least 90 minutes until the pellet become soluble.

2. Repeat step 1 if pellet is solid.

3. Add 250 μl of 6 M NaCl; shake carefully.

4. Centrifuge (3000 rpm) at 4°C for 15 minutes. Transfer the supernatant, with care, to a new tube.

5. Repeat steps 3 and 4.

6. Add 1 volume of isopropanol. Invert the tube until the DNA is visible. Wind the DNA onto a glass rod, press the DNA gently against the tube wall to drain excess isopropanol, and transfer the DNA to 300 μl of TE or double-distilled autoclaved H_2O.

7. Leave to dissolve at 4°C.

Extraction of DNA from 10 ml of Blood

8. Add 40 ml of cold (4°C) Triton lysis buffer.

9. Leave at 4°C for 30 minutes; turn the tube a few times during the incubation.

10. Centrifuge (3000 rpm) at 4°C for 30 minutes. Discard the supernatant.

11. Wash the pellet (nuclei) with 5–15 ml of 0.9% (0.15 mole/liter) NaCl.

12. Centrifuge 10 minutes at 2300 rpm.

13. Discard the supernatant. The nuclei can be stored at –20°C.

14. Add 3 ml of nuclei lysis buffer, shake well.

15. Add 25 μl of pronase (20 mg/ml) and 230 μl of 10% SDS, leave overnight shaking gently at room temperature.

16. Add 1 ml of 6 M NaCl; shake well.

17. Centrifuge (3000 rpm) at 4°C for 15 minutes. Transfer the supernatant, with care, to a new tube.

18. Repeat step 10.

19. Add 1 volume of isopropanol. Turn the tube upside-down until the DNA is visible. Wind the DNA onto a glass rod, press the DNA gently against the tube wall to drain excess isopropanol, and transfer the DNA to 400 μl of TE or double-distilled autoclaved H_2O.

20. Leave to dissolve at 4°C.

PROTOCOL 2 MULTIPLEX PCR AMPLIFICATION

MATERIALS

ENZYMES AND ENZYME BUFFERS

Taq polymerase

A large variety of *Taq* polymerases are available; we find that the enzyme (*Taq* polymerase) from Roche meets the majority of our needs. A hot-start polymerase should be used if the specificity of the PCR product needs to be increased. Use the *Taq* polymerase buffer supplied with the enzyme.

NUCLEIC ACIDS AND OLIGONUCLEOTIDES

DNA in H₂O

If DNA is in TE, MgCl₂ <!> should be added to the PCR mix. When selecting STR markers to analyze DNA purified from paraffin-embedded tissue, select markers between 78 and 250 bp, due to variability in the integrity of the DNA in the blocks.

Reference DNA can be purified from peripheral blood or nonmalignant tissue. For leukemia it is difficult to obtain a proper reference to the tumor DNA. We have analyzed DNA from blood samples obtained during the disease and after recovery.

Tumor DNA. It is difficult to avoid contamination of the tumor tissue with nonmalignant cells. Microdissection of the tumor tissue is one way to avoid contamination, but the process is time-consuming and the yield is low (see Chapter 11 on laser-assisted microdissection of DNA).

To ensure an informative LOH analysis, a large number of samples have to be analyzed. Therefore, the purification process must reliably and reproducibly extract quality DNA from a large number of samples. One way to evaluate the quality of the tumor DNA is to analyze a fraction of the samples for LOH using STR in chromosomal regions where LOH has already been established for the specific tumor type.

Primers

It is important to select the fluorophores carefully to ensure that they are detectable on the available electrophoresis instrument. For example, the green color TET cannot be used on ABI 3100 systems.

Stock solutions of the labeled primers may be stored at –20°C and working solutions at 4°C. Use black boxes to avoid exposure to light.

SPECIAL EQUIPMENT

Thermal cycler

METHOD

1. In a sterile 0.2-ml microfuge tube mix the following reagents;

DNA in H₂O	10–20 ng
Primers (each)*	1 pmole/liter**
dNTP solution	250 μM
Taq polymerase buffer	1×
Taq polymerase (Roche)	0.18 units
H₂O	to 6 μl***

   *One primer of each marker set is labeled with a fluorophore.
   **1.5–2.0 pmoles of primer may be optimal for some reactions.
   ***A final volume of 10 μl is used in PCR machines with a heated lid (no oil on top of the samples).

2. Perform PCR using the following general PCR amplification program:

Cycle number	Denaturation	Annealing*	Polymerization/Extension
1	4 min at 94°C	30 sec at 55°C	40 sec at 72°C
27	30 sec at 94°C	30 sec at 55°C	40 sec at 72°C
last cycle	30 sec at 94°C	30 sec at 55°C	6–10 min at 72°C

   *Adjust annealing temperature according to primers

3. Analyze 3 µl of the PCR product on an agarose gel. The products are analyzed by agarose gel electrophoresis when new primers are used. A few randomly chosen PCR products among 96 samples can be analyzed prior to the capillary electrophoresis.

4. If no product is seen after PCR amplifications at different annealing temperatures, search the databases for the CG content or specific CpG islands within the markers. A number of STR consist of only C and Gs, especially trinucleotide repeats. Add betaine and DMSO as described in Chapter 5 on PCR amplification of highly GC-rich regions.

INTERPRETATION AND EVALUATION OF RESULTS AFTER ELECTROPHORESIS

The electrophoresis results may be transferred to various programs for further analysis. Here, we describe the use of the ABI software of GenoTyper (Applied Biosystems) to analyze the PCR products from each STR allele and, finally, the LOH calculation in Excel sheets.

Electrophoresis can be carried out using either polyacrylamide gels with automated data collection (ABI 377 Applied Biosystems) or capillary electrophoresis (ABI 310–ABI 3700, Applied Biosystems). For capillary electrophoresis, 0.5–3 µl of PCR product, evaluated according to the agarose gel result, is mixed with 12.5 µl of formamide<!> and 0.5 µl of a fluorescent-labeled size standard (Applied Biosystems) prior to the electrophoresis.

GenoTyper

Results from the electrophoresis, presented in GeneScan (Applied Biosystems), are imported into the GenoTyper. Because the products and size markers may fluctuate slightly during the run of a large number of samples, it is important that a few single runs are selected uniformly among the samples and checked for a correct size marking. New size markers can be specified for each run.

For data analysis purposes, the published size range of each STR marker is used as a starting point. This range can be adjusted after the results have been examined. The software will label the two highest peaks of each PCR product with a peak height in relative fluorescence units, equivalent to the DNA content and with the DNA fragment length in base pairs. The height is used as the most convincing measurement of the DNA content of an allele (a peak) because it is difficult to specify an algorithm for calculating the specific peak spanning area, due to stutter peaks and nontemplate adenines added by the polymerase and their overlap with neighboring alleles. When loading and running the gel or the capillary electrophoresis, it is important to avoid overloading of the PCR product because the fluorescence detector will saturate and indicate a lower-than-actual signal. The peak height should be lower than 2500 units, and the tip of the peak should be sharp (see Fig. 2).

All results are examined by eye to ensure that the correct alleles are marked. For example, in cases with a considerable loss of one allele, a stutter peak may be misassigned as an allele. Additionally, the profile of the alleles may vary considerably, and some may be difficult to interpret, especially heterozygous alleles with a short mutual distance and overlap between stutter and allele peaks (see Fig. 2). In GenoTyper, all sample peaks from the specimens for each STR amplification are pictured below each other, facilitating the interpretation of the true allele. A homozygous sample may reveal the true profile of an allele (Fig. 2).

Depending on the tumor type, some samples may exhibit additional peaks in the tumor sample as compared with the wild type. These tumors may have lost the ability to repair

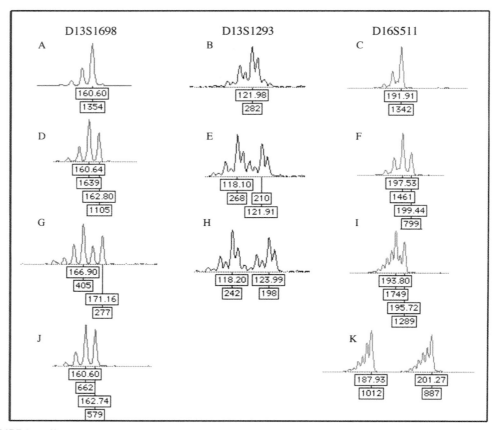

FIGURE 2. Different profiles of STRs. Comparison of homozygous (*A–C*) and heterozygous (*D–K*) profiles from PCR amplifications of the three different STR. J is a heterozygous sample with LOH. Length of the PCR product is specified in decimal numbers.

errors that occur during replication and should be discarded from further calculations (see Fig. 3).

The GenoTyper software will add "1" in the cells for peak height and length instead of the second allele from a homozygous sample. GenoTyper contains a macro, which can be adjusted to fulfill specific requirements of defining the number of STR (categories) and cut-off level for LOH. However, since we want to use our own definitions of the allele ratio and LOH/retention, we prefer to perform these calculations using an Excel spreadsheet, rather than the macro in the GenoTyper program.

For use in Excel, the samples must be sorted first according to the labeling dye, and then according to the sample names. The order on the vertical axis must be: sample 1 wild-type, sample 1 tumor, sample 2 wild type, and sample 2 tumor, etc. Sample name, peak height, and length are copied into a preprogrammed Excel spreadsheet, for further LOH calculations.

LOH Calculation

The principle of LOH calculation is described in Figure 4. Results from PCR amplification of the STR D16S511 on six consecutive samples of ovarian tumor and leukocyte DNA with variable levels of LOH are shown in Figure 5 (Hansen 2002). DNA fragment length and peak height from each of these samples are transferred to a preprogrammed Excel sheet, as shown in Figure 6.

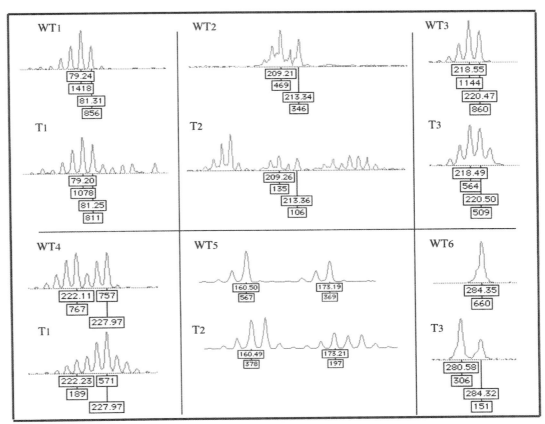

FIGURE 3. Replication error illustrated by PCR amplification of STR in two different ovarian tumor (T) and nontumor (WT) tissue pairs. The original alleles are labeled with peak height and PCR product length. One tumor has been found to carry a 1-nucleotide deletion in exon 7 of the DNA mismatch repair gene hMSH3, leading to a truncated protein (Arzimanoglou 2002).

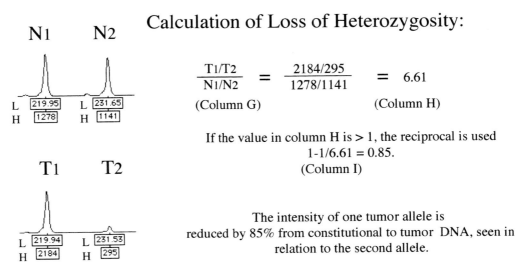

Calculation of Loss of Heterozygosity:

$$\frac{T_1/T_2}{N_1/N_2} = \frac{2184/295}{1278/1141} = 6.61$$

(Column G)　　　　　　　　　　　　(Column H)

If the value in column H is > 1, the reciprocal is used
1-1/6.61 = 0.85.
(Column I)

The intensity of one tumor allele is
reduced by 85% from constitutional to tumor DNA, seen in
relation to the second allele.

FIGURE 4. Calculation of LOH. N_1 and N_2 are wild-type alleles; T_1 and T_2 are tumor alleles from the same individual. L is the length of the PCR products in base pairs and H is the height in arbitrary fluorescence units. (Note that references to columns G, H, and I in this figure refer to Fig. 6.)

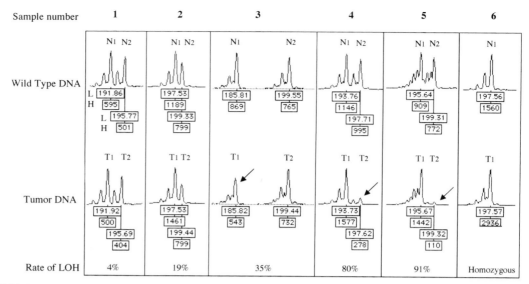

FIGURE 5. PCR amplification of D16S511 in six different samples of wild-type and ovarian tumor DNA. Samples 1–5 illustrate an increasing level of LOH, and sample 6 is homozygous. Numbers in boxes below the peaks are length (in decimal numbers) and heights of each peak. Arrows indicate affected alleles.

The formulas are inserted for the first set of data in row 3, where row 2 contains the blood cell data and row 3 contains the tumor data in 6 columns (A to F). The allele ratio (G2) is calculated for the WT (D2/F2) and Tumor (D3/F3). The retention will then be the ratio between these two figures (I3), but because it is not known which allele has a loss, if

Sample	Allele Ratio (G)	Ratio T/N (H)	Ratio N/T (I)	Retention (J)	Loss (K)
1 WT	D2/F2				
1 T	D3/F3	G2/G3	1/H3	MIN(H3,I3)	1-J3
2 WT	D4/F4				
2T	D5/F5	G4/G5	1/H5	MIN(H5,I5)	1-J5

0.18 level:
1 T: IF(AND(0.99>$K3,0.18<$K3),"Loss","Retention")

0.25 level:
1T: IF(AND(0.99>$K3,0.25<$K3),"Loss","Retention")

Homozygous?:
1 T: IF(OR(E3=1;C3=1),"Homoz?", "Het")

	A	B	C	D	E	F	G	H	I	J	K	L	M	N
1	Sample	Marker	Peak 1 (bp)	Allele1-ht	Peak 2 (bp)	Allele 2-ht	Allele Ratio	Ht Ration T/N	Ht Ration N/T	Retention	Loss	0.18 level	0.25 level	Homozygous?
2	1 WT	D16S511	191.86	595	195.77	501	1.19							
3	1 T	D16S511	191.92	500	195.69	404	1.24	0.96	1.04	0.96	0.04	Retention	Retention	Het
4	2 WT	D16S511	197.53	1189	199.33	799	1.49							
5	2 T	D16S511	197.53	1461	199.44	799	1.83	0.81	1.23	0.81	0.19	Loss	Retention	Het
6	3 WT	D16S511	185.81	869	199.55	765	1.14							
7	3 T	D16S511	185.82	543	199.44	732	0.74	1.53	0.65	0.65	0.35	Loss	Loss	Het
8	4 WT	D16S511	193.76	1146	197.71	995	1.15							
9	4 T	D16S511	193.73	1577	197.62	278	5.67	0.20	4.93	0.20	0.80	Loss	Loss	Het
10	5 WT	D16S511	195.64	909	199.31	772	1.18							
11	5 T	D16S511	195.67	1442	199.32	110	13.11	0.09	11.13	0.09	0.91	Loss	Loss	Het
12	6 WT	D16S511	197.56	1560	1	1	1560.00							
13	6 T	D16S511	197.57	2936	1	1	2936.00	0.53	1.88	0.53	0.47	Loss	Loss	Homoz?

FIGURE 6. Peak height and length of the six samples from B have been transferred into an Excel sheet. Algorithms for LOH calculation are shown above for columns G–N.

the ratio is greater than one, the reciprocal value must be used (J3). The real retention is thus the minor value (J3), and the loss is 1–J3 (K3).

To give a quick overview of the data, we have made simple formulas in columns L and M to indicate loss over 18% and 25%. In column N, the heterozygosity status of the samples is indicated. GenoTyper marks homozygosity with one in both the allele size (bp) as the peak height. In a few cases, the allele sizes have been changed in the tumor to reflect replication errors.

Cutoff Value of Selected Tumor Material

A number of factors can influence the calculation of LOH: e.g., the contamination of the sample by nontumor cells; the influence of stutter bands; uneven occurrence of chromosome-specific loss throughout the tumor, whether the LOH is an early or late event in tumor progression; and reproducibility of the PCR. The cut-off levels determining the border between LOH, allelic imbalance, and retention vary considerably among laboratories. The GenoTyper macro uses values <0.67 or >1.3 as an estimate of the rate of contamination by nontumor tissue. Others arbitrarily select values between 0.6 and 0.7, or even lower, based on a theoretical rationale. If the frequency of LOH is correlated to clinical prognostic factors, the use of too high a cutoff level can lead to the loss of important information on a subgroup of low-level LOH, and subsequent misinterpretation of the results.

Based on a repeated LOH calculation of 485 constitutional heterozygous genotypes from 20 different dinucleotide repeat loci, a cutoff value was established at a reduction of peak intensity by 0.84 (Skotheim et al. 2001). Results obtained by fluorescence and radioactively labeled PCRs, respectively, on the same panel of consecutive blood and tumor samples revealed that the fluorescence protocol was more sensitive than the radioactive approach (Liloglou et al. 2000).

A cutoff value should be determined for each type of tumor material used for LOH analysis. A histogram can be made by adding the number of samples within 2.5% or 5% LOH intervals distributed over the range of 0–100%. Figure 7 illustrates a histogram of LOH results from chromosome 13q in a panel of 65 ovarian tumors. A considerable number of samples with 100% allelic loss are specific for this material and not seen in our breast tumor panel, inset histogram (Fig. 7). The high columns to the left are samples with clear retention of both alleles. The column height drops considerably at 10% LOH and increases slightly to a more uniform level throughout the histogram. The level between 12.5% and 20% LOH is a gray zone between retention and allelic loss.

To perform this calculation using the Excel spreadsheet:

1. Copy columns A, C, E, and K to a new sheet. It is important that each sample is represented only once.

2. Sort all columns by column A, increasing.

3. Delete all rows with WT in column A.

4. The rest are sorted by column K.

5. Count the number of samples with a "LOH" between 0% and 2.49%, 2.5% and 4.99%, etc., using the macro/algorithm.

6. A histogram is formed using the results from the calculation and the LOH range from 0 to 100% divided into 2.5% intervals.

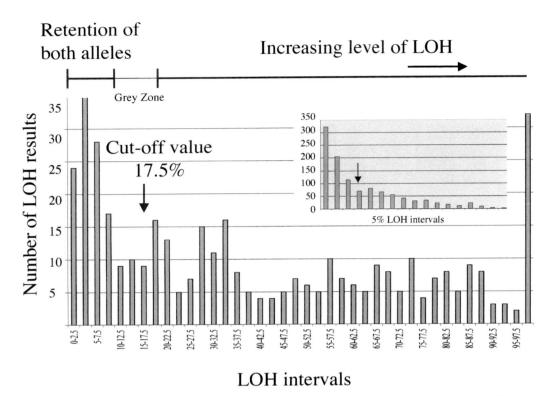

FIGURE 7. Histograms of calculated LOH ratios for different tumor types. For a panel of 65 ovarian carcinomas analyzed with 19 STR, the column depicts the number of heterozygous samples with LOH ratios divided into 2.5% intervals. (*Inset*) Distribution of 4800 LOH results from chromosome 14q in a panel of 265 breast tumors, divided by 5% intervals.

▶ TROUBLESHOOTING: Amplification

- Too many unspecific peaks or stutter bands appear after electrophoresis.
    1. Raise the annealing temperature; often one degree can improve the results.
    2. Use a hot-start polymerase.
    3. Move one or both primers. PCR results can be improved considerably by just shifting the position of the primer by 10–15 bp in the sequence.
    4. Add 1–2 M betaine to the PCR reaction.
- Heavy overrepresentation of some peaks compared with others from the same multiplex PCR.
    1. Increase the primer concentrations of the underrepresented PCR products.
    2. Decrease the primer concentrations of the overrepresented products.
    3. The addition of 1 M betaine and/or 10% DMSO<!> (v/v), may improve the specificity of the primer to template annealing.
- No PCR product.
    1. Decrease annealing temperature.
    2. If the DNA fragment contains large CpG islands or the overall CG contents constitutes more than 50% of the base pairs, one or more of the following steps may solve the problem:

a. Add 1 M betaine to the PCR.

b. Add 10% DMSO (v/v) the PCR.

c. Add both 1 M betaine and 10% DMSO (v/v) to the PCR.

d. Use a hot-start *Taq* polymerase.

e. Increase the initial denaturing step to 5 minutes at 96°C.

3. Check the pipettes; with these small volumes it is crucial that the pipetting is exact.

- The GenoTyper program.

  Some PCR products may appear in more than the original labeling color. The STR can be identified by the profile and the exact location of the peaks. Performing the PCR with a primer set labeled with another color can solve the problem.

ACKNOWLEDGMENTS

The Danish Medical Research Council, The Danish Cancer Society, and the Novo Nordisk Foundation have supported this work. Dr. L. G. Jensen is thanked for critical review of the manuscript, and E. Hein for excellent technical assistance.

REFERENCES

Arzimanoglou I.I., Hansen L.L., Chong D., Li Z., Psaroudi M.C., Dimitrakakis C., Jacovina A.T., Shevchuk M., Reid L., Hajjar K.A., Vassilaros S., Michalas S., Gilbert F., Chervenak F.A., and Barber H.R. 2002. Frequent LOH at hMLH1, a highly variable SNP in hMSH3, and negligible coding instability in ovarian cancer. *Anticancer Res.* **22:** 969–975.

Hansen L.L., Jensen L.L., Dimitrakakis C., Michalas S., Gilbert F., Barber H.R., Overgaard J., and Arzimanoglou I.I. 2002. Allelic imbalance in selected chromosomal regions in ovarian cancer. *Cancer Genet. Cytogenet.* **139:** 1–8.

Hansen L.L., Andersen J., Overgaard J., and Kruse T.A. 1998. Molecular genetic analysis of easily accessible breast tumour DNA, purified from the leftover from hormone receptor measurement. *APMIS* **106:** 371–377.

Liloglou T., Maloney P., Xinarianos G., Fear S., and Field J.K. 2000. Sensitivity and limitations of high throughput fluorescent microsatellite analysis for the detection of allelic imbalance: Application in lung tumors. *Int. J. Oncol.* **16:** 5–14.

Miller S.A., Dykes D.D., and Polesky H.F. 1988. A simple salting out procedure for extracting DNA from human nucleated cells. *Nucleic Acids Res.* **16:** 1215.

Skotheim R.I., Diep C.B., Kraggerud S.M., Jakobsen K.S., and Lothe R.A. 2001. Evaluation of loss of heterozygosity/allelic imbalance scoring in tumor DNA. *Cancer Genet. Cytogenet.* **127:** 64–70.

WWW RESOURCES

http://genome.cse.ucsc.edu/ UCSC Genome Bioinformatics site
http://www.ensembl.org/ Ensembl Genome Browser, The Wellcome Trust Sanger Institute
http://www.gdb.org/ The Genome Database (GDB)
http://www.ncbi.nlm.nih.gov/entrez/ National Center for Biotechnology Information (NCBI)

18 Single-strand Conformation Polymorphism Analysis

Anne E. Jedlicka[1] and Steven R. Kleeberger[2]

[1]The Johns Hopkins University Bloomberg School of Public Health, Department of Molecular Microbiology and Immunology, Baltimore, Maryland 21205; [2]The National Institute of Environmental Health Sciences, Laboratory of Pulmonary Pathobiology, Research Triangle Park, North Carolina 27709

Single-strand conformation polymorphism (SSCP) analysis is a technique first described by Orita et al. (1989) as a simple, powerful, and robust method to detect DNA sequence changes (single base substitutions) based on shifts in electrophoretic mobility. Used in combination with PCR, it was intended to be a method of discovery of variant DNA sequences, mutations, and polymorphisms. SSCP has been applied to genetic linkage studies, candidate gene evaluation, and diagnostics. Many mutations have been discovered by this method, including *RAS* mutations in lung carcinomas, somatic mutations in retinoblastoma (Rb) and p53 tumor suppressor genes in a variety of human cancers, and point mutations in the genes responsible for cystic fibrosis and neurofibromatosis (Yandell 1991).

SSCP analysis requires that the target sequence is simultaneously labeled and amplified, then heated to dissociate the strands, and resolved by nondenaturing polyacrylamide gel electrophoresis. Mutations in the target sequence are detected as altered mobility of the separated single strands in the autoradiogram compared to the normal, nonmutated sequence. Single-stranded DNA has a folded conformation that is stabilized by intrastrand interactions, and this conformation is dependent on sequence. Differences in sequence will alter conformation and therefore change electrophoretic mobility. Because of the high resolution of polyacrylamide gels, most conformational changes caused by subtle sequence differences can be detected (Hayashi 1991).

PCR-SSCP is perhaps the easiest to perform and one of the most sensitive PCR-based techniques for detecting mutations. Additional positive attributes of this application include minimal DNA requirement, convenience, low cost, and the potential to detect any sequence variation within the target sequence. In this chapter, we present a detailed protocol for high-throughput PCR-SSCP analysis that reflects these positive attributes and can be applied to any model organism.

METHOD OVERVIEW AND STRATEGY

PCR-SSCP analysis consists of the simultaneous labeling and amplification of a target sequence by PCR of genomic DNA, or cDNA, using labeled primers. The resulting amplicon is subsequently denatured and resolved on nondenaturing polyacrylamide gels, and mutations are detected as altered mobility of separated single strands in an autoradiogram. The following optimized procedure for SSCP is divided into detailed protocols as follows: (1) primer design and end-labeling, (2) PCR amplification, (3) nondenaturing gel electrophoresis, and (4) analysis and confirmation. These protocols include details for the preparation of a large population of individuals for analysis by multiple PCR-based techniques, and they are written with the assumption that the screened sequence variant is known.

Electrophoretic conditions are critical to the success of the SSCP assay. Under nondenaturing conditions, single-stranded DNA has a folded structure that is determined by intramolecular interactions and, therefore, by its sequence. A mutated sequence is detected as a change in mobility during polyacrylamide gel electrophoresis caused by its altered folded structure compared to the nonmutated form. The electrophoretic mobility is sensitive to the size and shape of the particle, but it is also strongly dependent on environmental conditions within the gel, such as temperature, and ion concentration. Therefore, multiple parameters are usually tested (Orita et al. 1989). The following protocol for nondenaturing gel electrophoresis is derived from a review by Hayashi (1991), which revealed that complementary single strands are better separated in gels with low cross-linking. The extent of cross-linking in a gel (%C) is a ratio of the percent concentration of bisacrylamide to the concentration of total acrylamide monomer. A gel of 5%C is the most rigid and has minimal pore size at any given total acrylamide concentration. A gel with a lower %C is softer, has increased pore size, and is more sensitive to molecular conformation. Gels at 1–2%C and 5–6% total acrylamide are commonly used in SSCP analysis. Hayashi (1991) also empirically determined that low concentrations of glycerol can improve separation of mutated sequences. This effect is perhaps due to its weak denaturing action on nucleic acids, which could partially open the folded structure so that more surface area of the molecule is exposed, thus allowing more chance for differential confinement of the variant strand. However, in rare cases, mobility shift of mutated sequences may be demonstrated only in the absence of glycerol.

Many reports of the sensitivity of SSCP for discovering polymorphisms have been published (Sheffield et al. 1993), but some mutations cannot be detected under the chosen conditions of electrophoresis. Therefore, false-negative results cannot be excluded (Hayashi 1994). Again, sequencing, particularly of aberrant bands, is very valuable for a thorough evaluation of regions or genes of interest in the absence of detected mutations by this technique, especially if it is used as a discovery method.

PROTOCOL 1   PRIMER DESIGN AND END-LABELING

A systematic analysis of the strengths and limitations of SSCP revealed a striking relationship between sensitivity and fragment size (Sheffield et al. 1993). Therefore, the size of a target region of amplification should be limited to a maximum of 200 bp. Oligonucleotide primers should be designed as 20-mers. Each primer of the pair should have a similar GC content and, therefore, similar estimated annealing temperature (T_m). As with all primers intended for amplification, it is best to avoid hairpins, loops, and self-annealing sequences within the primer. It is also advantageous to design the primers so that the polymorphism is centered within the amplified fragment, solely for the ease of confirmation by sequence analysis. This protocol uses very high-specific-activity isotope and includes a purification step to remove the unincorporated nucleotides, thereby reducing the amount of radioactivity in the overall reactions.

MATERIALS

BUFFERS, SOLUTIONS, AND REAGENTS
5x Kinase buffer
> 0.25 M Tris-HCl<!>, pH 9
> 0.05 M $MgCl_2$<!>
> 0.05 M dithiothreitol (DTT) <!>
> 0.25 mg/ml bovine serum albumin (BSA)
> Store at −20°C.

Sterile, double-distilled H_2O (ddH_2O)

TE buffer, pH 8.0
> 10 mM Tris-HCl<!>, pH 8.0
> 1 mM EDTA, pH 8.0

ENZYMES AND ENYZME BUFFERS
Polynucleotide kinase, 10 units/µl (Roche)

NUCLEIC ACIDS AND OLIGONUCLEOTIDES
Oligonucleotide primers

RADIOACTIVE COMPOUNDS
[γ-^{32}P]ATP <!>, 6000 Ci/mM (Perkin Elmer Life Sciences)

[γ-^{33}P]ATP may also be used. It is slightly less hazardous than [γ-^{32}P]ATP and has a longer half-life; however, it is more expensive.

CENTRIFUGES AND ROTORS
Centrifuge with swinging-bucket rotor and adapter for 15-ml conical tubes

SPECIAL EQUIPMENT
Acrylic shields, tube rack ("beta block"), and waste container
Beta block (e.g., 3008-1000 from USA Scientific)
Collection tube (1.5-ml, included with Quick Spin columns)
Cooling block
Conical tube, 15-ml
Geiger counter
Microfuge tube, 0.5-ml, sterile
Quick Spin columns (TE), G-25 Sephadex columns for radiolabeled DNA purification (Roche)
Water bath

METHOD

1. Dilute the primers to 10 µM with sterile, ddH_2O. The primers we purchase are synthesized on a 50 nM scale and are received in a lyophilized form. Resuspend the pellet in TE, pH 8.0, to be stored as a stock. Ten micromolar working dilutions should be prepared, aliquoted, and stored frozen at −20°C.

2. Select one primer to be end-labeled. Thaw the primer and 5x kinase buffer on ice. Keep the enzyme in a freezer until it is added to reaction, or use a bench-top cooling block that will maintain the enzyme at −20°C. Thaw the isotope behind an acrylic shield at room temperature.

3. On ice, in a sterile 0.5-ml microfuge tube, mix 3 μl of sterile ddH$_2$O, 5.6 μl of 5x kinase buffer, 12 μl of 10 μM primer, and 1.4 μl of polynucleotide kinase. Place the ice bucket behind an acrylic shield and add 6 μl of [γ-^{32}P]ATP, 6000 Ci/mM. Mix well, cap the tube, and incubate it at 37°C in a water bath for 1–2 hours.

4. During the final 10 minutes of incubation, prepare a Quick Spin column. Mix the column by inverting it repeatedly; remove the cap and tip cover, and place in a collection tube. Place column/collection tube assembly in a 15-ml conical tube and centrifuge in a tabletop centrifuge at 1100g for 2 minutes. Drain the collection tube and repeat the centrifugation. Transfer the packed column to a clean collection tube before loading the end-labeling reaction mix.

5. Upon completion of incubation, briefly spin the end-labeling reaction in a microcentrifuge. Add 52 μl of sterile ddH$_2$O, mix by pipetting, and transfer the full volume to the spin column for purification (i.e., removal of unincorporated nucleotide).

6. Centrifuge at 1100g for 4 minutes.

7. Remove the column from the collection tube and discard it in an appropriate radioactive-waste container. Place a cap on the collection tube containing the end-labeled primer. Proceed with the PCR protocol (Protocol 2) or store the primer at –20°C in a beta block.

PROTOCOL 2 TEMPLATE DNA PREPARATION, ORGANIZATION, AND PCR AMPLIFICATION

Genomic DNA may be extracted by a variety of methods, depending on the type (blood versus tissue) and quantity of starting material. Once isolated, all stock DNAs should be stored in TE, pH 8.0, in sterile Eppendorf tubes. After spectrophotometric quantitation, prepare 100 ng/μl "intermediate" dilutions with sterile ddH$_2$O, also in sterile Eppendorf tubes, from which "working" dilutions of DNA (20 ng/μl) may be prepared in 96-well plates. It is important to have caps and tubes that seal tightly to reduce sample loss during long-term storage, and to always centrifuge DNA plates before removing caps. (The Eppendorf 5810 and Beckman Allegra centrifuges have plate adapters for their swinging-bucket rotors.) Caps should be discarded when opening a plate for assay preparation and replaced with new caps when finished to prevent cross-contamination.

Include a negative control (no DNA, water only) on each DNA plate. For labeling and alignment of samples on films during analysis, it is helpful to have the negative control within the plate and not as the first or last sample. Also, for large cohorts consisting of multiple plates, use of multicolor tubes (in addition to careful labeling) can help to maintain sample organization.

When optimizing a new assay, select a subset of individuals for analysis before commencing with the entire cohort. The "intermediate" dilutions of DNA are useful for performing optimizations, repeats, and sequence confirmation reactions, so that volumes within the working plate are consistently maintained.

MATERIALS

▼ CAUTION

See Appendix for appropriate handling of materials marked with <!>.

BUFFERS, SOLUTIONS, AND REAGENTS

10x PCR buffer with 15 mM MgCl$_2$<!> (GeneAmp PCR Buffer I, Applied Biosystems)
Sterile ddH$_2$O

ENZYMES AND ENYZME BUFFERS
    Taq DNA polymerase, 5 units/μl (Applied Biosystems or Invitrogen)

NUCLEIC ACIDS AND OLIGONUCLEOTIDES
    Genomic DNA
    10 mM dNTPs (GeneAmp PCR dNTP blend, Applied Biosystems)
    Oligonucleotide primers (synthesized by Invitrogen, diluted to 10 μM)

CENTRIFUGES AND ROTORS
    Centrifuge with swinging-bucket rotor and microplate adapters

SPECIAL EQUIPMENT
    Acrylic shield
    Filter barrier tips
    Multichannel pipettors (Eppendorf or Apogent Discoveries)
    PCR machine, 96-well (Applied Biosystems GeneAmp 9700)
    Tray/retainer set and base, 96-well (Applied Biosystems)
    Tube caps, strips of 8 (USA Scientific or Applied Biosystems)
    Tubes, 0.2-ml strips of 8 (USA Scientific or Applied Biosystems)

METHOD

1. Thaw the dNTPs, 10x buffer, and unlabeled primer on ice. Keep *Taq* at –20°C until added to reaction cocktail. Thaw the labeled primer behind an acrylic shield.

2. Per reaction, make a cocktail consisting of:

sterile ddH$_2$O	6.95 μl
10x PCR buffer with MgCl$_2$	1.25 μl
10 mM dNTPs	0.25 μl
unlabeled 10 mM primer	0.5 μl
labeled primer	0.5 μl

   If, during optimization, the intensity of product is low, the amount of labeled primer may be increased and the volume of ddH$_2$O adjusted proportionally.

Taq DNA polymerase (0.25 units final concentration)	0.05 μl

   For a full 96-well plate, make a cocktail for 100.

3. Mix well, and aliquot into the PCR plate (9.5 μl/sample) on ice.

4. Add 3 μl of DNA (60 ng) with a multichannel pipettor and mix well.

5. Cap the tubes securely and place plate into preheated (94°C) PCR machine.

6. For products of this size (200 bp), amplify for 30 cycles. Cycling parameters are as follows: an initial denaturation of 94°C for 5 minutes; 30 cycles of 94°C for 30 seconds; an empirically determined annealing temperature (T_m) for 30 seconds; and 72°C extension for 30 seconds. The final extension step is performed for 7 minutes at 72°C. T_m may be calculated from primer sequence, utilizing the formula $2(A+T) + 4(G+C)$.

7. Upon completion of amplification, freeze the samples at –20°C, or proceed to Protocol 3.

PROTOCOL 3   GEL ELECTROPHORESIS

As previously mentioned, a gel with a lower %C is softer, has increased pore size, and is more sensitive to conformational changes. Furthermore, low concentrations of glycerol can improve separation of mutated sequences. This protocol reflects these observations using gels of low cross-linking percentage (2.67%) with 10% glycerol. It has been observed that, in some instances, a mobility shift of a mutated sequence will be observed only in the absence of glycerol. Furthermore, greater clarity of the bands may be found on a non-glycerol gel. Therefore, during optimization, PCR products should be run on both glycerol and non-glycerol gels, and resulting autoradiographic images should be compared. Test runs of samples on both types of gels can be achieved within a 24-hour period (see step 5). Once determined, the optimal gel type for a particular assay can then be used for all cohort analysis at that locus.

MATERIALS

BUFFERS, SOLUTIONS, AND REAGENTS

40% Acrylamide <!>/bis stock solution, 37.5:1 (BIO-RAD)

10% Ammonium persulfate <!>

Make 100 ml, and then store as 10-ml aliquots at –20°C. Screw-cap borosilicate scintillation vials make good storage containers.

Ethanol, 95%<!>

2x Loading ("stop") buffer

95% Formamide <!>

20 mM EDTA

0.05% Bromophenol Blue <!> (Sigma-Aldrich B5525)

0.05% Xylene Cyanol <!> (Sigma-Aldrich X4126)

Make 300 ml and then store in 15-ml aliquots in conical tubes at –20°C. For ease of addition to samples by multichannel pipettor, transfer an aliquot to a sterile reagent reservoir or trough (Eppendorf or Apogent Discoveries).

10x TBE buffer

Tris base <!>	108 g
Boric acid	55 g
EDTA	9.3 g
ddH$_2$O to 1 liter	

N,N,N´,N´-Tetramethylenediamine (TEMED)<!>

SafetyCoat NonToxic Glass Plate Coating<!> (*Caution:* flammable)

ENZYMES AND ENYZME BUFFERS

NUCLEIC ACIDS AND OLIGONUCLEOTIDES

PCRs from Protocol 2 <!> (Radioactive)

GELS

SSCP Gel mix with 10% glycerol (final %C=2.67%)

10x TBE	50 ml
40% acrylamide <!>(37.5:1)	75 ml
ddH$_2$O	325 ml

Filter-clarify with 0.45-micron ZapCap. Add 50 ml of Ultrapure Glycerol. Store at 4°C protected from light (foil-wrap the bottle).

SSCP Gel mix without glycerol (final %C+2.67%)

10x TBE	50 ml
40% acrylamide <!> (37.5:1)	75 ml
ddH$_2$O	375 ml

Filter-clarify with 0.45-micron ZapCap. Store at 4°C protected from light (foil-wrap the bottle).

MEDIA

SPECIAL EQUIPMENT

0.45-micron ZapCap bottle-top filters (Schleicher & Schuell)

108-well deep-well comb

> This is a square-well comb custom manufactured by The Gel Company, San Francisco, CA (www.gelcompany.com, 1-800-256-8596). This type of comb is better than a shark's tooth comb because it creates distinct, individual wells. Leakage or mixing of samples from well to well will not occur, as can happen with the shark's tooth wells, which can interfere with data interpretation.

3MM 46 x 57-cm chromatography paper (Whatman)

8-channel syringe gel loader (Hamilton)

> It should be noted that it has proven more cost-efficient to purchase model 84501, the 12-channel loader. By removing 4 syringes, 2 from each end, there will automatically be spare parts for maintenance. Syringes need to be replaced periodically, and ordering them individually is more expensive and may cause delays in experimentation, as they are frequently out of stock.

Acrylic shielding

Bottles, 125-ml

Conical tubes, 15 ml

Film cassettes with intensifying screens

GB002 Gel blot paper (Schleicher & Schuell)

Geiger counter

Gel Drying System (BioRad Model 583 with Hydrotech pump, or similar)

Kimwipes (or other lint-free wipe)

Kodak XAR5 14 x 17 film

Large binder clamps

Model S2 Sequencing apparatus (formerly from Life Technologies, now available from Whatman BioMetra) with 0.4-mm spacers and standard glass plates

Parafilm

Pipette, 50-ml

Power supply

Razor blade

Reagent reservoirs (Eppendorf or Apogent Discoveries)

SafetyCoat NonToxic Coating (JTBaker)<!> (*Caution:* flammable)

Saran Wrap

Syringe

Thermal Adhesive Sealing Film (Fisher Scientific) for covering 96-well plate during sample denaturation

Thermal cycler, or heat block

METHOD

Plate Assembly

1. Prior to assembly, the small plate should be treated with a silanizing agent such as SafetyCoat NonToxic Glass Plate Coating, according to the manufacturer's instructions.

2. Wash the plates with a soft sponge and nonabrasive detergent. The plates must be scrupulously clean and lint-free to reduce chance of bubble formation.

3. Rinse the plates with ddH$_2$O and wipe them with a large Kimwipe. Repeat the rinse with 95% ethanol, and wipe dry with Kimwipe.

   Water should "bead" on the small plate.

4. Place clean spacers on the large plate surface, place the small plate on top, and ensure that the foam blocks of spacers are flush against top of the small plate to avoid leaks.

5. Clamp the plates together along the sides (approximately four on each side). Clamps should not extend inward beyond the edge of the spacer.

6. Place the plate assembly horizontally but slightly elevated (a Styrofoam tube rack works nicely). Set out the comb and other materials for gel preparation.

Gel Preparation

1. Transfer 75 ml of gel mix into a 125-ml bottle. Add 37 µl of TEMED, cover with Parafilm, and shake vigorously. Add 705 µl of 10% ammonium persulfate, cover with Parafilm, and shake vigorously. De-gas the solution by vacuum.

2. Fill a 50-ml pipette with gel solution. Tilt the clamped plates (top of plates higher). Slowly and evenly dispense the gel solution by moving the pipette slowly from side to side across the top of the glass plates to dispense the gel solution between the plates. Lay the plates flat when gel reaches the bottom of plates. Immediately insert the teeth of the gel comb between the plates. The top of the wells should be level with the edge of glass plate.

3. Place additional clamps on sides of plates at comb edges.

4. Allow the gel to polymerize for at least 40 minutes.

Gel Setup

1. Remove all clamps.

2. Gently use a razor blade to loosen underneath the comb, then remove the comb slowly and evenly.

3. Place the plates into a gel box and fill the buffer chambers with 1x TBE.

4. Flush the wells with 1x TBE by syringe. If any wells are bent, use a flat pipette tip to straighten them.

5. No pre-run of gel is necessary; proceed with sample preparation and loading.

Sample Preparation and Electrophoresis

1. Thaw the PCRs from Protocol 2 if frozen and then briefly centrifuge to remove any condensation from tops and sides of tubes.

   The PCRs are radioactive; work with appropriate shielding.

2. Remove the caps carefully. Check gloves with Geiger counter and change if necessary.

   Use of a Kimwipe for cap removal reduces counts on gloves.

3. Add an equal volume (12.5 µl) of 2x Stop buffer to all samples with a multichannel pipettor. Mix well by pipetting. Cover the plate with Thermal Adhesive Sealing Film.

4. Denature the samples at 95°C for 5 minutes in a heat block or PCR machine, then place immediately on ice. Remove the sealing film and discard. Change gloves. Load the samples onto the gel using the multichannel syringe loader. The remaining samples may be stored at –20°C.

5. Attach the power cords to the gel box and power supply. Turn on the power supply.

   Caution: High voltage.

   Run the samples into the gel at ~40 W. If running a glycerol gel, reduce the power to 8 W and run for ~17 hours (overnight). If running a non-glycerol gel, reduce the power to 25 W and run for ~6 hours. Electrophorese until Xylene Cyanol (upper dye of loading buffer) has run approximately two-thirds through gel.

6. When the run is complete, turn the power off and disconnect cords from gel box. Remove the gel from gel box, and lay plates flat with the small plate up. Slide both spacers out and use a razor blade to pry top glass plate off.

7. Lay Whatman 3MM paper on the gel and press gently. Do not rub. Lay a piece of blotting paper on top and press gently. With one hand under lower glass plate and one hand on top of blotting paper, invert the assembly. Raise glass plate off gel surface. Trim excess paper around gel and cover with Saran Wrap.

8. Dry the gel for ~45 minutes at 80°C, transfer to a film cassette, and expose to X-ray film. (Exposure time can be determined during optimization.) Develop the film, label the samples, and analyze.

   Be sure to check all work areas and instruments with Geiger counter throughout protocols and at completion. Clean and decontaminate as necessary.

ANALYSIS, TROUBLESHOOTING, AND CONFIRMATION

An example of an SSCP assay we designed for a RANTES gene polymorphism (as described by Hajeer et al. [1999] for RANTES promoter position –403) is shown in Figure 1. Primers were designed to amplify a 200-bp fragment flanking the polymorphism, and SSCP analysis was performed on a large cohort. The optimized conditions utilized glycerol-containing gels. The heterozygotes and homozygotes are easily distinguishable; individuals 51, 238, and 179 are homozygous for the mutation (Fig. 1). At least one individual representative of each banding pattern was sequenced to verify that the aberrant pattern is due to the anticipated polymorphism. If the allele frequencies are published, the genotype may be extrapolated during initial analysis; however, it is best to confirm by direct sequencing.

An example of an SSCP assay we designed for a polymorphism in the Toll-like Receptor 4 gene (*TLR4*) is shown in Figure 2. *TLR4* has an A→G polymorphism at nucleotide 896, resulting in a missense Asp229Gly mutation (Arbour et al. 2000). Primers were designed to amplify a 200-bp fragment, which was resolved on glycerol-containing gels. Differentiation of the variant alleles was obvious, and they were subsequently confirmed by sequence analysis. For this verification, individuals were selected from the cohort representing each variant band, and a new PCR amplification was performed. The same primers were employed, although both primers were unlabeled. The resulting amplicon was purified by either gel electrophoresis methods (Low Melt Agarose [Invitrogen] followed by QiaQuick column purification [Qiagen]) or centrifugal filtration devices (Amicon MicroCon PCR filters [Millipore].) Quantitation was performed using the DNA Dipstick Kit (Invitrogen), and aliquots were sequenced using Applied Biosystems reagents and instrumentation. Representative sequence validation of the banding patterns seen on the SSCP gel (Fig. 2) is shown in Figure 3.

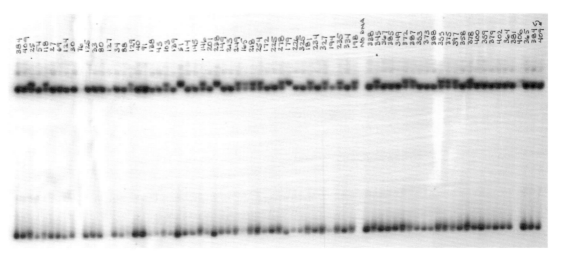

FIGURE 1. Autoradiograph of SSCP analysis of the polymorphism at –403 in the promoter region of the RANTES gene. The heterozygotes and homozygotes are easily distinguishable. Individuals 51, 238, and 179 are homozygous for the mutation.

Also within the *TLR4* gene, we designed an SSCP assay to screen a cohort for the polymorphism that results in a Thr399Ile missense mutation (Arbour et al. 2000). An autoradiograph of SSCP results for this assay indicated no aberrant banding in this region of known mutation (Fig. 4). The multiple bands present on the autoradiograph demonstrate multiple stable conformers of this particular amplicon; however, the polymorphism was not detected. The same sequence can have more than one stable conformation, resulting in multiple bands on the gel. The intensity of the bands may differ; however, the ratio of the bands will be constant across the gel, from individual to individual. The constant ratios allow them to be easily distinguished from the aberrant bands due to true sequence variants that do not appear at the same frequency across the gel.

This figure also demonstrates the limitations of the technique, because several of the individuals within the assayed region truly possess mutant alleles. This polymorphism cosegregates with the polymorphism that causes the Asp299Gly mutation; therefore, its presence within the population was anticipated. When it was not revealed by SSCP analysis, a different assay was created to detect and verify the polymorphism. The polymorphism would most likely have been resolved by redesigning and synthesizing the primers to adjust fragment size and location of polymorphism within the fragment. However, we chose to assess, by mutagenically separated PCR (MS-PCR), a variation of allele-specific PCR, which is a useful alternative technique for genotyping previously characterized substitution poly-

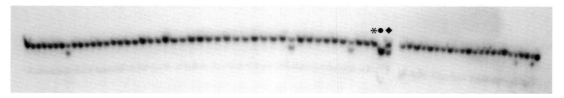

FIGURE 2. SSCP assay we designed for a previously described polymorphism within the *TLR4* gene. The asterisk denotes a wild-type individual, the diamond denotes a heterozygous individual, and the circle denotes an individual who is homozygous for the mutant allele.

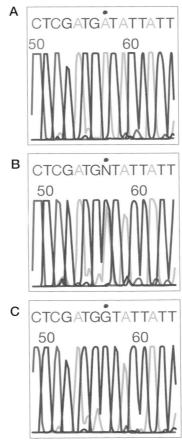

FIGURE 3. Sequence validation of the aberrant banding patterns found in SSCP analysis of the A→G polymorphism at nucleotide 896 of the *TLR4* gene resulting in the Asp299Gly mutation. (*A*) Sequence from an individual homozygous for the wild-type/normal A allele. (*B*) Representative heterozygote, as two overlapping peaks (A and G alleles) are called "N" by the sequencing software. (*C*) Sequence from an individual homozygous for the mutant G allele.

morphisms (Rust et al. 1993). For an alternative method of mutation detection, see Chapter 16.

ALTERNATIVE SSCP METHODS

Alternative methods for performing SSCP exist. In general, the amplification remains the same, but the detection methods vary. Nonlabeled primers may be used for amplification, and the samples are resolved by electrophoresis on smaller polyacrylamide gels and stained with silver stain or SYBR Green I, a fluorescent stain. The disadvantages include the potential hazard of such staining procedures, decreased resolution of bands, and reduced throughput due to lower sample capacity of gels. In addition, a good imaging system and special filter for the SYBR Green detection are required.

A more automated alternative is fragment analysis by fluorescent detection on an automated sequencer or capillary electrophoresis instrument. DNA fragments are amplified for this technique utilizing primers with 5′ fluorescent labels (e.g., HEX and FAM, synthesized by Applied Biosystems). The amplicons are denatured, then cooled and loaded onto a nondenaturing gel or polymer-filled capillary. Fragments containing a mutation display a dif-

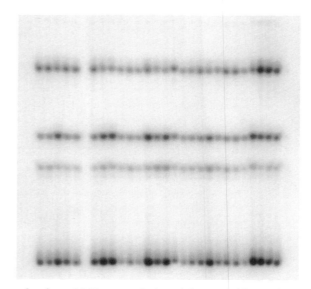

FIGURE 4. Autoradiograph of an SSCP assay designed for an additional polymorphism within the *TLR4* gene, one that results in a missense mutation changing amino acid 399 from threonine to isoleucine. This is an example of the limitations of the assay. The polymorphism is present in this cohort within this region; however, it is not detected by the assay.

ferent peak pattern in the electropherogram due to differential migration of the abnormal conformers. The advantages of this methodology include speed, reduced exposure to hazardous materials, long-term stability of reagents and PCR products, and automation of detection and analysis. A disadvantage is that the instrumentation and analysis software are very expensive. Furthermore, even if an institution has such an instrument in a core facility, it may be solely dedicated to sequencing and not available for fragment analysis applications or equipped with the appropriate software.

DISCUSSION AND SUMMARY

SSCP analysis, in combination with PCR, has been a robust and powerful technique for the detection of single-base variations in DNA. It is also relatively simple, quick, and inexpensive. Although predicted to some day be superseded by other techniques, it was a tremendous step forward at the time of its discovery. It became a prominent technique that led to enormous advances in our understanding of the human genome and the diseases that afflict us (Yandell 1991). The widespread use of PCR-SSCP resulted in the detection of mutations in genes responsible for various hereditary diseases, somatic mutations of oncogenes or tumor suppressor genes in cancer tissues, as well as polymorphisms useful for linkage mapping and association studies (Hayashi 1994).

Today, researchers are using such techniques for single nucleotide polymorphism (SNP) detection. SNPs are the most common genetic variation between individuals within a species, and they are responsible for phenotypic variation. SNP research continues to facilitate mapping studies and is leading to a better understanding of the genetic basis for complex diseases and individual variation in drug metabolism (pharmacogenetics). The National Center for Biotechnology Information (NCBI) has a major role in facilitating the identification and cataloging of SNPs through its creation and maintenance of the public SNP database (dbSNP) (http://www. ncbi.nlm.nih.gov/entrez/query.fcbi?db=snp). This powerful genetic tool may be accessed by the biomedical community worldwide and is intended to stimulate many areas of biological research, including the identification of the genetic components of disease. Records in dbSNP are cross-annotated with other resources and can be queried by many attributes, including gene name, chromosome location, and coding versus noncoding region. Also included are data about source, heterozygosity, and the method of discovery and validation. Many of the cataloged SNPs were discovered by SSCP.

Those who study human, mouse, and other well-characterized genomes have the luxury of these SNP databases, from which assays may be designed to screen target populations. Therefore, it might be assumed that SSCP is not as frequently used today, especially as it was originally intended for the discovery of polymorphisms. However, a perusal of current literature reveals that many researchers still use SSCP, and many still use SSCP for such discovery. Using the NCBI tool PubMed, a database to provide access to citations from biomedical literature, a query of the term "SSCP" limited to the dates January 1, 2002, to August 1, 2002, produced a list of 350 publications. These are publications by researchers who employ SSCP methods to perform screening, mapping, and evolution studies within less well-characterized species, such as *Oesophagostomum bufurcum* (a parasitic nematode), *Aedes aegypti* (yellow fever mosquito), and different breeds of dogs and horses (de Gruijter et al. 2002; Dekomien and Epplen 2002; Matiasovic et al. 2002; Severson et al. 2002). Even for those who select a SNP from a database for evaluation within a particular subpopulation, SSCP is still a valuable, easy, high-throughput, and relatively inexpensive tool for implementing and achieving such analysis. SSCP also has potential as a companion to microarray studies. During the complex data analysis of microarrays, variation of gene expression within groups of similar subjects may be explained by SNPs within certain genes. SSCP can be a useful tool to genotype the outliers of these subpopulations in array data sets.

In summary, SSCP analyses have great utility and impact and have facilitated a broad spectrum of discovery within the life sciences. In many laboratories, highly automated technology and high-throughput methods may be fiscally and practically unattainable for routine performance. Therefore, it is advantageous to have traditional, proven methods, such as SSCP, for molecular genetic research, including the more refined mapping of genes and the development of diagnostic and pharmacogenetic assays.

ACKNOWLEDGMENT

We thank Ms. Margaret V. Mintz for outstanding technical support, photographic assistance, and for sharing the joys of SSCP and other methods of candidate gene evaluation with us.

REFERENCES

Arbour N.C., Lorenz E., Schutte B.C., Zabner J., Kline J.N., Jones M., Frees K., Watt J.L., and Schwartz D.A. 2000. *TLR4* mutations are associated with endotoxin hyporesponsiveness in humans. *Nat. Genet.* **25:** 187–191.

de Gruijter J.M., Polderman A.M., Zhu X.Q., and Gasser R.B. 2002. Screening for haplotypic variability within *Oesophagostomum bifurcum* (Nemotoda) employing a single-strand conformation polymorphism approach. *Mol. Cell. Probes* **16:** 185–190.

Dekomien G. and Epplen J. 2002. Screening of the arrestin gene in dogs afflicted with generalized progressive retinal atrophy. *BMC Genet.* **3:** 12.

Hajeer A.H., al Sharif F., and Ollier W.E. 1999. A polymorphism at position –403 in the human RANTES promoter. *Eur. J. Immunogenet.* **26:** 375–376.

Hayashi K. 1991. PCR-SSCP: A Simple and sensitive method for detection of mutations in the genomic DNA. *PCR Methods Appl.* **1:** 34–38.

———. 1994. Manipulation of DNA by PCR. In *The polymerase chain reaction* (ed. K.B. Mullis et al.), pp. 3–13. Birkhäuser, Boston, Massachusetts.

Matiasovic J., Lukeszova L., and Horin P. 2002. Two bi-allelic single nucleotide polymorphisms within the promoter region of the horse tumour necrosis factor alpha gene. *Eur. J. Immunogenet.* **29:** 285–286.

Orita M., Suzuki Y., Sekiya T., and Hayashi K. 1989. Rapid and sensitive detection of point mutations and DNA polymorphisms using the polymerase chain reaction. *Genomics* **5:** 874–879.

Rust S., Funke H., and Assmann G. 1993. Mutagenically separated PCR (MS-PCR): A highly specific one step procedure for easy mutation detection. *Nucleic Acids Res.* **21:** 3623–3629.

Severson, D.W., Meece J.K., Lovin D.D., Saha G., and Morlais I. 2002. Linkage map organization of expressed sequence tagged sites in the mosquito, *Aedes aegypti. Insect Mol. Biol.* **11:** 371–378.

Sheffield V.C., Beck J.S., Kwitek A.E., Sandstrom D.W., and Stone E. 1993. The sensitivity of single-strand conformation polymorphism analysis for the detection of single base substitutions. *Genomics* **16:** 325–332.

Yandell D.W. 1991. SSCP analysis of PCR fragments. *Jpn. J. Cancer Res.* **82:** 1468–1469.

WWW RESOURCES

http://www.ncbi.nlm.nih.gov/entrez/query.fcbi?db=PubMed NCBI, National Library of Medicine, PubMed database

http://www.ncbi.nlm.nih.gov/entrez/query.fcbi?db=snp NCBI, National Library of Medicine, Single Nucleotide Polymorphism (SNP) database

www.gelcompany.com The Gel Company, San Francisco, California.

19   Reverse Transcriptase In Situ PCR

Gerard J. Nuovo

Ohio State University Medical Center, Columbus, Ohio 43210

Since the discovery of the structure of DNA in 1953, a great deal of effort has been devoted to detecting specific DNA target sequences. The first commonly used technique, filter (or dot-blot) hybridization, offered excellent sensitivity, being able to detect 1 copy per 100 cells. The later introduction of Southern blot hybridization further enhanced detection methods. However, the destructive nature of nucleic acid extraction, upon which these techniques rely, precluded localization of the target to its specific cell of origin. In many instances, this information is critical. For example, although PCR can be used routinely to detect HIV-1 DNA in the lymph nodes of seropositive, asymptomatic patients, it cannot be used to identify which of the many different cell types that reside in this site are reservoirs of the virus, prior to the development of AIDS.

During the last 15 years, direct cellular localization of a DNA or RNA target has been achieved routinely using in situ hybridization. Over the past 10 years, there have been dramatic improvements in the sensitivity of this technique, especially in the use of nonradioactive probes (Crum et al. 1988; Nuovo 1989, 1997; Nuovo and Richart 1989). However, despite these improvements, the usefulness of the technique has been limited by its relatively high detection threshold (~10 copies per cell; Nuovo 1993, 1994a). Although this is not a problem when detecting, for example, productive viral DNA infections, where a given cell may contain thousands of copies, there are common problems, such as the detection of latent viral infection or point mutations, which cannot be solved using in situ hybridization. Detection of low-copy events, of course, has become routine over the last 13 years with the widespread use of PCR. With the hot-start method, PCR can be used to detect one copy in a background of 1 μg of total cellular DNA (Erlich et al. 1991; Nuovo et al. 1991; Chou et al. 1992). However, PCR cannot be used to achieve sub-tissue localization of target sequences.

The last 10 years have seen the development of in situ PCR, a technology that combines the high sensitivity of PCR with the cell localization of in situ hybridization. This chapter presents protocols for in situ detection of PCR-amplified DNA and cDNA, and discusses some of its applications, especially as it relates to HIV-1 and AIDS.

PROTOCOL 1 REVERSE TRANSCRIPTASE IN SITU PCR

This section provides an abbreviated protocol for doing in situ PCR, specifically reverse transcriptase (RT) in situ PCR. Detailed information on the various steps is provided in the remainder of this chapter.

MATERIALS

BUFFERS AND SOLUTIONS

Nuclear Fast Red<!>
Formalin (v/v), 10%<!> buffered (Polyscientific)
Detection solution (0.1 M Tris-HCl<!>, pH 9.5, 0.1 M NaCl)
DEPC<!>-treated H_2O
Phosphate-buffered saline (PBS)(137 mM NaCl, 2.7 mM KCl, 10 mM Na_2HPO_4<!>)
2 mM KH_2PO_4<!>, pH 7.0, for use with cell cultures
Xylene<!> (for use with tissue samples)
Ethanol<!>
Bovine serum albumin (BSA), 2% (w/v)
SSC, 20x (3 M NaCl, 0.3 M sodium citrate, pH 7.0)
NBT/BCIP chromogen<!> (Enzo Clinical Labs)

ENZYMES AND ENZYME BUFFERS

Pepsin<!> (2 mg/ml) (Enzo Diagnostics)

> Prepare by adding 20 mg of pepsin, 9.5 ml of H_2O, and 0.5 ml of 2 N HCl<!>.

RNase-free DNase (Boehringer Mannheim)
Taq DNA polymerase Gold (Applied Biosystems, Foster City, CA)
RNase inhibitor (Applied Biosystems)
DNase buffer, 10x

> Prepare by adding 35 μl of sodium acetate (3 M, pH 3), 5 μl of $MgSO_4$ (1 M), and 60 μl of DEPC H_2O.

NUCLEIC ACIDS AND OLIGONUCLEOTIDES

Digoxigenin dUTP<!> (Boehringer Mannheim), 1 mM

ANTIBODIES

Antidigoxigenin conjugate (Boehringer Mannheim)
Appropriate PCR primers

SPECIAL EQUIPMENT

Silane-coated slides (Ventana)
In situ PCR cycler (Applied Biosystems)
Polypropylene sheet, to be used as coverslips (sold as autoclavable bags)
Incubators, preset to 37°C, 60°C
Humidity chamber (lidded box lined by water-saturated paper towel)

ADDITIONAL ITEMS

RT-PCR kit (Applied Biosystems)

> This kit includes the buffers, nucleotides, RNase inhibitor, manganese solution, rTth DNA polymerase, and control primers needed for RT-PCR.

Reagents and equipment for paraffin embedding and sectioning (for use with tissue samples)
Permount<!>

CELLS AND TISSUES

Appropriate tissue samples or cell culture

Many reagents, including the protease, buffers, washes, chromogen, probes, and counter-stains, are part of comprehensive in situ hybridization kits. These are now marketed by several biotechnology companies.

METHOD

Preparation of Tissue Sections and Cell Samples

1. Fix the tissue samples in 10% buffered formalin, preferably from 8 to 15 hours, and then embed in paraffin.

2. For cell cultures, wash the cells once with PBS added directly to the culture plate. Add 10% buffered formalin, let stand overnight, and then scrape the cells off the plate with a rubber policeman. Wash the cells twice in diethylpyrocarbonate (DEPC)-treated H_2O, and centrifuge at 2000 rpm for 3 minutes in a microcentrifuge. Resuspend the cells in 5 ml of DEPC-treated H_2O, and spot 50 µl (the ideal cell density is 2500–7500 cells) onto a slide, air-dry, and store at room temperature.

3. Place several (at least 3) 4-µm paraffin tissue sections or three cell suspensions on silane-coated slides.

   The silane coating is essential for cell adherence during the procedure.

4. Remove the paraffin from the tissue samples by placing the slides in fresh xylene for 5 minutes and then in 100% ethanol for 5 minutes; air-dry.

Protease Digestion

5. Bathe slides in pepsin solution for the desired incubation time at 37°C.

   See Tables 1 and 2 for protease incubation times. Proteases other than pepsin may be used for this digestion. For more information, see Protease Digestion, p. 257.

6. Inactivate the protease by washing the slide in DEPC-treated H_2O for 1 minute and then in 100% ethanol for 1 minute; air-dry.

   This simple wash step is sufficient to remove/inactivate pepsin; heat-inactivation is not required.

DNase Digestion (for RT In Situ PCR)

It is important to DNase-digest two of the three tissue sections/cell suspensions. The in situ PCR *positive control* is the section that is not treated with DNase. An intense signal in at least 50% of cells demonstrates that protease digestion time is optimal and that the other reagents are working properly; the signal represents DNA repair, genomic amplification, and mispriming. The in situ PCR *negative control* is the cells/tissue section that is DNase-digested and then incubated with irrelevant primers (i.e., primers that correspond to a target that could not be present in the tissue). The absence of signal demonstrates that the DNase digestion has rendered the native DNA template unavailable for DNA synthesis. The *test* is the section digested by DNase that is then treated with RT and PCR using the target-specific primers.

7. To two of the three sections add 1 µl of 10x DNase buffer, 1 µl of RNase-free DNase (10 units/µl), and 8 µl of DEPC-treated H_2O. Alternatively, RT/PCR buffer can be used in place of 10x DNase buffer.

8. Cover the solution with a polypropylene coverslip that is cut to the size of the sample to prevent drying. Place the slides in a humidity chamber at 37°C.

9. After overnight digestion, remove the coverslip, wash for 1 minute in DEPC H_2O, then 100% ethanol, and air-dry.

RT Step

10. In a sterile microcentrifuge tube, assemble the following RT-PCR reagents:

EZ buffer (Applied Biosystems)	10 µl
each of dATP, dCTP, dGTP, dTTP (stock of each nucleotide at 10 mM)	1.6 µl
Bovine serum albumin solution, 20% (w/v)	1.6 µl
DEPC-treated H_2O	15.1 µl
$MnCl_2$ (or Mn acetate) solution, 10 mM	12.4 µl
Sense primer (20 µM)	2.0 µl
Antisense primers (20 µM), 2.0 µl	
RNase inhibitor, 0.6 µl of digoxigenin dUTP (stock solution 1 mM; Applied Biosystems)	0.5 µl
rTth	2.0 µl

11. Add RT-PCR solution to one of the two sections treated with DNase.

12. To prevent drying, anchor the polypropylene coverslip with one small drop of nail polish and place the slide in an aluminum "boat" and cover with sterile mineral oil. Incubate on the block of the thermal cycler at 60°C for 30 minutes.

The Amplicover/Ampliclip method of Applied Biosystems can be used as an alternative to mineral oil.

PCR Step

For PCR in situ hybridization (for DNA targets), the hot-start manuever must be used. This can be achieved by withholding the DNA polymerase until the thermal cycler block reaches 60°C, but it is easier to add all the reagents at room temperature, including the *Taq* polymerase Gold. This polymerase contains an antibody that inhibits nonspecific polymerase activity. At the appropriate temperature, the anti-*Taq* antibody is eliminated and the polymerase is activated. The higher temperatures, in turn, prevent the nonspecific side reactions.

13. In a sterile microcentrifuge tube on ice, assemble the PCR reagents in the following order:

Taq polymerase buffer (Applied Biosystems)	5 µl
dNTP solution (each nucleotide at 10 mM stock)	9.0 µl
$MgCl_2$, 25 mM	8.0 µl
Bovine serum albumin solution, 2% (w/v)	1.6 µl
H_2O	22.4 µl
Sense primer, 20 µM	2.0 µl
Antisense primer, 20 µM	2.0 µl
Taq polymerase Gold (10 units/µl)	2.0 µl

14. Apply the PCR mix to each sample and perform the following thermal cycle series:

Cycle number	Denaturation	Annealing/Extension
1	3 min at 94°C*	
20**	1 min at 94°C	1.5 min at 60°C

*This initial denaturation step is used for RT in situ PCR and PCR in situ hybridization.

**If the newly formed cDNA is to be detected using a probe after RT and PCR rather than by direct labeling, increase the number of cycles to 30.

15. Remove the coverslip and nail polish from the slide. Wash for 5 minutes in xylene and 5 minutes in 100% ethanol. Allow the slide to air-dry.

Detection Step

16. Wash the slides for 15 minutes in a 0.2x SSC solution containing 0.2% (w/v) BSA at 60°C (for RT in situ PCR).

    For PCR in situ hybridization using full-length genomic probes of at least 80 bp in size, use the wash described above. For hybridization using oligoprobes of <45 bp, use 1x SSC containing 2% (w/v) BSA at 45°C for 10 minutes.

17. Remove the excess wash solution. To each slide, add 100 μl of a 0.1 M Tris-HCl, pH 7.5, 0.1 M NaCl solution containing the antidigoxigenin antibody (1:150 dilution) (Boehringer Mannheim). Incubate at 37°C for 30 minutes.

18. Wash the slides for 1 minute in detection solution and then incubate for 5–30 minutes at 37°C in detection solution containing NBT/BCIP chromogen, as specified in supplier's instructions (Enzo Clinical Labs). Shorter incubations can be used for direct incorporation of label into amplified DNA; longer incubations are required when using probes. Check the slides under a microscope and stop the reaction when the signal is strong.

    For PCR in situ hybridization, the probe–amplicon complex should be hybridized for 15 hours; digoxigenin-labeled probes are recommended for this purpose. It is important to use the longest probes possible (optimally 80–100 bp). Primer oligomerization does not appear to occur with PCR in situ hybridization, so full-length probes that include the region of the primers can be used. It is also important to label the probes by random priming (Nuovo 1997). If shorter probes are used, they must be at least 45 bp in size and labeled with the 3′ end-tailing method (Nuovo 1997).

19. Wash the slides in water for 1 minute, counterstain with nuclear fast red for 5 minutes, wash in water for 1 minute, then in 100% ethanol for 1 minute, and then in xylene for 1 minute. Mount coverslips using Permount. View under a microscope.

◗ TROUBLESHOOTING FOR RT IN SITU PCR

- If the positive control has >50% positive cells, and the negative control shows no positive cells, then the RT/DNase-treated section should show target-specific cDNA amplification.

- If the negative control shows positive cells, increase the protease digestion time. Insufficient protease digestion is the most common cause of problems with RT in situ PCR and is easily recognized using the two in situ PCR controls outlined above.

- If neither the negative nor the positive control shows positive cells, check the tissue morphology. If it is poorly preserved, reduce the protease digestion time. If morphology is well preserved, increase the protease digestion time.

DISCUSSION

Basics of In Situ PCR

Tissue Preparation

The correct choice of fixative is critical for success with in situ PCR. There are several fixatives that can be used for in situ hybridization, but their suitability for use in PCR is variable. In situ PCR can be performed on unfixed, frozen tissue or in tissue fixed in acetone, ethanol, or buffered formalin. Fixatives that include a heavy metal or picric acid cause rapid and extensive degradation of DNA and are incompatible with PCR (Greer et al. 1991). Fixatives such as acetone and ethanol function by denaturing proteins, but the most commonly used fixative in anatomic pathology is 10% buffered formalin, which cross-links proteins and nucleic acids. Tissues fixed for more than 8 hours in solutions that contain either a heavy metal or picric acid are not suitable for in situ hybridization, although shorter-term fixation can yield intense signals (Nuovo and Silverstein 1988).

Table 1 shows the results of a study comparing the utility of different fixatives for in situ PCR. The experiments used a cell line containing a known target sequence (*bcl-2* DNA), and fixatives that allowed a 100% detection rate were seen (Nuovo et al. 1993). Note that under the conditions defined in Table 1, only buffered formalin fixation, followed by protease digestion, gave a 100% detection rate of the target sequence. It is perhaps surprising that acetone or ethanol fixation did not produce 100% detection rates, given that they permit successful solution-phase PCR. Further investigation showed that the poor detection rate for acetone- or ethanol-fixed tissues was due not to inhibition of amplification, but to the migration of the PCR product out of the cell and into the amplification buffer. No such migration was detected in formalin-fixed samples. This does not mean that acetone- or ethanol-fixed cells should not be used for in situ PCR; however, the results obtained with these fixatives are less reliable than those obtained using formalin or its relatives (e.g., glutaraldehyde and paraformaldehyde). Formalin fixation appears to create a migration barrier that, under certain specified conditions, strongly limits the movement of the PCR product from its site of synthesis in the cell. It is reasonable to assume that this relates to the protein–DNA latticework that occurs as a secondary result of formalin fixation.

Because of the highly toxic nature of formalin, there has been a concerted effort to find alternatives to this fixative. Fixatives that do not contain formalin include Histochoice and Streck's solution. These chemicals can be used for in situ PCR with cell suspensions and tissue sections without the need for a protease digestion step. However, in our experience, these fixatives are less reliable than formalin-based reagents, with respect to both detection rate and precision of subcellular localization (Nuovo 1994a). Note that the use of formalin-containing fixatives can be problematic in combination with paraffin embedding (for more details, see Nonspecific Pathways during In Situ PCR, p. 258).

TABLE 1. Effect of fixation chemistry and duration on the detection of amplified *bcl-2* DNA in peripheral blood mononuclear cells

| NO PROTEASE DIGESTION (% positive cells) | | | | |
Fixation time	formalin	acetone	95% ethanol	Bouin's
5 min	5	2	14	0
15 hr	0	15	31	0
39 hr	not done	0	9	not done

| PROTEASE (pepsin) DIGESTION (% positive cells) | | | | | |
Fixation time	protease conditions	formalin	acetone	95% ethanol	Bouin's
5 min	2 mg/ml 12 min	0	0	0	0
5 min	20 µg/ml 1 min	1	0	0	—
5 min	20 µg/ml 2 min	35	0	0	—
5 min	20 µg/ml 5 min	0	0	0	—
15 hr	2 mg/ml 12 min	100	0	0	0

Protease Digestion

As shown by the data in Table 1, 100% detection of the target sequence was achieved only after prolonged formalin fixation *and* protease digestion. An expanded study, including tissues fixed in buffered formalin for up to 1 week, demonstrated that the signal strength in formalin-fixed tissues is strongly dependent on the duration of both formalin fixation and protease digestion (Nuovo et al. 1994a). These data are presented in Table 2. Note that the study used in situ PCR and not PCR in situ hybridization. For PCR in situ hybridization, using tissues fixed for 4 hours to several days, the optimal protease digestion time is from 20 to 30 minutes (Nuovo 1994a).

Several proteases have been extensively tested for in situ hybridization. These include proteinase K, pepsin, trypsin, trypsinogen, and pronase. All of these proteases are compatible with in situ amplification of DNA and cDNA, but trypsin and pepsin are preferred. These proteases can be inactivated simply by increasing the pH of the reaction to pH 7.0, and they do not have a tendency to over-digest, as does proteinase K. Overdigestion is indicated by the loss of distinct morphology and concomitant absence of signal (Fig. 1).

Inadequate protease digestion is the most common cause of unsuccessful in situ PCR. For both in situ PCR and RT-in situ PCR, the optimal length of protease treatment is dependent on the duration of formalin fixation. Optimal treatment is indicated by distinct morphology, concomitant with an intense signal in more than 50% of cells (3+). If the signal is less than 3+, the protease digestion times should be increased until 3+ is obtained. This important principle is illustrated in Figure 2. Note that the times listed in Table 1 were optimized for skin tissue. The times will vary depending on the tissue type used. Tissues such as lymph nodes, spleen, and kidney typically are more sensitive to protease digestion, as compared to brain, skeletal muscle, and skin. The use of longer formalin fixation times tends to allow a fixed protease digestion time to be used irrespective of cell type.

Amplifying Solution

For those doing PCR in situ hybridization (i.e., analyzing for a DNA target), it is important to note that the optimal $MgCl_2$ concentration is 4.5 mM, as illustrated in Figure 3, irrespective of the primer or target sequences. This relatively high and constant concentration of magnesium may be due to its partial sequestration on the glass slide and cellular proteins (Nuovo et al. 1993; Nuovo 1994a).

TABLE 2. Effect of protease digestion time on the primer-independent signal during in situ PCR as a function of the time of fixation in 10% buffered formalin

Fixation time	Protease[a] digestion time (min)								
	0	5	10	15	30	45	60	75	90
4 hr	0	1+	3+	2+	overdigested[b]				
6 hr	0	0	1+	3+	2+	overdigested[b]			
8 hr	0	0	0	0	1+	3+	—	—	—
15 hr	0	0	0	0	0	1+	1+	2+	3+
48 hr	0	0	0	0	0	0	1+	2+	3+
1 week	0	0	0	0	0	0	1+	1+	2+/3+[c]

The signal was scored as follows: 0, 1+ (<25% of cells positive), 2+ (25–50% cells positive), and 3+ (>50% of cells positive). Signal measurements were made without knowledge of the reaction conditions.
[a]Protease is pepsin (2 mg/ml) at room temperature.
[b]Overdigested refers to loss of tissue morphology with a concomitant loss of the in situ PCR signal.
[c]The 2+ signal was with pepsin; the 3+ signal with proteinase K digestion (1 mg/ml).

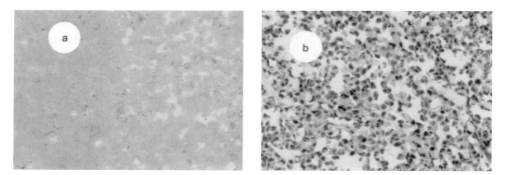

FIGURE 1. Overdigestion of tissue with protease. This tonsillar tissue was fixed in buffered formalin for 4 hours. No signal is evident with in situ PCR using *bcl-2* primers if the protease digestion time (pepsin 2 mg/ml) was 30 minutes (*a*); note the poor morphologic detail. If the digestion time was decreased to 10 minutes, an intense signal and good morphologic detail are evident (*b*).

The composition of the buffer (PCR buffer II, GeneAmp kit, Perkin-Elmer) and the concentration of the nucleotides (200 μM) and primers (1 μM) for in situ PCR are equivalent to those used for solution-phase PCR. The concentration of the DNA (*Taq*) polymerase (Perkin-Elmer or equivalent), however, should be increased to 15 units per 50 μl. This large amount of enzyme is thought to be necessary because the protein binds nonspecifically to the glass slide and plastic coverslip. If BSA is added to the reaction (final concentration 0.015%), the binding is inhibited, and strong signals can be obtained using only 1.5 units of enzyme (Nuovo et al. 1993; Nuovo 1994a). Similarly, higher concentrations of rTth must be used for RT in situ PCR as compared to solution-phase RT PCR.

Nonspecific Pathways during In Situ PCR

Studies have shown that primer-independent DNA synthesis occurs if paraffin-embedded fixed tissues are used for in situ PCR. The nonspecific signal results from the initiation of DNA synthesis at nicks in the DNA, formed during the 65°C paraffin-embedding process. Because of this background, target-specific direct incorporation of the reporter nucleotide into the amplicon is not possible for formalin-fixed, paraffin-embedded tissues for DNA targets.

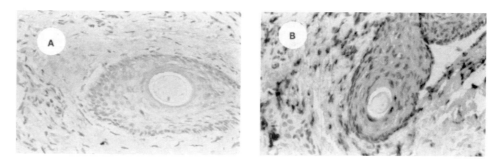

FIGURE 2. Relationship of protease digestion and formalin fixation times for successful in situ PCR. This skin biopsy was fixed in buffered formalin for 8 hours. No signal is evident with in situ PCR if the pepsin digestion time was 15 minutes (*A*). An intense signal is evident if the digestion time is increased to 45 minutes (*B*).

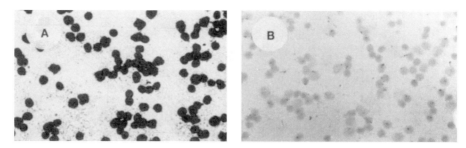

FIGURE 3. Importance of MgCl₂ concentration for successful in situ PCR. At an MgCl₂ concentration of 4.5 mM all peripheral blood mononuclear cells fixed for 15 hours in buffered formalin and digested have an intense signal using in situ PCR and *bcl-2* primers (*A*). No signal is evident if MgCl₂ is omitted (*B*). Similar poor results were obtained with magnesium concentrations of 1.5, 6.0, and 9.0 mM.

In Situ PCR

In situ PCR differs from PCR in situ hybridization (see Fig. 4) in the inclusion of a reporter molecule in the amplification step and, of course, the omission of the in situ hybridization

1. Place 3 paraffin embedded tissue sections on silane-coated slides, remove paraffin, protease digest for 20–30 min.

2. Place PCR reagent over 1 section; use others for − and + control. Place slide on thermal cycler.

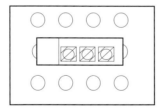

3. Do PCR for 30 cycles, wash in xylene and ETOH, then do in situ hybridization with internal oligoprobe.

4. Wash in 1XSSC with 0.2% BSA at 50°C for 10 minutes, then detect label on probe.

FIGURE 4. Graphic representation of the PCR in situ hybridization protocol.

step. A variety of reporter molecules may be used. Digoxigenin- and biotin-labeled nucleotides allow rapid and simple colorimetric detection of the amplified product by the use of the appropriate alkaline phosphatase-labeled conjugated antibodies (antidigoxigenin and streptavidin, respectively). In situ PCR is much quicker and simpler than PCR in situ hybridization. However, in our experience, it can be employed for DNA targets only when using frozen fixed tissues or cytospins that have not been heated. The power and widespread utility of in situ PCR are much more evident for RNA, where pretreatment overnight in a RNase-free DNase solution allows target-specific incorporation of the reporter molecule in the cDNA synthesized after the RT step.

RT In Situ PCR

The two steps of RT in situ PCR that differ from in situ PCR are the overnight digestion in RNase-free DNase that is performed after the protease digestion, and the RT step, prior to in situ PCR. After optimal protease digestion, a minimum of 7 hours of DNase treatment is needed to eradicate completely the nonspecific signal (Nuovo 1994a; Nuovo and Forde 1995).

An essential point for successful RT in situ PCR is evident from Tables 1 and 2. Specifically, there is a strong relationship between the protease digestion and formalin fixation times. Adequate protease digestion is defined as an intense signal in >50% of the cells and no signal after DNase digestion. It is important to realize that, with too little protease digestion, one usually sees a signal with DNase digestion that is stronger than that for the section not pretreated with DNase. Assuming that the nonspecific signal represents repair of DNA gaps, this DNase enhancement with suboptimal protease may represent the enlargement of these putative gaps under conditions where the DNase cannot completely destroy the DNA template. For RT in situ PCR, it is essential that the positive control (no DNase) and negative control (DNase, no RT) are performed on the same slide as the RT test. A successful run is defined by a strong signal in the positive control and no signal in the negative control (Fig. 5); note that ^{3}H is the reporter molecule. If a signal is seen with the negative control, the most likely reason is inadequate protease times. Another possible problem is no

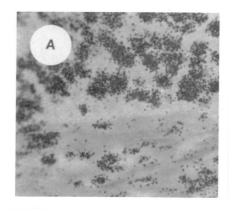 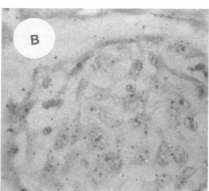 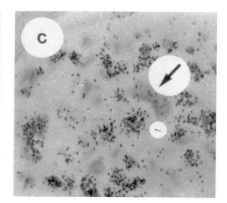

FIGURE 5. The importance of the positive and negative controls with RT in situ PCR. This cervical cancer was analyzed for MMP-92 expression using RT in situ PCR and direct incorporation of [^{3}H]dCTP. Note the intense nuclear signal with the positive control (no DNase, *A*) and the absence of signal with the negative control (DNase, no RT, *B*). These reactions were done on the same slide as the test and demonstrated that the protease and DNase digestions were adequate. On the same glass slide, the RT in situ PCR for MMP-9 was done. Many cells contained this mRNA; note the cytoplasmic localization of the signal (*C*; *small arrow*, cytoplasm; *large arrow*, nucleus).

signal with the positive control and poor morphology. This signifies overdigestion with protease and should prompt retesting at lower protease digestion times (Fig. 1).

The final important point about RT in situ PCR concerns the localization of the signal. The signal should localize to a different subcellular compartment(s) for the RNA and for the DNA in the positive control. For the DNA positive control, the signal should be pannuclear, as illustrated in Figure 6. Note the dramatically different patterns evident with various RNAs, suggesting different pathways from the nucleus to the cytoplasm.

UTILITY OF PCR IN SITU HYBRIDIZATION AND RT IN SITU PCR—PATHOGENESIS OF AIDS

It can be argued that the development of the methodology of in situ amplification of DNA and cDNA was due to AIDS. More specifically, the fact that HIV-1 DNA can integrate as one copy and remain transcriptionally silent (or nearly so) was the impetus to develop technologies that were capable of the routine in situ detection of one target copy. RT in situ PCR and PCR in situ hybridization have revealed the following about HIV-1-associated disease:

There is a massive, covert infection, primarily in the lymph nodes, by HIV-1 long before a person develops AIDS. On average, in the sites of infection, 30% of the CD4 cells are latently infected by the virus. The realization of this essential fact paved the way for aggressive anti-retroviral therapy, long before symptoms of AIDS were evident (Bagasra et al. 1992; Embretson et al. 1993; Nuovo et al. 1994c).

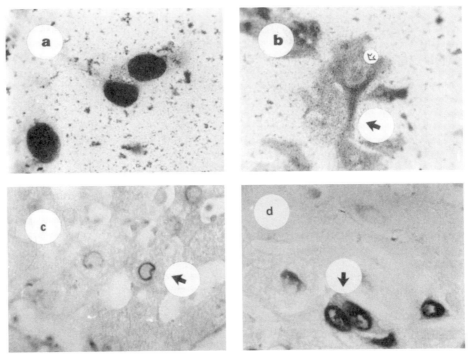

FIGURE 6. Different localizations of the signal with RT in situ PCR. The signal with the positive control, representing DNA repair, mispriming, and genomic DNA synthesis should localize to the nucleus, as is evident with this HT1080 cell line (*a*). The RNA-based signal should not be pan-nuclear. Variable patterns are seen. For example, a cytoplasmic signal is evident for MMP-92 in this cell line (*b*; *open arrow*, negative nucleus; *large arrow*, cytoplasmic signal). The signal for PCR-amplified hepatitis C RNA localizes to the junction of the nucleus and cytoplasm in the hepatocyte (*c*). A perinuclear and cytoplasmic signal is seen for the β chain of fibrinogen mRNA in trophoblasts (*d*; *arrow*, cytoplasm).

AIDS dementia is marked by productive HIV-1 infection of astrocytes, microglial cells, and neurons in the brain with concomitant up-regulation of a variety of cytokines in the area of HIV-1 infection but in the neighboring, noninfected cells (Nuovo and Alfieri 1996).

AIDS myopathy and cardiomyopathy involves (primarily) the recruitment of HIV-1-infected macrophages in the skeletal muscle and heart, respectively (Seidman et al. 1994; Cioc and Nuovo 2002).

SUMMARY

Many groups have published data generated by in situ detection of PCR-amplified DNA and cDNA. As with standard in situ hybridization or PCR, a number of variables can affect the results of in situ PCR. These include the type of fixative and duration of fixation, the duration of protease digestion, the composition of the amplifying solution, the size of the probe, and the stringency of the wash conditions. Investigators new to the field of in situ PCR should first try direct incorporation of the reporter molecule into paraffin-embedded tissue sections. Although nonspecific DNA synthesis occurs under these conditions, the experiment can be used to demonstrate successful DNA synthesis inside the nucleus and to demonstrate the importance of optimizing protease digestion for successful RT in situ PCR. In situ detection of PCR-amplified DNA and cDNA will have a strong impact on diverse fields of research, such as oncogenesis, embryology, RNA-trafficking, and detection of viral diseases. It has already significantly advanced our understanding of the pathogenesis of HIV-1 infection.

ACKNOWLEDGMENTS

The author thanks John Atwood, Phyllis MacConnell, Kim Rhatigan, Angella Forde, and Maricella Suarez for invaluable assistance.

REFERENCES

Bagasra O., Hauptman S.P., Lischner H.W., Sachs M., and Pomerantz R.J. 1992. Detection of human immunodeficiency virus type 1 provirus in mononuclear cells by in situ polymerase chain reaction. *N. Engl. J. Med.* **326:** 1385–1391.

Chou Q., Russell M., Birch D.E., Raymond J., and Bloch W. 1992. Prevention of pre-PCR mis-priming and primer dimerization improves low-copy-number amplifications. *Nucleic Acids Res.* **20:** 1717–1723.

Cioc A.M. and Nuovo G.J. 2002. Histologic and in situ viral findings in the myocardium in cases of sudden, unexpected death. *Mod. Pathol.* **15:** 914–922.

Crum C.P., Nuovo G.J., Friedman D., and Silverstein S.J. 1988. A comparison of biotin and isotype labeled ribonucleic acid probes for in situ detection of HPV 16 ribonucleic acid in genital precancers. *Lab. Invest.* **58:** 354–359.

Ehrlich H.A., Gelfand D., and Sninsky J.J. 1991. Recent advances in the polymerase chain reaction. *Science* **252:** 1643–1650.

Embretson J., Zupancic M., Ribas J.L., Burke A., Racz P., Tenner-Racz K., and Haase A.T. 1993. Massive covert infection of helper T lymphocytes and macrophages by HIV during the incubation period of AIDS. *Nature* **362:** 359–362.

Greer C.E., Peterson S.L., Kiviat N.B., and Manos M.M. 1991. PCR amplification from paraffin embedded tissues: Effects of fixative and fixative times. *Am. J. Clin. Pathol.* **95:** 117–124.

Nuovo G.J. 1989. A comparison of slot blot, Southern blot and in situ hybridization analysis for human papillomavirus DNA in genital tract lesion. *Obstet. Gynecol.* **74:** 673–677.

———. 1993. *Cytopathology of the cervix and vagina: An integrated approach.* Williams and Wilkins, Baltimore, Maryland.

————. 1997. *PCR in situ hybridization: Protocols and applications*, 3rd edition. Raven Press, New York.

Nuovo G.J. and Alfieri M.L. 1996. AIDS dementia is associated with massive, activated HIV–1 infection and concomitant expression of several cytokines. *Mol. Med.* **2:** 358–366.

Nuovo G.J. and Forde A. 1995. An improved system for reverse transcriptase in situ PCR. *J. Histotechnol.* **18:** 295–299.

Nuovo G.J. and Richart R.M. 1989. A comparison of different methodologies (biotin based and 35S based) for the detection of human papillomavirus DNA. *Lab. Invest.* **61:** 471–476.

Nuovo G.J. and Silverstein S.J. 1988. Comparison of formalin, buffered formalin, and Bouin's fixation on the detection of human papillomavirus DNA from genital lesions. *Lab. Invest.* **59:** 720–724.

Nuovo G.J., Gallery F., and MacConnell P. 1994a. Analysis of specific DNA synthesis during in situ PCR. *PCR Methods Appl.* **4:** 89–96.

Nuovo G.J., Gallery F., Hom R., MacConnell P., and Bloch W. 1993. Importance of different variables for optimizing in situ detection of PCR amplification. *PCR Methods Appl.* **2:** 305–312.

Nuovo G.J., Gallery F., MacConnell P., Becker J., and Bloch W. 1991. An improved technique for the detection of DNA by in situ hybridization after PCR-amplification. *Am. J. Pathol.* **139:** 1239–1244.

Nuovo G.J., Becker J., Simsir A., Margiotta M., and Shevchuck M. 1994b. *In situ* localization of PCR-amplified HIV-1 nucleic acids in the male genital tract. *Am. J. Pathol.* **144:** 1142–1148.

Nuovo G.J., Becker J., Margiotta M., Burke M., Fuhrer J., and Steigbigel R. 1994c. In situ detection of PCR-amplified HIV-1 nucleic acids in lymph nodes and peripheral blood in asymptomatic infection and advanced stage AIDS. *J. Acquir. Immune Defic. Syndr.* **7:** 916–923.

Seidman R., Peress N., and Nuovo G.J. 1994. In situ detection of PCR-amplified HIV-1 nucleic acids in skeletal muscle in patients with myopathy. *Mod. Pathol.* **7:** 369–375.

Sensitive and Fast Mutation Detection by Solid-phase Chemical Cleavage of Mismatches

Lise Lotte Hansen,[1] Just Justesen,[2] Andreana Lambrinakos,[3] and Richard G.H. Cotton[3]

[1]Department of Human Genetics, [2]Department of Molecular Biology, University of Aarhus, DK-8000 Aarhus C, Denmark; [3]Genomic Disorders Research Centre, St. Vincents Hospital, Fitzroy, Victoria 3065, Australia

Solid-phase chemical cleavage of mismatch (SpCCM), as illustrated in Figure 1, is a fast and sensitive method for the detection of mutations in genomic DNA. DNA fragments up to 2 kb in length can be analyzed in a single operation, determining the position (within 10–15 bp) and identity of the mismatched nucleotide. Almost all mutations can be located using this procedure (Cotton 1989; Ellis et al. 1998). Mutations can be detected even when the mutated DNA comprises up to 5% of the analyzed material (Hansen et al. 1996).

When searching for mutations in DNA from tumors or other sources that are contaminated with an excess of wild-type DNA, it is important to verify the results by DNA sequencing. The advantage of the SpCCM method is that both the type and the approximate position of mutation in the analyzed fragment can be determined. This allows results to be verified by sequencing a narrow well-defined region. Cleaved fragments can be detected by gel electrophoresis and autoradiography (using a radioactive label), or computer-controlled capillary electrophoresis (using a fluorescent label) and computer analysis. Automated capillary electrophoresis can be used to detect fluorescently labeled, cleaved fragments. The software provided with the equipment precisely calculates the length of the fragments, thus ensuring very reliable mutation detection.

All reactions, from the first amplification of the test DNA to the cleavage procedure, can be carried out in microtiter plates (or micro tubes in strips of eight). SpCCM can be automated using a workstation to carry out the reactions and an automated fluorescent detection-based DNA electrophoresis system, e.g., ABI-3100 from Applied Biosystems, to analyze the cleaved fragments.

Here we present two variations of SpCCM. The first uses a radioactively labeled probe and conventional gel electrophoresis, followed by autoradiography. The second method is dependent on an automated electrophoresis system with fluorescence detection, originally described by Roberts et al. (1997).

Both methods require wild-type and test DNAs to be amplified in the presence (for wild-type) or absence (for test) of a radioactive or fluorescent label. Amplification is per-

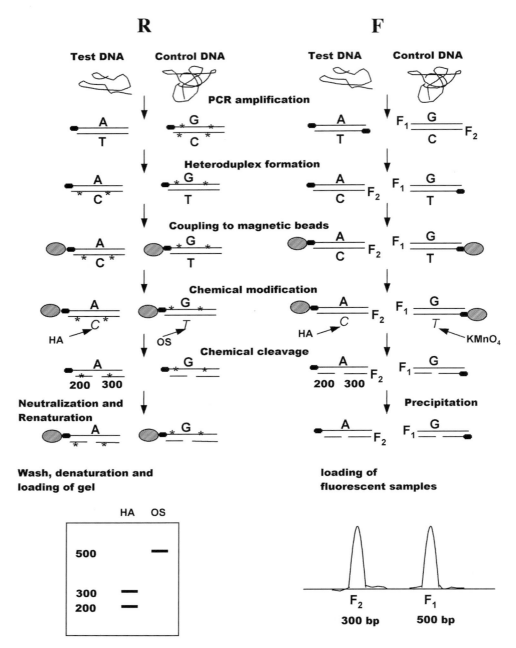

FIGURE 1. Schematic presentation of SpCCM with radioactive label (R) and fluorescence label (F). Test and control DNA is PCR-amplified with either biotin- or fluorophore-labeled primers (F_1 and F_2). [α-^{32}P]dCTP (H) is incorporated into the control DNA during amplification. Heteroduplexes are formed and attached to the streptavidin-coated magnetic beads. Chemical modification and cleavage are performed. The products are either reannealed and attached to new streptavidin-coated beads and washed (R), or precipitated with ethanol (F). In both R and F, the recovered DNA fragments are loaded onto denaturing polyacrylamide gels to resolve the size of the fragments. Modifications with hydroxylamine (HA) and osmium tetroxide (OS) are performed in separate reaction tubes but are illustrated on one DNA strand. Modified nucleotides are illustrated with letters in italics.

formed using two oligonucleotide primers, one of which is biotinylated. To establish the position of a mutation on a DNA fragment, it is very important to use two sets of primers for the amplification reactions. When using the radioactive method, the sense primer from one pair and the antisense primer from the other pair should be biotinylated. When using the fluorescent method, one primer from one pair should be biotinylated (sense or antisense) and the two primers from the other pair should each be labeled with a different fluorophore. This point is illustrated in Figure 2. The resulting amplicons are then combined to form heteroduplex molecules between the mismatched wild-type and test sequences. Once the heteroduplex is formed, streptavidin-coated magnetic beads, which bind to biotin, are used to capture the DNA. A magnet is used to collect the beads and attached heteroduplex molecules, thus avoiding tedious precipitations and loss of DNA during the procedure. The heteroduplexes are then split into two aliquots. One aliquot is treated with hydroxylamine and the other with osmium tetroxide. These treatments modify mismatched cytosines and thymines, respectively, and render the modified bases susceptible to cleavage by piperidine. Thymines and cytosines adjacent to the mismatch are also modified (Cotton et al. 1988; Cotton and Campbell 1989; Roberts et al. 1997; Lambrinakos et al. 1999). This results in the formation of more than multiple bands for a particular cleavage. The additional bands will be slightly smaller than the major bands. This difference would probably go unnoticed on an autoradiograph. The DNA fragments resulting from piperidine treatment are separated by capillary or polyacrylamide gel electrophoresis. Size analysis of the cleaved fragments reveals the position of the mutation, and comparison of the hydroxylamine- and osmium tetroxide-treated samples reveals the nature of the change.

PROTOCOL SOLID-PHASE CHEMICAL CLEAVAGE OF MISMATCHES

MATERIALS

BUFFERS, SOLUTIONS, AND REAGENTS

Annealing buffer, 2x: 1.2 M NaCl, 12 mM Tris-HCl<!>, pH 7.5, and 14 mM $MgCl_2$<!>. Autoclave and store at room temperature.

B &W (bind and wash) buffer: 10 mM Tris-HCl, pH 7.5, 1 mM EDTA, and 2.0 M NaCl. Autoclave and store at room temperature.

dNTP solution (containing all four dNTPs, each at 10 mM)

Formamide<!>dye: For 1 ml of loading dye, use 0.2 ml of blue dextran (50 mg/ml) and 0.8 ml of formamide. Use 1.5 µl of loading dye per sample.

Hydroxylamine<!>. Prepare hydroxylamine solution (4.2 M) in a fume hood as follows: Dissolve 1.39 g of solid hydroxylamine in 1.6 ml of H_2O in a glass test tube by shaking under hot tap water. Add 1 ml of diethylamine<!> dropwise and adjust the pH to ~6 by slowly adding up to 750 µl of diethylamine. To measure the exact pH, add two drops of hydroxylamine solution to 2 ml of H_2O. *Never* insert the electrode directly into an undiluted hydroxylamine solution. Store the solution for 7–10 days at 4°C or for 6 months at –20°C.

$KMnO_4$/TEAC, 100 mM $KMnO_4$<!> in 3 M TEAC (tetraethylammonium chloride<!>) (optional for method A)

Piperidine, 10%<!>

ENZYMES AND ENZYME BUFFERS

Taq DNA polymerase

Taq DNA polymerase buffer, 10x (as supplied with enzyme)

NUCLEIC ACIDS AND OLIGONUCLEOTIDES
Test and control genomic DNAs

SPECIAL EQUIPMENT
Magnetic stand suitable for use with chosen tube or microtiter plate system

Microtiter plates or microtubes of polyethylene or polypropylene (e.g., Advanced Technologies or Nunc)

Polycarbonate is not resistant to piperidine.

Shaker, for use at room temperature

Streptavidin-coated magnetic beads (e.g., Dynabeads M-280Norway)

Thermal cycler

METHOD A: RADIOACTIVELY LABELED REACTION

The following is an example of a protocol for preparing a radioactively labeled probe. Exon 5 from wild-type p53 tumor suppressor gene was used as the target DNA. The same procedure (minus the [α-^{32}P]dCTP) should be used to produce the test DNA amplicon.

ADDITIONAL MATERIALS

. . . .

BUFFERS, SOLUTIONS, AND REAGENTS
Formamide dye<!>

H_3PO_4<!>(1 M)

▼ CAUTION

See Appendix for appropriate handling of materials marked with <!>.

Osmium tetroxide<!>* (Sigma-Aldrich) buffer, prepared as follows: 100 mM Tris-HCl<!>, pH 7.7, 10 mM EDTA, and 15% pyridine<!>. In a fume hood, break an ampule containing 0.5 g of osmium tetroxide and place it in a glass bottle with 12.5 ml of double-distilled H_2O. Make sure that the lid is tight and well sealed by leaving the bottle in a container with a screw cap. Osmium tetroxide may be inactivated if left in plastic. Leave to dissolve for 2–3 days at 4°C wrapped in aluminum foil. Store at 4°C and dilute 1 in 5 in double-distilled H_2O before use. Osmium tetroxide solutions may be stored at –20°C for up to 6 months. A yellow color appears upon reaction with pyridine (from the buffer), and it indicates a successful reaction. The stock solution should be replaced when it turns green/grayish (usually after ~2 months).

Osmium tetroxide solution (1:5 in H_2O)

NUCLEIC ACIDS AND OLIGONUCLEOTIDES
5A: 5´-TTCAACTCTGTCTCCTTCCTCTTCC-3´

5B: 5´-CTGGGGACCCTGGGCAACC-3´

In one set of primers, the sense primer should be biotinylated; in the other, the antisense primer should be biotinylated.

RADIOACTIVE COMPOUNDS
[α-^{32}P]dCTP (50 µCi)<!>

SPECIAL EQUIPMENT
Geiger counter

Incubators, or water baths, preset to 37°C, 42°C, and 90°C

Magnet stand, precooled

ADDITIONAL ITEMS

> Equipment and reagents for standard polyacrylamide<!> gel electrophoresis (sequencing gel)
>
> Equipment and reagents for agarose gel electrophoresis, including ethidium bromide<!>
>
> Qiaquick spin columns, or equivalent, for purification of DNA from agarose gels
>
> > Osmium tetroxide can be replaced by a solution of 100 mM potassium permanganate ($KMnO_4$) in 3 M tetraethyl ammonium chloride (TEAC). The $KMnO_4$ solution must be made fresh before each use. The 3 M TEAC can be stored at −4°C for ~3 months.

METHOD

Preparation of the Uniformly Labeled Probe

1. In a sterile microfuge tube, assemble the following components on ice:

Genomic DNA	50–100 ng
Primer 5A	20 pmoles
Primer 5B	20 pmoles
dNTP solution (10 mM)	2.5 µl
[α-^{32}P]dCTP (25 µCi/µl)	2 µl
Taq polymerase (1 unit/µl)	1 µl
PCR buffer (10x)	10 µl
H_2O	to 100 µl

2. Carry out amplification as follows:

Cycle number	Denaturation	Annealing	Polymerization/Extension
1	2 min at 95°C		
30	30 sec at 94°C	45 sec at 65°C	45 sec at 72°C
1			10 min at 72°C, then cool to room temperature

3. Separate the products on a 2% (w/v) agarose gel. Excise the appropriate band and purify the DNA using a Qiaquick spin column, according to the manufacturer's instructions.

4. Resuspend the probe in H_2O to give 10,000 cpm/µl.

> If the DNA fragment has not been efficiently labeled, try repeating the amplification without the isotope. If this reaction produces a good-quality band on a gel, repeat the procedure, with isotope, using 2 µl of the product as a template.

Solid-phase Chemical Cleavage: Heteroduplex Formation

5. In a sterile microfuge tube, set up each of the annealing reactions as follows;

Probe DNA (10,000 cpm/µl)	5 µl
Test DNA	5 µl
Annealing buffer, 2x	10 µl

> Incubate for 5 minutes at 100°C, then for 1 hour at 42°C.

6. Collect the sample by centrifuging briefly, and add 20 µl of magnetic beads (prewashed according to the manufacturer's instructions) to each annealing reaction.

7. Incubate at room temperature with a gentle shaking (to keep the beads in suspension) for 15–30 minutes, depending on the length of the DNA fragments.

8. Place the tubes in the magnetic stand for 30 seconds. With the tubes still in the stand, remove the supernatant. Check the pellet with a Geiger counter to be sure that the

majority of the radioactive label is still attached to the beads. Note that a small amount of unbound radioactive DNA fragments will be present in the supernatant.

9. Remove the tubes and wash once with one volume of 2x B&W buffer.

10. Place the tubes in the magnetic stand, and remove the B&W buffer.

11. Resuspend the beads in 26 μl of H_2O.

Chemical Cleavage Reaction

12. For each annealing reaction, aliquot 6 μl of beads into four fresh tubes; two for the hydroxylamine treatment and two for the osmium tetroxide treatment.

> When a fragment is analyzed, at least for the first time (samples will often be analyzed several times), we recommend that the annealing reaction be split into four tubes, two for the HA and two for the osmium tetroxide treatment. The samples are treated with the chemicals for different periods of time (see steps 13 and 14). The nature of the mutation and the DNA fragment affect the optimum treatment time required to obtain a good clear result. Osmium tetroxide in particular has a tendency to modify other than mismatched thymines. Once the optimal incubation times have been established for the reactions, a single sample for each treatment can be used.

13. Add 20 μl of hydroxylamine solution to each of the hydroxylamine reaction tubes and incubate at 37°C for 10 and 30 minutes, respectively.

14. Add 2.5 μl of osmium tetroxide buffer and 15 μl of osmium tetroxide solution (1:5 in H_2O) to each of the osmium tetroxide reaction tubes, and incubate at 37°C for 1 and 5 minutes, respectively.

15. Place the tubes in the magnet for 30 seconds; then, with the tubes still in the stand, remove the supernatant. Check the supernatant with the Geiger counter; all counts should be in the pellet.

16. Remove the tubes from the magnetic stand and wash the beads once with 20 μl of 2x B&W buffer.

17. Return the tubes to the magnet and remove the supernatant.

18. Resuspend the beads in 50 μl of 10% piperidine and incubate for 30 minutes at 90°C.

> An unintended effect of piperidine cleavage is the separation of streptavidin-coated beads from the DNA. These beads are replaced at step 21.

19. Place the reactions on ice for 2–3 minutes.

20. To each reaction add 25 μl of 1 M H_3PO_4 and 8 μl of 10x *Taq* DNA polymerase buffer. Incubate the samples for 5 minutes at 100°C and cool slowly (~30 minutes) to 30°C.

21. Add 5 μl of fresh, prewashed magnetic beads to each sample. Incubate at room temperature with gentle shaking for 15–30 minutes.

22. Place the tubes in the magnet stand for 30 seconds; then, with the tubes still in the stand, remove the supernatant.

> **Important:** Use a Geiger counter to check that the radioactively labeled DNA fragments are still attached to the beads.

23. Remove the tube from the stand and wash the beads once with 20 μl of 2x B&W buffer.

24. Resuspend the beads in 5 μl of formamide dye and incubate for 3 minutes at 90°C. Place the tubes in a precooled magnet stand on ice, and load the supernatant onto a standard polyacrylamide sequencing gel. Once the gel is run, process and expose to X-ray film.

METHOD B: THE FLUORESCENT LABELED REACTION

As discussed in the introduction, for the accurate location of mutations, duplicate primer sets must be used for each reaction. The PCR primers used in fluorescent analysis are synthesized in duplicate: In the first set, both primers are labeled with biotin, and in the second set, each primer is labeled with a different fluorophore; see Figure 2.

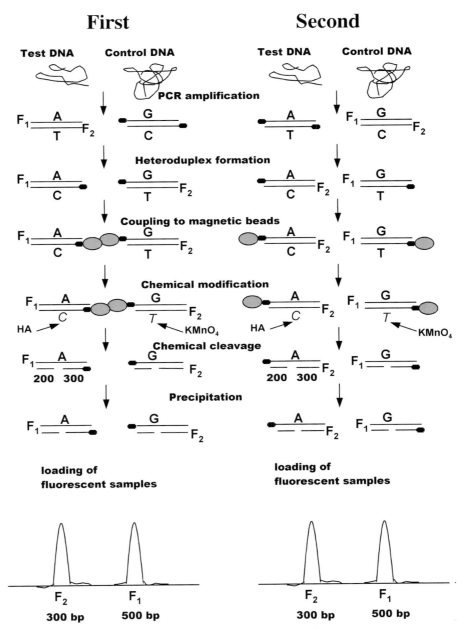

FIGURE 2. Illustration of the importance of performing reciprocal SpCCM experiments to ensure optimal mutation detection. First, the test DNA is labeled with two different fluorophores (F_1, F_2) on both 5′ ends, and the control DNA with biotin. Second, the control DNA is labeled with the fluorophores and the test DNA with biotin. In the first reaction, the 500-bp F_1-labeled DNA is cleaved by $KMnO_4$ and reveals a mispaired thymine 300 bp from F_2. In the second reaction, the 500-bp F_2-labeled fragment is cleaved by hydroxylamine (HA), revealing a mispaired cytosine in the control DNA 300 bp from F_2. Modified nucleotides are in italics.

The procedure presented here was optimized for the mouse β-globulin promoter (Youil et al. 1996), and the primers were labeled with biotin, or HEX and 6-FAM. The primers define a 547-bp amplicon (Youil et al. 1996; Lambrinakos et al. 1999).

ADDITIONAL MATERIALS

BUFFERS, SOLUTIONS, AND REAGENTS

TE (10 mM Tris-HCl<!>, pH 7.5, 1 mM EDTA, pH 8.0)
KMnO$_4$/TEAC<!> solution (1 mM KMnO$_4$, 3 M TEAC)
Glycogen (20 mg/ml)
Sodium acetate<!>, 0.6 M, pH 5.2
Ethanol<!>, ice-cold
Ethanol, 70%

NUCLEIC ACIDS AND OLIGONUCLEOTIDES

Two set of oligonucleotide primers : 5´-GCACGCGCTGGACGCGCAT; 5´-AGGTGCCCTTGAGGCTGTCC

One pair of primers to be biotinylated, the other pair to be labeled with fluorophores, e.g., HEX and 6-FAM.

SPECIAL EQUIPMENT

ABI 377, 310, 3100, or 3700 DNA sequencer
Dry ice (*optional*, see step 16)
Incubators, or water baths, preset to 65°C and 100°C

METHOD

Preparation of the Fluorescent Labeled Probe and Test DNA

1. In a sterile microfuge tube, assemble the following components on ice:

Genomic DNA	50–100 ng
Primer 5A*	20 pmoles
Primer 5B*	20 pmoles
dNTP solution (10 mM)	2.5 µl
Taq polymerase (1 unit/µl)	1 µl
PCR buffer (10x)	10 µl
H$_2$O	to 100 µl

   *First amplify the test DNA with fluorescently labeled primers (see Materials list) and amplify the wild-type DNA with the end-labeled biotin primers. In a second set of PCRs, switch the labelings; i.e., amplify the test DNA with end-labeled biotin primers and the wild-type DNA with the two different fluorophore-labeled primers.

   Use the following PCR conditions:

Cycle number	Denaturation	Annealing	Polymerization/Extension
30	30 sec at 94°C	45 sec at 65°C	45 sec at 72°C
1			10 min at 72°C, then cool to room temperature

2. Check the PCR product by analyzing a small aliquot on a 2% agarose gel. If a satisfactory product is obtained, set up a heteroduplex annealing reaction as follows (note that 40 µl is sufficient for both chemical treatments, KMnO$_4$ and hydroxylamine):

Wild-type amplicon (20 ng/µl)	5 µl
Test amplicon (20 ng/µl)	5 µl
Annealing buffer, 2x	20 µl
H_2O	to 40 µl

Incubate for 5 minutes at 100°C, and then for 1 hour at 42°C.

3. As a control, set up a homoduplex annealing reaction as follows:

Wild-type amplicon (biotinylated primers) (20 ng/µl)	5 µl
Wild-type amplicon (fluorescently labeled primers) (20 ng/µl)	5 µl
Annealing buffer, 2x	20 µl
H_2O	to 40 µl

Incubate for 5 minutes at 100°C, then for 80 minutes at 65°C. Cool slowly (overnight) to room temperature. Centrifuge briefly to collect the samples. Treat the homoduplex control sample exactly like the heteroduplex sample.

4. Prewash the streptavidin-coated beads according to the manufacturer's instructions and resuspend in 12 µl of B&W buffer.

Twelve µl of magnetic beads require three washes with 12 µl of 2x B&W buffer.

5. Add 6 µl of the magnetic bead suspension to each of the annealing reactions (homoduplex and heteroduplex). Incubate at room temperature for 15–30 minutes (depending on the length of the DNA fragments) with gentle shaking to keep the beads in suspension.

6. Place the tubes in the magnetic stand for 30 seconds. Then, with the tubes still in the stand, remove the supernatant.

7. Remove the tubes from the stand and wash the beads with 20 µl of TE. Return the tubes to the stand and remove the TE wash.

8. Resuspend each duplex sample in 12 µl of TE.

9. Divide each sample (heteroduplex and homoduplex) into two 6-µl aliquots.

Chemical Cleavage Reaction

10. Prepare one homoduplex and one heteroduplex sample for the $KMnO_4$ reaction by removing the TE in which they are suspended (use the magnetic stand).

At this point, each reaction should contain 75–100 ng of DNA bound to the magnetic beads.

11. Add 20 µl of 1 mM $KMnO_4$/3 M TEAC to each reaction and incubate at 25°C for 45 minutes.

12. Meanwhile, add 20 µl of hydroxylamine solution (4.2 M) to the remaining homoduplex and heteroduplex samples. Incubate at 37°C for 30 minutes.

13. Stop the reactions by placing the tubes in the magnetic stand and removing the supernatants.

14. Wash beads twice with 50 µl of TE.

15. Add 50 µl of 10% piperidine to *all four* tubes and incubate at 90°C for 30 minutes.

16. Place the tubes on ice and precipitate the DNA by adding 0.5 µl of glycogen (20 mg/ml), 50 µl of sodium acetate (0.6 M, pH 5.2), and 300 µl of ice-cold ethanol. Incubate at –20°C for 30–45 minutes or 10 minutes on dry ice.*

17. Centrifuge for 30 minutes at maximum speed in a microcentrifuge at 4°C. Remove the supernatants.

18. Wash the pellets with 180 µl of 70% ethanol.

19. Centrifuge the tubes for 10 minutes at maximum speed in a microcentrifuge and carefully remove the ethanol.

20. Redissolve the pellets in 1.5 µl of formamide dye and load onto an ABI-377 DNA Sequencer (Perkin Elmer) or add 12 µl of formamide and a size standard and load onto a capillary electrophoresis system. The cleaved fragments are detected via the fluorescence detection system in the instrument.

> * As an alternative to the DNA precipitation described in steps 16–19, add 5 µl of fresh, pre-washed magnetic beads to each sample. Incubate at room temperature with gentle shaking for 15–30 minutes. Place the tubes in the magnet stand for 30 seconds; then, with the tubes still in the stand, remove the supernatant. Proceed to step 20 above.

SENSITIVITY OF SpCCM

The following procedure was used to determine the sensitivity of SpCCM.

1. DNA from the breast cell line HMT-3522 at passage 45 was used as a wild-type control (Nielsen and Briand 1989).

2. The same cell line at passage 376 was used as the test DNA. The test DNA contained a mutation in exon 5 of the tumor suppressor gene p53, as well as an allelic loss including the second allele of p53 (Moyret et al. 1994). DNA from passages 45 and 376 was mixed in varying portions so that the resulting samples contained 0%, 5%, 10%, and 100% mutant DNA, respectively.

3. All DNA samples were PCR-amplified using the primers 5A-Biotin and 5B, which anneal to the introns flanking exon 5 of p53 (see Preparation of the Uniformly Labeled Probe, p. 269).

4. A heteroduplex was produced by annealing radioactively labeled DNA from the wild-type amplicon and each of the mutant amplicons.

5. The biotinylated sense strand of the amplicons was attached to magnetic beads and subjected to SpCCM. The mutation was detected, after 2 days of exposure to the X-ray film, in amplicons derived from samples containing 5% and 100% mutant DNA (Fig. 3C).

The p53 exon 5 is very GC-rich and has proved difficult to handle in the search for mutations by denaturing gradient gel electrophoresis (DGGE), probably due to the formation of secondary structures in the DNA (Beck et al. 1993). We repeated the experiment described above and performed traditional CCM (Cotton et al. 1988). Using this procedure, the mutation was detectable only in amplicons derived from samples containing 100% mutant DNA. A dense smear was seen at the top covering approximately one-third of the gel (Fig. 3A). We recommend the use of 1 M betaine in the PCR amplification of highly GC-rich DNA fragments. Some amplifications may be further enhanced by adding 5% DMSO.

No smear was ever seen in SpCCM reactions, probably because all nonspecific and unbound but labeled DNA is washed away before the chemical reactions and electrophoresis are performed (Fig. 3B,C).

◗ TROUBLESHOOTING

• *No signal appears on the film or the fluorescent gel.* It is important to keep the magnetic beads in solution, either by gentle shaking or by tapping the tube.

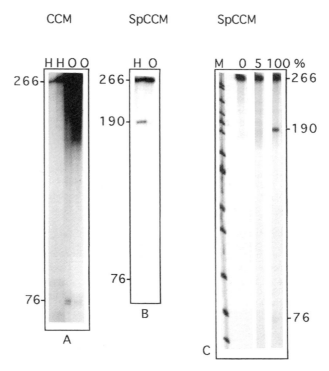

FIGURE 3. CCM and SpCCM performed on p53 exon 5, identifying the mutation H179D, which is a C →A substitution. (A) Heteroduplexes were formed between DNA from the breast cell line HMT-3522 passage 376 (with the p53 mutation, 376^m) labeled with [α-^32P]dCTP and DNA from the same cell line at passage 45 (wild-type p53, 45^wt) and subjected to CCM. The mismatched thymine present in the labeled strand was detected by osmium tetroxide (O). (B) SpCCM was performed on a heteroduplex formed between p53 exon 5 from 45^wt labeled with [α-^32P]dCTP and 376^m. The mutation, a mismatched C in the labeled DNA strand, was detected with hydroxylamine (H). (C) 45^wt and 376^m were mixed so that 376^m comprised 0%, 5%, and 100% of the total DNA amount. The mixes were PCR-amplified with the p53 exon-5 primers, and heteroduplexes were formed with 376^m labeled with [α-^32P]dCTP and subjected to SpCCM. The marker (M) is pBR322 digested with Msp1.

- *The microtiter plates were destroyed by the chemical treatment.* Microtiter plates made of polycarbonate are not resistant to piperidine. Use microtubes in strips of eight or microtiter plates made of polypropylene or polyethylene (Advanced Biotechnologies or Nunc).

- *A single high-molecular-weight band appears on the X-ray film.* If the specific activity of the probe is too low, relatively large quantities of the probe are added to the annealing reaction. This excess of probe DNA leads to the effective exclusion of nonradioactive DNA from the annealing reaction. Homoduplexes, which contain no mismatches, are formed in preference to heteroduplexes, and hence, only a single high-molecular-weight band is observed on the film.

- *No signal or no DNA on the gel.* This could be due to the use of insufficient DNA in the experiment. In this case, repeat the experiment using more DNA in each reaction. Alternatively, this problem may be caused by loss of DNA during the final precipitation step.

- *There are too many unspecific bands in each lane.* The explanation for this result is that the hydroxylamine or osmium tetroxide solutions are too old and the chemicals have attacked not only the mismatched, but all of the cytosines and thymidines, respectively.

Hydroxylamine should be replaced every 7–10 days. Osmium tetroxide develop a gray/green color when replacement is required. The problem bands are sequencing tracks of cytosines (hydroxylamine) or thymines (osmium tetroxide).

- *There are too many well-defined bands in each lane.* The probe may contain nonspecific DNA fragments arising from a contaminated PCR. Careful PCR and probe purification may relieve this problem. Note that this result would also be obtained if there were more than one mismatch between the probe and the test DNA. In this case, the sum of the observed fragment lengths should be equivalent to the length of the uncleaved fragment.

- *No cleavage is detected.* This result may be obtained because of poor resolution of large DNA fragments during polyacrylamide gel electrophoresis. If the mutation is very close to one end of the fragment, it may be difficult to see the difference in size between the cleaved and uncleaved fragments. Gradient gels may solve this problem.

- *For fluorescent gels, a low signal-to-noise ratio.* Reducing the background noise and increasing the mismatch signal may require optimizing the incubation times of the modifying reagents. Excessive cleavage of the $KMnO_4$/TEAC reaction can be reduced by decreasing the pH of piperidine from 10 to 7–8 (Lambrinakos et al. 1999), or by changing the piperidine incubation to 15 minutes at 60°C.

DISCUSSION

As a tool for detecting specific mutations, SpCCM offers the advantage of revealing both the position and the type of the mismatched nucleotide(s). DNA sequencing is the only alternative technique that can provide this information. SpCCM can detect all types of mismatched base pairs, since osmium tetroxide (or $KMnO_4$/TEAC) modifies mismatched thymine residues, and hydroxylamine modifies mismatched cytosine residues. Thymine and cytosine residues adjacent to a mismatch may also react to chemical modification (see Fig. 4).

Because the DNA amplicon is of known length and the analysis is performed using DNA sequencing technology, it is possible to determine which base is mismatched. The length of the cleaved fragments shows the position of the mismatch, and comparison of fragments obtained from osmium tetroxide treatment and hydroxylamine treatment reveals which residue (thymine or cytosine) is mismatched. In nonradioactive reactions, the specific fluorophore attached to the cleaved fragment allows the distance from the dye to the mutation to be determined and also indicates which strand is cleaved. A signal obtained by both hydroxylamine and osmium tetroxide (or $KMnO_4$/TEAC) for the same fragment indicates that both cytosines and thymines are mismatched, and that the mutation may be either a point mutation deletion or an insertion.

ACKNOWLEDGMENTS

The authors thank T.B. Christensen for excellent assistance, and Drs P. Briand and M.W. Madsen for kindly donating the human breast epithelial cell line HMT-3522. Dr. P. Guldberg is acknowledged for providing us with the primers and primer sequences for the p53 exon 5. This work was supported by The Danish Cancer Society and the Danish Medical Research Council.

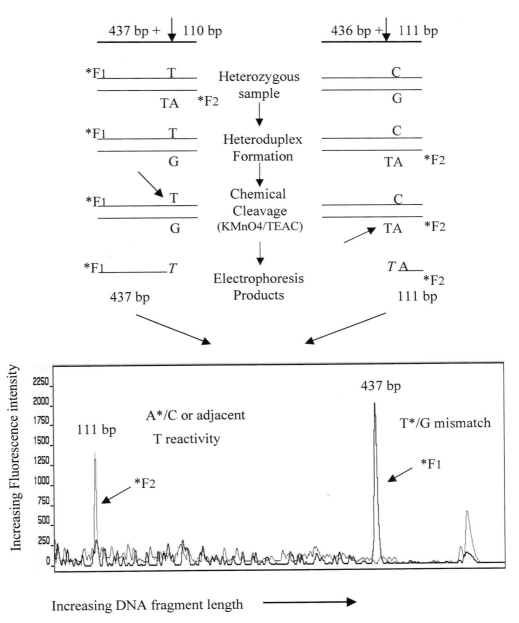

FIGURE 4. Fluorescent SpCCM was performed on a 547-bp heteroduplex generated from a mouse β-globulin promoter fragment, and the cleavage products were separated by electrophoresis on an ABI-377 DNA sequencer. The detection of a mismatched thymine is illustrated. KMnO$_4$/TEAC reacted strongly with a thymine*/guanine mismatch. Piperidine cleavage at the modified site produced a 437-bp fragment, which corresponds to the *F$_1$-labeled strand. Additionally, KMnO$_4$/TEAC reacted strongly with a matched thymine base adjacent to the adenine*/cytosine mismatch. Piperidine cleavage at this site produced an 111-bp fragment corresponding to the *F$_2$-labeled strand. The asterisk represents the fluorescence-labeled strands. The 5′ ends of the C/G strands are labeled with biotin (not shown).

REFERENCES

Beck J.S., Kwitek A.E., Cogen P.H., Metzger A.K., Duyk G.M., and Sheffield V.C. 1993. A denaturing gradient gel electrophoresis assay for sensitive detection of p53 mutations. *Hum. Genet.* **91:** 25–30.

Cotton R.G.H. 1989. Detection of single base changes in nucleic acids. *Biochem. J.* **263:** 1–10.

Cotton R.G. and Campbell R.D. 1989. Chemical reactivity of matched cytosine and thymine bases near mismatched and unmatched bases in a heteroduplex between DNA strands with multiple differences. *Nucleic Acids Res.* **17:** 4223–4233.

Cotton R.G.H., Rodrigues N.R., and Campbell R.D. 1988. Reactivity of cytosine and thymine in single-base-pair mismatches with hydroxylamine and osmium tetroxide and its application to the study of mutations. *Proc. Natl. Acad. Sci.* **85:** 4397–4401.

Ellis T.P., Humphrey K.E., Smith M.J., and Cotton R.G.H. 1998. Chemical cleavage of mismatch: A new look at an established method. *Hum. Mutat.* **11:** 345–353.

Hansen L.L., Justesen J., and Kruse T.A. 1996. Sensitive and fast mutation detection by solid-phase chemical cleavage. *Hum. Mutat.* **7:** 256–263.

Lambrinakos A., Humphrey K.E., Babon J.J., Ellis T.P., and Cotton R.G. 1999. Reactivity of potassium permanganate and tetraethylammonium chloride with mismatched bases and a simple mutation detection protocol. *Nucleic Acids Res.* **27:** 1866–1874.

Moyret C., Madsen M.W., Cooke J., Briand P., and Theillet C. 1994. Gradual selection of a cellular clone presenting a mutation at codon 179 of the p53 gene during establishment of the immortalized human breast epithelial cell line HMT-3522. *Exp. Cell Res.* **215:** 380–385.

Nielsen K.V. and Briand P. 1989. Cytogenetic analysis of in vitro karyotype evolution in a cell line established from nonmalignant human mammary epithelium. *Cancer Genet. Cytogenet.* **39:** 103–118.

Roberts E., Deeble V.J., Woods C.G., and Taylor G.R. 1997. Potassium permanganate and tetraethylammonium chloride are a safe and effective substitute for osmium tetroxide in solid-phase fluorescent chemical cleavage of mismatch. *Nucleic Acids Res.* **25:** 3377–3378.

Youil R., Kemper B., and Cotton R.G. 1996. Detection of 81 of 81 known mouse β-globin promoter mutations with T4 endonuclease VII—The EMC method. *Genomics* **32:** 431–435.

ANALYSIS OF DIFFERENTIAL GENE EXPRESSION BY PCR

One of the biggest challenges in biology is defining the shifts and changes in gene expression that are responsible for the alteration in physiology and phenotype. In this section, methods to determine differences in the patterns of genes expressed between two samples are presented. The standard PCR is designed for the amplification and detection of known genes by using gene-specific primers. In the protocols to examine differences in gene expression, the oligonucleotide primers are shorter in length than those routinely used in PCR and their sequence is random, rather than gene-specific, because no predictions of the differentially expressed genes are made. Microarray analysis is one of the methods routinely used to examine patterns of gene expression. In this case, PCR is very useful in preparing the templates for microarray, as described in "Applying PCR for Microarray-based Gene Expression Analysis" (p. 281), but is not used as the screening method per se. Applications and analysis of microarray technology are presented in depth in *DNA Microarrays: A Molecular Cloning Manual* (Bowtell and Sambrook 2003).

DIFFERENTIAL DISPLAY

The most successful method developed to date, differential display, takes full advantage of the amplification efficiency of PCR. As described in "Fluorescent mRNA Differential Display in Gene Expression Analysis" (p. 327), two mRNA populations are processed in parallel through RT-PCR using a fluorescent labeled oligo(dT) and a defined set of 13-nucleotide-long primers that contain sufficient variation to bind to and then amplify a number of cDNA sequences. The detection of differences in the two mRNA populations is performed on DNA sequencing gels. The method described here incorporates restriction enzyme recognition sequences into both primers to facilitate the cloning and further analysis of the differences of interest. From here the investigator clones and sequences the fragments, which correspond to the 3´ ends of the differentially expressed mRNAs, and then, using in silico techniques, determines the sequence of the entire gene product. Because all of the amplification reactions begin at the 3´ end of the RNA, this method produces one signal per mRNA and can detect differentially expressed splice variants of the gene. Therefore, careful experimental design must be used to characterize the changes in mRNA expression, and complementary techniques are needed to confirm the differential expression of the mRNA or the protein it encodes.

The drawbacks of this approach are twofold. First, because of the structure of the primers, multiple sets of amplifications must be performed to capture all the differences in mRNA populations. To assure 90–95% coverage of all the mRNAs expressed in cells, a minimum of 240 reactions is required. The second drawback is the tedious process of isolating and processing the differentially displayed bands. If there is a flaw in the experimental design and the mRNA populations are not optimal, significant amounts of time are wasted cloning and sequencing products that in fact do not arise from differentially expressed genes.

SUBTRACTIVE HYBRIDIZATION

"Identification of Differential Genes by Suppression Subtractive Hybridization" (p. 297) describes an alternative method of cloning and detecting these genes of interest. Unlike the method described above, this procedure can be performed using genomic DNA or mRNA. The critical steps in this procedure are the separate preparation of driver and target, or tester, DNA followed by hybridization, amplification, and physical separation of the DNA. This material can then be easily cloned. Additionally, the mirror orientation selection procedure, which eliminates a significant fraction of clones present in both populations, greatly increases the overall efficiency of the method. Because in this technique the high-abundance, shared sequences in a population are depleted, there is a good probability of identification of relevant low-abundance cDNA clones.

Like all techniques, this one has limitations as well. Sufficient starting material should be available for robust hybridizations and PCRs. The efficiency of some of the recovery steps drops dramatically with low concentrations of nucleic acid. Furthermore, there are a number of important controls to perform during the procedure that could tax the system if the template is limiting. Finally, once the library is constructed, a screening method needs to be used to detect the clones of interest. Great care must be taken here to prepare probes that will detect the low-abundance events, since these are most likely the genes of interest.

EXPERIMENTAL DESIGN

For both differential display and suppression subtractive hybridization, significant attention must be paid to the biology and handling of the specimens prior to isolation of the nucleic acid for analysis. First, everything possible must be done to avoid the introduction of trivial differences between the test and control material. This establishes an experiment that is focused on the change caused by the test variable. The avoidable sources of error could be as simple as differences in the feeding schedule, so that one culture is accidentally serum-starved, or subjected to inappropriate temperature shifts. If comparing tissue specimens from treated and control animals, it may be wise to pool tissues from several animals for each group so that consistent and significant changes are what are detected. The other major variable to consider is the time of harvest. If the phenotypic change under study is stable, such as a differentiation event or tumor versus normal tissue, timing is less critical, and either method should yield information. However, if the experiment is designed to detect a transient shift in gene expression that may be present for a relatively short period of time, the differential display method may be more practical. Even though differential display requires a large number of reactions, they can be batched and analyzed over time. Keep in mind that no matter which method is used, the quality of the result will be determined by the performance of the technique, and by the quality of the biology that generated the starting material.

REFERENCE

Bowtell D. and Sambrook J. 2003. *DNA microarrays: A molecular cloning manual.* Cold Spring Harbor Laboratory Press, Cold Spring Harbor, New York.

21 Applying PCR for Microarray-based Gene Expression Analysis

Joan M. Zakel

Clinical Sciences Group, Digene Corporation, Gaithersburg, Maryland 20878

Until the early 1990s, gene expression analysis was performed studying one gene at a time. Although several methods were available (i.e., northern blotting, primer extension, RNase protection assays), none addressed the multiplexing required for high-throughput gene expression analysis of multiple samples. With the introduction of DNA microarray technology, gene expression analysis has significantly advanced the field of genomics. It has made possible the simultaneous global study of entire genome sequences (Lander 1999). Gene expression analysis is the most obvious and widely used application of DNA microarrays. DNA microarrays are used to determine expression profiles in disease, progression and prognosis of disease states, evaluation of drug and therapeutic treatments, and in drug discovery and development. Additional applications involve mutational studies such as the analysis of single-nucleotide polymorphisms (SNPs), novel gene identification and validation studies, and genome sequencing (Marshall and Hodgson 1998). As microarray technology continues to advance and develop, its applications will continue to expand and benefit new research in the fields of genomics and proteomics.

PCR is an essential tool for the monitoring of gene expression and the fabrication of DNA microarrays. This chapter presents an overview of the principles and applications of PCR to microarray technology. Particular protocols for the optimization of large-scale PCR amplification and the purification of products from plasmid libraries are discussed. In addition, this chapter describes two distinct hybridization and detection protocols that are currently used for monitoring gene expression on the microarray platform. A two-color fluorescent detection method and Digene's application of Hybrid Capture technology for microarray detection are presented. Considerations for choosing an appropriate detection method are discussed briefly.

BASIC PRINCIPLES OF MICROARRAY-BASED ANALYSIS

In concise terms, a DNA microarray consists of an orderly set of DNA sequences (i.e., probes), each tethered to a solid support in a discrete, miniaturized location. Each DNA sequence that is tethered to the solid support represents a probe for an RNA molecule encoded by that gene in the sample (i.e., target). The DNA sequences on the microarray can comprise either PCR products or oligonucleotides. DNA microarrays use the same basic nucleic acid hybridization principles described by Southern (1975) with modifications to allow the fabrication of microarrays. Microarrays can be constructed upon porous membranes and nonporous surfaces, such as glass, silicon, plastic, or metal. The nonporous surfaces are preferred for most array applications because they are suitable for the scaleable manufacturing of quality microarrays, they utilize smaller reaction volumes, and they offer reduced background noise, facilitating nonradioactive, fluorescence-based detection methods (Shalon et al. 1996). In addition, the introduction and improvements of robotic microarray spotters and alternative printing techniques have advanced the manufacture of quality microarrays with a high density of spotted DNA elements (Schena et al. 1995; Okamoto et al. 2000). Therefore, the microarray platform allows massive parallel analyses of the expression levels of several thousands of genes.

FABRICATION OF MICROARRAYS

The fabrication of DNA microarrays requires multiple procedures and quality controls to ensure consistency within an arrayed grid and between microarrays. DNA microarrays can be divided into two general categories, in situ-synthesized arrays and spotted microarrays. These categories are determined by the manner in which the microarrays are manufactured.

In situ-synthesized arrays are composed of oligonucleotides. Affymetrix first pioneered this technology, in the early 1990s, by developing a method in which short oligonucleotides (25-mers) are synthesized directly on a silicon wafer using photolithographic masks (Fodor et al. 1991). The masks direct the synthesis of the oligonucleotides, one base at a time, by controlling the photoactiviation to produce 5´-hydroxy groups capable of reacting with another nucleoside (McGall et al. 1996). Rosetta Inpharmatics developed an additional in situ oligonucleotide synthesis based on ink-jet technology, which can generate oligonucleotide microarrays comprising 40- to 60-mers (Hughes et al. 2001).

Spotted DNA microarrays provide scientists with the flexibility to customize and fabricate their own microarrays for individual experiments. In contrast to in situ-synthesized arrays, either oligonucleotides or PCR products can be spotted onto a microarray surface, using a variety of attachment chemistries. The oligonucleotides usually range between 30 and 70 bp in length, whereas the PCR products range between 0.3 and 2.5 kb in length. For spotted microarray fabrication, researchers require a robotic arrayer, to control the print quality and quantity of spots on the slides, and a fluorescent scanner for data analysis. The researcher must also generate the material to spot on the array, through the purchase of oligonucleotides or the amplification of clones from a cDNA library (Petrik 2001). If the researcher is unable to make the required investment in money and time for the manufacture of spotted microarrays, numerous prespotted microarrays are available from commercial sources. This chapter concentrates on the protocols required for the generation of spotted microarrays using PCR products and the protocols for target hybridization and detection using these microarrays.

PROTOCOL 1   LARGE-SCALE PCR FOR GENERATING cDNA MICROARRAYS

This protocol describes the procedures required to generate a cDNA microarray using PCR amplification. First, the probes to be arrayed onto the slides must be selected. Probes can be selected from clone database resources such as Genbank or UniGene. Probes for full-length cDNA clones or expressed sequence tags (ESTs) can also be chosen from available cDNA libraries. Depending on the species of interest, sets of clones can be purchased from commercial sources. Most suppliers provide low-density bacterial cultures; therefore, these master clone sets need to be replicated into high-density working stocks. From these working stocks, the plasmids containing the clones are isolated for amplification. Following amplification, the PCR products are purified and quantified. The PCR products are then diluted in the appropriate spotting buffer and robotically printed onto glass slides. After printing, the slides are processed to couple the DNA products to the glass surface. The slides are then stored in a desiccator for future use. The protocol is presented in five stages:

Stage 1: Replica plating and preparation of working stocks

Stage 2: Isolation of plasmid templates

Stage 3: Amplification of clones and evaluation of PCR products

Stage 4: Purification and qualification of PCR products

Stage 5: Printing microarrays

The procedures for large-scale PCR amplification and purification from plasmid libraries were obtained, with permission, from the National Human Genome Research Institute Web site (http://research.nhgri.nih.gov/microarray/fabrication.html.)

STAGE 1: REPLICA PLATING AND PREPARATION OF WORKING STOCKS

MATERIALS

▼ CAUTION

See Appendix for appropriate handling of materials marked with <!>.

BUFFERS, SOLUTIONS, AND REAGENTS

Carbenicillin (100 µg/ml stock solution)

Ethanol (100%) <!>

Glycerol, 45% (v/v)

LB broth

1 liter of LB broth contains 10 g of tryptone, 5 g of yeast extract, 5 g of NaCl

Superbroth

1 liter of superbroth contains 32 g of tryptone, 20 g of yeast extract, 5 g of NaCl, 5 ml of 1 N NaOH<!>

BIOLOGICAL MOLECULES

Master set of clones or ESTs, sequence verified (e.g., gf211 release, Research Genetics)

OTHER EQUIPMENT

Centrifuge with a horizontal (swinging-bucket) rotor with a depth capacity of 6.2 cm for spinning microtiter plates and filtration plates (e.g., Sorvall SuperT 21, Sorvall)

Airpore Tape Sheets (Qiagen)

Bunsen burner

Incubator, preset to 37°C

Inoculating block, 96-pin (V&P Scientific)

Microtiter plates, deep 96-well, 1.0 ml, PP (VWR International)

Microtiter plates, U-bottomed 96-well (Corning)

Paper towels

Shaking incubator with holders for deep-well plates, 37°C

"Zip-lock" bag, 1 gallon (~4 liters)

METHOD

1. Incubate sealed master plates overnight at 37°C.

2. Prepare working copies of clone sets by labeling 96-well round- (U) bottomed plates and aliquoting 100 µl of LB broth containing 100 µg/ml carbenicillin in each well.

Use these working copies to preserve the master set of plates. Check to ensure that the clones are in a vector conferring ampicillin resistance, as is common with IMAGE clones.

3. To remove condensation from the plate sealers, spin the master plate in a horizontal microtiter plate rotor at 1000 rpm for 2 minutes.

4. Partially fill a small beaker with 100% ethanol. Dip the 96-pin replicating tool in the alcohol. Remove from the alcohol and flame the pins.

5. Following a brief cooling, dip the replicating tool into the master plate and then onto the working plates. Repeat as necessary for each plate to be inoculated.

6. Place the inoculated LB plates, with the lids on, into a 1-gallon "zip-lock" bag containing a moistened paper towel. Seal the bag and incubate at 37°C overnight.

7. Aliquot 1 ml of superbroth, containing 100 μg/ml carbenicillin, into each well of the deep-well plates.

    These plates will serve as the source of culture for template preparation.

8. The following day, use the replicating tool to inoculate the deep-well plates directly from the freshly grown LB plates.

9. Seal the deep-well plates with Qiagen Airpore Tape Sheets and place a plastic lid over the sheet. Incubate the plates in a 37°C shaking incubator at 200 rpm for 24 hours.

10. To any of the plates that are to be frozen (–80°C) for subsequent use as a culture source, add 50 μl of 45% (v/v) sterile glycerol to each well. Use the remaining plates for isolation of the plasmid templates.

STAGE 2: ISOLATION OF PLASMID TEMPLATES

MATERIALS

▼ CAUTION

See Appendix for appropriate handling of materials marked with <!>.

BUFFERS, SOLUTIONS, AND REAGENTS
    Ethanol, 70% <!>
    T low E buffer (10 mM Tris-HCl<!>, 0.1 mM EDTA, pH 8.0)

CENTRIFUGES AND ROTORS
    Centrifuge with a horizontal (swinging-bucket) rotor with a depth capacity of 6.2 cm
        for spinning microtiter plates and filtration plates (e.g., Sorvall SuperT 21, Sorvall)

SPECIAL EQUIPMENT
    Plate sealer
    Vortex mixer

VECTORS AND BACTERIAL STRAINS
    Bacterial cultures, as obtained in step 9 of Stage 1 above.

ADDITIONAL ITEMS
    Alkaline Lysis Miniprep kit, 96-well (Edge Biosystems)

METHOD

1. Warm the lysis buffer (Edge Biosystems Kit) to 37°C to dissolve the SDS.

2. Add 1 ml of RNase solution to 100 ml of resuspension buffer (Edge Biosystems Kit). Store at 4°C for up to 3 months.

3. Add 350 μl of bacterial culture to each well of the receiving plates from the Edge Biosystems kit. Place the filter plate on top and secure with tape.

4. Using a centrifuge equipped with a 96-well plate rotor, centrifuge the bacterial cultures in the deep-well plates at 1500*g* for 7 minutes.

5. Invert briefly and tap out excess medium onto a clean paper towel.

>   Do not delay or the pellets will loosen and may be lost when pouring off excess medium.

6. Resuspend each pellet in 100 μl of resuspension buffer and vortex until all the pellets are resuspended.

>   Inadequate resuspension results in clumps of cells that do not lyse in subsequent steps.

7. Add 100 μl of lysis buffer to each well. Mix gently, for 1 minute, by rocking the plates, to avoid shearing the bacterial chromosomal DNA.

8. Add 100 μl of precipitation buffer to each well. Mix for 1 minute.

9. Add 100 μl of neutralization buffer to each well and vortex for 20 seconds.

10. Transfer the contents of the deep-well plates to the filter plate/receiving plate stacks previously assembled.

11. Centrifiuge the stacked plates at 1500*g* for 12 minutes in a centrifuge equipped with a rotor for 96-well plates.

12. Remove and discard the filter plates. Decant the alcohol and the filtrate from the receiver plate. Blot the excess alcohol onto a clean paper towel.

13. Add 500 μl of 70% ethanol to each well. Decant immediately and blot the excess alcohol onto a clean paper towel.

14. Place plates in a clean drawer without their lids, cover with a clean paper towel, and dry overnight.

15. Resuspend each DNA pellet in 200 μl of T low E buffer. Seal top with plate sealer and allow the pellet to rehydrate at 4°C for at least 2 days. Long-term storage should be at –20°C.

STAGE 3: AMPLIFICATION OF CLONES AND EVALUATION OF PCR PRODUCTS

MATERIALS

▼ CAUTION

See Appendix for appropriate handling of materials marked with <!>.

BUFFERS, SOLUTIONS, AND REAGENTS

DEPC-treated H_2O<!>

TAE buffer, 1x (40 mM Tris-acetate<!>, 1 mM EDTA, pH~ 8.5)

DNA loading buffer (20% [w/v] Ficoll 400, 0.1 M Na_2EDTA, pH 8.0, 1.0% [w/v] sodium dodecyl sulfate<!>, 0.25% [w/v] bromophenol blue<!>, 0.25% [w/v] xylene cyanol<!>)

dATP solution, 100 mM (Pharmacia)

dGTP solution, 100 mM (Pharmacia)

dTTP solution, 100 mM (Pharmacia)

dCTP solution, 100 mM (Pharmacia)

ENZYMES AND ENZYME BUFFERS

AmpliTaq DNA polymerase (Perkin-Elmer)

GeneAmp PCR buffer, 10x (Applied Biosystems)

NUCLEIC ACIDS AND OLIGONUCLEOTIDES

Vector- or gene-specific PCR primers (1 mM)
100-bp DNA ladder (New England Biolabs)

GELS

Agarose gel (2%, w/v, in TAE buffer)

SPECIAL EQUIPMENT

Thin-wall PCR plate and Cycleseal PCR plate sealer (Robbins Scientific)
Electrophoresis apparatus with capacity for four 50-well combs
Thermal cycler

ADDITIONAL ITEMS

Equipment and reagents of agarose gel electrophoresis, including ethidium bromide <!>

METHOD

1. For each 96-well plate to be amplified, prepare a master PCR mixture containing the following ingredients:

PCR buffer, 10x (contains 15 mM MgCl$_2$)	1000 µl
dATP (100 mM)	20 µl
dGTP (100 mM)	20 µl
dCTP (100 mM)	20 µl
dTTP (100 mM)	20 µl
Primer 1 (1 mM)	5 µl
Primer 2 (1 mM)	5 µl
AmpliTaq polymerase (5 units/µl)	100 µl
DEPC-treated H$_2$O	8800 µl

2. Label 96-well PCR plates and aliquot 100 µl of PCR mix to each well.

   Gently tap plates to ensure that no air bubbles are trapped at the bottom of the wells.

3. Add 1 µl of purified plasmid template to each well. Cap or cover the plates.

4. Perform the following thermal cycle series:

Cycle number	Denaturation	Annealing	Polymerization/Extension
1	30 sec at 96°C		
25	30 sec at 94°C	30 sec at 55°C	150 sec at 72°C
1	30 sec at 94°C	30 sec at 55°C	5 min at 72°C, then cool to ambient temperature

   Store plates at 4°C while quality control gel analysis is performed.

5. Check the PCR products by agarose gel electrophoresis.

   Gel imaging provides a rough quantitation of the product while giving an excellent characterization of the composition of the product.

6. Combine 2 µl from each well with 2 µl of loading buffer.

7. Analyze the samples on 2% agarose gels with a 100-bp DNA ladder as a standard.

   Each lane should exhibit a single or dominant band with uniform brightness of the products between the lanes.

STAGE 4: PURIFICATION AND QUANTIFICATION OF PCR PRODUCTS

MATERIALS

BUFFERS, SOLUTIONS, AND REAGENTS

Ethanol/acetate mix <!> (95% ethanol, 5% 3 M sodium acetate, pH 6.0)

Ethanol, 70% <!>

SSC buffer, 20x (3 M NaCl, 0.1 M sodium citrate<!>•2H₂O, adjust pH to 7.0 with 1 M HCl<!>)

TAE buffer, 1x (40 mM Tris-acetate <!>, 1 mM EDTA, pH~8.5)

DNA loading buffer (20% Ficoll 400, 0.1 M Na₂EDTA, pH 8.0, 1.0% [w/v] sodium dodecyl sulfate <!>, 0.25% [w/v] bromophenol blue <!>, 0.25% [w/v] xylene cyanol <!>)

CENTRIFUGES AND ROTORS

Centrifuge with a horizontal swinging-bucket rotor with a depth capcacity of 6.2 cm for spinning microtiter plates and filtration plates (e.g., Sorvall SuperT 21, Sorvall)

NUCLEIC ACIDS AND OLIGONUCLEOTIDES

100-bp DNA ladder (New England Biolabs)

Double-stranded DNA standards ranging between 0, 50, 100, 250, and 500 µg/ml

Reference double-stranded DNA (0.5 µg/ml) (Invitrogen)

GELS

Agarose gel (2%, w/v, in TAE buffer)

SPECIAL EQUIPMENT

Microtiter plate, V-bottomed, 96-well (Corning)

Immunowash microtiter plate washer (BioRad)

Paper towels

Heat-sealable storage bags and heat sealer

65°C incubator

Electrophoresis apparatus with capacity for four 50-well combs

Electrophoresis power supply

96-well plates for fluorescent detection (Dynex)

Fluorometer (e.g., Perkin-Elmer Analytical Instruments)

ADDITIONAL ITEMS

FluoReporter Blue dsDNA Quantitation Kit (Molecular Probes)

METHOD

1. Fill 96-well V-bottomed plates with 200 µl per well of ethanol/acetate mix.

2. Transfer 100 µl per well of PCR product into the V-bottomed plates and mix by pipetting a volume of 75 µl per well, four times.

3. Place the plates in –80°C freezer for 1 hour or store overnight at –20°C.

4. Thaw the plates to reduce brittleness and melt any ice that may have formed in the wells.

5. Load the plates into a centrifuge with a microtiter plate rotor and centrifuge at 2600*g* for 40 minutes at 4°C.

6. Aspirate the supernatant from each well using an Immunowash plate washer.

   It is advisable to leave ~10–20 μl in the bottom of each well to avoid disturbing the pellet.

7. Add 200 μl of 70% ethanol to each well in the plate using the Immunowash plate washer.

8. Centrifuge plates at 2600*g* for 40 minutes at 4°C.

9. Aspirate the supernatant from each well using the Immunowash plate washer.

10. Cover the plates with a paper towel and allow the plates to dry overnight in a closed drawer.

Resuspension of PCR Products

11. Add 40 μl of 3x SSC to each well. Seal the plates with a foil sealer or appropriate seal, taking care to achieve a tight seal over each well.

    The buffer that PCR products are resuspended in is determined by the type of arrayer and slides that will be used in the manufacture of the microarrays.

12. Place the plates in heat-sealed bags with paper towels moistened with 3x SSC.

13. Place the bags in a 65°C incubator for 2 hours. Turn off the incubator and allow the plates to cool gradually.

    Cooling the plates in the incubator avoids condensation on the sealers.

14. Analyze 1 μl of the resuspended PCR products on a 2% agarose gel, as described previously in Stage 3.

    Adequate precipitation and resuspension will produce very intense bands.

15. Store the plates, containing the resuspended PCR products, at –20°C until use.

Quantification of PCR Products Using Fluorometry

16. Add 200 μl of fluor buffer (FluoReporter Blue dsDNA Quantitation Kit) to each well of a 96-well plate for fluorescent detection.

17. Add 1 μl of PCR product to each well in the fluorometry plate.

18. In the final row of the fluorometry plate, add 1 μl each of a series of double-stranded DNA (dsDNA) standards of 0, 50, 100, 250, and 500 μg/ml dsDNA.

19. Set the fluorometer for excitation at 346 nm and emission at 460 nm.

20. Test to see that the response for the standards is linear and reproducible from the range of 0 to 500 μg/ml dsDNA.

21. Calculate the concentration of dsDNA in the PCRs using the following equation after subtracting the average 0 μg/ml value from all other sample and control values:

$$\text{concentration of dsDNA}(\mu g/ml) = \frac{\text{PCR sample value}}{\text{average 100 } \mu g/ml \text{ value}} \times 100$$

22. On the basis of the above calculations, adjust the concentration of any PCR products that are out of the expected printing concentration range (0.1–0.5 μg/μl).

STAGE 5: PRINTING MICROARRAYS

Following the completion of the previous protocol, the researcher will have generated the necessary samples that will be used as the spotting material for the fabrication of microarrays. The protocol parameters for the printing of microarrays are determined by the type of arrayer and slides that will be used. Because there are several combinations of arrayers and slide chemistries that can be used for fabricating microarrays, a precise protocol for the spotting of PCR products is not presented. This is extensively covered elsewhere (Bowtell and Sambrook 2003).

HYBRIDIZATION AND DETECTION STRATEGIES FOR GENE EXPRESSION ANALYSIS ON MICROARRAYS

Once the microarrays have been manufactured, the optimal detection method must be chosen. There are several factors to consider when deciding on a detection strategy, including the nature and availability of the sample and time and resource constraints. Below is a discussion of the factors to consider when deciding between two-color fluorescent labeling protocols and direct detection of RNA on microarrays.

Two-color Fluorescent Detection

Two-color fluorescent detection allows the comparison of gene expression patterns between two cellular samples on one microarray. RNA is purified from two different samples (e.g., control vs. treated). The samples are each labeled using a different fluorescent dye. The labeled samples are then combined and hybridized to a single microarray (Shalon et al. 1996). The differential labeling allows the researcher to visualize the relative levels of gene expression in the two samples on one microarray. Thus, researchers are able to compare the expression profiles of two samples across thousands of genes (Fig. 1).

The obvious advantage of two-color fluorescent detection systems is that two samples can be compared to each other under identical conditions. Because both samples are labeled with a different fluorescent dye, they can be simultaneously hybridized and detected, under identical conditions, using the same microarray.

Although a two-color fluorescent detection method provides some advantages, it also has some disadvantages. This method requires the samples to be labeled by the incorporation of fluorescently labeled nucleotides into the cDNA. The reaction is catalyzed by a reverse transcriptase (e.g., Superscript II) and, most often, primed with an oligo-dT primer that targets polyadenylated RNA species. The efficiency of the reverse transcription reaction, and hence labeling, depends on the progression of the enzyme along the mRNA template. Progress can be impeded by the presence of secondary structures and incomplete poly(A) tails. In species lacking poly(A) tails, labeling must be performed by the less efficient method of random priming. Another problem is that the incorporation rates of the two fluorescent dyes differ. In the commonly used protocol, Amersham's Cy dyes are often used. The Cy3 dye is more readily incorporated than the more bulky Cy5 dye. Differences in labeling efficiency can lead to an inaccurate representation of the mRNAs in the samples. In an effort to normalize incorporation discrepancies, it is a common practice to perform follow-up experiments in which the fluorescent dyes are "swapped" between the samples (Bowtell and Sambrook 2003). The protocols for this fluorescent detection tech-

Two-Color Fluorescent Detection

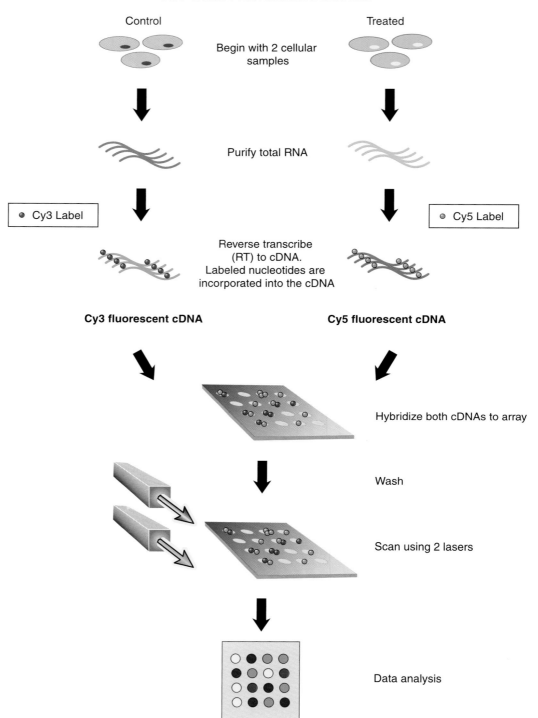

Total analysis time = 2–3 days

FIGURE 1. Principle of the procedures for a two-color fluorescent detection method for monitoring gene expression, using a microarray.

nique are not presented here, but they are extensively outlined in *DNA Microarrays: A Molecular Cloning Manual* (Bowtell and Sambrook 2003).

Direct Detection of RNA Using Hybrid Capture Technology

In contrast to the two-color fluorescent microarray detection methods, direct detection of RNA is possible through the use of Hybrid Capture technology, developed by Digene. Digene's HC (Hybrid Capture) *Express*Array Kit uses a proprietary antibody that binds specifically to RNA:DNA hybrids (Lazar et al. 1999) and a second, fluorescently labeled, antibody that detects the primary antibody. Total RNA is applied directly to a glass-spotted DNA microarray, which can be composed of cDNA, PCR products, or oligonucleotides of 70 bp in length or greater. The anti-RNA:DNA antibodies bind specifically to stable RNA:DNA hybrids and are visualized via a Cy3-labeled secondary antibody (Fig. 2A,B). Because the primary antibodies recognize only perfectly matched heteroduplex regions, the strength of the signal generated is proportional to the length of the hybrid formed.

Direct detection of gene expression using Digene's system provides a number of benefits. First, minimal sample preparation is required. Following the purification of total RNA, which can be performed using any reliable purification method (e.g., RNeasy Purification Kit from Qiagen or Strataprep Total RNA Purification Kit from Stratagene; see Chapter 10 or Sambrook and Russell 2001), the RNA can be directly hybridized to the array for detection and analysis. No RNA labeling is required, and thus the biases associated with this process are eliminated. This greatly reduces the time required for sample preparation and allows accurate monitoring of gene expression within samples. The stringency and specificity of direct detection are enhanced through increased hybridization temperatures (up to 75°C). Higher temperatures also allow hybridization times to be reduced to 3 hours. Using the HC *Express*Array Kit for direct detection of gene expression, the time from sample preparation to results is less than 8 hours, as compared with 2–3 days, using a two-color fluorescent detection method. However, because direct detection involves no sample labeling, only one sample can be detected per microarray. For comparative studies, it is necessary to normalize the detection between two microarrays (control and the treated conditions) using a panel of housekeeping genes. This must be done prior to calculating the relative fold change between two samples on separate microarrays. This normalization adjusts for any differences in signal or background that may be attributed to the individual slides or slide processing. Direct detection of RNA requires no enzymatic labeling of samples; thus, quantitative gene expression analysis, in terms of the number of transcripts per cell, is achievable. A standard curve can be generated by identifying a series of genes from an organism unrelated to the species of interest. These genes are then printed onto the array during fabrication. The corresponding gene transcripts are spiked into the target sample at known concentrations. Detection of these genes results in a standard curve, from which transcripts of genes of interest can be quantified.

(A) Principle of Hybrid Capture Technology

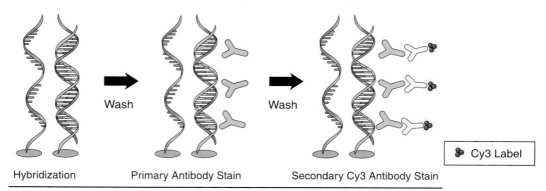

| Hybridization | Primary Antibody Stain | Secondary Cy3 Antibody Stain |

🔴 Cy3 Label

(B) Direct Detection Using the HC *Express*Array Kit

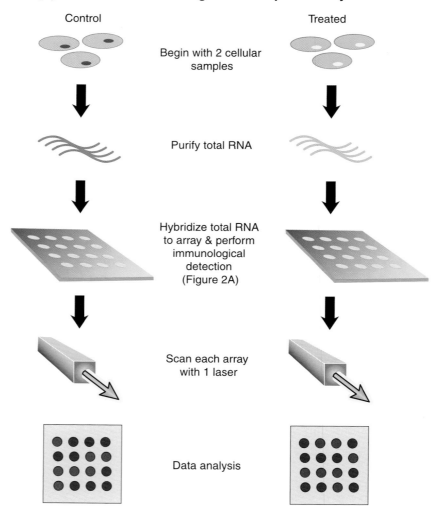

Control   Treated

Begin with 2 cellular samples

Purify total RNA

Hybridize total RNA to array & perform immunological detection (Figure 2A)

Scan each array with 1 laser

Data analysis

Total analysis time = 7–8 hours

FIGURE 2. (*A*) Overview of the immunological detection method of hybrid capture and its application for monitoring gene expression, using a microarray. (*B*) Principle of the procedures for direct detection of gene expression using the HC *Express*Array Kit.

PROTOCOL 2   HYBRIDIZATION AND DETECTION USING THE HC *EXPRESS*ARRAY KIT

MATERIALS

BUFFERS, SOLUTIONS, AND REAGENTS

Antibody dilution buffer*

Hybridization buffer*

Enhance buffer

> Add 50 ml of enhance buffer concentrate* <!> (contains sodium azide) into 950 ml of deionized H_2O and mix vigorously for 10 seconds.

PBS (137 mM NaCl, 2.7 mM KCl<!>, 4.3 mM Na_2HPO_4<!>•$7H_2O$, 1.4 mM KH_2PO_4<!>, pH ~7.2)

PBST

> Add 500 µl of Tween-20 to 1 liter of PBS and mix vigorously for 10 seconds.

Tween-20

Wash buffer

Add 50 ml of wash buffer concentrate*<!> (contains sodium azide) into 1.45 liters of deionized H_2O and mix vigorously for 10 seconds.

ANTIBODIES

Primary antibody*

Secondary antibody

Goat serum*<!> (contains sodium azide)

CENTRIFUGES AND ROTORS

Centrifuge with a swinging-bucket rotor

SPECIAL EQUIPMENT

Fluorescent array scanner with an excitation wavelength of 532 nm (for Cy3 detection)

Conical tubes, 50 ml (e.g., VWR)

Forceps

Hybridization cassettes (e.g., Corning)

LifterSlips (Erie Scientific) or equivalent

Paper towels

Microarray slide—glass-spotted, comprising cDNA, PCR products or oligonucleotides of 70 bp in length or greater

Water baths or humidified incubators, preset to 37°C, 55°C, and 75°C**

ADDITIONAL ITEMS

HC *Express*Array Kit (Digene Corporation)

> *Supplied with HC *Express*Array Kit (Digene).
> ** See step 8.

METHOD

1. Place the printed microarray slide into the hybridization cassette, spotted side up.

2. Carefully place a LifterSlip over the spotted area.

3. Add 10–20 µl of H_2O to each humidity well of the hybridization cassette.

4. In a microfuge tube, add 1–25 µl of total RNA (1–20 µg) to 25 µl of hybridization buffer provided in the kit. Add H_2O to bring the volume to 50 µl. Vortex the hybridization mixture for 10 seconds.

5. Denature the hybridization mix by incubating at 95°C for 2 minutes.

6. Immediately centrifuge the microfuge tube at >8000*g* for 5 seconds.

7. Quickly pipette the hybridization mix onto the edge of the LifterSlip on the glass array, and it will evenly disperse over the array.

8. Seal the hybridization cassette and incubate it in a water bath or humidified incubator for 3 hours at 75°C.

   75°C is recommended for cDNA arrays; 65°C should be used for oligonucleotide arrays.

9. Remove the hybridization cassette from the incubation chamber and place it at room temperature for 5 minutes.

10. Place the slide and LifterSlip into a 50-ml conical tube containing 35 ml of PBST until the LifterSlip slides off.

11. Remove the slide and transfer it to a second conical tube containing 35 ml of PBST. Wash the slide for 10 seconds using a shaking, vertical motion.

12. Using forceps, place the slide, spotted side up, onto a paper towel and allow it to air-dry for 1 minute. Return the slide to the hybridization cassette and place a fresh LifterSlip over the spotted area of the slide.

13. Add 5 µl of the primary antibody to 45 µl of the antibody dilution buffer to make the primary antibody staining solution.

14. Pipette the primary antibody staining solution onto the edge of the LifterSlip on the glass array, and it will evenly disperse over the array.

15. Seal the hybridization cassette and incubate in a humidified incubator for 1 hour at 37°C.

   Following this incubation, it is not necessary to allow the hybridization cassette to come to room temperature before proceeding to the wash step.

16. Repeat steps 10–12 using new 50-ml conical tubes with fresh PBST Buffer.

17. Prepare the secondary antibody staining solution by adding 5 µl of goat serum and 2 µl of secondary antibody to 43 µl of PBST buffer.

18. Pipette the secondary antibody staining solution onto the edge of the LifterSlip on the glass array, and it will evenly disperse over the array.

19. Seal the hybridization cassette and incubate again in a humidified incubator for 1 hour at 37°C. During this incubation, prewarm the enhance buffer to 55°C.

20. Place the slide and LifterSlip into a 50-ml conical tube containing 35 ml of wash buffer until the LifterSlip slides off.

21. Transfer the slide to a second conical tube containing 35 ml of PBST. Wash the slide for 10 seconds using a shaking, vertical motion.

22. Transfer the slide to a third conical tube containing 35 ml of prewarmed enhance buffer and wash again for 10 seconds.

   Extending the wash beyond 10 seconds will reduce the signal.

23. Transfer the slide to a fourth conical tube containing 35 ml of wash buffer. Wash for 10 seconds.

24. Finally, transfer the slide to a dry 50-ml conical tube and centrifuge in the tube, at 100–200*g* for 5–7 minutes, or until dry.

25. Scan the slide on an array scanner and analyze the data.

▶ TROUBLESHOOTING OF GENE EXPRESSION ANALYSIS USING MICROARRAYS

The success of microarray experiments is dependent on the quality of the reagents used. The two most important reagents are the microarray and the target.

- *Fabricating microarrays.* As previously stated, microarrays can be fabricated using PCR products generated from cDNA libraries. It is always best to begin with a "sequence-verified" cDNA library, although this does not guarantee accurate results. Studies have shown that following large-scale PCR amplification from sequence-verified cDNA libraries, only 79% of clones matched the original database (Taylor et al. 2001). For this reason, it is very important to re-verify PCR-amplified products that are to be used as microarray probes, prior to spotting. The root cause of such discrepancies lies in the multiple large-scale manipulations that are required for the fabrication of a high-density microarray. Because several steps in the fabrication procedure involve the manual transfer of clones, mistakes can be made quite easily. Although large-scale PCR has become a highly automated process, when the products are to be used as microarray probes, it is good practice to sequence the products. Consistency in the spotting of the PCR products is also an important factor, as is the quality of the slides on which the products are printed. High-quality robotic arrayers with uniform spot morphology must be used to manufacture arrays, and slides should provide a consistent surface upon which to attach the PCR products. Variability in the microarray manufacturing process can lead to problems when trying to replicate experiments between laboratories.

- *Preparing the target sample.* The second reagent that is important to a microarray experiment is the target sample. Again, the fewer manipulations needed before the target can be hybridized to the microarray, the better. For the two-color fluorescent detection method, the quality of the purified RNA will determine the quality of the labeled target. If genomic DNA contaminates the total RNA sample, it may become fluorescently labeled and hybridize to the microarray, thereby producing inaccurate results. For direct detection of total RNA using the HC *Express*Array Kit, the protocol is more flexible. Because of the specificity of the primary antibody for RNA:DNA hybrids, any remaining genomic DNA in the RNA sample will pass undetected. However, to maintain the accuracy of the experimental results, care should always be taken to ensure the highest-quality RNA sample.

- *Minimizing background fluorescence.* With any detection method, it is important to minimize background fluorescence. If the background is too high, weakly expressed genes may not be detected. If a cDNA labeling protocol is used, it may be necessary to perform additional sample purification steps to remove unincorporated labeled nucleotides. Alternatively, additional washes and some protocol optimization may be necessary to achieve minimal background signals.

- *General advice.* Finally, because there are several protocols for monitoring gene expression using microarrays, care should be taken to review each protocol thoroughly prior to use. It is critical that the temperatures for hybridization and wash procedures be adhered to. When performing any procedure underneath a coverslip, the researcher must be careful not to introduce air bubbles that might prevent uniform distribution of the sample or reagents over the microarray. General handling of the microarray should be carried out with extreme care. Thousands of DNA probes are present within a very small area and the surface must not be disturbed. Harsh handling, such as scratching or rubbing the surface, could damage the nucleic acids bound to the slide.

In summary, monitoring gene expression analysis using microarrays has been a significant advancement in the field of genomics. This development has been fueled by advancements in the fields of robotics, fluorescent scanners, and molecular biology techniques. With careful planning and consideration prior to the microarray experiment, a researcher can perform accurate and global analyses of gene expression.

ACKNOWLEDGMENTS

I thank Sheila Bingham, Joe Hernandez, Elena Katz, Jim Lazar, Richard Obiso, Donna Marie Seyfried, Fran Shay, Christina Strange, and Ha Thai for the commitment and effort they have put into the development of the HC *Express*Array Kit. I also thank Patti Mascone for her expert assistance with the chapter figures. In addition, I thank all the scientists whose efforts have contributed to the development and enhancement of the protocols presented in this chapter.

REFERENCES

Bowtell D. and Sambrook J. 2003. *DNA microarrays: A molecular cloning manual.* Cold Spring Harbor Laboratory Press, Cold Spring Harbor, New York.

Digene Corporation. 2001. HC *Express*Array™ Kit; Hybridization and detection for microarray gene expression analysis. Product package insert.

Fodor S.P., Read J.L., Pirrung J.C., Stryer L., Lu A.T., and Solas D. 1991. Light-directed, spatially addressable parallel chemical synthesis. *Science* **251:** 767–773.

Hughes T., Mao M., Jones A.R., Burchard J., Marton M.J., Shannon K.W., Lefkowitz S.M., Ziman M., Schelter J., Meyer M., Kobayashi S., Davis C., Dai H., He Y.D., Stephaniants S.B., Cavet G., Walker W.L., West A., Coffey E., Shoemaker D., Stoughton R., Blanchard A.P., Friend S.H., and Linsley P.S. 2001. Expression profiling using microarrays fabricated by an ink-jet oligonucleotide synthesizer. *Nat. Biotechnol.* **19:** 342–347.

Lander E.S. 1999. Array of hope. *Nat. Genet.* **21:** 3–4.

Lazar J.G., Cullen A., Mielzynska I., Meijide M., and Lorincz A. 1999. Hybrid Capture®: A sensitive signal amplification-based chemiluminescent test for the detection and quantitation of human viral and bacterial pathogens. *J. Clin. Ligand Assay* **22:** 139–151.

Marshall A. and Hodgson J. 1998. DNA chips: An array of possibilities. *Nat. Biotechnol.* **16:** 27–31.

McGall G., Labadie J., Brock P., Wallraff G., Nguyen T., and Hinsberg W. 1996. Light-directed synthesis of high-density oligonucleotide arrays using semiconductor photoresists. *Proc. Natl. Acad. Sci.* **93:** 13555–13560.

Okamoto T., Suzuki T., and Yamamoto N. 2000. Microarray fabrication with covalent attachment of DNA using bubble jet technology. *Nat. Biotechnol.* **18:** 438–441.

Petrik J. 2001. Microarray technology: The future of blood testing? *Vox Sanguinis* **80:** 1–11.

Sambrook J. and Russell D.W. 2001. *Molecular cloning : A laboratory manual,* 3rd edition. Cold Spring Harbor Laboratory Press, Cold Spring Harbor, New York

Schena M., Shalon D., Davis R.W., and Brown P.O. 1995. Quantitative monitoring of gene expression patterns with a complementary DNA microarray. *Science* **270:** 467–470.

Shalon D., Smith S.J., and Brown P.O. 1996. A DNA microarray system for analyzing complex DNA samples using a two-color fluorescent probe hybridization. *Genome Res.* **6:** 639–645.

Southern E.M. 1975. Detection of specific sequences among DNA fragments separated by gel electrophoresis. *J. Mol. Biol.* **98:** 503–517.

Taylor E., Cogdell D., Coombes L., Hu L., Ramdas L., Tabor A., Hamilton S., and Zhang W. 2001. Sequence verification as quality-control step for production of microarrays. *BioTechniques* **31:** 62–65.

WWW RESOURCES

http://research.nhgri.nih.gov/microarray/fabrication.html National Human Genome Research Institute, Microarray project, protocols.

22 Identification of Differential Genes by Suppression Subtractive Hybridization

Denis V. Rebrikov

Laboratory of Genes for Regeneration, Institute of Bioorganic Chemistry, Moscow, Russia 117997

Suppression subtractive hybridization (SSH) is one of the most powerful and popular methods for generating subtracted cDNA or genomic DNA libraries (Lukyanov et al. 1994; Diatchenko et al. 1996; Gurskaya et al. 1996; Jin et al. 1997; Akopyants et al. 1998). This powerful technique can be used to compare two mRNA populations and obtain cDNAs representing genes that are either overexpressed or exclusively expressed in one population as compared to another. The technique can also be used for comparison of genomic DNA populations.

The SSH method is based on a suppression PCR effect, introduced by Sergey Lukyanov (Lukyanov et al. 1994). A key feature of the method is simultaneous normalization and subtraction steps. The normalization step equalizes the abundance of DNA fragments within the target population, and the subtraction step excludes sequences that are common to the two populations being compared (Gurskaya et al. 1996). SSH eliminates any intermediate steps demanding the physical separation of single-stranded (ss) and double-stranded (ds) DNAs, it requires only one round of subtractive hybridization, and it can achieve a >1000-fold enrichment for differentially presented DNA fragments.

The level of enrichment of a particular DNA depends greatly on its original abundance, the ratio of its concentration in the samples being subtracted, and the number of other differentially presented genes. Other factors, such as the complexity of a starting material, hybridization time, and ratio of two samples being subtracted, play a very important role in the success of SSH for any given application. For instance, the high complexity of eukaryotic genomic DNA makes SSH very difficult. Some cDNA subtractions can also be very challenging due to the nature of the starting samples. Subtracted libraries generated from complex samples may contain very high background. An especially challenging problem is the generation of so-called "false-positive" clones that show a differential signal in a primary screening procedure but are not confirmed by subsequent detailed analysis. To overcome this problem, a simple procedure called mirror orientation selection (MOS) can be used to decrease substantially the number of background clones (Rebrikov et al. 2000).

In this chapter, we describe an improved version of the SSH technique for generating subtracted cDNA or genomic DNA libraries. Detailed protocols for cDNA synthesis, subtractive hybridization, PCR amplification, cloning, and differential screening analysis are provided. The MOS procedure is also described. Finally, some examples of SSH- and MOS-subtracted libraries are presented.

OVERVIEW

A brief overview of the SSH strategy is shown in Figure 1. SSH includes several steps. The most critical step is the choice of cells or tissue for comparison. The two samples should be related, with stable expression of the phenotypic difference of interest. If the two samples are too distantly related, a large number of irrelevant, differentially expressed genes will be identified.

The DNA population in which specific fragments are to be found is called the tester. The reference DNA population is called the driver. The generation of tester and driver DNAs begins with either poly(A)⁺ mRNA, followed by conversion to cDNA, or genomic DNA. After isolation and synthesis, the tester and driver DNAs are digested with a four-base-cutting restriction enzyme (*Rsa*I in this protocol) yielding blunt ends. The enzyme should be selected according to the GC-content of the target organism. Choose a restriction enzyme that will yield blunt-ended fragments averaging around 0.5–1 kb. The tester DNA is then subdivided into two portions, and each portion is ligated to a different pseudo-double-stranded (ds) adapter (adapter 1 [Ad1] and adapter 2R [Ad2R]). These adapters will serve as primer-binding sites for PCR amplification in later steps. Tester and driver DNA are then incubated and treated so that only appropriate adapter-labeled molecules will be amplified during the PCR steps (Fig. 2).

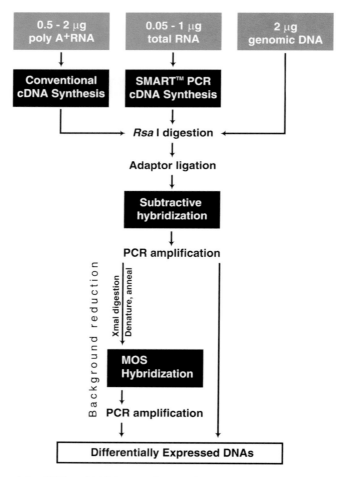

FIGURE 1. Overview of the SSH and MOS procedures. The DNA in which specific sequences are to be found is called "tester" and the reference DNA is called "driver."

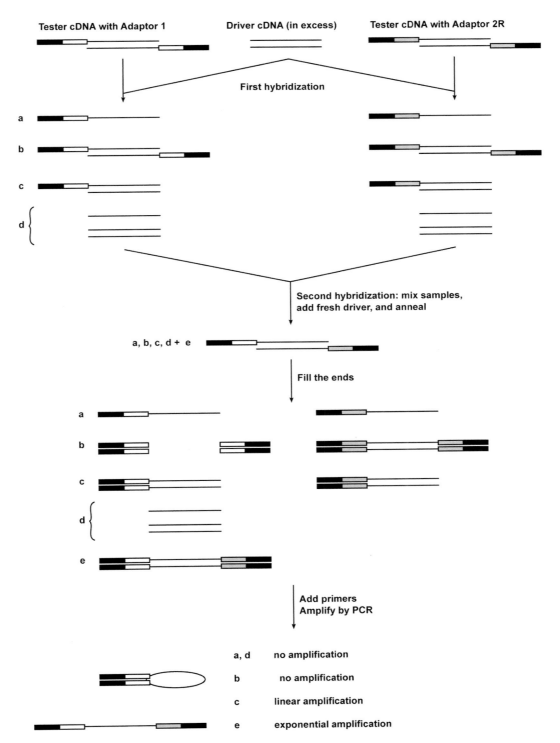

FIGURE 2. Schematic diagram of SSH procedure.

High background in the SSH-generated subtracted library can be reduced using MOS. The MOS technique is based on the rationale that after PCR amplification, during SSH, background molecules will be present in one orientation only, relative to the adapter sequences. Genuine SSH clones will be present in both sequence orientations (Rebrikov et al. 2000), as detailed in Figure 3.

Oligonucleotides for SSH

The following oligonucleotides are used throughout this chapter at a concentration of 10 μM. Whenever possible, oligonucleotides should be gel-purified.

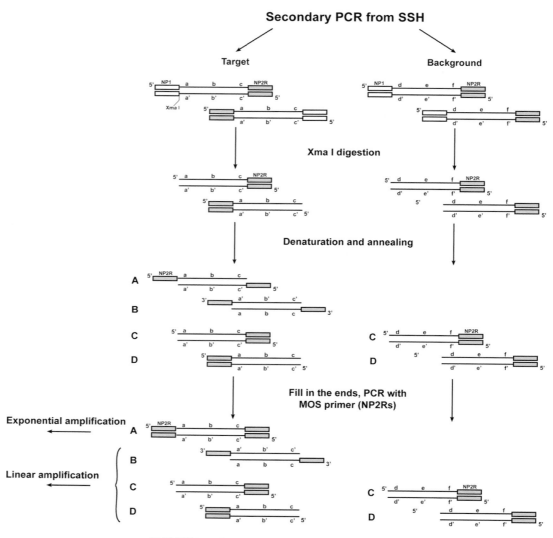

FIGURE 3. Schematic diagram of MOS procedure.

cDNA synthesis primer	5´-TTTTGTACAAGCTT$_{30}$-3´
Ad1	5´-CTAATACGACTCACTATAGGGCTCGAGCGGCCGCCCGGGCAGGT-3´
	3´-GGCCCGTCCA-5´
Ad2R	5´-CTAATACGACTCACTATAGGGCAGCGTGGTCGCGGCCGAGGT-3´
	3´-GCCGGCTCCA-5´
PCR primer 1 (P1)	5´-CTAATACGACTCACTATAGGGC-3´
Nested primer 1 (NP1)	5´-TCGAGCGGCCGCCCGGGCAGGT-3´
Nested primer 2R (NP2R)	5´-AGCGTGGTCGCGGCCGAGGT-3´
Nested primer 1s (NP1s)	5´-GCCGCCCGGGCAGGT-3´
MOS PCR primer (NP2Rs)	5´-GGTCGCGGCCGAGGT-3´
Blocking solution	A mixture of the cDNA synthesis primer, nested primers (NP1 and NP2R), and their respective complementary oligonucleotides (2 mg/ml each)

Buffers and Enzymes Used in SSH

Complementary DNA (cDNA) synthesis was performed by the method of Gubler and Hoffmann (Gubler and Hoffman 1983) using the AMV reverse transcriptase (20 units/μl; Invitrogen, San Diego, CA). Second-strand synthesis used a cocktail of DNA polymerase I, *E. coli* DNA ligase, T4 DNA polymerase (New England Biolabs, Beverly, MA), and RNase H (Epicentre Technologies, Madison, WI). *Rsa*I and T4 DNA ligase were also from New England Biolabs.

The subtractive hybridization buffer should be constructed as a 4x hybridization buffer stock: 4 M NaCl, 200 mM HEPES, pH 8.3, and 4 mM cetyltrimethyl ammonium bromide (CTAB). The final 1x hybridization mix is diluted with dilution buffer of 20 mM HEPES-HCl, pH 8.3, 50 mM NaCl, and 0.2 mM EDTA.

PCR amplification reactions are performed using a mixture of KlenTaq-1 and DeepVent DNA polymerases (New England Bio Labs, Beverly, MA) and TaqStart Antibody (Clontech, Palo Alto, CA) with 10x PCR buffer (40 mM Tricine-KOH, pH 9.2 at 22°C, 3.5 mM magnesium acetate, 10 mM potassium acetate, and 75 mg/ml BSA). Alternatively, *Taq* DNA polymerase alone can be used, but some additional cycles may be needed in both the primary and secondary PCR.

Please note that the cycling parameters in this protocol have been optimized using the MJ Research PTC-200 DNA Thermal Cycler. For a different type of thermal cycler, the cycling parameters must be optimized. It is not possible to use this protocol with water bath thermal cyclers because there is no PCR suppression effect there.

It is recommended that subtraction hybridizations be performed in both forward and reverse directions. We also recommend performing self-subtractions, with both tester and driver prepared from the same DNA sample, as a control to determine subtraction efficiency. These controls should yield little, if any, PCR product after amplification. For systems where there is experimentally induced overexpression of a known set of genes, such as viral infections, it is extremely important to add these known sequences to the driver DNA sample. For example, when searching for p53-up-regulated genes in a p53-overexpressing cell line, add *Rsa*I-digested p53 cDNA to the *Rsa*I-digested driver sample to a level of 10% of the total driver DNA concentration. This must be done after the preparation of the adapter-ligated tester sample. Addition of this exogenous DNA before *Rsa*I digestion will cause disproportional representation of this material in the samples. All chemical reagents were obtained from Sigma Chemical (St. Louis, MO).

PROTOCOL 1   PREPARATION OF SUBTRACTED cDNA OR GENOMIC DNA LIBRARY

STAGE 1: RNA AND DNA ISOLATION

A total of 2 μg of genomic DNA or cDNA is required per bidirectional subtraction. Most methods for isolation of RNA and genomic DNA are appropriate for the application presented here (Sambrook et al. 1989; Chomczynski and Sacchi 1987; Farrell 1993). Whenever possible, samples being compared should be purified side by side using the same reagents and protocol. If genomic DNA is used as the starting material, proceed to *Rsa*I digestion (Stage 3 of this protocol). If RNA is used as the starting material, proceed to cDNA synthesis (Stage 2 of this protocol).

STAGE 2: cDNA SYNTHESIS

MATERIALS

▼ CAUTION

See Appendix for appropriate handling of materials marked with <!>.

BUFFERS, SOLUTIONS, AND REAGENTS
dNTP solution (10 mM)
DTT <!>, 0.1M
EDTA, 0.2 M
First-strand buffer, 5x
Mineral oil

> SSH technology demands very accurate PCR. In the author's experience, mineral oil is necessary to maintain this accuracy. Heated lids should only be used for large-scale PCR screening when samples might be more easily analyzed in the absence of oil. Thus, I strongly advise against the use of heated lids until Protocol 5.

Second-strand buffer, 5x
TN buffer (10 mM Tris-HCl<!>, pH 7.6, 10 mM NaCl)

ENZYMES AND ENZYME BUFFERS
AMV reverse transcriptase (20 units/μl)

NUCLEIC ACIDS AND OLIGONUCLEOTIDES
cDNA synthesis primer (10 μM)
Poly(A)$^+$ RNA

RADIOACTIVE COMPOUNDS
Optional: [α-^{32}P]dCTP (10 mCi/ml, 3000 Ci/mmole)<!>

SPECIAL EQUIPMENT
Thermal cycler, or water baths preset to 16°C, 42°C, and 72°C

ADDITIONAL ITEMS
Reagents for phenol:chloroform <!> extraction and ethanol <!> precipitation

First-strand cDNA Synthesis

1. For each tester and driver sample, combine the following components in a sterile 0.5-ml microcentrifuge tube:

Poly(A)$^+$ RNA (2 μg)	2–4 μl
cDNA synthesis primer (10 μM)	1 μl
H$_2$O	to 5 μl

2. Overlay samples with one drop of mineral oil and incubate at 72°C in a thermal cycler or H$_2$O bath for 2 minutes.

3. Cool at room temperature for 2 minutes.

4. To each reaction tube, add the following:

First-strand buffer, 5x (see p. 301)	2 μl
dNTP solution (10 mM each)	1 μl
H$_2$O*	0.5 μl
DTT, 0.1 M	0.5 μl
AMV reverse transcriptase (20 units/μl)	1 μl

*Optional: To monitor the progress of cDNA synthesis, dilute 0.5 μl of [α-^{32}P]dCTP<!> (10 mCi/ml, 3000 Ci/mmole) in 9 μl of H$_2$O and replace the H$_2$O above with 0.5 μl of the diluted label.

5. Gently vortex and briefly centrifuge the tubes.

6. Incubate the tubes at 42°C for 90 minutes in a thermal cycler or H$_2$O bath.

7. Place the tubes on ice to terminate first-strand cDNA synthesis, and immediately proceed to second-strand cDNA synthesis.

Second-strand cDNA Synthesis

8. Add the following components (previously cooled on ice) to the first-strand synthesis reaction tubes:

Sterile H$_2$O	48.4 μl
Second-strand buffer, 5x (see p. 301)	16.0 μl
dNTP solution (10 mM each)	1.6 μl
Second-strand enzyme cocktail, 20x	4.0 μl

9. Mix the contents and briefly spin the tubes. The final volume should be 80 μl.

10. Incubate the tubes at 16°C (water bath or thermal cycler) for 2 hours.

11. Add 2 μl (6 units) of T4 DNA polymerase. Mix contents well.

12. Incubate the tube at 16°C for a further 30 minutes in a water bath or a thermal cycler.

13. Add 4 μl of 0.2 M EDTA to terminate second-strand synthesis.

14. Perform phenol:chloroform extraction and ethanol precipitation.

15. Redissolve pellet in 50 μl of TN buffer.

16. Transfer 6 μl to a fresh microcentrifuge tube. Store this sample at −20°C until after *Rsa*I digestion.

This sample will be used for agarose gel electrophoresis to estimate yield and size range of the ds cDNA products synthesized.

STAGE 3: *Rsa*I DIGESTION

Perform the following procedure with each experimental ds tester and driver DNA. This step generates shorter, blunt-ended dsDNA fragments optimal for subtractive hybridization.

▼ CAUTION

See Appendix for appropriate handling of materials marked with <!>.

MATERIALS

. . . .

BUFFERS, SOLUTIONS, AND REAGENTS

EDTA, 0.2 M

TN buffer (10 mM Tris-HCl<!>, pH 7.6, 10 mM NaCl)

ENZYMES AND ENZYME BUFFERS
> *Rsa*I (10 units/µl)
> *Rsa*I restriction buffer, 10x

NUCLEIC ACIDS AND OLIGONUCLEOTIDES
> Genomic DNA, obtained from Stage 1, or cDNA obtained from Stage 2, 2 µg per bidirectional subtraction

SPECIAL EQUIPMENT
> Water bath, preset to 37°C

ADDITIONAL ITEMS
> Reagents for phenol:chloroform<!> extraction and ethanol<!> precipitation
> Reagents and equipment for agarose gel electrophoresis<!>

METHOD

1. Add the following reagents to the tube from Protocol 2, step 8 above or to genomic DNA (gDNA) sample:

ds cDNA or genomic DNA	43.5 µl
*Rsa*I restriction buffer, 10x	5.0 µl
*Rsa*I (10 units/µl)	1.5 µl

2. Mix and incubate for 2–4 hours at 37°C.

3. To check the efficiency of *Rsa*I digestion, analyze 5 µl of the digestion on a 2% agarose gel, alongside undigested DNA (Protocol 2, Stage 2, step 8 or Protocol 1 for genomic DNA).

   Continue the digestion during electrophoresis and terminate the reaction only after you are satisfied with the results of the analysis.

4. Terminate the digestion by adding 2.5 µl of 0.2 M EDTA. Extract the digestion with phenol:chloroform and collect the DNA by ethanol precipitation.

   The use of glycogen or any type of coprecipitants at this stage is not recommended. These agents may increase the viscosity of the DNA solution and inhibit later DNA hybridization. The use of silica matrix-based PCR purification systems should also be avoided.

6. Dissolve each pellet in 6 µl of TN buffer (not H$_2$O) and store at –20°C. Driver DNA preparation is now complete.

STAGE 4: ADAPTER LIGATION

It is strongly recommended that subtractions be performed in both directions for each tester/driver DNA pair. Forward subtraction is designed to enrich for differentially presented molecules present in the tester but not in the driver; reverse subtraction is designed to enrich for differentially presented sequences present in the driver but not in the tester. The availability of such forward- and reverse-subtracted DNAs will be useful for differential screening of the resulting subtracted tester DNA library (Protocol 6).

The tester DNAs are ligated separately to Ad1 (Tester 1-1 and 2-1) and Ad2R (Tester 1-2 and 2-2). A third ligation of both adapters Ad1 and Ad2R to the tester DNAs (unsubtracted tester control 1-c and 2-c) should be performed and used as a negative control for subtraction. The adapters are *not* ligated to the driver DNA.

MATERIALS

BUFFERS, SOLUTIONS, AND REAGENTS
ATP, 3 mM
EDTA, 0.2 M
Mineral oil
TN buffer (10 mM Tris-HCl <!>, pH 7.6, 10 mM NaCl)

ENZYMES AND ENZYME BUFFERS
Ligation buffer, 10x
T4 DNA ligase (400 units/μl)

NUCLEIC ACIDS AND OLIGONUCLEOTIDES
Adapter Ad1 (10 μM)
Adapter Ad2R (10 μM)
*Rsa*I-digested tester DNA

SPECIAL EQUIPMENT
Water bath, preset to 37°C

ADDITIONAL ITEMS
Reagents for phenol:chloroform<!> extraction and ethanol<!> precipitation
Reagents and equipment for agarose gel electrophoresis<!>

METHOD

1. Dilute 1 μl of each *Rsa*I-digested tester DNA from the above section with 5 μl of TN buffer.

2. Prepare a master ligation mix of the following components for each reaction:

H_2O	2 μl
Ligation buffer, 5x	2 μl
ATP (3 mM)	1 μl
T4 DNA ligase (400 units/μl)	1 μl

3. For each tester DNA mixture, combine the following reagents in a 0.5-ml microcentrifuge tube in the order shown. Pipette the solution up and down to mix thoroughly.

	Tube #	
Component	Tester 1-1	Tester 1-2
Diluted tester cDNA	2 μl	2 μl
Adapter Ad1 (10 μM)	2 μl	—
Adapter Ad2R (10 μM)	—	2 μl
Master ligation mix	6 μl	6 μl
Final volume	10 μl	10 μl

4. In a fresh microcentrifuge tube, mix 2 μl of Tester 1-1 and 2 μl of Tester 1-2. This is the unsubtracted tester control 1-c. Do the same for each tester DNA sample.

   After ligation, approximately one-third of the DNA molecules in each unsubtracted tester control tube will be ligated to two different adapters and will be suitable for exponential PCR amplification with adapter-derived primers.

5. Overlay samples with one drop of mineral oil, centrifuge the tubes briefly and incubate at 16°C overnight.

6. Stop the ligation reaction by adding 1 μl of 0.2 M EDTA.

7. Heat samples at 72°C for 5 minutes to inactivate the ligase.

8. Remove 1 μl from each unsubtracted tester control (1-c, 2-c ...) and dilute into 1 ml of TN buffer. These samples will be used for PCR amplification (Protocol 3).

    Preparation of experimental adapter-ligated tester DNAs 1-1 and 1-2 is now complete.

9. Perform a ligation efficiency test before proceeding to the next section.

LIGATION EFFICIENCY TEST

MATERIALS

BUFFERS, SOLUTIONS, AND REAGENTS
dNTP solution (containing all four dNTPs, each at 10 mM)
Mineral oil

ENZYMES AND ENZYME BUFFERS
PCR buffer, 10x (40 mM Tricine-KOH, pH 9.2 at 22°C, 3.5 mM magnesium acetate, 10 mM potassium acetate, 75 mg/ml BSA, or as supplied by the manufacturer)
Polymerase mixture, 50x (Advantage 2, Clontech, or equivalent)

NUCLEIC ACIDS AND OLIGONUCLEOTIDES
PCR Primer P1
Unsubtracted control samples (Protocol 1, Stage 4, ligation efficiency test)

SPECIAL EQUIPMENT
Thermal cycler

ADDITIONAL ITEMS
Reagents and equipment for agarose gel electrophoresis<!>

METHOD

1. Aliquot 1 μl of each undiluted unsubtracted control sample into a 0.5-ml PCR tube.

2. Prepare a master mix for all of the reaction tubes. Combine the reagents as follows:

H$_2$O	19.5 μl
PCR buffer, 10x	2.5 μl
dNTP solution (10 mM each)	0.5 μl
PCR Primer P1	1.0 μl
Polymerase mixture, 50x	0.5 μl
Total volume	24.0 μl

3. Mix and briefly centrifuge the tubes.

4. Aliquot 24 μl of master mix into each of the reaction tubes prepared in step 1.

5. Overlay with one drop of mineral oil.

6. Incubate the reaction mixture in a thermal cycler at 72°C for 5 minutes to extend the adapters.

7. Immediately commence thermal cycling as follows:

Cycle number	Denaturation	Annealing	Polymerization/Extension
15	10 sec at 95°C	10 sec at 66°C	1.5 min at 72°C

8. Analyze 4 µl from each tube on a 2% agarose gel.

> **Important:** This PCR product should have a similar pattern to that of *Rsa*I-digested DNA. If no PCR products are visible after 15 cycles, perform 3 more cycles and again analyze the PCR product. If no PCR products are visible after 21 cycles, the activity of the polymerase mixture needs to be examined. If there is no problem with polymerase mixture or other PCR reagents, the ligation reaction should be repeated with fresh samples before proceeding to the next step.

PROTOCOL 2 SUBTRACTIVE HYBRIDIZATION

MATERIALS

BUFFERS, SOLUTIONS, AND REAGENTS

Dilution buffer (20 mM HEPES-HCl, pH 8.3, 50 mM NaCl, 0.2 mM EDTA)

Hybridization buffer stock 4x: 4 M NaCl, 200 mM HEPES, pH 8.3, 4 mM cetyltrimethyl ammonium bromide (CTAB)<!>. The final 1x hybridization mix is diluted with dilution buffer.

Mineral oil

NUCLEIC ACIDS AND OLIGONUCLEOTIDES

Ad1-ligated Tester 1-1 (Protocol 1, Stage 4, step 8)

Ad2R-ligated Tester 1-2 (Protocol 1, Stage 4, step 8)

Driver DNA (Protocol 1, Stage 3, step 6)

*Rsa*I-digested driver DNA (Protocol 1, Stage 3, step 6)

SPECIAL EQUIPMENT

Two thermal cyclers (see steps 11 and 12)

METHOD

First Hybridization

1. For each tester sample, combine the reagents in the following order:

	Hybridization 1-1	Hybridization 1-2
*Rsa*I-digested driver DNA	1.5 µl	1.5 µl
Ad1-ligated Tester 1-1	1.5 µl	—
Ad2R-ligated Tester 1-2	—	1.5 µl
Hybridization buffer, 4x	1.0 µl	1.0 µl
Final volume	4.0 µl	4.0 µl

2. Overlay with one drop of mineral oil and centrifuge briefly.

3. Incubate samples in a thermal cycler at 98°C for 1.5 minutes.

4. Incubate samples for the time listed below, based on the sample type at 68°C, and then proceed immediately to the second hybridization.

Sample type	First hybridization time
Bacterial genome subtraction	1–3 hours
Eukaryotic genome subtraction	3–5 hours
cDNA subtraction	7–12 hours

The protocol uses 15 ng of ligated tester DNA and 450 ng of driver DNA. The ratio of driver to tester can be changed if a different subtraction efficiency is desired.

Second Hybridization

9. Repeat the following steps for each experimental driver DNA.

Place the following reagents in a sterile 0.5-µl microcentrifuge tube:

Component	Amount per reaction
Driver DNA (Protocol 3, step 7)	1 µl
Hybridization buffer, 4x	1 µl
Sterile H$_2$O	2 µl

10. Place 1 µl of this mixture in a 0.5-ml microcentrifuge tube and overlay it with one drop of mineral oil.

11. Incubate in a thermal cycler at 98°C for 1.5 minutes.

12. Remove the tube of freshly denatured driver and immediately place in a second thermal cycler equilibrated to 68°C.

    Using a second thermal cycler (or second block) ensures rapid equilibration of specimens.

13. Then add hybridized sample 1-1 and hybridized sample 1-2 (from first hybridization) in that order.

    This ensures that the two hybridization samples are mixed only in the presence of excess of freshly denatured driver DNA.

14. Incubate the hybridization reaction at 68°C overnight.

15. Add 100 µl of dilution buffer to the tube and mix well by pipetting.

16. Incubate in a thermal cycler at 72°C for 7 minutes.

17. Store hybridization reaction at –20°C.

PROTOCOL 3   PCR AMPLIFICATION OF DIFFERENTIALLY PRESENTED DNAs

Differentially presented DNAs are selectively amplified during the reactions described in this protocol. Each experiment should have at least four reactions: (1) subtracted tester DNAs, (2) unsubtracted tester control (1-c), (3) reverse-subtracted tester DNAs, and (4) unsubtracted control for the reverse subtraction (2-c).

MATERIALS

....

BUFFERS, SOLUTIONS, AND REAGENTS

dNTP solution (containing all four dNTPs, each at 10 mM)

ENZYMES AND ENZYME BUFFERS

PCR reaction buffer, 10x (see p. 306)
Polymerase mixture, 50x (Advantage 2, Clontech, or equivalent)

NUCLEIC ACIDS AND OLIGONUCLEOTIDES

DNA sample (i.e., each subtracted sample from Protocol 2, step 16)
Nested PCR primer NP1 (10 μM)
Nested PCR primer NP2R (10 μM)
PCR primer P1 (10 μM)
Unsubtracted tester control from Protocol 1, Stage 4, step 8

SPECIAL EQUIPMENT

Thermal cycler, preset to 72°C

ADDITIONAL ITEMS

Equipment and reagents required for 2% agarose gel electrophoresis<!>

METHOD

Primary PCR

1. Aliquot 1 μl of each diluted DNA sample (i.e., each subtracted sample from Protocol 2, step 16, and the corresponding diluted unsubtracted tester control from Protocol 1, Stage 4, step 7) into an appropriately labeled tube. Frozen samples should be thawed and incubated at 72°C and then mixed well by pipetting prior to use.

2. Prepare a master mix for all of the primary PCR tubes plus one additional tube. For each reaction combine the reagents in the order shown:

Sterile H_2O	19.5 μl
PCR reaction buffer, 10x	2.5 μl
dNTP solution (10 mM)	0.5 μl
PCR primer P1 (10 μM)	1.0 μl
50x polymerase mixture	0.5 μl
Total volume	24.0 μl

3. Aliquot 24 μl of master mix into each reaction tube prepared in step 1 above.

4. Overlay with one drop of mineral oil.

5. Incubate the reaction mixture in a thermal cycler at 74°C for 5 minutes to extend the adapters. This step "fills in" the missing strand of the adapters and thus creates binding sites for the PCR primers. Do not remove the samples from the thermal cycler.

6. Immediately commence thermal cycling as follows:

Cycle number	Denaturation	Annealing	Polymerization/Extension
26	10 sec at 95°C	10 sec at 66°C	1.5 min at 72°C

7. Analyze 4 μl from each tube on a 2.0% agarose TAE gel.

For some complicated subtractions (with complex tissues or eukaryotic genomes), we recommend performing primary PCR in two steps. This procedure may significantly reduce background. First, perform primary PCR as described in this protocol. Then perform another primary PCR as described in Protocol 4, Stage 1, step 10. Then proceed to secondary PCR (Protocol 3).

Secondary PCR

8. Dilute 2 μl of each primary PCR mixture in 38 μl of H_2O.

9. Aliquot 1 μl of each diluted primary PCR product mixture from step 1 into an appropriately labeled tube.

10. Prepare a master mix for the secondary PCR samples plus one additional reaction by combining the reagents in the following order:

Sterile H$_2$O	18.5 μl
PCR reaction buffer, 10x	2.5 μl
Nested PCR primer NP1 (10 μM)	1.0 μl
Nested PCR primer NP2R (10 μM)	1.0 μl
dNTP solution (10 mM)	0.5 μl
50x polymerase mixture	0.5 μl
Total volume	24.0 μl

11. Aliquot 24 μl of master mix into each reaction tube from step 2.

12. Overlay with one drop of mineral oil.

13. Immediately commence thermal cycling as follows:

Cycle number	Denaturation	Annealing	Polymerization/Extension
10–12	10 sec at 95°C	10 sec at 68°C	1.5 min at 72°C

14. Analyze 4 μl from each reaction on 2.0% agarose.

15. Store reaction products at –20°C. The PCR mixture is now enriched for differentially presented DNAs.

If MOS is not required, proceed to the section Cloning of Subtracted DNAs.

PROTOCOL 4 MIRROR ORIENTATION SELECTION

The major drawback of SSH is the presence of background clones that represent nondifferentially expressed DNA species in the subtracted libraries. In some cases, the number of background clones may considerably exceed the number of target clones. To overcome this problem, we recommend MOS—a simple procedure that substantially decreases the number of background clones in the libraries generated by SSH. We recommend the use of MOS in the following cases:

• When the percentage of differentially expressed clones found during differential screening is very low (for example, 1–5%). The MOS procedure can increase the number of differential clones up to 10-fold.

• When most of the differentially expressed clones found are false positives. The MOS procedure decreases the portion of false-positive clones severalfold.

• When the primary PCR in SSH requires more than 30 cycles (but no more than 36 cycles) to generate visible PCR product. In this case, the problems described in the previous two items will usually appear.

• If the complexity of tester and driver samples is very great or if the difference in gene expression between tester and driver is very small. In this case, plan to perform MOS from the beginning of the experiment. After subtractive hybridization, perform PCR amplification using the following protocol instead of the procedure in Protocol 3.

• If the SSH subtracted library has already been made and found, upon differential screening, to contain high background, the option to perform MOS on the SSH-generated library should be considered. The hybridization mix generated in Protocol 2, step 17, can be used for PCR amplification using the following protocol.

STAGE 1: PCR AMPLIFICATION FOR MOS AND *XMA*I DIGESTION

MATERIALS

BUFFERS, SOLUTIONS, AND REAGENTS
EDTA, 0.2 M
Mineral oil
TN buffer (10 mM Tris-HCl<!>, pH 7.6, 10 mM NaCl)

ENZYMES AND ENZYME BUFFERS
PCR buffer, 10x (40 mM Tricine-KOH<!>, pH 9.2 at 22°C, 3.5 mM magnesium acetate<!>, 10 mM potassium acetate, 75 mg/ml BSA, or as supplied by the manufacturer)
Polymerase mixture, 50x (Advantage 2, Clontech, or equivalent)
*Xma*I (10 units/µl)
*Xma*I restriction buffer, 10x

NUCLEIC ACIDS AND OLIGONUCLEOTIDES
DNA (Protocol 10, step 10)
dNTP solution (containing all four dNTPs, each at 10 mM)
Each diluted second hybridization (from Protocol 2, step 17)
PCR Primer NP1
PCR Primer NP2R
PCR Primer P1

SPECIAL EQUIPMENT
Thermal cycler
Water baths, preset to 37°C and 74°C

ADDITIONAL ITEMS
Equipment and reagents for agarose gel electrophoresis<!>
Equipment and reagents for phenol:chloroform <!> extraction and ethanol <!> precipitation

METHOD

Primary PCR Amplification for MOS

1. Transfer 10 µl of each diluted second hybridization (from Protocol 2, step 17) into appropriately labeled tubes.

2. Prepare a master mix for the primary PCR-1 as follows. This recipe is sufficient for one reaction. Scale up the volumes as required. For each reaction, combine the reagents as follows:

Component	Amount per reaction
Sterile H₂O	92.5 µl
10x PCR buffer	12.5 µl
dNTP solution (10 mM each)	2.5 µl
PCR Primer P1	5.0 µl
50x polymerase mixture	2.5 µl
Total volume	115 µl

3. Mix well and briefly centrifuge the tube.

4. Aliquot 115 µl of master mix into each reaction tube from step 1.

5. Aliquot 125 μl of final mix into five 0.5-μl PCR tubes (25 μl per tube).

6. Overlay with one drop of mineral oil.

7. Incubate the reaction mixture in a thermal cycler at 74°C for 5 minutes to extend the adapters, then immediately begin thermal cycling as follows:

Cycle number	Denaturation	Annealing	Polymerization/Extension
x*	10 sec at 95°C	10 sec at 66°C	1.5 min at 72°C

*The recommended number of primary PCR-1 cycles for MOS is the number of SSH primary PCR cycles minus 2. For example, if the SSH primary PCR was visible on agarose/gel after 31 cycles, you will need 31 − 2 = 29 cycles of primary PCR1 for MOS.

8. Combine 2 μl of each of the 5 primary PCR-1 products in one tube and add 390 μl of H_2O.

9. Aliquot 1 μl of each diluted primary PCR-1 product mixture from step 8 into an appropriately labeled PCR tube.

10. Prepare master mix for primary PCR-2 as follows:

Component	Amount per reaction
Sterile H_2O	19.5 μl
10x PCR buffer	2.5 μl
dNTP solution (10 mM each)	0.5 μl
PCR Primer P1	1.0 μl
50x polymerase mixture	0.5 μl
Total volume	24 μl

11. Mix well and briefly centrifuge the tube.

12. Aliquot 24 μl of master mix into each reaction tube from step 9.

13. Overlay with one drop of mineral oil.

14. Immediately commence thermal cycling as follows:

Cycle number	Denaturation	Annealing	Polymerization/Extension
10–12	30 sec at 95°C	30 sec at 66°C	1.5 min at 72°C

15. Analyze 4 μl from each reaction on a 2.0% agarose gel.

Secondary PCR for MOS

16. Dilute 2 μl of each primary PCR-2 mixture generated in step 14, above, in 38 μl of H_2O.

17. Aliquot 2 μl of each diluted primary PCR-2 product mixture into an appropriately labeled tube.

18. Prepare a master mix for secondary PCR as follows. The recipe is sufficient for one reaction. Scale up the volumes as required. For each reaction, combine the reagents as follows:

Component	Amount per reaction
Sterile H_2O	37.0 μl
PCR buffer, 10x	5.0 μl
dNTP solution (10 mM each)	1.0 μl
PCR Primer NP1	2.0 μl
PCR Primer NP2R	2.0 μl
50x polymerase mixture	1.0 μl
Total volume	48.0 μl

19. Mix well and briefly centrifuge the tube.

20. Aliquot 48 µl of master mix into each reaction tube from step 2.

21. Overlay with one drop of mineral oil.

22. Immediately commence thermal cycling for 10–12 cycles at 95°C for 10 sec, 68°C for 10 sec, and 72°C for 1.5 minutes.

23. Analyze 4 µl from each tube on a 2% agarose<!> gel.

24. Purify secondary PCR product by phenol/chloroform extraction and ethanol precipitation.

25. Dissolve the pellet in 20–40 µl of TN buffer up to concentration 20 ng/µl of DNA.

26. Analyze 2 µl of purified PCR product on a 2% agarose/ethidium bromide gel.

27. Dilute 1 µl of purified PCR product from step 10 in 1.6 ml of H_2O (this will be the undigested control).

28. Store reaction products at –20°C.

XmaI Digestion

29. Place the following reagents in the tube:

Component	Amount per reaction
Sterile H_2O	12 µl
10x XmaI restriction buffer	2 µl
DNA (Protocol 10, step 10)	5 µl
XmaI (10 units/µl)	1 µl

30. Mix and incubate at 37°C for 2 hours.

31. Add 2 µl of 0.2 M EDTA to terminate the reaction.

32. Incubate at 74°C for 5 minutes to inactivate enzyme.

33. Store at –20°C.

STAGE 2: MOS HYBRIDIZATION

MATERIALS

▼ CAUTION

See Appendix for appropriate handling of materials marked with <!>.

BUFFERS, SOLUTIONS, AND REAGENTS
Dilution buffer (20 mM HEPES-HCl, pH 8.3, 50 mM NaCl, 0.2 mM EDTA)
Hybridization buffer stock 4x: 4 M NaCl, 200 mM HEPES, pH 8.3, 4 mM cetyltrimethyl ammonium bromide (CTAB)<!>. The final 1x hybridization mix is diluted with dilution buffer.
Mineral oil

NUCLEIC ACIDS AND OLIGONUCLEOTIDES
XmaI-digested DNA (Protocol 11, step 5)

SPECIAL EQUIPMENT
Thermal cycler

METHOD

1. Combine the following reagents in a 1.5-ml microfuge tube:

Component	Amount per reaction
H$_2$O	2 µl
*Xma*I-digested DNA (Protocol 11, step 5)	1 µl
4x Hybridization buffer	1 µl

2. Place 2 µl of this mixture in a 0.5-ml microcentrifuge tube and overlay with one drop of mineral oil.

3. Incubate in a thermal cycler at 98°C for 1.5 minutes.

4. Incubate in a thermal cycler at 68°C for 3 hours.

5. Add 200 µl of dilution buffer to the tube and mix well by pipetting.

6. Heat in a thermal cycler at 70°C for 7 minutes.

7. Store at –20°C.

STAGE 3: MOS PCR AMPLIFICATION

MATERIALS

BUFFERS, SOLUTIONS, AND REAGENTS
dNTP solution (containing all four dNTPs, each at 10 mM)
Mineral oil

▼ CAUTION

See Appendix for appropriate handling of materials marked with <!>.

ENZYMES AND ENZYME BUFFERS
PCR buffer, 10x (40 mM Tricine-KOH<!>, pH 9.2 at 22°C, 3.5 mM magnesium acetate<!>, 10 mM potassium acetate, 75 mg/ml BSA, or as supplied by the manufacturer)
Polymerase mixture, 50x (Advantage 2, Clontech, or equivalent)

NUCLEIC ACIDS AND OLIGONUCLEOTIDES
Diluted DNA sample (after hybridization and the corresponding undigested control, Protocol 4, Stage 2, step 6 and Protocol 4, Stage 1, step 26)
MOS PCR Primer (NP2Rs)

SPECIAL EQUIPMENT
Thermal cycler

ADDITIONAL REAGENTS
Equipment and reagents for agarose gel electrophoresis

METHOD

1. Prepare a master mix for all MOS PCRs as follows:

Component	Amount per reaction
Sterile H$_2$O	19.5 µl
10x PCR buffer	2.5 µl
dNTP solution (10 mM each)	0.5 µl
MOS PCR Primer (NP2Rs)	1.0 µl
50x polymerase mixture	0.5 µl
Total volume	24.0 µl

2. Add 1 µl of each diluted DNA sample (after hybridization and the corresponding Undigested control) to an appropriately labeled tube containing 24 µl of master mix.

3. Overlay with one drop of mineral oil.

4. Incubate the reaction mix in a thermal cycler at 74°C for 5 minutes to extend the adapters. (Do not remove the samples from the thermal cycler.)

5. Immediately commence thermal cycling as follows:

Cycle number	Denaturation	Annealing	Polymerization/Extension
19	10 sec at 95°C	10 sec at 62°C	1.5 min at 72°C

6. Analyze 4 μl from each tube agarose gel electrophoresis.

7. Store reaction products at –20°C.

Cloning of Subtracted DNAs

Once a subtracted sample has been confirmed to be enriched in DNAs derived from differentially presented genes, the PCR products (from Protocol 3, secondary PCR amplification or from Protocol 4, Stage 3, step 7) can be subcloned using several conventional cloning techniques. See Chapters 27 and 28. It is recommended that cloning reactions begin with either 3 μl of the secondary PCR product (Protocol 3, step 15) or MOS PCR product (Protocol 4, Stage 3, step 4).

For site-specific cloning, using the adapter sequences described here, cleave at the *Eag*I (*Not*I) and *Xma*I (*Sma*I, *Srf*I) sites and then ligate the products into an appropriate plasmid vector. Keep in mind that some or all of these sites might also be present in the DNA fragments. The number of independent colonies obtained for each library depends on the number of differentially expressed genes, as well as the subtraction and subcloning efficiencies. In general, 500 colonies can be initially arrayed and studied. Additional colonies can be obtained easily by performing further subclonings of the secondary PCR products listed above.

Differential Screening of the Subtracted DNA Library

Two approaches can be utilized for differential screening of the arrayed subtracted DNA clones: colony dot blots and PCR-based DNA dot blots. For colony dot blots, bacterial colonies are spotted on nylon filters, grown on antibiotic plates, and processed for colony hybridization. This method is cheaper and more convenient, but it is less sensitive and gives a higher background than PCR-based DNA dot blots. The DNA dot blot approach is highly recommended (Protocol 5).

PROTOCOL 5   PCR-BASED DNA DOT BLOT

For high-throughput screening, a 96-well or 384-well format PCR from one of several thermal cycler manufacturers is recommended. Alternatively, single tubes can be used.

STAGE 1: AMPLIFICATION OF DNA INSERTS BY PCR

MATERIALS

BUFFERS, SOLUTIONS, AND REAGENTS

dNTP solution (containing all four dNTPs, each at 10 mM)

ENZYMES AND ENZYME BUFFERS

10x PCR buffer (40 mM Tricine-KOH<!>, pH 9.2 at 22°C, 3.5 mM magnesium acetate<!>, 10 mM potassium acetate, 75 mg/ml BSA, or as supplied by the manufacturer)

Polymerase mixture, 50x (Advantage 2, Clontech, or equivalent)

NUCLEIC ACIDS AND OLIGONUCLEOTIDES

Nested Primer NP1*

Nested Primer NP2R*

Alternatively, primers flanking the insertion site of the vector can be used in PCR amplification of the inserts.

MEDIA

LB-amp liquid medium

For 1 liter, dissolve 10 g of Bacto-tryptone, 5 g of Bacto-yeast extract, and 10 g of NaCl in 950 ml of deionized H_2O. Adjust the pH to 7.0 with 5 N NaOH<!>. Make up the volume to 1 liter with deionized H_2O. Sterilize by autoclaving for 20 minutes at 15 lb/in² on liquid cycle. When autoclaved medium has cooled, add ampicillin to a final concentration of 50 μg/ml.

SPECIAL EQUIPMENT

Thermal cycler

At this stage, the PCR is in plate format, and the precision required earlier is no longer required and it is not necessary to use mineral oil.

Tubes or reaction plate

ADDITIONAL ITEMS

Equipment and reagents for agarose gel electrophoresis<!>

CELLS AND TISSUES

cDNA library, grown as colonies on LB agar plates in an appropriate bacterial strain

METHOD

1. Randomly pick 96 or 384 white bacterial colonies from the library.

2. Grow each colony in 150 μl of LB-amp medium in a plate at 37°C for at least 4 hours (up to overnight) with gentle shaking.

3. Prepare a master mix for 100 or 400 PCRs:

Component	Amount per reaction
Sterile H_2O	15.0 μl
10x PCR buffer	2.0 μl
dNTP solution (10 mM)	0.4 μl
Nested Primer NP1*	0.6 μl
Nested Primer NP2R*	0.6 μl
50x polymerase mixture	0.4 μl
Total volume	19.0 μl

*Alternatively, primers flanking the insertion site of the vector can be used in PCR amplification of the inserts.

4. Aliquot 19 μl of the master mix into each tube or well of the reaction plate.

5. Transfer 1 μl of each bacterial culture (from step 2, above) to each tube or well containing master mix.

6. Perform PCR in an oil-free thermal cycler with the following conditions:

Cycle number	Denaturation	Annealing	Polymerization/Extension
1	2 min at 95°C		
22	10 sec at 95°C	10 sec at 68°C	3 min at 72°C

7. Analyze 4 μl from each reaction on a 2.0% agarose gel.

STAGE 2: DOT BLOT ANALYSIS

MATERIALS

BUFFERS, SOLUTIONS, AND REAGENTS
0.6 M NaOH<!> (freshly made or at least freshly diluted from concentrated stock)
SSC, 20x (for 1 liter:175.3 g of NaCl and 88.2 g of sodium citrate<!>, pH 7.0)
0.5 M Tris-HCl<!>, pH 7.5

NUCLEIC ACIDS AND OLIGONUCLEOTIDES
PCR product

SPECIAL EQUIPMENT
Microtiter dish
Nylon membrane
UV crosslinking device<!> (such as Stratagene's UV Stratalinker), or 68°C oven
96-well or 384-well replicator

METHOD

1. For each PCR, combine 5 μl of the PCR product and 5 μl of 0.6 M NaOH.

2. Transfer 1–2 μl of each mixture to a nylon membrane. This can be accomplished by dipping a 96-well or 384-well replicator in the corresponding wells of a microtiter dish used in the PCR amplification and spotting it onto a dry nylon filter. Make at least two identical filters for hybridization with subtracted and reverse-subtracted probes (Protocol 1, Stage 3).

   We highly recommend that you make four identical filters. Two of the filters will be hybridized to forward- and reverse-subtracted DNAs and the other two can be hybridized to initial cDNAs or genomic DNAs.

3. Neutralize the blots for 2–4 minutes in 0.5 M Tris-HCl, pH 7.5, and wash in 2x SSC.

4. Immobilize DNA on the membrane using a UV crosslinking device (such as Stratagene's UV Stratalinker), or bake the blots for 4 hours at 68°C.

PROTOCOL 6   DIFFERENTIAL HYBRIDIZATION WITH TESTER AND DRIVER DNA PROBES

The following four probes are used for differential screening hybridization:

1. Tester-specific subtracted probe (forward-subtracted probe)

2. Driver-specific subtracted probe (reverse-subtracted probe)

3. cDNA probe synthesized directly from tester mRNA (or tester genomic DNA)

4. cDNA probe synthesized directly from driver mRNA (or driver genomic DNA)

   The first two probes are the secondary PCR products (Protocol 3, step 15) of the subtracted DNA pool. The last two cDNA probes can be synthesized from the tester and driver poly(A)$^+$ RNA. They can be used as either single-stranded or double-stranded cDNA probes (Protocol 1, Stage 2). Alternatively, unsubtracted tester and driver DNA (Protocol 3, step 15) or preamplified cDNA from total RNA (Chenchik et al. 1998) can be used if enough poly(A)$^+$ RNA is not available. If you have made the MOS-subtracted library, you can still screen it using the same probes. Do not use MOS PCR products (Protocol 4, Stage 3, step 6) for probe preparation.

MATERIALS

BUFFERS, SOLUTIONS AND REAGENTS

High-stringency wash buffer (0.2x SSC, 0.5% SDS<!> [w/v]) prewarmed to 68°C (*optional*, see step 9)

Low-stringency wash buffer (2x SSC, 0.5% SDS [w/v]) prewarmed to 68°C (*optional*, see step 9)

SDS, 0.5% (*optional*, see step 9)

SSC, 20x (for 1 liter: 175.3 g of NaCl and 88.2 g of sodium citrate<!>, pH 7.0)

NUCLEIC ACIDS AND OLIGONUCLEOTIDES

Blocking solution (containing 2 mg/ml unpurified NP1, NP2R, cDNA synthesis primers [in case of cDNA] and their complementary oligonucleotides)

Sheared salmon sperm DNA (10 mg/ml)

PROBES

Purified tester and driver probes labeled by random primer labeling (at least 10^7 cpm per 100 ng of subtracted DNA)<!> (*optional*, see step 9)

SPECIAL EQUIPMENT

Boiling water bath

Hybridization oven, preset to 72°C

ADDITIONAL ITEMS

ExpressHyb Hybridization Kit (Clontech)

Reagents and equipment for autoradiography

METHOD

1. Label tester and driver DNA probes by random-primer labeling using a commercially available kit. The hybridization conditions given here have been optimized for Clontech's ExpressHyb solution; the optimal hybridization conditions for other systems should be determined empirically.

2. Prepare a sufficient volume of prehybridization solution for each membrane.

   a. Combine 50 µl of 20x SSC, 50 µl of sheared salmon sperm DNA (10 mg/ml), and 10 µl of blocking solution (containing 2 mg/ml of unpurified NP1, NP2R, cDNA synthesis primers [in case of cDNA] and their complementary oligonucleotides).

   b. Boil the blocking solution for 5 minutes, then chill on ice.

   c. Combine the blocking solution with 5 ml of ExpressHyb Hybridization Solution (Clontech).

3. Place each membrane in the prehybridization solution prepared in step 2. Prehybridize for 40–60 minutes with continuous agitation at 72°C.

It is important to add blocking solution to the prehybridization solution. Because subtracted probes contain the same adapter sequences as arrayed clones, these probes hybridize to all arrayed clones regardless of their sequences.

4. Prepare hybridization probes.

   a. Mix 50 µl of 20x SSC, 50 µl of sheared salmon sperm DNA (10 mg/ml), 10 µl of blocking solution, and purified probe (at least 10^7 cpm per 100 ng of subtracted DNA). Make sure the specific activities of each probe are approximately equal.

   b. Boil the probe for 5 minutes, then chill on ice.

   c. Add the probe to the prehybridization solution.

5. Hybridize overnight with continuous agitation at 72°C.

6. Prepare low-stringency (2x SSC, 0.5% SDS) and high-stringency (0.2x SSC, 0.5% SDS) washing buffers and warm them to 68°C.

7. Wash the membranes with low-stringency buffer (4 x 20 minutes at 68°C), and then wash with high-stringency buffer (2 x 20 minutes at 68°C).

8. Perform autoradiography.

9. If desired, remove probes from the membranes by boiling for 7 minutes in 0.5% SDS.

   Blots can typically be reused at least five times.
   To minimize hybridization background, store the membranes at –20°C in Saran Wrap when they are not in use.

RESULTS AND DISCUSSION

The utility of SSH for comparison has been shown for (1) cDNA subtraction of FACS (fluorescence-activated cell sorter) sorted mammalian stem cells and differentiated progenitors; (2) subtraction of genomic DNA from different bacterial strains; and (3) subtraction of genomic DNA of two closely related strains of freshwater planaria *Girardia tigrina* (*Platyhelminthes, Turbellaria, Tricladida*).

Subtractive Cloning of Mammalian cDNA

Subtracted libraries for mouse stem cells (ST) and differentiated progenitor cells (PR) were constructed using the SSH protocol (A. Terskikh et al., in prep.).To create the ST subtracted library, RNA from 25,000 FACS-multiparameter double-sorted mouse stem cells served as the tester and RNA from 25,000 FACS-differentiated progenitor cells served as the driver. cDNAs were synthesized and amplified as described by Chenchik et al. (1998). The ST subtracted library was thus enriched for genes up-regulated in stem cells and down-regulated in progenitor cells. To create the PR subtracted library, the opposite combination of tester and driver was taken, with the resulting library enriched for the genes down-regulated in stem cells and up-regulated in progenitor cells. Each subtracted cDNA pool was then cloned into the pUC-based cloning vector.

We randomly picked 576 white colonies (6 x 96-well plates) from each subtracted library. Equal amounts of PCR-amplified inserts from the ST subtracted library and PR subtracted library were arrayed on nylon membranes for screening. Duplicate membranes were prepared from each 96-well plate. Each membrane was hybridized with either ST- or PR-subtracted cDNA. The genes that were enriched in one subtracted library, but not in the other, yielded a significantly higher hybridization signal (Fig. 4). These genes were strong candidates for genuine differential expression.

These results reveal the following types of clones:

1. Clones hybridizing to one probe but not to the other. These clones correspond to the differentially presented cDNA, subject to confirmation by northern blot analysis.

2. Clones hybridizing to both subtracted probes with the same efficiency. These clones do not correspond to the differentially presented cDNA; this is background.

3. Clones hybridizing to both subtracted probes with different hybridization efficiencies. This difference may be a result of random fluctuation and usually does not represent differential clones for genomic DNA subtractions. However, in the case of cDNA subtraction, these clones may represent transcripts that are differentially expressed, but closely related. Such clones should be further analyzed by northern blot or RT-PCR.

4. Clones that do not appear to hybridize to either probe. These clones may not contain DNA inserts or may be present at very low concentration in the subtracted probe. In most cases, they do not represent differentially presented sequences.

The ST subtracted library was found to contain 13% differentially expressed clones, whereas the PR subtracted library was found to contain 2% differentially expressed clones. Differential expression of the selected candidates was confirmed by virtual northern blot analysis (Franz et al. 1999) (data not shown). Virtual northern blot analysis showed no false-positive clones in either library.

Table 1 shows a summary of the differential screening results and the sequence analysis of clones isolated from the ST and PR subtracted libraries. Of 37 nonredundant clones isolated from both subtracted libraries, 11 represented known genes, 18 corresponded to mouse ESTs, and 8 corresponded to new genes.

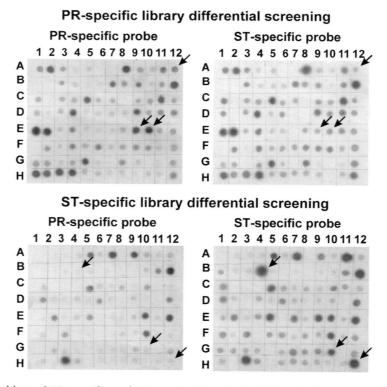

FIGURE 4. Dot blots of PR-specific and ST-specific libraries hybridized with DNA probes made from forward-subtracted DNA and reverse-subtracted DNA. Arrows indicate possible differential clones.

TABLE 1. Summary of ST- and PR-specific libraries

Library	Number of screened clones	Number of positive clones	Number of nonredundant sequences	Homology with known genes			
				exact mouse/rat match	mouse EST match	non-mouse match	no homology
ST cell-specific	576	77	29	7	15	1	6
PR cell-specific	576	14	8	3	3	0	2

Subtractive Cloning of Bacterial Genes

To demonstrate the potential of the SSH technique for bacterial genome subtractions, we performed a comparison of two *Staphylococcus aureus* methicillin-resistant and methicillin-sensitive bacterial strains (D. Rebrikov et al., in prep.). The bacterial genomes of these two strains have been completely sequenced and the subtracted library provides an excellent reference for testing the ability of SSH to recover all strain-specific gene content.

Whole-genome sequences of two related *S. aureus* strains (29213 and ZW) were performed using a random, shotgun approach at 99%. Methicillin-resistant *S. aureus* strain ZW (isolated at the University of Chicago) was used as tester, and *S. aureus* strain 29213 (methicillin-sensitive strain, ATCC number 29213) was used as the driver. Based on a comparison of the genome sequences of ZW and 29213 strains, 79 loci were identified as containing ZW-unique genes. These were found to be largely the result of two classes of events: integration of phage DNA and gene transfer from a related bacterial species. Indeed, the

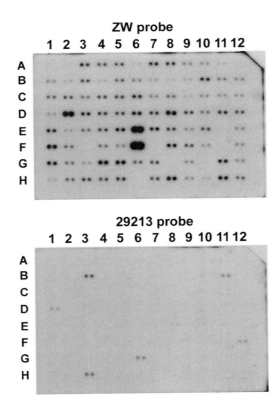

FIGURE 5. Dot blots of ZW-specific library hybridized with DNA probes made from ZW and 29213 *Rsa*I-digested genomic DNAs.

remarkable ability of *S. aureus* to acquire useful genes from various organisms has been reported previously (Kuroda et al. 2001).

The SSH approach was applied to the same *S. aureus* strains. Differential screening of the ZW-specific SSH-generated library has shown that 95% of clones present were ZW-specific (Fig. 5). A total of 289 randomly chosen clones were selected for sequencing and analysis. Only 11 sequences were found to be present in both the ZW and 29213 strains, representing less than 4% of the subtracted library. Another 278 clones represented 88 ZW-specific loci. These loci were subdivided into four classes:

1. Clones containing fragments unique to the tester (ZW) genome (69 loci)

2. Clones containing fragments of paralogous genes representing approximately 50–60% homology between the 29213 and ZW strains (6 loci)

3. Clones containing fragments of a transposon with a different adjustment area in the 29213 and ZW genomes (8 loci)

4. Clones containing ZW fragments where one or both *Rsa*I sites were found to have been lost in the 29213 strain as a result of point mutations (5 loci)

Figure 6 shows the identification of ZW-specific loci as a function of the number of sequenced clones. The curve begins to plateau as more products represent loci that have already been identified. A total of 75% of all ZW-specific loci were identified after 120 clones were sequenced, with the percentage reaching 100% after 250 clones had been sequenced.

The SSH approach thus yielded a subtracted library containing 69 ZW-unique genes, identified using the shotgun sequencing approach. Ten loci that were expected to be ZW-specific based on whole-genome sequencing were not found in the ZW-specific SSH-subtracted library. Of these ten loci, we chose four loci and examined the strain specificity of each by Southern blot analysis. All four of the examined fragments showed uniform distribution between the strains (data not shown).

This result demonstrates that uniform coverage of a bacterial genome is more effective when performed by SSH. There is no particular advantage for recovering certain fragments over others. More importantly, these results demonstrate that the sequencing of only 250 DNA clones from a SSH-subtracted library provides the same information, in terms of strain-specific differences, as the complete sequencing of two comparative bacterial genomes.

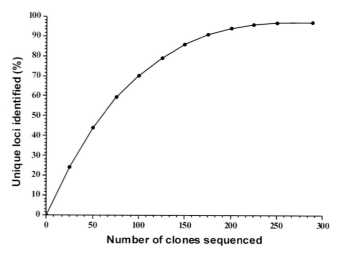

FIGURE 6. Percentage of unique loci identified by SSH with increasing number of sequenced clones.

Subtractive Cloning of Eukaryotic Genes

To illustrate the utility of combining SSH and MOS for eukaryotic genome comparison, we describe our efforts to isolate genes that are present in one freshwater planaria strain but are absent in another (Rebrikov et al. 2002). In this study, we used two closely related strains of freshwater planaria *G. tigrina* that reproduce in different ways. One strain has exclusively asexual reproduction, the other reproduces both sexually and asexually. We compared the genomes of the two strains of *G. tigrina* to search for genetic determinants of asexuality.

Total DNA from these strains was purified using the procedure described by Chomczynski and Sacchi (1987), modified for DNA preparation. The SSH and MOS combination was used to isolate genes that are differentially present in each planaria strain. Forward subtraction was performed using asexual DNA as tester and sexual DNA as driver (AB subtraction), and the forward-subtracted DNA was enriched for DNA fragments specific to the asexual strain of freshwater planaria. Reverse-subtracted DNA (BA subtraction) was enriched for DNA fragments specific to the sexual planaria strain. Self-subtractions were performed for both DNA samples to get a quick idea of the subtraction efficiency. Subsequent MOS-PCR analysis confirmed that the self-subtractions (as well as undigested controls) require more PCR cycles to generate visible PCR product, indicating that the subtraction had been successful (Fig. 7).

We anticipated that the differences between tester and driver DNA would be small, so we proceeded with a differential screening procedure (data not shown). Eighty-six randomly selected clones from each (forward- and reverse-subtracted) library were arrayed

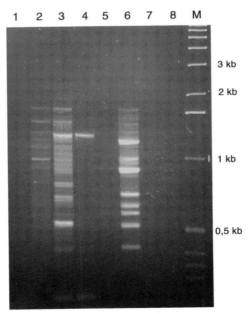

FIGURE 7. Typical results of the experimental and control self-subtractions. The MOS-PCR product of the AB and BA subtractions contains a lot of unique DNA fragments. (*Lane M*) 1-kb DNA ladder (Gibco BRL), (*lane 1*) MOS PCR product of undigested control of BB subtraction, (*lane 2*) MOS PCR product of BB subtraction, (*lane 3*) MOS PCR product of BA subtraction, (*lane 4*) MOS PCR product of of undigested control of BA subtraction, (*lane 5*) MOS PCR product of undigested control of AB subtraction, (*lane 6*) MOS PCR product of AB subtraction, (*lane 7*) MOS PCR product of AA subtraction, (*lane 8*) MOS PCR product of undigested control of AA subtraction.

(DNA dot blot) on nylon membranes. DNA dot blots were hybridized to probes prepared from the subtracted and reverse-subtracted libraries.

Differential screening revealed that ~60% and 30% of the clones in the AB and BA libraries, respectively, were strain-specific. Most of the nondifferential DNA sequences were identified as the mariner element, ~7000 copies of which were present in each compared genome (Garcia-Fernandez et al. 1993). About 50% of the asexual-specific clones turned out to be a novel extrachromosomal DNA-containing virus-like element. Several unique strain-specific genes were identified as lectins.

CONCLUSION

We have used SSH in our laboratory for more than 10 years in studies of regeneration and development on various types of model organisms (including freshwater planaria regeneration, *Xenopus laevis* development, and mammalian brain cortex development). We also use SSH for the analysis of strain-specific genes in bacteria with different characteristics. During these studies, a large number of differentially regulated and differentially presented genes have been identified, including transcriptional regulation factors and restriction modification enzymes (see http://www.ibch.ru:8101/~lgr/pub.htm).

To obtain the maximum data from a cDNA or genomic DNA subtraction experiment, it is important to achieve the highest efficiency of subtraction. The power of SSH subtraction makes it possible to achieve a level of 90–95% differentially expressed clones in the cDNA-subtracted library (Diatchenko et al. 1996; Zuber et al. 2000). In cases where differentially expressed clones represent the majority of the clones in the subtracted library, the time-consuming process of differential screening can be omitted. Whenever possible, the researcher should consider designing the experiment to yield the highest level of difference between the tester and driver RNA populations, possibly by choosing the time point with the highest fold induction of control gene, for example. In the case of a sample comprising a mixed cell population, homogeneity can often be achieved by fine dissection of fixed or frozen tissues and/or by cell sorting. However, when this is not possible, the MOS procedure should be applied.

Typically, it is necessary to analyze 500–1000 clones from a subtracted library to ensure that genes representing low-abundance transcripts are not lost. Sequence data from various studies show that the majority of the clones will be picked repeatedly, two to six times, indicating a degree of redundancy. This finding confirms the high level of normalization of SSH libraries, suggesting that the libraries contain both high- and low-abundance differentially presented DNAs.

REFERENCES

Akopyants N.S., Fradkov A., Diatchenko L., Hill J.E., Siebert P.D., Lukyanov S.A., Sverdlov E.D., and Berg D.E. 1998. PCR-based subtractive hybridization and differences in gene content among strains of *Helicobacter pylori*. *Proc. Natl. Acad. Sci.* **95:** 13108–13113.

Chenchik A., Zhu Y.Y., Diatchenko L., Li R., Hill J., and Siebert P.D. 1998. Generation and use of high-quality cDNA from small amounts of RNA by SMART PCR. In *Gene cloning and analysis by RT-PCR* (eds. P.D. Siebert and J.W. Larrick), pp. 305–319. BioTechniques Press, Natick, Massachusetts.

Chomczynski P. and Sacchi N. 1987. Single-step method of RNA isolation by acid guanidinium thiocyanate-phenol-chlorophorm extraction. *Anal. Biochem.* **162:** 156–159.

Diatchenko L., Lau Y.F., Campbell A.P., Chenchik A., Moqadam F., Huang B., Lukyanov S., Lukyanov K., Gurskaya N., Sverdlov E.D., and Siebert P.D. 1996. Suppression subtractive hybridization: A method for generating differentially regulated or tissue-specific cDNA probes and libraries. *Proc. Natl. Acad. Sci.* **93:** 6025–6030.

Farrell R.E. Jr. 1993. *RNA methodologies: A guide for isolation and characterization.* Academic Press, San Diego, California.

Franz O., Bruchhaus I.I., and Roeder T. 1999. Verification of differential gene transcription using virtual northern blotting. *Nucleic Acids Res.* **27:** e3.

Garcia-Fernandez J., Marfany G., Baguna J., and Salo E. 1993. Infiltration of mariner elements. *Nature* **364:** 109–110.

Gubler U. and Hoffman B.J. 1983. A simple and very efficient method for generating cDNA libraries. *Gene* **25:** 263–269.

Gurskaya N.G., Diatchenko L., Chenchik A., Siebert P.D., Khaspekov G.L., Lukyanov K.A., Vagner L.L., Ermolaeva O.D., Lukyanov S.A., and Sverdlov E.D. 1996. Equalizing cDNA subtraction based on selective suppression of polymerase chain reaction: Cloning of Jurkat cell transcripts induced by phytohemaglutinin and phorbol 12-myristate 13-acetate. *Anal. Biochem.* **240:** 90–97.

Jin H., Cheng X., Diatchenko L., Siebert P.D., and Huang C.C. 1997. Differential screening of a subtracted cDNA library: A method to search for genes preferentially expressed in multiple tissues. *BioTechniques* **23:** 1084–1086. [Au: Not cited in text-delete ref or add citation]

Kuroda M., Ohta T., Uchiyama I., Baba T., Yuzawa H., Kobayashi I., Cui L., Oguchi A., Aoki K., Nagai Y., Lian J., Ito T., Kanamori M., Matsumaru H., Maruyama A., Murakami H., Hosoyama A., Mizutani-Ui Y., Takahashi N.K., Sawano T., Inoue R., Kaito C., Sekimizu K., Hirakawa H., Kuhara S., Goto S., Yabuzaki J., Kanehisa M., Yamashita A., Oshima K., Furuya K., Yoshino C., Shiba T., Hattori M., Ogasawara N., Hayashi H., and Hiramatsu K. 2001. Whole genome sequencing of methicillin-resistant *Staphylococcus aureus*. *Lancet* **357:** 1225–1240.

Lukyanov S.A., Gurskaya N.G., Lukyanov K.A., Tarabykin V.S., and Sverdlov E.D. 1994. Highly efficient subtractive hybridization of cDNA. *J. Bioorg. Chem.* **20:** 386–388.

Rebrikov D.V., Bulina M.E., Bogdanova E.A., Vagner L.L., and Lukyanov S.A. 2002. Complete genome sequence of a novel extrachromosomal virus-like element identified in planarian *Girardia tigrina*. *BMC Genomics* **3:** 15.

Rebrikov D.V., Britanova O.V., Gurskaya N.G., Lukyanov K.A., Tarabykin V.S., and Lukyanov S.A. 2000. Mirror orientation selection (MOS): A method for eliminating false positive clones from libraries generated by suppression subtractive hybridization. *Nucleic Acids Res.* **28:** e90.

Sambrook J., Fritsch E.F., and Maniatis T. 1989. *Molecular cloning: A laboratory manual*, 2nd edition. Cold Spring Harbor Laboratory Press, Cold Spring Harbor, New York.

Zuber J., Tchernitsa O.I., Hinzmann B., Schmitz A.C., Grips M., Hellriegel M., Sers C., Rosenthal A., and Schafer R. 2000. A genome-wide survey of RAS transformation targets. *Nat. Genet.* **24:** 144–152.

WWW RESOURCE

http://www.ibch.ru:8101/~lgr/pub.htm Russian Academy of Science, Shemyakin & Ovchinikov Institute of Bioorganic Chemistry, Laboratory of Genes for Regeneration Publications 1994–2002.

23 Fluorescent mRNA Differential Display in Gene Expression Analysis

Jeffrey S. Fisher, Jonathan Meade, Jamie Walden, Zhen Guo, and Peng Liang[1]

GenHunter Corporation, Nashville, Tennessee and [1]Vanderbilt University, Nashville, Tennessee 37232

How can a fertilized egg, containing a complete set of genes unique to a species, give rise to many different cell types that will ultimately organize into the different tissues and organs that define each specific organism? This has been one of the most intriguing questions in biology. Even the complete sequencing of many genomes, from a few thousand base pairs for bacteria to over 3 billion base pairs for human, has yet to provide answers to the mystery of life. Of the estimated 30,000–35,000 genes embedded in the human genome, only 10–15% are "turned on" (expressed as mRNAs for protein synthesis) at any given time in each of our cells. Thus, interpretation of the genomic instruction will have to rely on tools that can allow us to determine when and where a gene is to be turned on or off as a cell divides, differentiates, and ages. These tools are also powerful for detecting when and where the seemingly precise genomic instruction goes awry, as in many disease states such as cancer. Differential display (DD) technology is a major tool for interpreting the genomic information (Liang and Pardee 1992).

DIFFERENTIAL DISPLAY

DD technology continues to be one of the most reliable methods for gene expression analysis available to biomedical researchers. Since its invention in 1992, the number of publications using DD has exploded to more than 3800, outnumbering the combined total of publications using competitive methodologies such as DNA microarrays, serial analysis of gene expression (SAGE), and representational difference analysis (RDA). The rapid and successful adoption of DD has been largely attributed to its simplicity, which ensures a higher probability of success and few artifactual differences caused by experimental errors. Essentially, starting from the RNA samples being compared, only two steps—reverse transcription and PCR—are needed before the results are analyzed on a gel matrix. No second-strand DNA synthesis, purification of cDNA, restriction enzyme digestion or adapter primer ligation, probe labeling and normalization, hybridization, or washing steps are required. Each of these steps could introduce and amplify errors or lead to the loss of mRNAs being detected.

FLUORESCENT DIFFERENTIAL DISPLAY

Fluorescent differential display (FDD) (Fig. 1), also referred to as FDDRT-PCR in PCR nomenclature (Bauer et al. 1993; Liang et al. 1995), takes advantage of three of the most simple, powerful, and commonly used molecular biological methods: RT-PCR, DNA sequencing gel electrophoresis, and cDNA cloning (Liang and Pardee 1992; Cho et al. 2001). FDD is an improvement over conventional DD technology because it makes use of fluorochrome-labeled primers rather than radioactively labeled primers in the FDDRT-PCRs. Thus, the hazards traditionally associated with isotopic DD are eliminated, allowing easier and faster acquisition of gene expression data. As with other differential gene expression technologies, FDD may use any of a variety of confirmation techniques, including, but not limited to, northern blot analysis, reverse northern blot analysis, and quantitative RT-PCR (qRT-PCR).

Primer Design and Amplification

FDD methodology starts with the harvesting of total RNA from the cells of interest. After first-strand cDNA synthesis by reverse transcription, the messenger RNAs (mRNAs) within the total RNA population are used as the templates for FDD-PCR. The current methodology makes use of three anchor primers (that target the polyadenylation site of eukaryotic mRNA) of the type H-T_{11}M, where H is a sequential HindIII restriction site (AAGCTT) and M can be G, C, or A. The sequential HindIII restriction site was added to the anchor primer design to make the primers longer and more efficient in annealing to the targeted poly(A) site, as well as improving downstream applications such as cDNA cloning. Using the current anchor primer design, the cDNA populations are subsequently divided into three subpopulations that, in theory, each represent one-third of the potential mRNA expressed in the cell at any given time (Liang et al. 1994). Previous work indicated using anchor primers of the type T_{11}VN, where V can be A, G, or C, and N can be any of the four nucleotides, as well as anchors of the type T_{12}MN, where M is a degenerate mixture of A, G, or C, and N is any of the four nucleotides (Liang and Pardee 1992). Both of these primer designs result in larger subfractions of the mRNA population (12 for type T_{11}VN and 4 for T_{12}MN), and, in combination with the arbitrary primers utilized in subsequent PCRs, unnecessarily increases the amount of FDD-PCRs for the same level of gene coverage versus the H-T_{11}M primer design.

The next step in FDD is the amplification of the cDNA subpopulations employing a combination of a fluorescently labeled anchor primer (FH-T_{11}M) and a set of second primers that are arbitrary and short in length (Table 1). The design of the arbitrary 13-mers (H-AP) utilized in FDD technology includes a 7-bp backbone of random base combinations, and, similar to that of the anchor primer, a sequential HindIII restriction site (AAGCTT) for more efficient primer annealing and easier downstream manipulation of the cDNA (Liang et al. 1994). The primers used in FDD represent a random selection from over 16,000 (4^7) base-pair combinations. Additionally, the length of an arbitrary primer is designed so that, by probability, each will recognize 50–100 mRNAs under a given PCR condition (Liang et al. 1994). As a result, mRNA 3′ termini defined by any given pair of anchored primer and arbitrary primer combination are amplified and then displayed by denaturing polyacrylamide gel electrophoresis. A mathematical model of estimated gene coverage using various combinations of anchor and arbitrary primers was developed shortly after the advent of DD technology (Liang et al. 1994). This mathematical model indicates that ~240

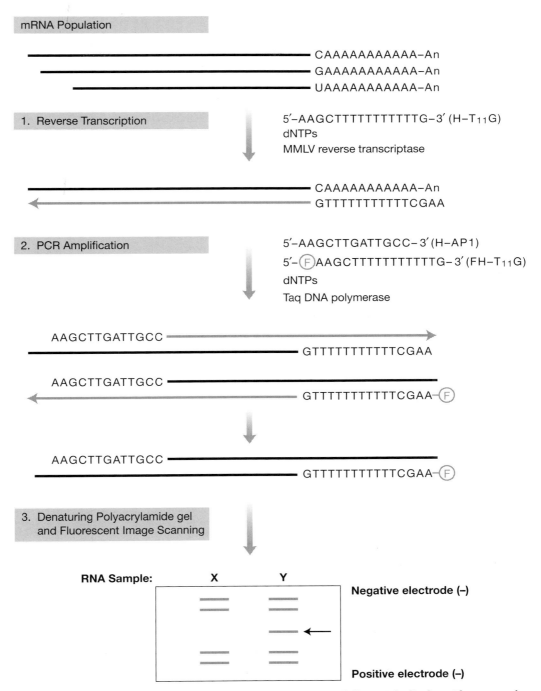

FIGURE 1. Schematic representation of fluorescent mRNA differential display. Three one-base anchored oligo(dT) primers with 5′ *Hin*dIII sites are used in combination with a series of arbitrary 13-mers (also containing 5′ *Hin*dIII sites) to reverse-transcribe (I) and amplify (II) the mRNAs from a cell. Fluorescently labeled anchor primers are used in FDD-PCR (III).

primer combinations need to be used to approach the level of estimated genome-wide screening for eukaryotes (>95%). Fewer primer combinations are expected to provide lower estimated gene coverage (Fig. 2).

TABLE 1. List of 13-mer upstream primers (H-AP) used for FDDRT-PCR

No.	Sequence (5′ to 3′)
H-AP1	AAGCTTGAT TGCC-
H-AP2	AAGCTTCGA CTGT-
H-AP3	AAGCTTTGG TCAG
H-AP4	AAGCTTCTCAACG
H-AP5	AAGCTTAGTAGGC
H-AP6	AAGCTTGCACCAT
H-AP7	AAGCTTAACGAGG
H-AP8	AAGCTTTTACCGC
H-AP9	AAGCTTCATTCCG
H-AP10	AAGCTTCCACGTA
H-AP11	AAGCTTCGGGTAA
H-AP12	AAGCTTGAGTGCT
H-AP13	AAGCTTCGGCATA
H-AP14	AAGCTTGGAGCTT
H-AP15	AAGCTTACGCAAC
H-AP16	AAGCTTTAGAGCG
H-AP17	AAGCTTACCAGGT
H-AP18	AAGCTTAGAGGCA
H-AP19	AAGCTTATCGCTC
H-AP20	AAGCTTGTTGTGC
H-AP21	AAGCTTTCTCTGG
H-AP22	AAGCTTTTGATCC
H-AP23	AAGCTTGGCTATG
H-AP24	AAGCTTCACTAGC

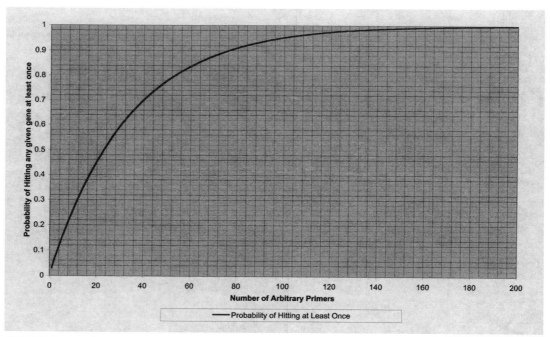

FIGURE 2. Graphic representation of the estimated level of gene coverage using FDD technology. Based on published calculations (Liang et al. 1994), the potential of hitting a specific mRNA increases with an increase in number of arbitrary 13-mers used in FDD-PCRs. For ~95% gene coverage, a minimum of 240 reactions, using the three anchored oligo(dT)s in combination with 80 arbitrary 13-mers, must be performed.

Analysis of DDRT-PCR Products

Gel electrophoresis can be performed with sequencing gels (Liang and Pardee 1992; Liang et al. 1993), nondenaturing polyacrylamide gels (Bauer et al. 1993), or agarose gels (Hsu et al. 1993; Sokolov and Prockop 1994). Sequencing gels are recommended here because of their ability to accommodate a larger number of reactions, thereby reducing the number of gels that must be used for FDD analysis. Compared to agarose gels, the use of sequencing gels (less specifically polyacrylamide) results in better band resolution due to increased sensitivity, and also allows easier and more efficient recovery of genes. Side-by-side comparison of resulting cDNA patterns between or among relevant RNA samples reveals differences in the gene expression profile for each sample. Because the resulting cDNAs are fluorescently labeled, the use of a fluorescent imager scanner is required for this technology. We recommend using the FMBio Series (Miraibio, Alameda, CA) for digital acquisition of the cDNA profiles. Other fluorescent imagers, such as the Typhoon 8600 (Molecular Dynamics, Piscataway, NJ), are also compatible with FDD.

Cloning and Reamplification of Products

After completion of the gene expression profiles by electrophoresis, the next step is to begin characterization of the potential differentially expressed genes of interest. After excising the bands from the gel matrix, each cDNA must be reamplified with the same primer combination as the original FDD-PCR and under the same reaction conditions. The reamplification fragments are subsequently cloned into a suitable vector. We recommend the PCR-TRAP Cloning System (GenHunter Corporation, Nashville, TN) because it is designed specifically for cloning genes that use the primer designs of FDD previously mentioned. Because of the potential for more than one distinct cDNA being contained within an excised band, more than one colony should be screened for the correctly sized insert. Furthermore, if the screening results indicate that more than one cDNA is present in the colony population, each of the different cDNAs should then be further characterized.

Confirmation of Results

Characterization of each potential gene begins with sequencing of the cloned cDNAs of interest, with the results giving an indication of whether the cDNA is a known or unknown sequence. Additionally, one must be sure that the characterized sequences are actually differentially regulated and not artifactual sequences. Although there are several assays to choose from, each with its own distinct advantages and disadvantages, we suggest the northern blot analysis, which is considered the gold standard for gene expression confirmation. The sensitivity of northern analysis is a distinct advantage over other confirmation methods, as both high- and low-level mRNA expression can be validated with this standard assay. Although it may be labor-intensive and time-consuming, northern blot analysis is by far the most accepted tool for confirmation.

PROTOCOLS

The FDD approach to differential gene expression analysis does not require any prior knowledge in mRNA sequences, thus making the gene screening systematic and nonbiased. Depending on the desired gene coverage, FDD methodology enables quicker results when compared to traditional isotopic DD or other DD-related technologies, yet ensures more reliable results when compared to microarray or other competing, non-DD technologies. Because each researcher will complete each protocol in a different time frame based on the success of that step, a time line for completing the entire methodology is dependent on many variables. However, if diligence is used and errors kept to a minimum, the researcher should be able to complete the entire process (from RNA extraction to confirmation of gene expression) within 3–5 days. The following protocols introduce the beginner to gene expression analysis using FDD technology.

PROTOCOL 1   RNA EXTRACTION AND PURIFICATION

Although FDD takes advantage of the polyadenylation (poly[A]$^+$) site of eukaryotic mRNA, poly(A)$^+$ RNA is not recommended because of possible contaminating oligonucleotides in the reverse transcription reactions, thus increasing the incidence of false positives. To this end, as well as to verify the integrity of the RNA after extraction, total RNA is suggested for FDD analysis. See Chapter 10 for alternate RNA extraction protocols.

MATERIALS

▼ CAUTION

See Appendix for appropriate handling of materials marked with <!>.

BUFFERS, SOLUTIONS, AND REAGENTS
  Chloroform<!>
  Diethyl pyrocarbonate-(DEPC) <!>-treated H$_2$O (GenHunter R105)
  Ethanol <!>
  70% Ethanol wash solution (prepared in DEPC-treated H$_2$O)
  Isopropanol <!>
  Phenol-guanidium<!> monophasic solution [RNApure, GenHunter P501-P503, is recommended]
  Phosphate-buffered saline (PBS)

CENTRIFUGES AND ROTORS
  Microfuge

SPECIAL EQUIPMENT
  50-ml conical tube
  1.5-ml microfuge tubes
  1000-ml pipette
  100- to 150-mm plates
  P10 pipette
  Polytron Homogenizer (for RNA extraction from tissue)

METHOD

STEP 1: EXTRACTION OF RNA

METHOD 1: Extraction of RNA from Tissue Cultures
  1. If the cells are attached to the flask or plate, pour off the medium. Set the plate on ice.
  2. If the cells are in suspension, spin down the cells and remove the medium.
  3. Rinse the cells with 10–20 ml of cold phosphate-buffered saline (PBS).

4. Pour off the PBS and remove the residual PBS with a 1000-μl pipette.

5. Add 2 ml of phenol-guanidium monophasic solution (RNApure) per 100- to 150-mm plate to lyse the cells (spread the solution by shaking the plate). This volume is sufficient for 1–10 million cells.

6. Let the plate sit on ice for 10 minutes.

7. Pipette the lysate into two labeled 1.5-ml microfuge tubes.

METHOD 2: Extraction of RNA from Tissues

1. Add at least 2 ml of phenol-guanidium monophasic solution (RNApure) to the tissue in a 50-ml conical tube on ice. Ideally, the volume ratio of tissue to phenol solution should be at least 1:10.

2. Homogenize the tissue with a Polytron Homogenizer until the tissue is dispersed.

3. Let the tube sit on ice for 10 minutes.

4. Transfer 1-ml aliquots of the lysate into labeled 1.5-ml centrifuge tubes.

METHOD 3: Extraction of RNA from Blood

1. Spin down blood products and remove the plasma.

2. Follow the instructions for Extraction of RNA from Tissues, above.

STEP 2: PURIFICATION OF RNA

1. Add 150 μl of chloroform per milliliter of lysate. Vortex for 10 minutes

   The Protocol can be stopped here by placing the lysates at –80°C.

2. Centrifuge the tubes at 4°C with maximum speed for 10 minutes.

3. Carefully remove the upper phase into a clean, labeled 1.5-ml centrifuge tube.

4. Add an equal volume of isopropanol and let the tube sit on ice for 10 minutes. Then mix vigorously or vortex for 30 seconds.

5. Centrifuge the mixture for 10 minutes at 4°C at maximum speed.

6. Rinse the RNA pellet with 1 ml of cold 70% ethanol (in DEPC-treated H_2O). Centrifuge for 2 minutes at 4°C at maximum speed.

7. Remove the ethanol. Spin briefly and remove the residual wash solution with a pipette.

8. Resuspend the RNA in 50 μl of DEPC-treated H_2O.

   Caution: Do not use SDS in resuspension if using RNA for any PCR application.

9. Measure the concentration by taking 1 μl of the RNA (using a P10 pipette) and diluting with 1 ml of H_2O. Read at 260 nanometers. Note that 1 OD_{260} = 40 μg.

PROTOCOL 2   REMOVAL OF GENOMIC DNA FROM TOTAL RNA

For the purposes of FDD gene expression analysis, as well as any other RNA-based gene expression technologies, contaminating genomic DNA must be removed before single-strand cDNA synthesis by reverse transcription and subsequent PCRs. If left unchecked, the contaminating DNA could potentially anneal to the primers within the FDDRT-PCRs, thus causing amplification of DNA sequences and leading to a high false-positive rate. Therefore, the following protocol for removal of the contaminating genomic DNA is ultimately the most important procedure in preventing any irregularities or artifacts during the FDDRT-PCRs. DNase removal of contaminating genomic DNA is also presented in Chapter 10.

MATERIALS

▼ CAUTION

See Appendix for appropriate handling of materials marked with <!>.

BUFFERS, SOLUTIONS, AND REAGENTS

Denaturing (formaldehyde<!>) agarose gel (see step 18 for preparation)
- 1 g of agarose
- 83 ml of distilled H_2O
- 12.3 M formaldehyde <!>

Distilled H_2O

Ethanol, 100% <!>

Ethanol, 70% in DEPC-treated H_2O<!>

12.3 M (37%) formaldehyde<!>, pH >4.0

10x MOPS buffer:
- 0.01 M ethylenediamine tetraacetic acid (EDTA)
- 0.2 M MOPS <!>
- 0.05 M sodium acetate<!>

Phenol/chloroform (3:1) solution:
- 10 ml of chloroform <!>
- 30 ml of melted phenol <!>
- 10 ml of Tris-HCl<!>, pH 7.0

Running buffer (see step 18f)
- 10 ml of 10x MOPS<!>
- 900 ml of distilled H_2O to 1x concentration

NUCLEIC ACIDS AND OLIGONUCLEOTIDES

RNA from Protocol 1

CENTRIFUGES AND ROTORS

Microfuge (we recommend MicroSpin 24, Sorvall Instruments, Wilmington, DE)

SPECIAL EQUIPMENT

1.5-ml centrifuge tubes
Gel casting plate
Microwave oven

ADDITIONAL ITEMS

MessageClean DNA Removal Kit (GenHunter M601), which includes a 10x Reaction buffer, RNase-free DNase I (10 units/μl), 3 M sodium acetate, diethyl pyrocarbonate (DEPC)-treated H_2O, and RNA Loading Mix. <!>

METHOD

DNase I Digestion of Total RNA

1. If necessary, dilute desired amount of RNA to be digested (maximum of 50 μg) with DEPC-treated H_2O to a maximum volume of 50 μl.

2. In a 1.5-ml centrifuge tube, add the following *in order* (total reaction volume is 56.7 μl):

Total RNA	50 μl
10x Reaction buffer	5.7 μl
DNase I (10 units/μl)	1.0 μl

3. Mix gently and incubate at 37°C for 30 minutes.

Extraction and Ethanol Precipitation of DNA-free RNA

4. Prepare the phenol/chloroform solution by melting crystalline phenol at 65°C in the supplied glass storage container.

5. Add 30 ml of melted phenol to 10 ml of chloroform and mix well.

6. Add 10 ml of Tris-HCl, pH 7.0, and mix well. Allow the saturation phase to form before using.

7. Add 40 µl of the phenol/chloroform solution to each DNase I reaction in the 1.5-ml centrifuge tube. Vortex for 30 seconds.

8. Let the solution sit on ice for 10 minutes.

9. Centrifuge at maximum speed for 5 minutes at 4°C.

10. Collect the upper phase and place it in a clean, labeled 1.5-ml microfuge tube.

11. Add 5 µl of 3 M sodium acetate and 200 µl of 100% ethanol. Mix well.

12. Let the mixture sit for at least 1 hour at –80°C.

13. Centrifuge the mixture at 4°C for 10 minutes at maximum speed to pellet the RNA.

14. Carefully remove the supernatant and rinse the RNA pellet with 0.5 ml of 70% ethanol (in DEPC-treated H_2O). Do not disturb the pellet.

15. Centrifuge the pellet for 5 minutes at 4°C at maximum speed and remove the supernatant. Centrifuge again briefly, removing the residual liquid without disturbing the RNA pellet.

16. Resuspend the RNA in 10–20 µl of DEPC-treated H_2O.

RNA Quantification and Integrity Verification

17. Quantitate by OD_{260} after 1:1000 dilution of the DNA-free RNA sample with distilled H_2O.

18. Prepare the denaturing (formaldehyde) agarose gel by the following protocol.

   a. Add the following to a microwave-safe container:

10x MOPS	10 ml
agarose	1 g
distilled H_2O	83 ml

   b. Microwave for ~3 minutes or until agarose is melted.

   c. Let agarose cool to at least 50°C.

   d. Add 7 ml of a 12.3 M (37%) formaldehyde solution. Gently mix.

   e. Pour into prepared gel casting plate and replace gel comb.

   f. Running buffer (1 liter) is made by diluting 100 ml of 10x MOPS with 900 ml of distilled H_2O to a 1x concentration. Cover the agarose gel with running buffer.

19. Check the integrity of the RNA by resolving 2–3 µg of both pre-DNase and post-DNase RNA samples on a 7% formaldehyde agarose gel with RNA loading mix by the following protocol.

   a. Add 1–10 µl (2–3 µg) of RNA to 20 µl of RNA loading mix in a labeled 1.5-ml microfuge tube. Mix well.

   b. Incubate at 65°C for 10 minutes.

c. Centrifuge sample briefly to collect condensation.

d. Put samples on ice for 5 minutes.

e. Load the entire amount onto a RNA gel.

f. Run at 50–60 V for ~45 minutes or until resolution of the ribosomal subunits is achieved.

20. The resulting ribosomal subunits will appear as crisp bands on the denaturing gel. RNA that has been degraded by the extraction or DNase treatment will appear as a smear. The size of the ribosomal subunits depends on the organism used for analysis.

PROTOCOL 3 SINGLE-STRAND cDNA SYNTHESIS BY REVERSE TRANSCRIPTION

The volume of the cDNA synthesis reaction depends on the number of anchor–arbitrary primer combinations being performed. For 24-primer combinations and 2 samples, the following protocol will be applicable. For more samples and/or increased primer combinations, adjust the volumes in the master mix accordingly.

MATERIALS

NUCLEIC ACIDS AND OLIGONUCLEOTIDES
RNA from Protocol 2

A separate RT master mix for each individual H-T_{11}M primer is constructed:

distilled H_2O	28.2 µl
5x RT buffer	12.0 µl
2.5 mM dNTP mix	4.8 µl
H-T_1M primer (where M = G, A, or C)	6.0 µl

SPECIAL EQUIPMENT
0.5-ml microfuge tube
0.2-ml thin-walled PCR tube (GenHunter T101)
Thermocycler (GeneAmp PCR System 9600 [Perkin-Elmer Corporation, Norwalk, CT] or Mastercycler [Eppendorf Scientific, Westbury, NY] recommended)

ADDITIONAL ITEMS

▼ CAUTION

See Appendix for appropriate handling of materials marked with <!>.

RNAspectra Fluorescent mRNA Differential Display System (GenHunter F501-F510 & R501-R510), which includes distilled H_2O, 5x RT buffer [125 mM Tris-HCl <!>, pH 8.3, 188 mM KCl <!>, 7.5 mM $MgCl_2$ <!>, and 25 mM dithiothreitol (DTT) <!>, FDD dNTP mix (2.5 mM), anchor primers (H-T_{11}M) (2 µM), and MMLV reverse transcriptase (100 units/µl).

METHOD

cDNA Synthesis

1. Prepare three reaction tubes (one for each of the three H-T_{11}M primers) as follows:

   a. Aliquot 42.5 µl of the RT master mix into a labeled, RNase-free, 0.2-ml thin-walled PCR tube. Remember to prepare a separate RT reaction for each individual anchor primer (H-T_{11}M) used!

2. Dilute the RNA to a final concentration of 0.1 μg/μl with DEPC-treated H$_2$O and mix thoroughly. Place on ice.

3. Add 5.0 μl of the diluted RNA (0.1 μg/μl) to the 0.5-ml tube and mix thoroughly.

4. Program your thermocycler as follows: 65°C for 5 min. → 37°C for 60 min. → 75°C for 5 min. → 4°C soak.

5. Place the tubes on the thermocycler and begin the program.

6. After the tubes have been at 37°C for 10 minutes, pause the thermocycler and add 2.5 μl of MMLV reverse transcriptase to each tube. Quickly mix well before continuing incubation.

7. At the end of the reverse transcription, spin the tube briefly at maximum speed to collect condensation.

8. Set the tubes on ice or store at –20°C for later use.

PROTOCOL 4   FDD PCR

This protocol is designed for 24 PCRs utilizing the three cDNA subpopulations (RT reactions using H-T$_{11}$M primers from Protocol 3) with a combination of fluorescent dye-labeled anchor primers (FH-T$_{11}$M) and 24 upstream arbitrary primers (H-AP). For a complete, genome-wide screening, 240 primer combinations must be completed. Therefore, this protocol will need to be repeated 10 times using varying anchor–arbitrary primer combinations.

MATERIALS

BUFFERS, SOLUTIONS, AND REAGENTS
Mineral oil (*optional*)

NUCLEIC ACIDS AND OLIGONUCLEOTIDES
cDNA populatons from Protocol 3

SPECIAL EQUIPMENT
Thermocycler (GeneAmp PCR System 9600 [Perkin-Elmer] or Mastercycler [Eppendorf Scientific] recommended)
0.2-ml thin-walled PCR tubes

▼ CAUTION

See Appendix for appropriate handling of materials marked with <!>.

ADDITIONAL ITEMS
RNAspectra Fluorescent mRNA Differential Display System (GenHunter F501-F510 & R501-R510), including distilled H$_2$O, 10x PCR buffer (100 mM Tris-HCl <!>, pH 8.4, 500 mM KCl <!>, 15 mM MgCl$_2$ <!>, and 0.01% gelatin), FDD dNTP mix (2.5 mM), fluorescent anchor primers (F-H-T$_{11}$M) (2 μM), and arbitrary primers (H-AP) (2 μM)

METHOD

1. Prepare a separate FDD-PCR master mix for each individual F-H-T$_{11}$M anchor primer in separate 1.5-ml centrifuge tubes:

distilled H$_2$O	91.8 µl
10x PCR buffer	18.0 µl
FDD dNTP mix	14.4 µl
FH-T$_{11}$M primer (where M = G, A, or C)	18.0 µl
Taq DNA polymerase	1.8 µl
(Qiagen 201207 Valencia, CA)	

2. Program your thermocycler to 94°C for 15 seconds → 40°C for 2 minutes → 72°C for 60 seconds → 40 cycles → 4°C soak. If you are not using the recommended thermocyclers, adjust the denaturation (94°C) time to 30 seconds.

3. Into individually labeled, 0.2-ml thin-walled PCR tubes, place 16 µl of the corresponding FDD-PCR master mix. For example, label 8 tubes with combinations of the anchor primer F-H-T$_{11}$G with arbitrary primer H-AP 1-8. Do this with all 24 primer combinations, including F-H-T$_{11}$A with H-AP 1-8 and F-H-T$_{11}$C with H-AP 1-8.

4. Add 2.0 µl of the corresponding cDNA (RT reaction from Protocol 3) to the 0.2-ml PCR tube.

5. Add 2.0 µl of the corresponding H-AP to the appropriate 0.2 ml PCR tube. Mix well. The total reaction volume will be 20 µl. Add 25 µl of mineral oil if needed.

6. Place the 0.2-ml PCR tubes on the thermocycler and begin the program (see step 2). Once completed, store reactions at –20° C in the dark.

PROTOCOL 5   SEQUENCING GEL ELECTROPHORESIS

For ease of use, the Sequagel 6 Ready-To-Use 6% Sequencing Gel (National Diagnostics EC-836 Atlanta, GA) is recommended for denaturing gel electrophoresis. However, a general protocol is given here for the 6% denaturing polyacrylamide gel that is recommended for resolution of cDNA profiles.

MATERIALS

BUFFERS, SOLUTIONS, AND REAGENTS
   10% Ammonium persulfate (APS) <!>
   6% Denaturing gel solution

38% acrylamide/2% bisacrylamide solution <!>	9 ml
urea (ultrapure) <!>	25.2 g
10x TBE	6 ml
distilled H$_2$O	24 ml

   Ethanol, 50% <!>
   FDD Loading Dye (RNAspectra Fluorescent mRNA Differential Display System, GenHunter F501-F510 & R501-R510, or F201)
   FDD locator dye (GenHunter F202 & R202)
   10x TBE
      20 mM disodium ethylenediamine tetraacetic acid (Na$_2$EDTA)
      0.89 M Tris-borate, pH 8.3

NUCLEIC ACIDS AND OLIGONUCLEOTIDES
    FDDRT-PCRs from Protocol 4

SPECIAL EQUIPMENT
    Fluorescence imager (we recommend the FMBio Series [Miraibio])
    Gel electrophoresis equipment
    1.5-ml microfuge tubes
    Printer for creating template
    Razor

METHOD

1. Use 60 ml of the denaturing gel solution for a 45 x 28 x 0.04-cm gel.

2. Add 0.5 ml of the 10% APS solution and mix thoroughly.

3. Pour the gel into the sequencing gel cast and let it polymerize.

4. Flush the urea from the gel wells and pre-run the sequencing gel in 1x TBE buffer for 30 minutes.

    Caution: It is crucial that all urea is removed from the wells before loading samples.

5. Add 3.5 µl of each FDDRT-PCR from Protocol 4 with 2 µl of FDD loading dye and incubate at 80°C for 2 minutes immediately before loading onto the gel.

6. Electrophorese for 2–3 hours at 60 watts constant power (voltage not to exceed 1700) until the xylene dye (the slower-moving dye) reaches the bottom of the gel. If voltage exceeds 1700, lower the wattage.

7. Turn off the power supply and remove the plates from the gel apparatus. Clean the outside of the glass plates very well with H_2O and 50% ethanol.

8. Scan the gel on a fluorescence imager with an appropriate filter following the manufacturer's instructions.

9. Print a real-size image of the gel scan on appropriately sized paper using a quality ink jet or laser printer. This printed image will be used as the template to excise differentially expressed cDNAs. FDD locator dye with its combination of fluorescent and visible dyes should be used to align the gel with the printed template for band excision before scanning.

10. Excise the band with a razor or other sharp utensil and place it in a 1.5-ml microfuge tube.

PROTOCOL 6   REAMPLIFICATION, CLONING, AND SEQUENCING OF DIFFERENTIALLY EXPRESSED cDNAs

Following excision of potential differentially expressed cDNAs from the acrylamide gel matrix, the cDNAs will be reamplified using the same anchor–arbitrary primer combinations and reaction conditions as the initial PCRs. The reamplification products can then be cloned and sequenced for further characterization.

MATERIALS

BUFFERS, SOLUTIONS AND REAGENTS
    1.5% Agarose gel with ethidium bromide<!> (see step 9)

Distilled H$_2$O
DNA loading dye
    30% glycerol
    70% distilled H$_2$O
    0.003% bromophenol blue <!>
    0.003% Xylene cyanol FF <!>(GenHunter S403)
    Ethidium bromide solution (0.5 µg/ml) <!>
Distilled H$_2$O
Reamplification mix (see step 7)

NUCLEIC ACIDS AND OLIGONUCLEOTIDES
Arbitrary primer (H-AP) (2 mM)
cDNAs (excised band) from Protocol 5
dNTP mix (250 µM) (GenHunter S501)
Unlabeled anchor primer (H-T$_{11}$M) (2 µM)
Vector-specific primers (Lseq/Rseq [GenHunter]).

ENZYMES AND ENZYME BUFFERS
10x PCR buffer
    100 mM Tris-HCl<!>, pH 8.4
    500 mM KCl<!>
    15 mM MgCl$_2$<!>
0.01% gelatin
Taq DNA polymerase (Qiagen 201207)

SPECIAL EQUIPMENT
Electrophoresis equipment
1.5-ml Microfuge tubes
Parafilm
Thermocycler
0.2-ml Thin-walled PCR tubes (GenHunter T101)
UV transilluminator
Microwave oven

ADDITIONAL ITEMS
PCR-TRAP Cloning System (GenHunter P404), which includes insert-ready PCR-TRAP cloning vector, T4 DNA ligase (200 units/µl), distilled H$_2$O, 10x ligase buffer (500 mM Tris-HCl, pH 7.8, 100 mM MgCl$_2$, 100 mM DTT<!>, 10 mM ATP, 500 µg/ml BSA), and GH-competent
RNAspectra Fluorescent mRNA Differential Display System (GenHunter F501-F510 & R501-R510, or S201)

CELLS
GH-competent cells

SPECIAL ITEMS
Agar plates containing antibiotic (consult cloning system recommendations for specific antibiotic and protocol)

METHOD

1. Add 1 ml of distilled H$_2$O to the excised band from Protocol 5 in a 1.5-ml microfuge tube.

2. Let the band soak for 30 minutes; vortex occasionally.

3. Remove the distilled H_2O (avoiding the band) and replace with 50 µl of fresh distilled H_2O.

4. Boil the tightly closed tube (covered with Parafilm) for 15 minutes to elute the cDNA from the gel slice.

5. Centrifuge for 1 minute to collect the condensation and pellet the gel.

6. Transfer the supernatant to a fresh 1.5-ml microfuge tube. Discard the tube with the gel slice.

7. Reamplification of the cDNA should be accomplished by the following in a 0.2-ml thin-walled PCR tube:

distilled H_2O	22.6 µl
10x PCR buffer	4.0 µl
dNTP mix (250 µM)	1.0 µl
H-AP primer (2 µM)	4.0 µl
H-T$_{11}$M (2 µM)	4.0 µl
cDNA template	4.0 µl
Taq DNA polymerase	0.4 µl

8. Place the reamplification reactions on the thermocycler and perform under the initial PCR conditions (see step 1, FDD-PCR Protocol 4).

9. Make a 1.5% agarose gel stained with ethidium bromide:

   a. Add 1.5 mg of ultrapure electrophoresis-grade agarose into a glass flask.

   b. Add 1x TAE to 100 ml volume.

   c. Place agarose in microwave until melted.

   d. Let cool and add 3 µl of 1:1 ethidium bromide solution.

   e. Mix well and pour into gel apparatus.

   f. Replace comb and let gel solidify.

   g. Use 1x TAE solution as running buffer (amount depends on gel apparatus).

10. Add 30 µl of the reamplification reaction to 5 µl of DNA loading dye in a 0.5-ml microfuge tube. Load the entire volume onto the 1.5% agarose gel.

11. Electrophorese at 70 V for ~45 minutes.

12. Confirm the cDNA reamplification by visualizing gel using a UV transilluminator.

13. Clone the differentially expressed cDNAs into recommended PCR-TRAP cloning vector, or other suitable cloning vector, following the manufacturer's protocol.

14. If using the PCR-TRAP Cloning System, sequencing can be performed employing vector-specific primers. If you are using another cloning vector, consult the manufacturer's guidelines for sequencing instructions.

PROTOCOL 7 NORTHERN BLOT ANALYSIS

To confirm differential expression of the selected cDNAs, we suggest using northern blot analysis rather than other confirmation techniques such as reverse northern hybridization (Zhang et al. 1996) or quantitative RT-PCR. The northern blot technique is technically simple and straightforward in approach, and requires no manipulation of the RNA sequences from which differential gene expression has been detected. Additionally, northern blot analysis is the most accepted confirmation technique for differential gene expression, often being referred to as "the gold standard" of gene expression confirmation assays. If the recommended PCR-TRAP cloning vector is used, the probe template is produced by a PCR of the cDNA construct within the cloning vector. The required primers are supplied with the cloning system. Additionally, we suggest using the HotPrime cDNA Labeling Kit (GenHunter) for more efficient priming as compared to random priming, because it is specifically designed to efficiently label cDNA probes isolated from differential display for northern blot analysis. This method makes use of decamers, rather than the traditional hexamers used in random priming, and also incorporates the anchored oligo(dT) primer (H-T_{11}M) into the labeling buffer to ensure full-length antisense cDNA probe labeling. These improvements greatly increase the chance for signal detection on the northern blot analysis.

MATERIALS

▼ CAUTION

See Appendix for appropriate handling of materials marked with <!>.

BUFFERS, SOLUTIONS, AND REAGENTS

1.5% Agarose gel with ethidium bromide staining <!>
Colony lysis buffer (PCR-TRAP Cloning System, GenHunter P404 or L102)
Denhardt's solution (500 ml)

Ficoll<!>	5 g
polyvinylpyrrolidone<!>	5 g
BSA (Pentax Fraction V)	5 g
distilled H_2O	up to 500 ml

Distilled H_2O
12.3 M (37%) formaldehyde<!>, pH >4.0
Formamide prehybridization/hybridization solution (GenHunter ML1). If preparing yourself, use the following protocol (for 500 ml):

20x saline-sodium phosphate-EDTA (SSPE)	125 ml
50x Denhardt's solution	50 ml
20% sodium dodecyl sulfate (SDS) <!>	2.5 ml
formamide <!>	250 ml
distilled H_2O	up to 500 ml

Mix well, aliquot into smaller volumes, and store at –20°C until use.
HotPrime cDNA Labeling Kit (GenHunter H50), which includes KlenowDNA polymerase (1 unit/μl), 10x labeling buffer, dNTP (–dATP) or dNTP (–dCTP) (500 μM), stop buffer, and distilled H_2O.

Mineral oil (*optional*)
10x MOPS buffer:
    0.2 M MOPS <!>
    0.05 sodium acetate <!>
    0.01 M EDTA
10x PCR Buffer (RNAspectra Differential Display System, F501-F510 & R501-R510, PCR-TRAP Cloning System, GenHunter, P404, or S201)
20x Saline-sodium citrate (SSC)
    3 M NaCl
    0.3 M trisodium citrate•2H_2O
    Adjust pH to 7.0 with 1 M HCl <!>.
1x SSC, 0.1% SDS<!> (w/v)
0.25x SSC, 0.1% SDS<!> (w/v)

20x SSPE
   3 M NaCl
   0.1 M NaH$_2$PO$_4$ (dibasic)<!>
   0.02 M EDTA
Stop buffer (0.1 M EDTA, pH 8.0)

NUCLEIC ACIDS AND OLIGONUCLEOTIDES

[α-^{32}P]dATP (3000 curies/μM)<!>
dNTP (250 μM) (PCR-TRAP Cloning System, GenHunter P404 or S501)
Lgh/Rgh Primers (2 μM) (PCR-TRAP Cloning System, GenHunter P404 or L201 & L202)
Salmon sperm DNA (GenHunter ML2)

ENZYMES AND ENZYME BUFFERS

Taq DNA polymerase (Qiagen 201207)

SPECIAL EQUIPMENT

Centrifuge
3-mm filter paper sheets
Intensifying screen
1.5-ml microfuge tube
Nitrocellulose or nylon membrane
Pipette
Scintillation counter
Sephadex G50 column (Roche Applied Science 1814419, Indianapolis, IN)
Single-emulsion scientific imaging film (we recommend Kodak Biomax MS, No. 8715187 [Kodak-Eastman, Rochester, NY])
Thermocycler
0.2-ml thin-walled PCR tube
UV-transparent plastic wrap (we recommend Glad Cling Wrap [The Glad Products Company, Oakland, CA])

ADDITIONAL ITEMS

QIAEX II Gel Extraction Kit (Qiagen 20021)

METHOD

Generation of cDNA Probes

1. On the bottom of each clone plate, number each tetracycline-resistant colony chosen for analysis (at least 5 are recommended per plate).

2. Aliquot 50 μl of colony lysis buffer into a 1.5-ml microfuge tube.

3. Pick each colony with a clean pipette tip, trying not to get too much (a tiny amount that can be seen by the eye is usually enough). Transfer the cells into the appropriately labeled 1.5-ml microfuge tube.

4. Incubate the tubes in boiling water for 10 minutes.

5. Centrifuge at maximum speed for 2 minutes at room temperature to pellet the cellular debris. Transfer the supernatant to a clean 1.5-ml microfuge tube. Discard the pelleted debris.

6. Use the lysate immediately for PCR analysis or store at –20°C until use.

7. For the colony PCR, add the following to a 0.2-ml thin-walled PCR tube (it is highly recommended that a core mix containing all components except the colony lysate be prepared to minimize pipetting errors):

distilled H$_2$O	10.2 μl
10x PCR buffer	2.0 μl
dNTP mix (250 μM)	1.6 μl
Lgh primer	2.0 μl
Rgh primer	2.0 μl
colony lysate	2.0 μl
Taq DNA polymerase	0.2 μl

Mix reaction well and add 30 μl of mineral oil if needed.

8. Program the thermocycler as follows: 94°C for 30 seconds → 52°C for 40 seconds → 72°C for 1 minute → 30 cycles → 72°C for 5 minutes (extension) → 4°C incubation.

9. Run all 20 μl of the PCR product on a 1.5% agarose gel with ethidium bromide staining. This not only serves to generate cDNA probes for northern blot analysis, but also indicates which cDNA clones contain the appropriate insert when using the PCR-TRAP Cloning System (see Protocol 6).

10. Purify each band from the agrose gel using a QIAEX II Gel Extraction Kit. Use the resulting purified fragment as a template to generate a probe using the HotPrime DNA Labeling Kit.

11. If a different vector was used for cloning, consult manufacturer's suggestions for generating cDNA probes for northern blot hybridization.

Labeling of cDNA Probes

12. If using the recommended HotPrime DNA Labeling Kit, thaw all components completely and immediately set them on ice.

13. Set up the following reaction in a 1.5-ml microfuge tube with a locking cap (so the cap will not loosen during boiling):

distilled H$_2$O	11 μl
10x labeling buffer	3 μl
DNA template to be labeled (10–50 ng)	7 μl

14. Incubate the mixture in a boiling water bath for 10 minutes.

15. Quickly chill the tubes on ice. Spin the tubes briefly to collect the condensation.

16. To the reaction, add the following *in order*:

dNTP (–dATP) (500 μM)*	3 μl
[α-^{32}P]dATP (3000 Ci/mM)<!>*	5 μl
Klenow DNA polymerase	1 μl

*If using [α-^{32}P]dCTP instead of [α-^{32}P]dATP, substitute dNTP (–dCTP) for dNTP (–dATP).

17. Incubate the mixture for 20 minutes at room temperature, followed by incubation at 37°C for an additional 10 minutes.

18. Add 6 μl of the stop buffer and mix well.

19. Purify the labeled probe with a Sephadex G50 column. Collect the purified probe in a 1.5-ml microfuge tube with a lock-on cap. Count 1 μl of labeled probe in a scintillation counter. A total of 10 million or more cpm can be obtained for most of the labeled DNA probes.

Probe Hybridization

20. Protocols for the preparation of a denaturing agarose (RNA) gel, including sample loading and electrophoresis conditions, and RNA transfer to nitrocellulose or nylon membrane, are described in Ausubel et al. (1995) and Sambrook and Russell (2001).

21. If the prehybridization buffer has been stored at –20°C, thaw at 37°C for 20 minutes.

22. Denature the salmon sperm DNA by incubating for 10 minutes in a boiling water bath.

23. Add the salmon sperm DNA (to a final concentration of 100–200 µg/ml) in the prehybridization solution. Mix well.

24. Use 5 ml of prehybridization solution or enough to cover the membrane. Prehybridize at 42°C for at least 4 hours.

25. Denature the purified probe in a 1.5-ml microfuge tube with a lock-on cap (otherwise the cap may loosen) by boiling for 10 minutes in a water bath.

26. Chill on ice for 2 minutes.

27. Spin down the condensation and add the probe directly to the prehybridization solution.

28. Hybridize overnight.

29. Carefully decant the radioactive hybridization and dispose of in an appropriate container for radioactive waste.

30. Wash with 1x SSC containing 0.1% SDS *twice* at room temperature, each time disposing of wash solution in an appropriate container.

31. Wash for 15–20 minutes with 0.25x SSC containing 0.1% SDS *prewarmed* to the final washing temperature of 50–55°C.

Blot Exposure

32. Blot the membrane dry with 3-mm filter paper and cover using UV-transparent plastic wrap

33. Expose the blot to single emulsion with an intensifying screen at –70°C for best signal detection.

DISCUSSION

The development of FDD represents a marked improvement over conventional DD with respect to safety and high-throughput capability. In the development of FDD, it was crucial that the new platform have similar sensitivity to traditional DD with isotopic labeling, as well as other advantages that would make the platform a viable and improved alternative to the established DD methodology. Such improvements as elimination of radioactivity, less time-consuming assay methodology, and digital data acquisition were goals successfully reached by the establishment of the FDD platform.

To test the sensitivity of the new fluorescence platform, DNA-free total RNA from normal and *ras* oncogene-transformed rat embryo fibroblasts was used for DDRT-PCRs utilizing the anchored oligo(dT), $HT_{11}G$, and two arbitrary primers, H-AP6 and H-AP29. The reactions were a comparison between isotopic DD, using ^{33}P-labeled α-dATP, and FDD, employing a fluorescent dye-labeled one-base anchored primer, under identical PCR conditions. Initially, use of the fluorescent oligo(dT) in DD-PCRs resulted in lower fluorescent signals when carried out under the same conditions as conventional DD. However, because the dNTP concentration for conventional DD is intentionally kept low to increase the efficiency of incorporation of radioactively labeled nucleotides into amplified cDNAs, the dNTP concentration was increased to compensate for the initial low signal using the fluorescent anchored oligo(dT). Such optimized FDD was shown to be essentially identical in both sensitivity and reproducibility to that of conventional DD (Fig. 3) (Cho et al. 2001).

The elimination of isotopic primer labeling has allowed researchers to explore gene expression analysis without exposure to potentially dangerous radioactive compounds. Combined with robotics and digital data analysis, FDD has been shown to be accurate and amenable to high throughput (Liang 2000; Cho et al. 2001). Elimination of manual reac-

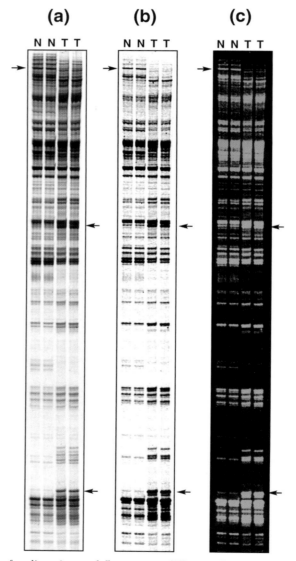

FIGURE 3. Comparison of radioactive and fluorescent differential display. DNA-free RNA from normal (lanes *N*) and *ras* oncogene-transformed (lanes *T*) rat embryo fibroblasts were compared in duplicate by either conventional differential display with ^{33}P-labeled [α]dATP or FDD with fluorescein-labeled anchor primer under identical PCR conditions. The autoradiogram (*a*) and fluorescent images in gray scale (*b*) or color (*c*) were compared in sensitivity and reproducibility as indicated. Reproducible differences were marked by arrows. The anchored primer, H-T$_{11}$G, was used in combination with the arbitrary 13-mer, H-AP29.

tion setup, through the use of a robotic liquid dispenser, not only ensures reproducibility by reduction of pipetting errors, but, in combination with the elimination of conventional DD autoradiography, also decreases the amount of time required for a differential gene expression screening. This allows researchers to accomplish their research expression goals in as little as half the time needed to complete the same screening using isotopic DD.

In assessing the sensitivity of the fluorescence platform of DD technology, previous results have shown that use of a fluorescent-dye-labeled anchored oligo(dT) resulted in equal signal sensitivity as compared to conventional DD (Cho et al. 2001). This conservation of sensitivity of the FDD platform allows an equal chance of detecting differentially expressed genes as conventional DD, but with the methodological improvements of FDD with regard to safety, time, and digital data acquisition.

In terms of high-throughput capabilities, the optimized FDD technology is now able to compete with other gene expression tools such as DNA microarray technology. The simplistic design of FDD has an inherent advantage over microarrays; microarray technology also has additional disadvantages such as probe sensitivity, nonlinearity in signal detection (Ramdas et al. 2001), probe cross-hybridization due to homologous cDNA sequences (Richmond et al. 1999), and data management (Gibbs 2001). Thus, FDD represents a truly remarkable achievement in gene expression technology.

REFERENCES

Ausubel F., Brent R., Kingston R.E., Moore D.D., Seidman J.G., Smith J.A., and Struhl K., eds. 1995. *Short protocols in molecular biology,* 3rd edition. Wiley, New York.

Bauer D., Muller H., Reich J., Riedel H., Ahrenkiel V., Warthoe P., and Strauss M. 1993. Identification of differentially expressed mRNA species by an improved display techniqe (DDRT-PCR). *Nucleic Acids Res.* **21:** 4272–4280.

Cho Y.-J., Meade J.D., Walden J.C., Chen X., Guo Z., and Liang P. 2001. Multicolor fluorescent differential display. *BioTechniques* **30:** 562–572.

Gibbs W.W. 2001. Shrinking to enormity: DNA microarrays are reshaping basic biology – but scientists fear that they may soon drown in data. *Sci. Am.* **284:** 33–34.

Hsu D.K., Donohue P.J., Alberts G.F., and Winkles J.A. 1993. Fibroblast growth factor-1 induces phosphofructokinase, fatty acid synthase and Ca (2+)-ATPase mRNA expression in NIH 3T3 cells. *Biochem. Biophys. Res. Commun.* **197:** 1483–1491.

Liang P. 2000. Gene discovery using differential display. *Gen. Eng. News* **20:** 37.

Liang P. and Pardee A.B. 1992. Differential display of eukaryotic messenger RNA by means of the polymerase chain reaction. *Science* **257:** 967–971.

Liang P., Averboukh L., and Pardee A.B. 1993. Distrubution and cloning of eukaryotic mRNAs by means of differential display: Refinements and optimization. *Nucleic Acids Res.* **21:** 3269–3275.

———. 1994. Method of differential display. In *Methods in molecular genetics* (Ed. K.W. Adolph). Academic Press, San Diego. pp. 3–16.

Liang P., Bauer D., Averboukh L., Warthoe P., Rohrwild M., Muller H., Strauss M., and Pardee A.B. 1995. Analysis of altered gene expression by differential display. *Methods Enzymol.* **254:** 304–321.

Ramdas L., Coombes K.R., Baggerly K., Abruzzo L., Highsmith W.E., Krogmann T., Hamilton S.R., and Zhang W. 2001. Sources of nonlinearity in cDNA microarray expression measurements. *Genome Biol.* **2:** research0047.1-0047.7.

Richmond C.S., Glasner J.D., Mau R., Jin H., and Blattner F.R.1999. Genome-wide expression profiling in *Escherichia coli* K-12. *Nucleic Acids Res.* **27:** 3821–3835.

Sambrook J. and Russell D.W. 2001. *Molecular cloning: A laboratory manual,* 3rd edition. Cold Spring Harbor Laboratory Press, Cold Spring Harbor, New York.

Sokolov B.P. and Prockop D.J. 1994. A rapid and simple PCR-based method for isolation of cDNAs from differentially expressed genes. *Nucleic Acids Res.* **22:** 4009–4015.

Zhang H., Zhang R., and Liang P. 1996. Differential screening of gene expression difference enriched by differential display. *Nucleic Acids Res.* **24:** 2454–2455.

WWW RESOURCE

http://www.ncbi.nlm.nih.gov/entrez/query.fcgi?db=PubMed NCBI, National Library of Medicine, PubMed search database.

Traditionally, researchers have screened cDNA libraries to obtain clones for further evaluation. In this section, we present PCR methods for library construction, screening and isolation of full-length clones.

SCREENING BY PCR

Detailed in "A PCR-based Method for Screening Libraries" (p. 351) are the techniques to prepare phage pools for screening by PCR, as well as methods for DNA release and hints on primer design. When screening libraries by PCR, the time and effort necessary to isolate the desired clone are significantly reduced in comparison to screening with labeled nucleic acid probes. The single most important step in this protocol is selection of the primers. The PCR primers must be designed to amplify specifically and efficiently a region of the gene of interest between 100 and 1000 bp, ideally in the range of 200–500 bp. Smaller products are more difficult to resolve accurately on gels and sometimes are not as robust. Primers used in screening amplifications should not anneal to *E. coli* and phage DNA to prevent background amplification. Additionally, it is important to know the lower limit of detection of the primer pair designed for the screening using a dilution series of target DNA in carrier nucleic acid. The range of copies tested should be between 10,000 and 0.1 copies per PCR. Both the template DNA and the carrier must be of a length to assure that viscosity problems do not prevent the complete mixing of the DNA solutions during the construction of the dilution series.

RACE

A common problem with traditional library screening methods is that the isolated cDNA may not be of full length. In general, because of the manner in which the cDNA libraries are constructed, the obtained cDNA clone lacks the 5′ end corresponding to the amino-terminal region of the protein. Alternatively, the clone may have arisen from a differential selection method and is lacking both 3′ and 5′ sequence. In all of these cases, sequence-specific primers can be designed and used in combination with a second primer designed to hybridize to added nucleotides (such as in the 5′ RACE protocol) or based on the common poly(A) tail of mRNAs (such as in the 3′ RACE method). The protocols in "Beyond Classic RACE" (p. 359) describe one-sided techniques that are designed to yield full-length cDNAs when starting from limited nucleotide or amino acid sequence. This chapter discusses the different variations of RACE that have been developed. Protocols for classic 3′ and 5′ RACE,

RNA ligase-mediated RACE, and cap-switching RACE are provided. Additionally, there is extensive discussion comparing and contrasting these different methodologies.

PEPTIDE LIBRARY CONSTRUCTION

PCR is also useful when constructing, analyzing, and screening recombinant DNA libraries. In "Synthesis and Analysis of DNA Libraries Using PCR" (p. 391), a method is presented for the construction of a library consisting of random 12-amino-acid-long peptides. This type of library is very useful for the mapping of protein–protein interactions, epitopes for antibody binding, and detection of high-affinity peptides that bind cellular factors such as receptors and DNA-binding proteins.

Success in library construction is defined as the generation of a pool of clones that accurately reflects the nucleic acid composition of the starting material. As an analytical tool, PCR can determine the accuracy of the individual steps in library construction. This is important in assuring that the full complexity of the starting material is captured. A method for this kind of analysis is presented in "Synthesis and Analysis of DNA Libraries Using PCR" (p. 391). No matter which technique for screening is devised, success in obtaining the desired clone resides heavily in the quality of the library to be screened.

24 PCR-based Method for Screening Libraries

David I. Israel

PRAECIS Pharmaceuticals Incorporated, Waltham, Massachusetts 02451-1420

One of the fundamental techniques of molecular biology is the isolation of a rare clone from a complex mixture of DNA sequences contained within a library. The isolation of cDNA and genomic clones from highly complex libraries is often an early step in studies aimed at understanding basic biological processes as well as in applied biological research. In a typical genomic library with an average insert size of 20,000 bp from an organism with a haploid genome size of 2×10^9 bp, the occurrence of a single-copy gene will be approximately $1/10^5$. Likewise, for cDNA clones within a highly complex library derived from a tissue or cell line that expresses many different genes, a particular clone may occur with a similarly low frequency. Screening of libraries of high complexity by techniques such as filter hybridization (Benton and Davis 1977) or expression cloning (Wong et al. 1985) is therefore a labor-intensive and time-consuming process because of the large number of clones that must be screened to obtain the clone of interest.

PCR results in the amplification of a given nucleic acid sequence by many orders of magnitude (Saiki et al. 1985). When applied to the screening of highly complex DNA libraries contained within either bacteriophage or plasmid vectors, PCR offers the opportunity to identify rare DNA sequences in complex mixtures of molecular clones by increasing the abundance of a particular sequence, thereby allowing the easy identification of a particular clone in a portion of the library. This is accomplished by subdividing the original library into pools of decreased complexity and screening each pool or groups of pools for a given DNA sequence (Fig 1). A pool that contains the desired clone is subsequently subdivided into smaller pools, each of which is screened using the same PCR protocol that was used for the primary screen. After several cycles of subdividing and screening, the initially rare clone is greatly enriched and can be easily obtained as a pure clone (Israel 1993).

A method for screening highly complex DNA libraries using PCR is described in this chapter. This method allows a library of high complexity to be screened in a short time and provides an alternative to more traditional and time-consuming screening methods that entail plaque or colony hybridization, or methods that require the expression of a functional gene product. The main advantages of this screening technique are its speed, sensitivity, and ease. Some providers of molecular biology reagents, such as MRC geneservice, have PCR-ready library pools available for screening. In addition, the method as described below requires more than one oligonucleotide to anneal correctly to the template DNA and/or PCR product, thereby providing a high degree of stringency for a true positive signal.

Precise sequence information from the target gene is required for the design of specific and efficient PCR primers. Because the complete genomes of human, mouse, and several other species have now been sequenced, this information can be used to design primers for virtually any gene in many species. In some cases, such as using sequences from one species to clone the corresponding gene from a different species whose genome is not yet sequenced, the precise DNA sequence within the target clone may not be known. Hybridization conditions of reduced stringency or using mixtures of degenerate oligonucleotide probes are commonly employed in these circumstances when screening by plaque or colony hybridization (Goeddel et al. 1980; Toole et al. 1984). Using annealing conditions of reduced stringency or mixtures of degenerate oligonucleotide primers may present technical difficulties when using PCR, because the occurrence of false-positive signals increases as the annealing specificity is decreased. In addition, because of the exquisite sensitivity of PCR, careful laboratory technique must be practiced diligently to avoid cross-contamination of samples that can result in false positives. However, with reliable sequence information for primer synthesis and careful experimental design and technique, this method provides an efficient means to screen libraries with a high probability of success.

PROTOCOL   PCR-BASED SCREENING OF DNA LIBRARIES

MATERIALS

BUFFERS, SOLUTIONS, AND REAGENTS
Glass-distilled H_2O

ENZYMES AND ENZYME BUFFERS
PCR cocktail (can be frozen in aliquots): For a 1-ml reaction cocktail, include:
    200 nmoles each dNTP
    1x *Vent* buffer, or equivalent
    2.5 μmoles $MgCl_2$ <!>
    2 nmoles each primer
PCR master mix (should be prepared fresh for each reaction)
    1 μl of *Vent* DNA polymerase, or equivalent
    99 μl of PCR cocktail

▼ CAUTION

See Appendix for appropriate handling of materials marked with <!>.

NUCLEIC ACIDS AND OLIGONUCLEOTIDES
2 PCR primer oligonucleotides
1 hybridization oligonucleotide
Positive control for PCR: 10 ng of total genomic DNA or an aliquot of the starting library that yields a positive PCR signal (see step 5)

MEDIA
L broth: for 1 liter in H_2O:
    10 g of Bacto-tryptone
    5 g of Bacto-yeast extract
    10 g of NaCl
    Adjust the pH to 7.5 with NaOH<!>.
SM: For 1 liter in H_2O:
    5.8 g of NaCl
    2 g of $MgSO_4$<!>
    50 ml of 1 M Tris-HCl<!>, pH 7.5

5 ml of 2% gelatin
L broth containing 10 mM MgSO$_4$

SPECIAL EQUIPMENT
Agarose gel electrophoresis supplies, including ethidium bromide <!>

ADDITIONAL ITEMS
96-well U-bottomed plates (Corning Costar)
Polyester sealing tape (Nalge Nunc International)
Materials for plaque hybridization

VECTORS AND BACTERIAL STRAINS
DNA library in phage vector
Bacterial strain to propagate library <!>

METHOD

Before screening the DNA library, several parameters should be investigated to establish efficient experimental conditions. These include

- *PCR conditions.* Use the primers that will be used to screen the library, and vary annealing temperature and cycle number to optimize the specificity and yield of the product. The source of template can be either the library to be screened (10^6 phage particles) or 10 ng of total genomic DNA. During the PCR, phage DNA is released and serves as template. Therefore, phage DNA does not need to be purified prior to the reaction. Typical PCR primers (16–24 nucleotides long, 50% G+C content) yielding a product 0.1–1.0 kb in length can be used. If necessary (for example, when using primers from different exons that span a large intron[s]), target sequences may be greater than 1 kb apart, but the yield of product may be lower.

- *Determination of the frequency of the gene in the library.* Using the PCR conditions established above, titer the library by varying the amount of input phage. The minimum number of phage that yields a PCR product is the experimentally determined frequency of the gene in the library. Genomic libraries with an average insert size of 20 kb should have a complexity of greater than 10^5 to assure that the target gene is present at least once. Use of 10^4–10^6 phage as template from a typical library of high complexity should indicate the frequency of the gene in the library. The number of input phage that contains one or two copies of the target gene should be used in the screen of the library.

Once the PCR conditions and gene frequency have been determined, the library can be screened as follows:

1. Mix 0.5 ml of a fresh overnight bacterial culture grown in L broth with 0.5 ml of SM and add phage containing the library. Incubate for 20 minutes at room temperature.

2. Add 20 ml of L broth containing 10 mM MgSO$_4$ and dispense 100 μl/well in a 96-well plate in an 8 x 8 matrix. Seal the plate with polyester sealing tape and incubate for 5–6 hours at 37°C, shaking at 225 rpm. This allows amplification of the phage within the subpool of library. Phage titers should increase to ~10^9/ml after amplification.

   Approximately one-third of the culture is aliquoted into the 96-well plate. The number of input phage used in step 1 should take this into account.

3. Combine phage from 8 wells across a row or 8 wells down a column (25 μl/well) using a multiwell pipette (see Discussion for alternative formats). Special care must be taken at this step when removing the plate sealer and when pipetting to avoid cross-contamination of samples. Brief centrifugation of plates should help clear liquid from the plate

sealer if cross-contamination is a problem. Reseal the plate with fresh polyester sealing tape and store at 4°C for up to 1 month.

4. Dilute pooled phage 1:1 with glass-distilled H_2O. The phage are now ready to use as PCR templates.

5. Perform PCR using the conditions established above by adding 0.5 μl of PCR template (pooled phage) to 24.5 μl of PCR master mix. Each experiment should contain a negative control (no template) and positive control (i.e., 10 ng of total genomic DNA or an aliquot of the starting library known to yield a positive signal).

6. Analyze the PCR products by agarose gel electrophoresis. Stain the gel with ethidium bromide and photograph (see Sambrook and Russell 2001).

7. Dry the gel in vacuo at 70°C, denature the DNA, then hybridize the oligonucleotide probe (end-labeled with [^{32}P]phosphate) directly to the dried gel using standard DNA hybridization conditions. Wash, and perform autoradiography. (Technical details for this step can be found in Israel [1993].) This step is optional if the specific PCR product can be readily visualized by ethidium bromide staining, as discussed below. Hybridization can also be performed after the transfer of DNA to a nitrocellulose or nylon filter using standard techniques.

> The data from step 6 and/or step 7 should allow the identification of a subpool of the library containing the gene of interest (see Discussion). The primary screen is now complete, and the gene within the positive subpool is now enriched relative to the starting library. Subsequent screening cycles are iterations of steps 1–7 and can be performed after titration of the phage in the amplified subpool.

8. Determine the phage titer from the positive well by plaque formation (see Sambrook and Russell 2001).

9. Initiate the next round of screening by infecting bacteria with ~30-fold fewer phage than were used in the previous round of screening.

10. Repeat steps 2–9.

DISCUSSION

For a screen that yields a single positive clone, the positive well within a row is located at the column that also yields a positive signal (Fig. 1). For screens that yield two or more positive clones, a second PCR using phage from individual wells within the positive column or row will definitively locate the positive well(s). Alternative formats for screening libraries are discussed below and in Figure 1.

Depending on the purity and yield of the specific PCR product, the data from the agarose gel may be sufficient to identify positive pools. For greater sensitivity and specificity, a radioactive or fluorescent oligonucleotide probe specific for sequences between the PCR primers can be hybridized to the PCR products in the gel, as detailed in step 7. This step is optional, particularly at the secondary and tertiary stages of screening. Figure 2A shows the analysis of a primary screen by agarose gel electrophoresis visualized by ethidium bromide staining. Because of both the complexity of the PCR products and the low amount of specific product, hybridization is required to identify positive pools (Fig. 2B). As the target gene becomes more enriched at the secondary and tertiary levels of screening, the abundance and purity of the PCR product increases (compare Fig. 2A, lane 2 to Fig. 3A, lane 9). When this occurs, the hybridization step becomes dispensable. Once pure phage containing the target gene have been isolated, the correct PCR product predominates over side products (Fig. 3).

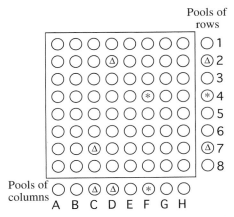

FIGURE 1. Formats for library plating and PCR screening. The figure shows the schematic division of a library into 64 subpools where the target gene occurs once (*) or three times (* and Δ). In a typical primary screen, each well is seeded with 1000 phage in an 8 x 8 format. A portion of the library with complexity of 64,000 is therefore subdivided into 64 subpools, each containing 1000 independent phage. After amplification of phage, PCR can be performed in a number of formats. (1) Pooling strategy: Eight wells are pooled across rows and down columns. PCR is then performed on the 16 pools of phage. For a single positive (*), pool F and pool 4 will yield the correct PCR product, identifying well 4F as the source of the positive clone. However, for three occurrences of the target gene (* and Δ), pools C, D, F, 2, 4, and 7 will all be positive. The precise identification of the positive well(s) therefore requires a second PCR using individual wells within a positive pool. (2) Two-step strategy: Phage are pooled in only one direction, yielding 8 pools which are used as template in the first PCR. The 8 individual wells within a positive pool are then analyzed in a second PCR. For a single positive, the pooling strategy is more efficient, because the 16 samples are analyzed in a single PCR. For multiple positives, the two-step format is more efficient because it requires less pooling and the analysis of fewer samples. (3) Analysis of individual wells: Phage from single wells are used as template. No pooling is required, and although the number of individual samples is higher (64 versus 16 for the pooling and two-step strategies), a single experiment should unambiguously identify positive wells. This strategy becomes more efficient as the number of positive clones in the screen increases, and if PCR is performed in a 96-well format.

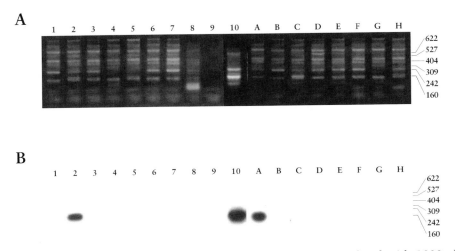

FIGURE 2. PCR screening of pooled phage. Sixty-four wells were inoculated with 1000 phage/well and screened for the murine M-CSF gene (Israel 1993) using the pooling strategy. The agarose gel in A was vacuum-dried and used for direct hybridization to an internal M-CSF oligonucleotide. (A) Ethidium bromide staining of PCR products; (B) hybridization to M-CSF-specific oligonucleotide. The templates for each reaction were: (lanes 1–8) pools of rows; (lane 9) no template; (lane 10) 10 ng of mouse genomic DNA; (lanes A–H) pools of columns. The migration of MspI-digested pBR322 is indicated on the top right (bp).

The frequency of the gene after any round of screening can be estimated by PCR titration of the positive subpool as described in the protocol or by plaque hybridization. This frequency should increase at each round of screening and can be used to check the effectiveness of the enrichment procedure. Once the gene is sufficiently enriched, the clone can be obtained as a pure plaque by performing PCR on individually picked plaques or by plaque hybridization.

Several modifications of this technique can be made in certain situations to make cloning more efficient or for other applications. These include:

- *Changing the format of sublibrary plating or pooling.* Pooling in an 8 x 8 matrix is one of many formats that can be used with this technique. At one extreme, the pooling of phage can be dispensed with, and PCR can be performed using phage from individual wells. This approach is now more feasible with increasingly sophisticated laboratory automation, and with the availability of 96-well and 384-well PCR plates and thermal cyclers, as phage amplification and PCR can both be performed using multiwell pipettes and the same plate formats. The PCR product still needs to be analyzed by gel electrophoresis, however, and the number of individual samples will be much higher if the pooling strategy is not employed. An intermediate approach is to pool amplified phage in one dimension (i.e., only columns or only rows), perform PCR, and then use phage from individual wells within a positive pool for a second PCR. The advantage of this approach is that it requires only eight PCR analyses for the pools and eight more PCR analyses for individual wells, and yields the exact location of the positive sample. This strategy, however, requires two PCR protocols to be done serially, adding time to the overall procedure.

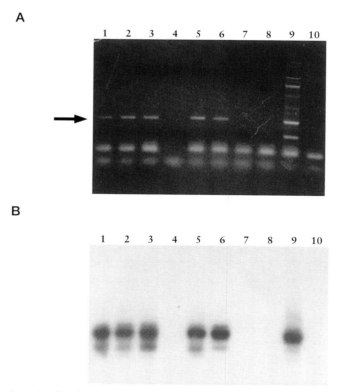

FIGURE 3. PCR screening of individual plaques. (*A*) Phage from a positive well after the tertiary screen were picked from random plaques (lanes *1-8*) and used as template. The arrow indicates the correct M-CSF PCR product. (Lane *9*) Phage from the tertiary positive well; (lane *10*) no template. (*B*) Hybridization to the M-CSF-specific oligonucleotide.

- *Screening for genes within non-phage vectors.* DNA libraries, particularly cDNA libraries, are often contained in plasmid vectors. The PCR screening technique described here has been used to clone a cDNA gene from a library within a plasmid vector (Israel 1993). To screen plasmid libraries, the plasmid-bearing bacteria should be allowed to grow for 16 hours in the multiwell plate. This assures that the bacterial number will be sufficiently high to allow representation of all the clones in the pool during the PCR. The other steps of analysis are as described for phage libraries. Pooled cDNA and genomic libraries in YAC vectors can similarly be screened using PCR, with appropriate modifications for amplification of YACs in yeast.

- *Priming from the vector.* When limited sequence information is available, or when purposefully screening for a clone that contains a particular sequence toward one end of the insert, one of the primers can be complementary to sequences within the vector (Jansen et al. 1989). This modification may yield a higher amount of incorrect product because one PCR primer will anneal to all phage within the library. In addition, the amount of PCR product will be greatly decreased if the amplified sequence is greater than 2 kb. As a consequence, the apparent frequency of a particular clone in a library will be lower. The description of methods for obtaining long PCR products (Cheng et al. 1994) makes this approach more feasible.

The basic PCR protocol for screening highly complex DNA libraries, and several modifications, have been described in this chapter. This strategy for screening libraries should consistently yield positive results with the use of good reagents and careful technique.

REFERENCES

Benton W.D. and Davis R.W. 1977. Screening λgt recombinant clones by hybridization to single plaques in situ. *Science* **196:** 180–182.

Cheng S., Fockler C., Barnes W.M., and Higuchi R. 1994. Effective amplification of long targets from cloned inserts and human genomic DNA. *Proc. Natl. Acad. Sci.* **91:** 5695–5699.

Goeddel D.V., Yelverton E., Ullrich A., Heyneker H.L., Miozzari G., Holmes W., Sceburg P.H., Dull T., May L., Stebbing N., Crea R., Maeda S., McCandliss M., Sloma A., Tabor J.M., Gross M., Familletti P.C., and Pestka S. 1980. Human leukocyte interferon produced by *E. coli* is biologically active. *Nature* **287:** 411–416.

Israel D.J. 1993. A PCR-based method for high stringency screening of DNA libraries. *Nucleic Acids Res.* **21:** 2627–2631.

Jansen R., Kalousek F., Fenton W.A., Rosenberg L.E., and Ledley F.D. 1989. Cloning of full-length methylmalonyl-CoA mutase from a cDNA library using the polymerase chain reaction. *Genomics* **4:** 198.

Saiki R.K., Scharf S., Faloona F., Mullis K.B., Horn G.T., Erlich H.A., and Arnheim N. 1985. Enzymatic amplification of β-globin genomic sequences and restriction site analysis for diagnosis of sickle cell anemia. *Science* **230:** 1350–1354.

Sambrook J. and Russell D.W. 2001. *Molecular cloning: A laboratory manual*, 3rd edition. Cold Spring Harbor Laboratory Press, Cold Spring Harbor, New York.

Toole J.J., Knopf J.L., Wozney J.M., Sultzman L.A., Buecker J.L., Pittman D.D., Kaufman R.J., Brown E., Shoemaker C., Orr E.C., Amphlett G.W. Foster W.B., Coe M.L., Knutson G.J., Fass D.N., and Hewick R.M. 1984. Molecular cloning of a cDNA encoding human antihaemophic factor. *Nature* **312:** 342–347.

Wong G.G., Witek J.S., Temple P.A., Wilens K.M., Leary A.C., Luxenberg D.P., Jones S.S., Brown E.L., Kay R.M., Orr E.C., Shoemaker C., Golde D.W., Kaufman R.J., Hewick R.M., Wang E.A., and Clark S.C. 1985. Human GM-CSF: Molecular cloning of the complementary DNA and purification of the natural and recombinant proteins. *Science* **228:** 810–815.

25 Beyond Classic RACE (Rapid Amplification of cDNA Ends)

Michael A. Frohman

Department of Pharmacology and the Center for Developmental Genetics, SUNY at Stony Brook, New York 11794-5140

Most attempts to identify and isolate a novel cDNA result in the acquisition of clones that represent only a part of the mRNA's sequence. However, once the partial sequence has been identified, the remainder of the transcript can often be obtained from the ever-growing collections of sequenced genomes and high-quality cDNA libraries. Unfortunately, for less well-characterized organisms, or for low-abundance cDNAs in well-characterized organisms, such information is frequently unavailable, particularly at the 5′ end of the transcript. The missing sequence (cDNA ends) can be cloned by PCR, using a technique variously called rapid amplification of cDNA ends (RACE) (Frohman et al. 1988), anchored PCR (Loh et al. 1989), or one-sided PCR (Fig. 1) (Ohara et al. 1989). Since the initial reports describing this technique, many laboratories and companies have developed significant improvements on the basic approach (Frohman 1989, 1993, 1994; Frohman and Martin 1989; Dumas et al. 1991; Fritz et al. 1991; Borson et al. 1992; Jain et al. 1992; Rashtchian et al. 1992; Schuster et al. 1992; Bertling et al. 1993; Monstein et al. 1993; Templeton et al. 1993).

This chapter describes three different versions of RACE: the relatively simple classic RACE and two more powerful protocols that are frequently, although not always, better suited for the task. The latter protocols evolved through the work of a number of laboratories (Tessier et al. 1986; Mandl et al. 1991; Volloch et al. 1991; Brock et al. 1992; Bertrand et al. 1993; Fromont-Racine et al. 1993; Liu and Gorovsky 1993; Sallie 1993; Schmidt and Mueller 1999; Schramm et al. 2000). Commercial RACE kits are available from many companies, including Invitrogen (RLM-RACE) and Clontech (SMART-RACE). The kits are convenient, but the commercial methods are not as powerful as the versions described in this chapter.

PRINCIPLES

Why use PCR (RACE) instead of screening (additional) cDNA libraries? RACE cloning is advantageous for several reasons: First, it takes weeks to go through the process of screening cDNA libraries, obtaining individual cDNA clones, and analyzing the clones to determine whether the missing sequence is present; using PCR, such information can be generated within a few days. As a result, it becomes practical to spend time optimizing RNA preparation and/or reverse transcription conditions to produce full-length cDNAs. Second, whereas library screens generally produce very few cDNA clones, RACE can be used to generate an essentially unlimited number of independent clones. The availability of large numbers of clones provides confirmation of nucleotide sequence and allows the isolation of unusual transcripts that may be alternately spliced or that begin at infrequently used promoters.

CLASSIC RACE

PCR is used to amplify partial cDNAs representing the region between a single point in a mRNA transcript and its 3′ or 5′ end (Figs. 1 and 2). For the technique to work, a short internal stretch of sequence from the mRNA of interest must already be known. Using this sequence, gene-specific primers are designed to extend in the direction of the missing sequence. Extension of the partial cDNAs, from the unknown end of the message back to the known region, is achieved using primers that anneal to the preexisting poly(A) tail (3′ end) or an appended homopolymer tail or linker (5′ end). Using RACE, enrichments on the order of 10^6–10^7 can be obtained. As a result, relatively pure cDNA "ends" are generated. These can be easily cloned and rapidly characterized using conventional techniques (Frohman et al. 1988).

To generate "3′ end" partial cDNA clones, mRNA is reverse-transcribed using a "hybrid" primer (Q_{total}, Q_T) that consists of two mixed bases (GATC/GAC followed by $[T]_{17}$) and a unique 35-base oligonucleotide sequence (Q_I-Q_O; see Fig. 2a,c). Amplification is then performed using a primer containing part of this sequence (Q_{outer}, Q_O), (which now binds to each cDNA at its 3′ end) and a primer derived from the gene of interest, GSP1 (gene-specific primer 1). A second set of amplification cycles is then carried out using "nested" primers (Q_{inner}, Q_I, and GSP2) to enhance the specificity of the amplification.

To generate "5′ end" partial cDNA clones, the first-strand products are generated by reverse transcription (primer extension) from a known gene-specific primer (GSP-RT; Fig. 2b). Then, a poly(A) tail is appended using terminal deoxynucleotidyltransferase (Tdt) and dATP. Amplification is carried out using three primers; (1) the hybrid primer Q_T to form the second strand of cDNA; (2) the Q_O primer; and (3) a gene-specific primer upstream of the one used for reverse transcription. Finally, a second set of PCR cycles is carried out using nested primers (Q_I and GSP2) to enhance specificity (Frohman and Martin 1989).

RACE VARIATIONS

The most technically challenging step in classic 5′ RACE is to cajole reverse transcriptase to copy the mRNA of interest in its entirety into first-strand cDNA. Prematurely terminated first-strand cDNAs are a problem because they are tailed by terminal transferase just as effectively as full-length cDNAs. Populations of cDNA that are composed largely of prematurely terminated first strands will result primarily in the amplification and recovery of cDNA ends that are also incomplete (Fig. 3a). This problem is encountered routinely for vertebrate genes, which tend to be GC-rich at their 5′ ends, a condition that hinders reverse

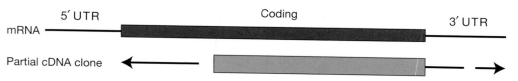

FIGURE 1. Schematic representation of the setting in which classic RACE is used. The figure shows a mRNA for which only a partial, internal cDNA is available.

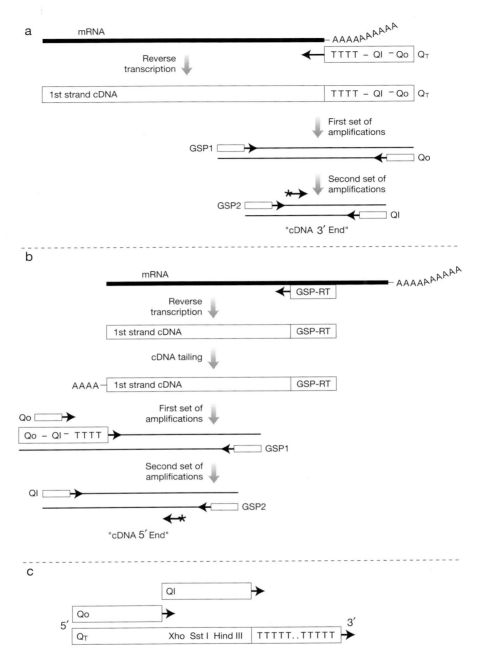

FIGURE 2. Schematic representation of classic RACE. See text for details. (GSP1)Gene-specific primer 1; (GSP2) gene-specific primer 2; (GSP-RT) gene-specific primer, used for reverse transcription; (*→) GSP-Hyb/Seq; i.e., a gene-specific primer for use in hybridization and sequencing reactions. (*a*) Amplification of 3′ partial cDNA ends. (*b*) Amplification of 5′ partial cDNA ends. (*c*) Schematic representation of the primers used in classic RACE. The 52-nt Q_T primer (5′ Q_O-Q_I-TTTT 3′) contains a 17-nt oligo-(dT) sequence at the 3′ end followed by a 35-nt sequence encoding *Hin*dIII, *Sst*I, and *Xho*I recognition sites. The Q_I and Q_O primers overlap by 1 nt; the Q_I primer contains all three of the restriction enzyme recognition sites. Optionally, two additional nucleotides can be added to the 3′ end of Q_T to force it to bind to the junction of the cDNA and the poly(A) tail: (G, A, or C, followed by G, A, T, or C). Primers:

Q_T: 5′- CCAGTGAGCAGAGTGACGAGGACTCGAGCTCAAGCTTTTTTTTTTTTTTTTT-3′
Q_O: 5′- CCAGTGAGCAGAGTGACG-3′
Q_I: 5′-GAGGACTCGAGCTCAAGC-3′

transcription. A number of laboratories and companies have developed steps or protocols addressing the problem of GC-rich sequences (Tessier et al. 1986; Mandl et al. 1991; Volloch et al. 1991; Brock et al. 1992; Bertrand et al. 1993; Fromont-Racine et al. 1993; Liu and Gorovsky 1993; Sallie 1993; Schmidt and Mueller 1999; Schramm et al. 2000); the protocols described here are composites, adapted from the cited reports and commercial products, and they have been used by a number of laboratories.

New RACE

New RACE, a variation of RNA ligase-mediated-RACE (RLM-RACE; Liu and Gorovsky 1993), departs from classic RACE in that the "anchor" primer is attached to the 5′ end of the mRNA *before* the reverse transcription step; hence the anchor sequence becomes incorporated into the first-strand cDNA if, and only if, the reverse transcription proceeds through the entire length of the mRNA of interest (and through the relatively short anchor sequence), as shown in Figure 3b.

Before beginning new RACE (Fig. 4), the mRNA is subjected to a dephosphorylation step using calf intestinal phosphatase (CIP). This step does not affect full-length mRNAs, which have methyl-G caps at their termini, but it does dephosphorylate degraded mRNAs, which are uncapped at their termini (Volloch et al. 1991). Thus, degraded RNAs are excluded from the ligation step later in the protocol. After the dephosphorylation step, the full-length mRNAs are treated with tobacco acid pyrophosphatase (TAP). This removes the cap structure but preserves the active phosphorylated 5′ termini (Mandl et al. 1991; Fromont-Racine et al. 1993). Using T4 RNA ligase, the full-length mRNA is then ligated to a short synthetic RNA oligonucleotide that has been generated by in vitro transcription of a linearized plasmid (Tessier et al. 1986). The RNA oligonucleotide–mRNA hybrids are then

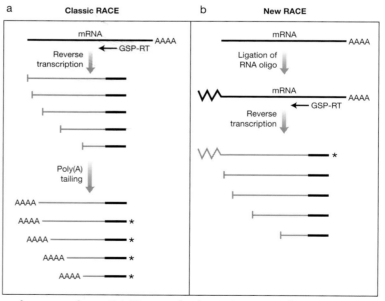

FIGURE 3. The advantage of new RACE over classic RACE. (*a*) In classic RACE, premature termination in the reverse transcription step results in polyadenylation of less-than-full-length first-strand cDNAs, all of which can be amplified using PCR to generate less-than-full-length cDNA 5′ ends. * indicates cDNAs ends that will be amplified in the subsequent PCR. (*b*) In new RACE, less-than-full-length cDNAs are also created, but only full-length molecules are terminated by the RNA oligonucleotide (the anchor sequence) and hence amplified in the subsequent PCR.

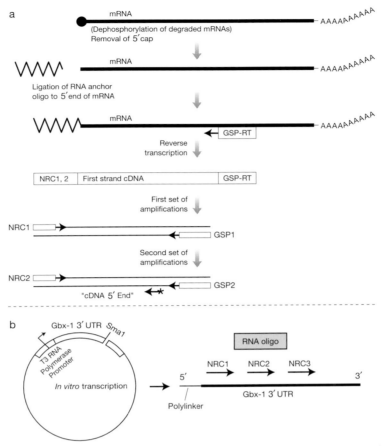

FIGURE 4. A schematic representation of new RACE. See text for details. See Fig. 1 for a description of GSP1, GSP2, and GSP-RT. See below for details of primers NRC1, -2, and -3. (*a*) Amplification of 5′ partial cDNA ends. (*b*) In vitro synthesis of the RNA oligonucleotide used for ligation in new RACE and schematic representation of the corresponding required primers. A 132-nt RNA oligonucleotide is produced by in vitro transcription of the plasmid depicted using T3 RNA polymerase. Primers NRC1, -2, and -3 are derived from the sequence of the oligonucleotide but do not encode restriction sites. To assist in the cloning of cDNA ends, the sequence "ATCG" is added to the 5′ end of NRC2, as described in the cloning section of the text.

reverse-transcribed using a gene-specific primer, thus creating the first-strand cDNA. Finally, the 5′ cDNA ends are amplified in two nested PCRs using additional gene-specific primers and primers derived from the sequence of the RNA oligonucleotide.

New RACE can also be used to generate 3′ cDNA ends (Mandl et al. 1991; Volloch et al. 1991), and it is particularly useful for non-polyadenylated RNAs. In brief, cytoplasmic RNA is dephosphorylated and ligated to a short synthetic RNA oligonucleotide, as described above. Although ligation of the oligonucleotide to the 5′ end of the RNA was emphasized above, RNA oligonucleotides actually ligate to both ends of cytoplasmic RNAs. For the reverse transcription step, a primer derived from the RNA oligonucleotide sequence is used (e.g., the reverse complement of NRC-3; Fig. 4). Reverse transcription of the RNA oligonucleotides that happen to be ligated to the 3′ end of the cytoplasmic RNAs results in the creation of cDNAs that have the RNA oligonucleotide sequence appended to their 3′ end. Gene-specific primers, oriented in the 5′→3′ direction, and new RACE primers, e.g., the reverse complements of NRC-2 and NRC-1 (Fig. 4), are used in nested PCRs to amplify the 3′ ends.

Cap-switching RACE

A simpler method for the selective amplification of full-length cDNA ends involves the addition of an adapter during reverse transcription. This method takes advantage of the propensity of Moloney murine leukemia virus reverse transcriptase (MMLV RT) to append 2–4 cytosines to the 3′ end of newly synthesized cDNA strands. The additional residues are added when the enzyme reaches the 5′ cap structure at the end of the mRNA template (Schmidt and Mueller 1999; Schramm et al. 2000). In the presence of a primer terminating in multiple Gs at its 3′ end, annealing and then complementary copying of the sequence

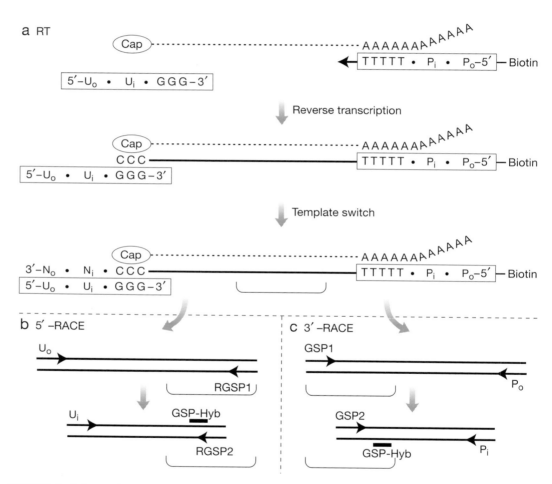

FIGURE 5. Schematic representation of cap-switching RACE. (*a*) Reverse transcription, template switch, and incorporation of adapter sequences at the 3′-end of first strand of cDNA. Biotin-labeled primer P_{total} is used to initiate reverse transcription through hybridization of the poly(dT) tract with the mRNA poly(A) tail. After reaching the 5′ end of the mRNA, oligo(dC) is added by reverse transcriptase in a cap-dependent manner. Then, through template switch via base-pairing between the oligo(dC) and the oligo(dG) at the end of cap finder Adapter, the reverse complementary sequence of the cap finder primer is incorporated into the first strand of the cDNA. Dotted line indicates mRNA; solid line, cDNA; rectangle, primer. The bracket indicates the known region. (*b*) 5′-RACE. The first round of PCR uses primer U_o and RGSP1 (reverse gene-specific primer 1), the second round, U_i and RGSP2. GSP-Hyb is also within the known region, and it can be used to confirm the authenticity of the RACE product. (*c*) 3′-RACE. (Reprinted, with permission, from Zhang 2003.)

of the annealed oligonucleotide take place, thus adding a linker sequence to the cDNA terminus (Fig. 5). Because the template-independent addition of cytosines is cap-dependent, the oligonucleotide is ligated only to full-length cDNA ends. This method is simpler than classic and new RACE, because it has fewer steps; however, the presence of the dG-terminating ("switch") primer can lead to problems if it binds to C-rich sequences in the mRNA of interest.

LIMITATIONS OF COMMERCIAL SYSTEMS

Commercial systems are geared toward the construction of universal pools of full-length cDNAs; for example, see Figure 5, in which all of the mRNAs in the starting material are converted to cDNA. The value of this approach is that, in theory, a single reverse transcription pool can be used to obtain the 5′ end of any transcript. In contrast, noncommercial versions of RACE have emphasized the use of a gene-specific primer to generate the first-strand cDNA templates. Although it lacks universality, the latter approach is more powerful because the reverse transcription reaction starts closer to the 5′ end of the transcript and the relative frequency of the desired cDNA is increased >50-fold in the resulting pool. This enrichment increases the chances of the desired 5′ end being present in sufficient quantity to be amplified using standard PCR methods.

Which approach should investigators choose? For the one-time user, or for those with limited molecular biology experience, the most practical approach would be to obtain a commercial system and, if possible, a pre-made pool of reverse-transcribed cDNAs. Pools representing many human tissues are available. Failing that, use of a gene-specific reverse transcription primer in conjunction with the commercial kits will overcome the limitation described above. For this purpose, the Clontech cap-finding Smart system is easier to use than the Invitrogen RLM system, although both have been reported to work well by many investigators. However, because commercial kits are relatively expensive, investigators who plan to use RACE regularly will achieve substantial savings if they prepare the reagents themselves.

PROTOCOL 1   3′-END cDNA AMPLIFICATION USING CLASSIC RACE

Reagents

The materials required for this procedure can be purchased, along with the appropriate 5x or 10x enzyme reaction buffers, from most major suppliers. SuperScript II reverse transcriptase and RNase H can be obtained from Invitrogen, RNasin from Promega Biotech, *Hercules* heat-stable Hot-Start DNA polymerase from Stratagene, and Tdt from either Invitrogen or Boehringer Mannheim. Enzymes are used as directed by the suppliers. Any of many different heat-stable DNA polymerases can be used, although investigators should look for robust amplification (as opposed to fidelity or cost). Expand from Boehringer Mannheim is another good choice. Oligonucleotide primer sequences are listed in the legend to Figure 1. Primers can be used "crude" except for Q_T, which should be purified to ensure that it is uniformly full length (this can be requested as part of the commercial service from most companies that synthesize primers). dNTPs can be purchased as 100 mM solutions from many suppliers.

This protocol is split into two stages:
Stage 1: Reverse transcription to generate cDNA templates
Stage 2: Amplification

STAGE 1: REVERSE TRANSCRIPTION TO GENERATE cDNA TEMPLATES

MATERIALS

BUFFERS, SOLUTIONS, AND REAGENTS
dNTP solution (containing all four dNTPs, each at 10 mM)
Dithiothreitol (DTT)<!> (0.1 M)
TE (10 mM Tris-HCl<!>, pH 7.5, 1 mM EDTA, pH 8.0)

ENZYMES AND BUFFERS
Reverse transcription buffer, 5x (as supplied by manufacturer)
RNase H
RNasin
SuperScript II reverse transcriptase (Invitrogen)

NUCLEIC ACIDS AND OLIGONUCLEOTIDES
Poly(A)$^+$ RNA, or total RNA
Q_T primer (100 ng/µl)

SPECIAL EQUIPMENT
Water baths or heating blocks preset to 37°C, 42°C, 50°C, 70°C, 80°C

METHOD

1. In a sterile microfuge tube, assemble the following transcription components on ice:

Reverse transcription buffer, 5x	4 µl
dNTP solution (10 mM)	1 µl
DTT (0.1 M)	2 µl
Q_T primer (100 ng/µl)	0.5 µl
RNasin (40 units/µl)	0.25 µl

2. In a separate tube, add 1 µg of poly(A)$^+$ RNA, or 5 µg of total RNA, to 13 µl of H$_2$O. Incubate at 80°C for 3 minutes, cool rapidly on ice, and centrifuge for 5 seconds in a microcentrifuge.

   Poly(A)$^+$ RNA is used in preference to total RNA for reverse transcription to reduce background, but it is unnecessary to prepare it if only total RNA is available.

3. Add the RNA to the reverse transcription components. Then add 1 µl (200 units) of SuperScript II reverse transcriptase, and incubate for 5 minutes at room temperature, 1 hour at 42°C, and 10 minutes at 50°C.

4. Inactivate the reverse transcriptase by incubating at 70°C for 15 minutes. Centrifuge for 5 seconds in a microcentrifuge.

5. Destroy RNA template by adding 0.75 µl (1.5 units) of RNase H and incubating at 37°C for 20 minutes.

6. Dilute the reaction mixture to 1 ml with TE and store at 4°C. (This is the 3′ end cDNA pool.)

COMMENTS

An important factor in the generation of full-length 3′ end partial cDNAs is the stringency of the reverse transcription reaction. Historically, reverse transcription reactions were carried out at relatively low temperatures (37–42°C), using a vast excess of primer (~1/2 the mass of the mRNA, which represents an ~30:1 molar ratio). Under these low-stringency conditions, a stretch of A residues as short as 6–8 nucleotides will suffice as a binding site for an oligonucleotide(dT)-tailed primer. This binding can result in the initiation of cDNA synthesis at sites upstream of the poly(A) tail, leading to truncation of the desired amplification product. One should be suspicious that this has occurred if a canonical polyadenylation signal sequence (Wickens and Stephenson 1984) is not found near the 3´ end of the cDNAs generated. The problem can be minimized by controlling two parameters: primer concentration and reaction temperature. The primer concentration can be reduced dramatically without significantly decreasing the yield of cDNA (Coleclough 1987). At lower concentrations, the primer will bind preferentially to the longest A-rich stretches present (i.e., the poly[A] tail). The quantity recommended in the protocol above represents a good starting point and can be reduced 5-fold further if significant truncation is observed.

In the protocol described above, the incubation temperature is raised slowly to encourage reverse transcription to proceed through regions of difficult secondary structure. Since the half-life of reverse transcriptase rapidly decreases as temperature increases, the reaction cannot be carried out at elevated temperatures in its entirety. In theory, the problem of difficult secondary structure (and nonspecific reverse transcription) could be ameliorated by using heat-stable reverse transcriptases, which are available from several suppliers. However, the authors have never been able to attribute improvements to the use of these enzymes, which are merely promiscuous DNA polymerases.

STAGE 2: REVERSE TRANSCRIPTION TO GENERATE cDNA TEMPLATES

MATERIALS

▼ CAUTION

See Appendix for appropriate handling of materials marked with <!>.

BUFFERS, SOLUTIONS, AND REAGENTS
dNTP solution (containing all four dNTPs, each at 10 mM)
TE (10 mM Tris-HCl<!>, pH 7.5, 1 mM EDTA, pH 8.0)

ENZYMES AND BUFFERS
Hercules Hot-Start polymerase (Stratagene)
Hercules Hot-Start polymerase buffer (10x)

NUCLEIC ACIDS AND OLIGONUCLEOTIDES
3´ cDNA pool obtained in Stage 1 above
Oligonucleotide primers Q_O, Q_I, GSP1, and GSP2 (25 pmoles/µl) (see Fig. 2 for sequence of Q_O and Q_I)

SPECIAL EQUIPMENT
Programmable thermal cycler

METHOD

First round:

1. In a sterile 0.2-ml microfuge tube mix the following reagents:
   Hercules Hot-Start polymerase buffer (10x)   5 µl
   dNTP solution (10 mM)   1.0 µl
   Hercules Hot-Start polymerase   2.5 units
   H_2O to 50 µl

2. Add a 1-µl aliquot of the 3´ end cDNA pool (obtained in step 6 above) and 25 pmoles each of primers GSP1 and Q_O.

3. Mix and heat in a DNA thermal cycler for 5 minutes at 98°C to denature the first-strand products and to activate the polymerase. Cool to the appropriate annealing temperature (56–68°C) for 2 minutes. Extend the cDNAs at 72°C for 40 minutes.

4. Carry out 30 cycles of amplification as follows:

Cycle number	Denaturation	Annealing	Polymerization/Extension
29	10 sec at 94°C	10 sec at 52–68°C	3 min at 72°C
Last cycle	10 sec at 94°C	10 sec at 52–68°C	15 min at 72°C, then cool to room temperature

Second round:

5. Dilute a portion of the amplification products from the first round 1:20 in TE.

6. Amplify 1 μl of the diluted material with primers GSP2 and Q_I using the procedure described above, but eliminate the initial 2-minute annealing step and the 72°C, 40-minute extension step.

COMMENTS

Use of "Hot Start" PCR (via any of several approaches) is vital to minimize nonspecific amplification. An annealing temperature close to the effective T_m of the primers should be used. The Q_I and Q_O primers work well at 64°C under the PCR conditions recommended here, although these were optimized using legacy machines. Conditions may vary depending on the PCR machine used. Gene-specific primers of similar length and GC content should be chosen. Computer programs to assist in the selection of primers are widely available and should be used (e.g., http://www-genome.wi.mit.edu/cgi-bin/primer/primer3_www.cgi/). Extension should be carried out for 1 minute for every 1000 bp of the expected product. If the expected product length is unknown, extension for 3–4 minutes is a good starting point.

Very little substrate is required for the PCR. A 1-μg amount of poly(A)+ RNA typically contains ~5 $\times$ 10^7 copies of each low-abundance transcript. The PCR described here works optimally when 10^3–10^5 templates (of the desired cDNA) are present in the starting mixture; thus, as little as 0.002% of the reverse transcription mixture suffices for the reaction! Addition of too much starting material will lead to production of large amounts of nonspecific product and should be avoided. The RACE technique is particularly sensitive to this problem, because every cDNA in the mixture, desired and undesired, contains a binding site for the Q_O and Q_I primers.

It was found empirically that allowing extra extension time (40 minutes) during the first amplification round (when the second strand of cDNA is created) can result in increased yields of the specific product, relative to background amplification. In particular, the extra extension time frequently leads to increased yields of long cDNAs versus short cDNAs when specific cDNA ends of multiple lengths were present (Frohman et al. 1988). Prior treatment of cDNA templates with RNA hydrolysis, or a combination of RNase H and RNase A, occasionally improves the efficiency of specific cDNA amplification.

PROTOCOL 2   5′ END cDNA AMPLIFICATION USING CLASSIC RACE

This protocol is split into three stages:

Stage 1: Reverse transcription to generate cDNA templates
Stage 2: Appending a poly(A) tail to first-strand cDNA products
Stage 3: Amplification

STAGE 1: REVERSE TRANSCRIPTION TO GENERATE cDNA TEMPLATES

MATERIALS

BUFFERS, SOLUTIONS, AND REAGENTS

dNTP solution (containing all four dNTPs, each at 10 mM)
DTT<!> (0.1 M)
TE (10 mM Tris-HCl<!>, pH 7.5, 1 mM EDTA, pH 8.0)

ENZYMES AND BUFFERS

Reverse transcription buffer, 5x (as supplied by manufacturer)
RNase H
RNasin
SuperScript II reverse transcriptase (Invitrogen)

NUCLEIC ACIDS AND OLIGONUCLEOTIDES

GSP-RT primer (100 ng/µl)
Poly(A)$^+$ RNA, or total RNA

SPECIAL EQUIPMENT

Water baths or heating blocks preset to 37°C, 42°C, 50°C, 70°C, 80°C

METHOD

1. In a sterile microfuge tube, assemble the following transcription components on ice:

Reverse transcription buffer, 5x	4 µl
dNTPs (containing all four dNTPs, each at 10 mM)	1 µl
DTT, 0.1 M	2 µl
RNasin (40 units/µl)	0.25 µl

2. In a separate sterile microfuge tube, mix 0.5 µl of GSP-RT primer (100 ng/µl) and 1 µg of poly(A)$^+$ RNA, or 5 µg of total RNA, with 13 µl of H$_2$O. Incubate at 80°C for 3 minutes, cool rapidly on ice, and centrifuge for 5 seconds in a microcentrifuge.

3. Add the RNA/primer mix to the reverse transcription components. Then add 1 µl (200 units) of SuperScript II reverse transcriptase. Mix gently, and incubate for 1 hour at 42°C, and then 10 minutes at 50°C.

4. Inactivate the reverse transcriptase by incubating at 70°C for 15 minutes. Centrifuge for 5 seconds in a microcentrifuge.

5. Destroy RNA template by adding 0.75 µl (1.5 units) of RNase H to tube and incubating at 37°C for 20 minutes.

6. Dilute the reaction mixture to 400 µl with TE and store at 4°C (this is the 5′ end non-tailed cDNA pool).

COMMENTS

Many of the comments in Protocol 1, stage 1 (reverse transcribing 3′-end partial cDNAs) also apply here and should be noted. However, in 5′-end cDNA amplification, the efficiency of cDNA extension is critically important. In the 5′ procedure, each specific cDNA, no matter how short, is tailed and becomes a potential target for amplification (Fig. 2a). Thus, the quality of the final PCR products directly reflects that of the reverse transcription reaction. The length of the first-strand cDNA can be maximized by using clean, intact RNA, and by selecting a reverse transcriptase primer that anneals near the 5´ end of region of known sequence. Improvements can also be made, at least in theory, by using a combination of SuperScript II and heat-stable reverse transcriptase at multiple temperatures. At elevated temperatures the amount of secondary structure encountered in GC-rich regions of the mRNA should be reduced.

STAGE 2: APPENDING A POLY(A) TAIL TO FIRST-STRAND cDNA PRODUCTS

MATERIALS

▼ CAUTION

See Appendix for appropriate handling of materials marked with <!>.

BUFFERS, SOLUTIONS, AND REAGENTS
 $CoCl_2$<!> (25 mM)
 dATP solution (1 mM)
 TE (10 mM Tris-HCl<!>, pH 7.5, 1 mM EDTA, pH 8.0)

ENZYMES AND ENZYME BUFFERS
 Terminal deoxynucleotidyltransferase (Tdt)
 Tailing buffer, 5x (125 mM Tris-HCl<!>, pH 6.6, 1 M potassium cacodylate<!>, 1250 µg/ml BSA)

SPECIAL EQUIPMENT
 Microcon-100 filter (Millipore) or equivalent
 Water bath or heating blocks preset to 37°C, 65°C

METHOD

1. Remove excess primer from the 5´-end nontailed cDNA pool (obtained in Stage 1, step 2, above) using Microcon-100 spin filters (Millipore) or an equivalent product, according to the manufacturer's instructions. Wash the material by spin filtration twice more using TE, according to manufacturer's instructions. The final volume recovered should not exceed 15 µl. Adjust volume to 15 µl using H_2O.

2. Add 4 µl of 5x tailing buffer and 10 units of Tdt.

3. Incubate for 5 minutes at 37°C and then for 5 minutes at 65°C.

4. Dilute to 500 µl with TE and store at 4°C (this is the 5´-end tailed cDNA pool).

COMMENTS

To attach a known sequence to the 5´ end of the first-strand cDNA, a homopolymeric tail is appended using Tdt. It is preferable to add poly(A) tails rather than poly(C) tails for a number of reasons. First, the 3´-end strategy is based on the naturally occurring poly(A) tail; adding a poly(A) tail to the 5´ end allows the same adapter primer to be used for both ends, thus simplifying the protocol and reducing cost. Second, because AT binding is weaker than GC binding, longer stretches of A residues (~2x) are required before the oligo(dT)-tailed Q_T primer will bind to the template. Internal poly(A) tracts are rare, so the chance of nonspecific binding and the production of truncated amplification products is reduced. Third, vertebrate coding sequences and 5´ untranslated regions tend to be biased toward GC residues; thus, use of a poly(A) tail further decreases the likelihood of inappropriate amplification.

Unlike many other applications that use homopolymeric tails, the actual length of the tail added here is unimportant, as long as it exceeds 17 nucleotides. This is because the oligo(dT)-tailed primer binds at the junction of the appended poly(A) tail and the cDNA transcript. The conditions described above result in the addition of 30–400 A residues.

STAGE 3: AMPLIFICATION

MATERIALS

BUFFERS, SOLUTIONS, AND REAGENTS
dNTP solution (containing all four dNTPs, each at 10 mM)
TE (10 mM Tris-HCl<!>, pH 7.5, 1 mM EDTA, pH 8.0)

ENZYMES AND ENZYME BUFFERS
Hercules Hot-Start polymerase (Stratagene)
Hercules Hot-Start polymerase buffer (10x)

NUCLEIC ACIDS AND OLIGONUCLEOTIDES
5′-end tailed cDNA pool (obtained in Stage 2 above)
Oligonucleotide primers Q_o, Q_T and GSP1 (see Fig. 2 for sequence of Q_o and Q_T)

SPECIAL EQUIPMENT
Programmable thermal cycler

METHOD

First round:

1. In a sterile 0.2-ml microfuge tube, mix the following reagents:

Hercules Hot-Start polymerase buffer (10x)	5 µl
dNTP solution (10 mM)	1.0 µl
Hercules Hot-Start polymerase	2.5 units
H_2O	to 50 µl

2. Add a 1-µl aliquot of the 5′-end tailed cDNA pool (obtained in step 4 above) and 25 pmoles each of primers GSP1, Q_o (shown in Fig. 2b), and Q_T.

3. Mix and heat in a DNA thermal cycler for 5 minutes at 98°C to denature the first-strand products and to activate the polymerase. Cool to 48°C for 2 minutes. Extend the cDNAs at 72°C for 40 minutes.

4. Carry out 30 cycles of amplification using a step program as follows:

Cycle number	Denaturation	Annealing	Polymerization/Extension
29	10 sec at 94°C	10 sec at 52–68°C	3 min at 72°C
Last cycle	10 sec at 94°C	10 sec at 52–68°C	15 min at 72°C, then cool to room temperature

Second round:

5. Dilute a portion of the amplification products from the first round 1:20 in TE.

6. Amplify 1 µl of the diluted material with primers GSP2 and Q_I using the procedure described above, but eliminate the initial 2-minute annealing step and the 72°C, 40-minute extension step.

> ## COMMENTS
>
> Many of the remarks made above in the section on amplifying 3´ end partial cDNAs also apply here and should be noted. There is, however, one major difference. The annealing temperature in the first step (48°C) is lower than that used in successive cycles (52–68°C). This is because cDNA synthesis during the first round depends on the interaction of the appended poly(A) tail and the oligo(dT)-tailed Q_T primer. In all subsequent rounds, amplification can proceed using the Q_O primer, which is composed of ~60% GC and which can anneal to its complementary target at a much higher temperature.

PROTOCOL 3 5´ END cDNA AMPLIFICATION USING NEW RACE

Reagents

See the classic RACE section above for sources of some of the required materials. Calf intestinal phosphatase (CIP) and proteinase K can be purchased from Boehringer Mannheim; tobacco acid pyrophosphatase (TAP) and a cost-effective RNA transcription kit can be purchased along with the appropriate 10x enzyme reaction buffers from Epicentre. T4 RNA ligase can be purchased from New England Biolabs or Boehringer Mannheim. Note that 10x T4 RNA ligase buffers supplied by some manufacturers contain too much ATP (Tessier et al. 1986). Check the composition of any commercially supplied 10x buffer and make your own if it contains more than 1 mM ATP (final 1x concentration should be 0.1 mM), as described below.

The procedure detailed below uses relatively large amounts of RNA and can be scaled down if RNA quantities are limited. If large quantities of RNA are used, it is practical to sacrifice an aliquot at each stage of the experiment to check for sample degradation. This can be done by agarose gel electrophoresis. Samples in which no detectable degradation has occurred can be stored indefinitely for future experiments. The protocol is split into six stages:

Stage 1: Dephosphorylation of degraded RNAs
Stage 2: Decapping of intact RNAs
Stage 3: Preparation of RNA oligonucleotide
Stage 4: RNA oligonucleotide–cellular RNA ligation
Stage 5: Reverse transcription
Stage 6: Amplification

STAGE 1: DEPHOSPHORYLATION OF DEGRADED RNAs

MATERIALS

▼ CAUTION
See Appendix for appropriate handling of materials marked with <!>.

BUFFERS, SOLUTIONS, AND REAGENTS
Sodium acetate <!>, 3 M, pH 5.2
Ethanol <!>
Dithiothreitol (DTT)<!> (0.1 M)
Phenol/chloroform<!> (1:1, v/v)

ENZYMES AND ENZYME BUFFERS
Phosphatase buffer, 10x

Proteinase K<!>
RNasin
Calf intestinal phosphatase (CIP)

NUCLEIC ACIDS AND OLIGONUCLEOTIDES
RNA sample

SPECIAL EQUIPMENT
Water baths or heating blocks preset to 37°C, 50°C

ADDITIONAL ITEMS
Equipment and reagents for agarose gel electrophoresis, including ethidium bromide <!>

METHOD

In general, follow the manufacturer's recommendations for use of the phosphatase.

1. In a sterile microfuge tube, combine the following reagents;

RNA	50 µg
10x phosphatase buffer	5 µl
DTT (100 mM)	0.5 µl
RNasin (40 units/µl)	1.25 µl
CIP (1 unit/µl)	3.5 µl
H$_2$O	to 50 µl

2. Incubate the reaction at 50°C for 1 hour.

3. Add proteinase K to 50 µg/ml and incubate at 37°C for 30 minutes.

4. Extract the reaction with a mixture of phenol/chloroform, extract again with chloroform, and precipitate the RNA using 1/10 volume of 3 M sodium acetate, pH 5.2, and 2.5 volumes of ethanol. Resuspend the RNA in 43.6 µl of H$_2$O.

5. Analyze 2 µg (1.6 µl) of RNA on a 1% agarose gel (TAE) adjacent to a lane containing 2 µg of the original RNA preparation, stain the gel with ethidium bromide, and visually confirm that the RNA has remained intact during the dephosphorylation step.

STAGE 2: DECAPPING OF INTACT RNAs

▼ CAUTION

See Appendix for appropriate handling of materials marked with <!>.

MATERIALS

BUFFERS, SOLUTIONS, AND REAGENTS
ATP (100 mM)
Chloroform<!>
Sodium acetate<!>, 3 M, pH 5.2
Ethanol<!>
Phenol/chloroform<!> (1:1, v/v)
TE (10 mM Tris-HCl<!>, pH 7.5, 1 mM EDTA, pH 8.0)

ENZYMES AND ENZYME BUFFERS
RNasin
Tobacco acid pyrophosphatase (TAP), 5 units/µl
TAP buffer, 10x

NUCLEIC ACIDS AND OLIGONUCLEOTIDES
RNA sample, recovered from Stage 1 above

SPECIAL EQUIPMENT
Water bath or heating block preset to 37°C

ADDITIONAL ITEMS
Equipment and reagents for agarose gel electrophoresis, including ethidium bromide <!>

METHOD

....

1. In a sterile microfuge tube mix the following reagents:

RNA (as obtained in Stage 5 above)	38 µg
TAP Buffer (10x)	5 µl
RNasin (40 units/µl)	1.25 µl
ATP (100 mM)	1 µl
TAP (5 units/µl)	1 µl
H_2O	to 50 µl

2. Incubate the reaction at 37°C for 1 hour, and then add 200 µl of TE.

3. Extract the reaction with a mixture of phenol/chloroform, extract again with chloroform, and precipitate the RNA using 1/10th volume of 3 M sodium acetate, pH 5.2, and 2.5 volumes of ethanol. Resuspend the RNA in 40 µl of H_2O.

4. Analyze 2 µg of RNA on a 1% agarose gel (TAE) adjacent to a lane containing 2 µg of the original RNA preparation, stain the gel with ethidium bromide, and visually confirm that the RNA remained intact during the decapping stage.

COMMENT

Most protocols call for large amounts of TAP. The enzyme is very expensive and it is not necessary to use more than the amount recommended in this protocol!

STAGE 3: PREPARATION OF RNA OLIGONUCLEOTIDE

Choose a plasmid that can be linearized at a site ~100 bp downstream from a T7 or T3 RNA polymerase site (see Fig. 3b). Ideally, use a plasmid containing some insert cloned into the first polylinker site, because primers made from palindromic polylinker DNA do not perform well in PCR. A tried and tested construct is pBS-SK-GBX-1-3′UTR, which contains the 3′ untranslated region (UTR) of the mouse gene *Gbx-1* (Frohman et al. 1993) cloned into the *Sst*I site of pBS-SK (Stratagene). This can be linearized with *Sma*I and transcribed with T3 RNA polymerase to produce a 132-nucleotide RNA oligonucleotide. All but 17 of these nucleotides are from *Gbx-1*. Note that adenosines are the best "acceptors" for the 3′ end of the RNA oligonucleotide to ligate to the 5′ end of its target, if an appropriate restriction site can be found. The primers subsequently used for amplification are all derived from the *Gbx-1* 3′UTR sequence. The *Gbx-1* NRC primer sequences and plasmid are available from the Frohman lab upon request.

Carry out a test transcription to make sure that everything is working; then scale up. The oligonucleotide can be stored at –80°C indefinitely for future experiments, and it is important to synthesize enough oligonucleotide to allow for losses due to purification and spot checks along the way.

MATERIALS

BUFFERS, SOLUTIONS, AND REAGENTS
Chloroform<!>
Ethanol<!>
Diethyl pyrocarbonate (DEPC)<!>-treated H_2O
rUTP solution, 10 mM
rATP solution, 10 mM
rCTP solution, 10 mM
rGTP solution, 10 mM
DTT<!> (0.1 M)
Phenol/chloroform (1:1, v/v)<!>
Sodium acetate<!>, 3 M, pH 5.2
TE (10 mM Tris-HCl<!>, pH 7.5, 1 mM EDTA, pH 8.0)

ENZYMES AND ENZYME BUFFERS
Appropriate restriction enzymes and buffers
Pancreatic DNase I (RNase-free)
Proteinase K
DNA-dependent RNA polymerase
RNasin
Transcription buffer, 5x (as supplied by the manufacturer)

NUCLEIC ACIDS AND OLIGONUCLEOTIDES
Plasmid template DNA

SPECIAL EQUIPMENT
Microcon filter of appropriate size (see step 8)
Water bath or heating block preset to 37°C

ADDITIONAL ITEMS
Equipment and reagents for TAE agarose gel electrophoresis, including ethidium bromide<!>

METHOD

1. Linearize 25 µg of the plasmid that is to be transcribed (the plasmid should be reasonably free of RNases) by restriction digestion using the appropriate enzymes and buffers.

2. Treat the digestion reaction with 50 µg/ml proteinase K for 30 minutes at 37°C, extract twice with phenol/chloroform, once with chloroform, and collect the DNA by standard ethanol precipitation.

3. Redissolve the template DNA in 25 µl of TE, pH 8.0. This will give a final concentration of ~1 µg/µl.

4. In a sterile microfuge tube, at room temperature, mix transcription reagents in the following order:

	Test scale	Preparative scale
DEPC-treated H_2O	4 µl	80 µl
Transcription buffer (5x)	2 µl	40 µl
DTT (0.1 M)	1 µl	20 µl
rUTP (10 mM)	0.5 µl	10 µl
rATP (10 mM)	0.5 µl	10 µl
rCTP (10 mM)	0.5 µl	10 µl
rGTP (10 mM)	0.5 µl	10 µl
Linearized DNA (1 µg/µl)	0.5 µl	10 µl
RNasin (40 units/ul)	0.25 µl	5 µl
RNA polymerase (20 units/µl)	0.25 µl	5 µl

Incubate at 37°C for 1 hour.

5. After transcription, remove the DNA template by adding 0.5 µl of DNase (RNase-free) for every 20 µl of reaction volume. Incubate at 37°C for 10 minutes.

6. Check the oligonucleotide product by analyzing 5 µl of the test or preparative reaction on a TAE agarose gel (1%). Expect to see a diffuse band at about the right size for the expected product (or a bit smaller) in addition to some smearing up and down the gel.

7. Purify the oligonucleotide by extracting with phenol/chloroform and then chloroform. Rinse three times with H_2O and then pass though a Microcon (Millipore) spin filter (prerinsed with H_2O).

> Microcon 30 spin filters have a cutoff size of 60 nucleotides, and Microcon 100 spin filters have a cutoff size of 300 nucleotides. Microcon 10 spin filters are probably most appropriate if the oligonucleotide is smaller than 100 nucleotides, and Microcon 30 spin filters for anything larger.

8. Analyze a second aliquot of the oligonucleotide on a TAE agarose gel (1%) to check the integrity and concentration of the sample. Store in aliquots at –80°C indefinitely.

STAGE 4: RNA OLIGONUCLEOTIDE–CELLULAR RNA LIGATION

MATERIALS

BUFFERS, SOLUTIONS, AND REAGENTS
ATP, 2 mM

ENZYMES AND ENZYME BUFFERS
Ligation buffer, 10x (500 mM Tris-HCl<!>, pH 7.9, 100 mM $MgCl_2$<!>, 20 mM DTT<!>, 1 mg/ml BSA)
RNasin
T4 RNA ligase

▼ CAUTION

See Appendix for appropriate handling of materials marked with <!>.

NUCLEIC ACIDS AND OLIGONUCLEOTIDES
RNA oligonucleotide, as prepared in Stage 3 above
RNA sample, TAP-treated and untreated

SPECIAL EQUIPMENT
Microcon-100 filter
Water bath or heating block preset to 17°C

ADDITIONAL ITEMS
Equipment and reagents for agarose gel electrophoresis, including ethidium bromide <!>

METHOD

1. Set up two sterile microfuge tubes: one with TAP-treated cellular RNA, the other with untreated cellular RNA.

Ligation buffer (10x)	3 µl
RNasin (40 units/µl)	0.75 µl
RNA oligonucleotide	4 µg*
TAP-treated (or untreated) RNA	10 µg
ATP (2 mM)	1.5 µl
T4 RNA ligase (20 units/µl)	1.5 µl
H$_2$O	to 30 µl

    *3–6 molar excess over target cellular RNA

2. Incubate for 16 hours or overnight at 17°C.

3. Purify the ligation product using Microcon 100 spin filtration (three times in H$_2$O; pre-rinse filter with RNase-free H$_2$O). The volume recovered should not exceed 20 µl.

4. Check integrity of the ligated RNA by analyzing one-third of the product on a TAE agarose gel (1%). It should be little changed by ligation.

STAGE 5: REVERSE TRANSCRIPTION

MATERIALS

BUFFERS, SOLUTIONS, AND REAGENTS
dNTP solution (containing all four dNTPs, each at 10 mM)
DTT<!>, 0.1 M
TE (10 mM Tris-HCl<!>, pH 7.5, 1 mM EDTA, pH 8.0)

ENZYMES AND ENZYME BUFFERS
Reverse transcription buffer, 5x (as supplied by the manufacturer)
RNase H
RNasin
SuperScript II reverse transcriptase (Invitrogen)

NUCLEIC ACIDS AND OLIGONUCLEOTIDES
Anti-sense-specific primer (20 ng/µl), or random hexamers (50 ng/µl)
Oligo-ligated RNA, as prepared in Stage 5 above

SPECIAL EQUIPMENT
Water baths or heating blocks, preset to 37°C, 42°C, 50°C, 70°C

METHOD

1. In a sterile microfuge tube, assemble the following transcription components on ice:

Reverse transcription buffer, 5x	4 µl
dNTP solution (10 mM)	1 µl
DTT (0.1 M)	2 µl
RNasin (40 units/µl)	0.25 µl

2. In a separate tube, add 20 ng of antisense-specific primer, or 50 ng of random hexamers, to the remaining RNA (~6.7 μg) in 13 μl of H_2O. Incubate at 80°C for 3 minutes, cool rapidly on ice, and centrifuge for 5 seconds in a microcentrifuge.

3. Add the RNA/primer mix to the reverse transcription components. Then add 1 μl (200 units) of SuperScript II reverse transcriptase. Incubate for 1 hour at 42°C, and 10 minutes at 50°C. If using random hexamers, insert a room-temperature, 10-minute incubation period after mixing everything together.

4. Inactivate the reverse transcriptase by incubating at 70°C for 15 minutes. Centrifuge for 5 seconds in a microcentrifuge.

5. Destroy the RNA template by adding 0.75 μl (1.5 units) of RNase H. Incubate at 37°C for 20 minutes.

6. Dilute the reaction mixture to 100 μl with TE (10 mM Tris-HCl, pH 7.5, 1 mM EDTA) and store at 4°C (this is the 5′ end oligonucleotide-cDNA pool).

STAGE 6: AMPLIFICATION

MATERIALS

BUFFERS, SOLUTIONS, AND REAGENTS
TE (10 mM Tris-HCl<!>, pH 7.5, 1 mM EDTA, pH 8.0)
DMSO<!>

ENZYMES AND ENZYME BUFFERS
dNTP solution (containing all four dNTPs, each at 10 mM)
Taq polymerase buffer (10x)

NUCLEIC ACIDS AND OLIGONUCLEOTIDES
5′ end oligonucleotide-cDNA pool (obtained in Stage 6 above)
Oligonucleotide primers GSP1, GSP2, NRC1, and NCR2 (see Fig. 4 for details of primers NRC1 and NRC2)

SPECIAL EQUIPMENT
Programmable thermal cycler

METHOD

First round:

1. In a sterile 0.2-ml microfuge tube mix the following reagents:

Taq polymerase buffer (10x)	5 μl
dNTP solution (10 mM)	1.0 μl
DMSO	5 μl
H_2O	to 50 μl

3. Add a 1-μl aliquot of the 5′-end oligonucleotide–cDNA pool and 25 pmoles each of primers GSP1 and NRC1.

4. Heat in a DNA thermal cycler for 5 minutes at 98°C to denature the first-strand products and to activate the polymerase. Cool to the appropriate annealing temperature (56–68°C) for 2 minutes. Extend the cDNAs at 72°C for 40 minutes.

5. Carry out 35 cycles of amplification using a step program as follows:

Cycle number	Denaturation	Annealing	Polymerization/Extension
29	10 sec at 94°C	10 sec at 52–60°C	3 min at 72°C
Last cycle	10 sec at 94°C	10 sec at 52–60°C	15 min at 72°C, then cool to room temperature

Second round:

6. Dilute a portion of the amplification products from the first round 1:20 in TE.

7. Amplify 1 µl of the diluted material with primers GSP2 and NRC2 using the procedure described above, but eliminate the initial 2-minute annealing step and the 72°C, 40-minute extension step.

COMMENTS

Many of the remarks made above in the classic RACE section on amplifying 5′-end partial cDNAs also apply here and should be noted.

PROTOCOL 4   CAP-SWITCHING RACE

In the classic RACE protocol, first-strand products are cleaned up using Microcon filters. In the method presented here, a biotinylated primer is used to facilitate first-strand purification. The clean-up techniques are interchangeable. Similarly, the use of gene-specific primers for the initiation of reverse transcription (as described above) is interchangeable with the use of an oligonucleotide-dT-based primer as described here.

Reagents

The success of 5′ RACE relies on the incorporation of the cap finder adapter sequence at the beginning of the cDNA. This step depends on the addition of extra oligonucleotide(dC) to the end of the first-strand cDNA. It has been shown that the presence of Mn^{++} ions in the reverse transcription buffer greatly enhances the efficiency of this step (Schmidt and Mueller 1999).

STAGE 1: REVERSE TRANSCRIPTION TO GENERATE cDNA TEMPLATES

MATERIALS

> ▼ CAUTION
>
> *See Appendix for appropriate handling of materials marked with <!>.*

BUFFERS, SOLUTIONS, AND REAGENTS

dNTP solution (containing all four dNTPs, each at 10 mM)

TE (10 mM Tris-HCl<!>, pH 7.5, 1 mM EDTA, pH 8.0), prewarmed to 50°C

ENZYMES AND ENZYME BUFFERS

Reverse transcription buffer, 5x (250 mM Tris-HCl<!>, pH 8.3 at 45°C, 30 mM $MgCl_2$<!>, 10 mM $MnCl_2$<!>, 50 mM dithiothreitol<!>, 1 mg/ml BSA)

RNasin

SuperScript II reverse transcriptase (Invitrogen)

NUCLEIC ACIDS AND OLIGONUCLEOTIDES

Biotin <!>-labeled primer P_{total} (Biotin-labeled primers can be ordered from Invitrogen.) See Figure 6 for sequence of P_{total}.

Cap-finder adapter primer (10 mM). See Figure 6 for sequence of cap-finder adapter primer.

Poly(A)$^+$ RNA

SPECIAL EQUIPMENT

Streptavidin magnetic beads (e.g., Streptavidin MagneSphere particles from Promega)

Magnetic separator (e.g., MagneSphere Magnetic Separation Stand from Promega)

Water baths or heating blocks preset to 42°C, 45°C, 50°C, 70°C, 80°C

METHOD

1. In a sterile microfuge tube, assemble the following transcription components on ice:

Reverse transcription buffer, 5x	4 µl
dNTP solution (10 mM)	0.4 µl
CapFinder adapter primer (10 mM)	1 µl
RNasin (40 units/µl)	0.25 µl

2. In a separate tube, mix 1 µg of poly(A)$^+$ RNA and 10 pmoles of P_{total} primer in 12 µl of H$_2$O. Incubate for 3 minutes at 80°C, cool rapidly on ice, and centrifuge for 5 seconds in a microcentrifuge. Combine with the reverse transcription components from step 1.

3. Add 1 µl (200 units) of SuperScript II reverse transcriptase, and incubate for 5 minutes at room temperature, 30 minutes at 42°C, 30 minutes at 45°C, and 10 minutes at 50°C.

4. Inactivate reverse transcriptase by incubating for 15 minutes at 70°C. Separate the first-strand cDNA using magnetic streptavidin beads; follow the manufacturer´s instructions

```
CapFinder: 5'-TGGTTGCCATAAGCGGATCATCGGGAGGAGAAACGGG-3'
Uo:        5'-TGGTTGCCATAAGCGGATC-3'
Ui:                            5'-TCATCGGGAGGAGAAACGG-3'

Ptotal: 3'-G/A/C(T)17CTATCGCTCCGCTCGCAAGGTTTGGGTCAGGTTGGTTT-5'
Pi:                  3'-CTATCGCTCCGCTCGCAAG-5'
Po:                                    3'-GGTTTGGGTCAGGTTGGTTT-5


For convenience: The 5'->3' sequences of the P primers:

Ptotal: 5'-TTTGGTTGGACTGGGTTTGGAACGCTCGCCTCGCTATC(T)17C/A/G-3'
Po:        5'-TTTGGTTGGACTGGGTTTGG-3'
Pi:                          5'-GGAACGCTCGCCTCGCTATC -3'
```

FIGURE 6. Primer sequences and their relationship. Gene-specific primers are not included. The cap finder sequence was selected from the *Yersinia pestis* genome sequence. BLASTN search results using it can be retrieved with request ID (RID) 1009985097-16592-24994. P_{total} is biotin-labeled at the 5´ end. This is a "lock-docking" degenerate primer that actually consists of three primers with different (A, G, or C) nucleotides at its 3´ end. Its RID is 1009987520-16411-27741. (Reprinted, with permission, from Zhang 2003.)

and use about five times the binding capacity required to complex the amount of biotinylated primer used. Wash with TE at 50°C three times to eliminate excess CapFinder adapters (according to manufacturer's instructions).

5. Dilute the reaction mixture to 0.5 ml with TE, and store at 4°C (this is the cDNA pool).

STAGE 2: AMPLIFICATION OF THE cDNA

MATERIALS

BUFFERS, SOLUTIONS, AND REAGENTS
dNTP solution (containing all four dNTPs, each at 10 mM)

ENZYMES AND ENZYME BUFFERS
Hercules Hot-Start polymerase (Stratagene)
Hercules Hot-Start polymerase buffer (10x)

> Other hot-start PCR systems can also be used; e.g., Expand High Fidelity (Roche).

NUCLEIC ACIDS AND OLIGONUCLEOTIDES
Biotinylated cDNA pool (obtained in Stage 1 above)
Oligonucleotide primers GSP1 and P_o (for 3´ RACE), or RGSP1 and U_o (for 5´ RACE). See Figure 6 for sequences of P_o and U_o.

SPECIAL EQUIPMENT
Programmable thermal cycler

ADDITIONAL ITEMS
If a second round of amplification is necessary, the following reagents are required for steps 5 and 6:
Primers GSP2 and P_i (for 3´ RACE), or RGSP2 and U_i (for 5´ RACE). See Figure 6 for sequences of P_i and U_i.
TE (10 mM Tris-HCl<!>, pH 7.5, 1 mM EDTA, pH 8.0)

▼ CAUTION

See Appendix for appropriate handling of materials marked with <!>.

METHOD

First round:

1. In a sterile 0.2-ml microfuge tube mix the following reagents:

Hercules Hot-Start polymerase buffer (10x)	5 µl
dNTP solution (10 mM)	1.0 µl
Hercules Hot-Start polymerase	2.5 units
H_2O	to 50 µl

2. Add a 1-µl aliquot of the cDNA pool (resuspend the beads well) and 25 pmoles of each of primers GSP1 and P_o (for 3´ RACE), or RGSP1 and U_o (for 5´ RACE).

3. Heat the mixture in the thermal cycler at 98°C for 5 minutes to denature the first-strand products and the streptavidin and to activate the polymerase. Incubate at appropriate annealing temperature for 2 minutes. Extend the cDNAs at 72°C for 40 minutes.

> It is not necessary to keep the magnetic streptavidin beads in suspension during these incubations because the biotin should not interact with the denatured streptavidin. It has been reported that styrene beads have the advantage of being smaller and remain in suspension without agitation (Schramm et al. 2000).

4. Carry out 30 cycles of amplification using a step program as follows:

Cycle number	Denaturation	Annealing	Polymerization/Extension
29	10 sec at 94°C	10 sec at 52–68°C	3 min at 72°C
Last cycle	10 sec at 94°C	10 sec at 52–68°C	15 min at 72°C, then cool to room temperature

The extension time at 72°C must be adjusted according to the length of the product expected and the processivity of the polymerase used.

Second round (if necessary):

5. Dilute a portion of the amplification products from the first round 1:20 in TE.

6. Amplify 1 μl of the diluted material with primers GSP2 and P_I (for 3′ RACE), or RGSP2 and U_i (for 5′ RACE) using the first-round procedure, but eliminate the initial 2-minute annealing step and the 72°C, 40-minute extension step.

ANALYSIS OF AMPLIFICATION PRODUCTS

The production of specific partial cDNAs by the RACE protocol can be assessed using Southern blot hybridization analysis. After the second set of amplification cycles, analyze the first- and second-set reaction products on an agarose gel (1%), stained with ethidium bromide. Using standard procedures, transfer the nucleic acids to a nylon membrane and hybridize with a labeled oligomer or gene fragment derived from the amplification template (e.g., GSP-Hyb/Seq in Fig. 1a,b). Gene-specific partial cDNA ends should be detected easily. Yields of the desired product, relative to nonspecific amplified cDNA in the first-round products, should vary from <1% of the amplified material to nearly 100%. The yield will depend largely on the stringency of the amplification reaction, the amplification efficiency of the specific cDNA end, and the relative abundance of the specific transcript within the mRNA source. In the second set of amplification cycles, ~100% of the cDNA detected by ethidium bromide staining should represent specific product. If no specific hybridization is observed, then troubleshooting steps should be initiated (see below).

Use the information gained from this analysis to optimize the RACE procedure. If significant yields of nonspecific products are amplified at the expense of specific product, annealing temperatures can be increased gradually (~2° at a time) at each stage of the procedure until the background products are lost. Alternatively, the "touchdown PCR" procedure (Chapter 4) can be used to optimize the annealing temperature of a reaction without trial and error (Don et al. 1991). Optimizing the annealing temperature is also recommended if multiple species of specific products are observed, which may indicate that truncation of specific products is occurring. If multiple species of specific products continue to be observed after the reverse transcription and amplification reactions have been fully optimized, then the possibility should be entertained that alternate splicing or promoter use is occurring.

For classic RACE only: If an almost continuous smear of specific products is observed up to a specific size limit after 5′-end amplification, this suggests that the polymerase is pausing during reverse transcription. To maximize the length of cDNA ends, the amplification mixture can be enriched for full- or near-full-length product using gel electrophoresis. The longest products can be isolated from the gel using standard procedures and then reamplified for a limited number of cycles.

For new RACE only: Expect to see one or two extra nucleotides inserted between the RNA oligonucleotide 3′ end and the 5′ end of the gene of interest. These come from the transcription step using T7, T3, or SP6, which tends to exceed the template by a base or two (template-independent transcription).

If the genomic sequence of interest is available, look for promoter elements such as TATA, CCAAT, and initiator element (Inr) sites at or around your candidate transcription site: You should usually be able to find either a TATA or an Inr.

FURTHER ANALYSIS AND USE OF RACE PRODUCTS

Cloning

RACE products can be cloned in the same way as other PCR products.

Option 1: To clone the cDNA ends directly from the amplification reaction (or after gel purification, which is recommended), it is convenient to use a "T" vector. These vectors exploit the tendency of thermostable DNA polymerases to add a single, unpaired 3´ adenosyl residue to the end of PCR products. The vectors carry restriction sites that, upon digestion, generate a single 3´ T overhang on both strands. Such vectors are available commercially (e.g., Invitrogen's "TOPO TA Cloning Kit") or can be easily and cheaply prepared in the laboratory. Note that not all heat-stable DNA polymerases generate the single-base overhang (e.g., *Pfu* does not).

Option 2: The classic RACE Q_I primer encodes *Hin*dIII, *Sst*I, and *Xho*I restriction enzyme sites. These sites allow products to be cloned efficiently into linearized vectors carrying appropriate "sticky" or "blunt" ends. It is important to "polish" the amplification products with Klenow enzyme or T4 DNA polymerase and to separate them from residual *Taq* polymerase and dNTPs before carrying out restriction enzyme digestions. If cloning attempts are unsuccessful, the problem may be caused by the presence of additional restriction sites in the unknown portion of the sequence. A simpler strategy is to append a restriction site (*not Hin*dIII, *Sst*I, or *Xho*I) to the 5´ end of the GSP2 primer, allowing the creation of "sticky" ends at both ends of the amplified product.

Option 3: An alternative approach is to modify the ends of the primers to allow the creation of overhanging ends using (1) T4 DNA polymerase to chew back a few nucleotides from the amplified product in a controlled manner and (2) Klenow enzyme (or Sequenase) to partially fill in restriction enzyme-digested overhanging ends on the vector, as shown below in Figure 7; for another conceptual variation, see Rashtchian et al. (1992).

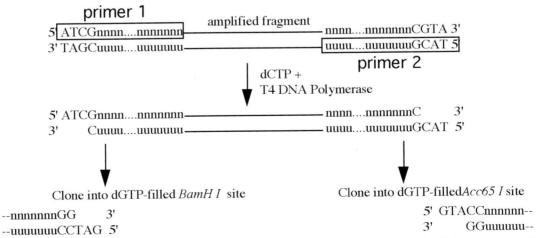

FIGURE 7. Safe cloning scheme. Note: *Acc*65I recognizes the same sequence as *Kpn*I, but creates a 5´ rather than a 3´ overhang. (Adapted from Stoker 1990; Iwahana et al. 1994.)

The advantages of this last approach are that (1) it eliminates the possibility that the restriction enzymes chosen for the cloning step will cleave the cDNA end in the unknown region; (2) vector dephosphorylation is not required since vector self-ligation is no longer possible; (3) since vector self-ligation is not possible, insert phosphorylation (and polishing) is not necessary; (4) since inserts are not "polished" or phosphorylated, the problem of insert multimerization and fusion clones does not arise. The procedure is also more reliable than "TA" cloning.

Procedure for Option 3

Insert Preparation

1. Select a pair of restriction enzymes for which "half-sites" can be appended to PCR primers. Choose sites that can be chewed back to form the appropriate overhangs using T4 DNA polymerase, as shown for BamHI and Acc65I, in Figure 7. NotI and EagI are compatible with BamHI and Acc65I, respectively. For RACE cloning, add "ATCG" to the 5′ end of Q_I or NRC-2, and add "TACG" to the 5′ end of GSP2. Carry out PCR as usual.

2. After PCR, remove the polymerase, excess primers, and nucleotides from the DNA products using any of several "PCR clean-up kits" (e.g., Qiagen's QIAquik mini-PCR purification kit).

3. In a sterile microfuge tube on ice, assemble the following reagents;

Purified PCR product	8.5 µl
Selected dNTP(s) (e.g., dCTP) (20 mM stock)	0.1 µl
T4 DNA polymerase buffer (10x)	1 µl
T4 DNA polymerase	1–2 units
H$_2$O	to 10 µl

4. Incubate for 15 minutes at 12°C, then inactivate the T4 DNA polymerase by incubating for 10 minutes at 75°C. *Optional:* If sufficient DNA is available, it can be gel-purified at this stage.

5. Prepare vector (e.g., pGem-7ZF [Promega] or pBluescript [Stratagene]) by digesting it with the appropriate restriction enzymes (e.g., Acc65I and BamHI) under optimal conditions, in a volume of 10 µl.

6. After restriction digestion, add a 10-µl mixture containing the selected dNTP(s) (e.g., dGTP) at a final concentration of 0.4 mM, 1 µl of the restriction buffer used for digestion, 0.5 µl of Klenow, and 0.25 µl of Sequenase.

7. Incubate for 15 minutes at 37°C, then inactivate the polymerases by incubating for 10 minutes at 75°C.

8. Gel-purify the linearized vector fragment using standard techniques.

9. Set up the ligation using standard procedures and equal molar amounts of vector and insert.

For additional information on cloning PCR products, see Section 7, Cloning PCR Products.

SEQUENCING

RACE products can be sequenced directly on a population level using a variety of protocols, including cycle sequencing. Sequencing can be performed using either the gene-specific primer or the Q_I primer at the unknown end. Note that the Q_I primer will yield useful sequence only if all the cDNAs initiate at the same nucleotide, which is not always the case.

Hybridization Probes

RACE products are generally pure enough to be used as probes for RNA and DNA blot analyses. Note, however, that small amounts of contaminating nonspecific cDNAs will always be present. It is also possible to include a T7 RNA polymerase promoter in one or both primer sequences and to use the RACE products in in vitro transcription reactions to produce RNA probes (Frohman and Martin 1989). Primers encoding the T7 RNA polymerase promoter sequence do not appear to function as amplification primers as efficiently as the ones listed in Figure 1 (personal observation). Thus, as a general rule, the T7 RNA polymerase promoter sequence should not be incorporated into RACE primers.

Construction of Full-length cDNAs

It is possible to use the RACE protocol to create overlapping 5′ and 3′ cDNA ends that can later, through judicious choice of restriction enzyme sites, be joined through subcloning to form a full-length cDNA. It is also possible to use the sequence information gained from analysis of the 5′ and 3′ cDNA ends to make new primers, representing the extreme 5′ and 3′ ends of the cDNA. The new primers can be used to amplify a de novo copy of a full-length cDNA, directly from the "3′ end cDNA pool" (see Protocol 1, Stage 1, step 6.) Despite the added expense of making two additional primers, there are several reasons that the second approach is preferred.

First, the PCR conditions required for efficient RACE (use of DMSO and elevated concentrations of nucleotides) cause a higher than necessary error rate by the thermostable polymerase. This means that potentially large numbers of clones have to be analyzed to identify sequences without mutations. In contrast, two specific primers from the extreme ends of the cDNA can be used under inefficient, but low-error-rate conditions (Eckert and Kunkel 1990) for a minimal number of cycles to amplify a new cDNA that is likely to be free of mutations. Second, because convenient restriction sites are often not available, subcloning RACE products can be difficult. Third, this approach allows the synthetic poly(A) tail (if present) to be removed from the 5′ end of the cDNA. The presence of homopolymer tails has been reported to inhibit translation. Finally, if alternate promoters, splicing, and polyadenylation signal sequences are present, multiple 5′ and 3′ ends can arise. This makes it possible to join two cDNA "halves" that would never be found together in vivo. Employing primers from the extreme ends of the cDNA, as described here, confirms that the resulting amplified cDNA represents a mRNA that is actually present in the starting population.

▶ TROUBLESHOOTING

Problems with Reverse Transcription and Prior Steps
- *Damaged RNA.* Examine integrity of the 18S and 28S ribosomal RNA bands using polyacrylamide gel electrophoresis (1% formaldehyde minigel). If the ribosomal bands are not sharp, discard the RNA preparation.

- *Contaminants.* Ensure that the RNA preparation is free of agents that might inhibit reverse transcription, e.g., LiCl and SDS. See Sambrook and Russell (2001) for advice on the optimization of reverse transcription reactions.

- *Poor-quality reagents.* To monitor reverse transcription of the RNA, add 20 μCi of [^{32}P]dCTP to the reaction and separate the newly created cDNAs using gel electrophoresis. Wrap the gel in cling film and expose it to X-ray film. Accurate estimates of cDNA size are best made using alkaline agarose gels, but a simple 1% agarose minigel will suffice to confirm that reverse transcription took place and that cDNAs of reasonable length were generated. Note that adding [^{32}P]dCTP to the reverse transcription reaction results in the detection of cDNAs synthesized both through the specific priming of mRNA and through RNA self-priming. When a gene-specific primer is used to prime transcription (5′-end RACE) or when total RNA is used as a template, the majority of the labeled cDNA will actually have been generated from RNA self-priming. To monitor extension of the reverse transcriptase primer, label the primer using T4 DNA kinase and [γ-^{32}P]ATP, prior to use. Much longer exposure times will be required to detect the labeled primer-extension products than when [^{32}P]dCTP is added to the reaction.

To monitor reverse transcription of the gene of interest, one may attempt to amplify an internal fragment of the gene containing a region derived from two or more exons, if sufficient sequence information is available.

Problems with Tailing

- *Poor-quality reagents.* The quality of reagents can be tested using the following procedure:

    a. Using the protocol given above (Protocol 2, Stage 2), append a poly(A) tail to 100 ng of a 100- to 300-bp DNA fragment for 30 minutes.

    b. In a separate tube, under the same conditions, mock-tail the same fragment (add everything but the Tdt).

    c. Analyze both samples on a 1% agarose minigel. The mock-tailed fragment should run as a tight band. The tailed fragment should have increased in size by 20–200 bp and should run as a diffuse band that trails off into higher-molecular-weight products; i.e., smears upward. If this is not the case, replace reagents.

- *Experimental control:* Mock-tail 25% of the cDNA pool (add everything but the Tdt). Dilute to the same final concentration as the tailed cDNA pool and analyze the samples on an agarose minigel. Both samples will contain amplification products, but the banding patterns of those products will vary. In general, the untailed sample will give rise to discrete bands after the first set of cycles, and the tailed templates will typically produce a broad smear of amplified cDNA with some individual bands. If the samples are different, this confirms that tailing took place and that the oligo(dT)-tailed QT primer is annealing effectively to the tailed cDNA during PCR. The observation of specific products unique to the tailed template indicates that these products are being synthesized from an A-tailed cDNA terminal, rather than from an A-rich sequence in or near the gene of interest.

Problems with Amplification

- *No product.* If no products are observed after 30 cycles in the first set of amplifications, add fresh *Taq* polymerase and carry out an additional 15 rounds of amplification (extra enzyme is not necessary if the entire set of 45 cycles is carried out without interruption at cycle 30). If efficient amplification is taking place, a product will be observed after a total of 45 cycles. If no product is observed, test the integrity of reagents by carrying out a control PCR using known templates and primers.

- *Smeared product from the bottom of the gel to the loading well.* Too many cycles, or too much starting material.

- *Nonspecific amplification, but no specific amplification.* Check the sequence of cDNA and primers. If all are correct, examine the primers (using computer program) for secondary structure and self-annealing problems. Consider ordering new primers. Determine whether too much template is being added, or whether the choice of annealing temperatures could be improved. Alternatively, secondary structure in the template may be blocking amplification. Consider adding formamide (Sarker et al. 1990) or ^{7}aza-GTP (in a 1:3 ratio with dGTP) to the reaction to assist polymerization. ^{7}aza-GTP can also be added to the reverse transcription reaction.

- *The last few nucleotides of the 5′ end sequence do not match the corresponding genomic sequence.* Note that reverse transcriptase, T7, and T3 RNA polymerase can add a few extra template-independent nucleotides.

- *Inappropriate templates.* To determine whether the amplification products observed are being generated from cDNA or whether they derive from residual genomic DNA or contaminating plasmids, pretreat an aliquot of the RNA with RNase A.

SUMMARY

The RACE protocol offers several advantages over conventional library screening to obtain additional sequence for cDNAs already partially cloned. RACE is cheaper, much faster, requires very small amounts of primary material, and provides rapid feedback on the generation of the desired product. Information regarding alternate promoters, splicing, and polyadenylation signal sequences can be obtained, and a judicious choice of primers (e.g., within an alternately spliced exon) can be used to amplify a subpopulation of cDNAs from a gene for which the transcription pattern is complex. Furthermore, differentially spliced or initiated transcripts can be separated by electrophoresis and cloned separately, and essentially unlimited numbers of independent clones can be generated to examine rare events. Finally, for 5′ end amplification, the ability of reverse transcriptase to extend cDNAs all the way to the ends of the mRNAs is greatly increased because a primer extension library is created instead of a general purpose one.

ACKNOWLEDGMENTS

The author thanks Yue Zhang and the publisher for permission to use and adapt material from "RACE all the way to the end" (Zhang 2003).

REFERENCES

Bertling W.M., Beier F., and Reichenberger E. 1993. Determination of 5′ ends of specific mRNAs by DNA ligase-dependent amplification. *PCR Methods Appl.* **3:** 5–99.

Bertrand E., Fromont-Racine M., Pictet R., and Grange T. 1993. Visualization of the interaction of a regulatory protein with RNA in vivo. *Proc. Natl. Acad. Sci.* **90:** 496–3500.

Borson N.D., Salo W.L., and Drewes L.R. 1992. A lock-docking oligonucleotide(dT) primer for 5′ and 3′ RACE PCR. *PCR Methods Appl.* **2:** 144–148.

Brock K.V., Deng R., and Riblet S.M. 1992. Nucleotide sequencing of 5′ and 3′ termini of bovine viral diarrhea virus by RNA ligation and PCR. *J. Virol. Methods* **38:** 39–46.

Coleclough C. 1987. Use of primer-restriction end adapters in cDNA cloning. *Methods Enzymol.* **154:** 64–83.

Don R.H., Cox P.T., Wainwright B.J., Baker K., and Mattick J.S. 1991. 'Touchdown' PCR to circumvent spurious priming during gene amplification. *Nucleic Acids Res.* **19:** 4008.

Dumas J.B., Edwards M., Delort J., and Mallet J. 1991. Oligonucleotide deoxyribonucleotide ligation to single-stranded cDNAs: A new tool for cloning 5′ ends of mRNAs and for constructing cDNA libraries by in vitro amplification. *Nucleic Acids Res.* **19:** 5227–5233.

Eckert K.A. and Kunkel T.A. 1990. High fidelity DNA synthesis by the *Thermus aquaticus* DNA polymerase. *Nucleic Acids Res.* **18:** 3739–3745.

Fritz J.D., Greaser M.L., and Wolff J.A. 1991. A novel 3′ extension technique using random primers in RNA-PCR. *Nucleic Acids Res.* **119:** 3747.

Frohman M.A. 1989. Creating full-length cDNAs from small fragments of genes: Amplification of rare transcripts using a single gene-specific oligonucleotide primer. In *PCR protocols and applications: A laboratory manual.* (ed. M. Innis et al.), pp. 28–38. Academic Press, New York.

———. 1993. Rapid amplification of cDNA for generation of full-length cDNA ends: Thermal RACE. *Methods Enzymol.* **218:** 340–356.

———. 1994. Cloning PCR products. In *The polymerase chain reaction (*ed. K.B. Mullis et al.), pp. 14-37. Birkhäuser, Boston, Massachusetts.

Frohman M.A. and Martin G.R. 1989. Rapid amplification of cDNA ends using nested primers. *Technique* **1:** 165–173.

Frohman M.A., Dush M.K., and Martin G.R. 1988. Rapid production of full-length cDNAs from rare transcripts by amplification using a single gene-specific oligonucleotide primer. *Proc. Natl. Acad. Sci.* **85:** 8998–9002.

Frohman M.A., Dickinson M.E., Hogan B.L.M., and Martin G.R. 1993. Localization of two new and related homeobox-containing genes to chromosomes 1 and 5, near the phenotypically similar mutant loci *dominant hemimelia* (*Dh*) and *hemimelic extra-toes* (*Hx*). *Mouse Genome* **91:** 323–325.

Fromont-Racine M., Bertrand E., Pictet R., and Grange T. 1993. A highly sensitive method for mapping the 5′ termini of mRNAs. *Nucleic Acids Res.* **21:** 1683–1684.

Iwahana H., Mizusawa N., Ii S., Yoshimoto K., and Itakura M. 1994. An end-trimming method to amplify adjacent cDNA fragments by PCR. *BioTechniques* **16:** 94–98.

Jain R., Gomer R.H., and Murtagh J.J.J. 1992. Increasing specificity from the PCR-RACE technique. *BioTechniques* **12:** 58–59.

Liu X. and Gorovsky M.A. 1993. Mapping the 5′ and 3′ ends of tetrahymena-thermophila mRNAs using RNA ligase mediated amplification of cDNA ends (RLM-RACE). *Nucleic Acids Res.* **21:** 4954–4960.

Loh E.L., Elliott J.F., Cwirla S., Lanier L.L., and Davis M.M. 1989. Polymerase chain reaction with single sided specificity: Analysis of T cell receptor delta chain. *Science* **243:** 217–220.

Mandl C.W., Heinz F.X., Puchhammer-Stockl E., and Kunz C. 1991. Sequencing the termini of capped viral RNA by 5′-3′ ligation and PCR. *BioTechniques* **10:** 484–486.

Monstein H.J., Thorup J.U., Folkesson R., Johnsen A.H., and Rehfeld J.F. 1993. cDNA deduced procionin: Structure and expression in protochordates resemble that of procholecystokinin in mammals. *FEBS Lett.* **331:** 60–64.

Ohara O., Dorit R.I., and Gilbert W. 1989. One-sided PCR: The amplification of cDNA. *Proc. Natl. Acad. Sci.* **86:** 5673–5677.

Rashtchian A., Buchman G.W., Schuster D.M., and Berninger M.S. 1992. Uracil DNA glycosylase-mediated cloning of PCR-amplified DNA: Application to genomic and cDNA cloning. *Anal. Biochem.* **206:** 91–97.

Sallie R. 1993. Characterization of the extreme 5′ ends of RNA molecules by RNA ligation-PCR. *PCR Methods Appl.* **3:** 54–56.

Sambrook J. and Russell D.W. 2001. *Molecular cloning: A laboratory manual,* 3rd edition. Cold Spring Harbor Laboratory Press, Cold Spring Harbor, New York.

Sarker G., Kapelner S., and Sommer S.S. 1990. Formamide can dramatically improve the specificity of PCR. *Nucleic Acids Res.* **18:** 7465.

Schmidt W.M. and Mueller M.W. 1999. CapSelect: A highly sensitive method for 5′ CAP-dependent enrichment of full-length cDNA in PCR-mediated analysis of mRNAs. *Nucleic Acids Res.* **27:** e31.

Schramm G., Bruchhaus I., and Roeder T. 2000. A simple and reliable 5′-RACE approach. *Nucleic Acids Res.* **28:** e96.

Schuster D.M., Buchman G.W., and Rastchian A. 1992. A simple and efficient method for amplification of cDNA ends using 5′ RACE. *Focus* **14:** 46–52.

Stoker A.W. 1990. Cloning of PCR products after defined cohesive termini are created with T4 DNA polymerase. *Nucleic Acids Res.* **18:** 4290.

Templeton N.S., Urcelay E., and Safer B. 1993. Reducing artifact and increasing the yield of specific DNA target fragments during PCR-RACE or anchor PCR. *BioTechniques* **15:** 48–50.

Tessier D.C., Brousseau R., and Vernet T. 1986. Ligation of single-stranded oligonucleotidedeoxyribonucleotides by T4 RNA ligase. *Anal. Biochem.* **158:** 171–178.

Volloch V., Schweizer B., Zhang X., and Rits S. 1991. Identification of negative-strand complements to cytochrome oxidase subunit III RNA in *Trypanosoma brucei*. *Biochemistry*. **88:** 10671–10675.

Wickens M. and Stephenson P. 1984. Role of the conserved AAUAAA sequence: Four AAUAAA point mutants prevent mRNA 3′end formation. *Science* **226:** 1045–1050.

Zhang Y. 2003. Rapid amplification of cDNA ends. *Methods Mol. Biol.* **221:** 13–24.

Synthesis and Analysis of DNA Libraries Using PCR

John L. Joyal, Jinwei Jiang, and David I. Israel

PRAECIS Pharmaceuticals, Waltham, Massachusetts 02451

High-quality DNA libraries that contain a diverse collection of recombinant molecules are critical reagents for many experimental protocols in both basic and applied research. cDNA and genomic libraries are often the source materials used when cloning genes. DNA libraries have tremendous utility for the investigation of protein function and protein–protein interactions (Chien et al. 1991), as well as in gene expression analysis techniques such as SAGE (Velculescu et al. 1995). They are also useful in protein engineering applications, where improvements in protein function are selected from gene expression libraries that contain variation introduced by mutagenesis or recombination (Stemmer 1994). Synthetic gene libraries encoded by oligonucleotides in phage display (Scott and Smith 1990) and other expression systems are finding increased utility in drug discovery, in functional genomics (Xu et al. 2001), and in the discovery of peptides for use as affinity reagents (Sato et al. 2002).

Biologically expressed random peptide libraries have been described in several systems, including phage display vectors (Scott and Smith 1990), yeast and mammalian two-hybrid systems (Chien et al. 1991; Fearon et al. 1992), and as fusion proteins in both bacterial and mammalian cell expression systems (Pelletier et al. 1999). Library insert size, library complexity, and amino acid usage are easily controlled in these systems, as the libraries are encoded by synthetic DNA oligonucleotides that can be precisely designed and synthesized using standard oligonucleotide chemistries. The complexity of random peptide libraries is often limited by bacterial transformation efficiencies and typically approaches 10^9 members. Such libraries have many applications in functional genomics, peptide interaction mapping, antibody epitope mapping, discovery of peptide affinity reagents, and discovery of synthetic ligands for receptor proteins.

Construction of DNA libraries is a multistep, technically demanding task, and the ultimate value of a library is often limited by library complexity and/or quality. Construction of a typical library requires purification of the library DNA, enzymatic modification of both the ends of the library DNA and an appropriate vector system to make them compatible with ligation, ligation of the library DNA into the vector, and transformation of the ligation product into bacteria. Inefficiency at any step in the library construction process can result in a library of poor quality and, therefore, limited use.

PCR is a powerful tool for synthesizing and analyzing DNA, and it has several applications in the synthesis and analysis of DNA libraries. As a synthetic tool, PCR enables the production of high-complexity libraries using limiting amounts of input nucleic acid. As an analytical tool, PCR enables the rapid characterization of many intermediary reaction products and allows the researcher to proceed confidently to the next step of library construction with well-characterized reactants. Finally, PCR enables rapid and sensitive characterization of the final DNA library. In this chapter, we describe the use of PCR as both a preparative and analytical tool for synthesizing high-quality, complex DNA libraries.

PROTOCOL 1   USE OF PCR TO PREPARE A DOUBLE-STRANDED DNA LIBRARY ENCODING RANDOM PEPTIDES

PCR can serve two important functions in synthesis of a library derived from synthetic oligonucleotides. First, a large amount of PCR product can be generated by amplifying a small amount of diverse library template with primers that anneal to the fixed flanking sequence. Second, PCR synthesis is an efficient method for generating a complex set of complementary strands of library DNA. Most random libraries derived from synthetic oligonucleotides have vast potential molecular diversity; in the example presented here, an NNK library that encodes 12 amino acids has greater than 10^{18} potential sequences. Annealing of complementary strands produced by oligonucleotide chemistry, therefore, is not a plausible method for generating duplex library DNA.

In the example presented here, a random peptide library is encoded by a mixture of synthetic oligonucleotides (Fig. 1A), which are flanked by the PCR primer-binding sites. The PCR primers are synthesized with a 5′-biotin modification, which permits removal of the DNA termini after digestion with restriction endonucleases. PCR is used to make single-stranded library oligonucleotides into double-stranded DNA (Fig. 1B).

The PCR will yield equal amounts of each of the complementary strands of library DNA. However, if either the primers or dNTPs are consumed during the PCR, or if the enzymatic activity of the thermostable polymerase is lost before the end of the reaction, complementary strands will denature and may anneal to mismatched DNA (Fig. 1B). To assure that the final product is authentic complementary double-stranded DNA, and not mismatched DNA, a second single-cycle PCR is performed upon addition of fresh polymerase, dNTPs, and the two primers, with equal stoichiometry to the library product. The products of the single-cycle polymerase reaction are not subjected to high-temperature denaturation that occurs with conventional PCRs, thereby assuring that the final products are correctly matched, duplex DNA (Fig. 1B).

MATERIALS

▼ **CAUTION**

See Appendix for appropriate handling of materials marked with <!>.

BUFFERS, SOLUTIONS, AND REAGENTS
   ddH$_2$O
   10 mM Tris-HCl<!>, pH 8.0

ENZYMES AND ENZYME BUFFERS
   Restriction enzymes and buffers (see step 9)
   10x Vent Buffer
   Vent DNA polymerase

NUCLEIC ACIDS AND OLIGONUCLEOTIDES
   Biotin-labeled PCR primers, at least 15 bp in length, with 5′-biotin modification (see Fig. 1)

NNK library encoding 12 amino acids. This library is designed as shown in Figure 1A. The synthetic strand contains a random peptide library that is flanked on both the 5′ and 3′ ends by constant regions that serve as primer-binding sites for PCR and also contain restriction enzyme recognition sites for cloning library DNA into a suitable vector system. The flanking sequences should be at least 15 bp in length and should be designed with three criteria in mind: (1) They should be efficient PCR primer-binding sites; (2) they should include restriction enzyme sites for subsequent cloning steps (*option:* the restriction enzyme sites may be in the PCR primer 5′ of the overlap with the library oligonucleotide, rather than within the library oligonucleotide itself); and (3) they should not include codons that interfere with library function, as the DNA sequence in one or both flanks is likely to be represented as amino acids in the peptide library. During the synthesis of the coding

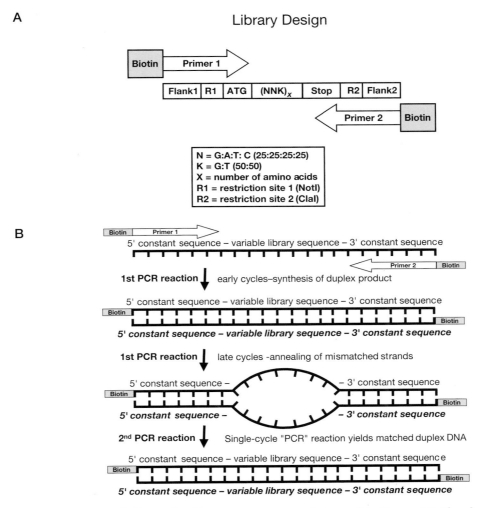

FIGURE 1. Design of oligonucleotides to generate a random peptide library. (*A*) The characteristics of the random peptide library template and primers are shown. Both PCR primers contain a 5′ biotin modification as indicated. The 5′ and 3′ constant sequences (Flank1/R1 and Flank2/R2) are different from each other, allowing specific priming on either side of the variable library sequence. (*B*) The PCR produces duplex DNA during early cycles, but can also produce mismatched strands during later cycles when denaturation/annealing occurs without polymerization. A second single-cycle reaction with fresh reagents produces fully matched duplex DNA as shown in the bottom of the figure.

region of the library, equal amounts of the 4 nucleotides are present at the first two positions of each codon, and a 1:1 ratio of G + T is present at the third position (Fig. 1A). The exclusion of A and C nucleotides at the third position decreases the occurrence of stop codons but still allows codons for all 20 amino acids.
dNTPs, 20 mM stocks of each

ADDITIONAL ITEMS

Agarose gel electrophoresis supplies, including ethidium bromide<!> (see step 11)
QIAprep Miniprep silica column (QIAGEN)
Streptavidin beads (Dynal)

METHOD

First Round of Polymerization: PCR Amplification of the Library

1. Assemble the following reaction to amplify the single-strand synthetic oligonucleotides into sufficient starting material for library construction. The amounts given here are for a single 100-μl reaction. To generate the library at a large scale, prepare 1 ml of reactants and aliquot into ten individual PCR tubes.
   10 pmoles of library oligonucleotides
   2 units of Vent DNA polymerase
   100 pmoles of each biotin-labeled PCR primer
   10 μl of 10x Vent buffer
   1 μl each of dATP, dCTP, dTTP, and dGTP
   Bring to a final volume of 100 μl with ddH$_2$O.

2. Preheat the mixture in a thermal cycler for 20 seconds at 94°C.

3. Perform 30 cycles of PCR with the following parameters:
   94°C for 10 seconds
   55°C for 10 seconds (or appropriate annealing temperature for primers being used)
   72°C for 15 seconds

4. Cool the reactions to 4°C.

Second Round of Polymerization: Synthesis of Matched Duplex Library DNA

5. Add 10 μl of 10x reaction buffer, 100 pmoles of each primer, and 2 units of Vent DNA polymerase to each 100-μl first-round reaction product.

6. Perform a single cycle of 94°C for 30 seconds, 55°C for 30 seconds, 72°C for 15 seconds.

7. Cool the reactions to 4°C.

8. Purify the second-round polymerization products on a QIAprep column, and elute in 10 mM Tris, pH 8.0.

9. Digest the purified product (Fig. 2, lane 1) with the appropriate restriction enzymes, then ethanol-precipitate the DNA and dissolve in ddH$_2$O (see Sambrook and Russell 2001). This digestion releases the biotinylated DNA ends from the library core.

10. Use streptavidin beads to remove the biotinylated DNA ends (Korn et al. 1992). This step increases the yield of the desired ligation product, as the ends can ligate onto either library insert or the expression vector. The streptavidin affinity step also removes any undigested or partially digested PCR product (Fig. 2, lane 3 vs. lane 4).

11. Analyze aliquots of the reaction intermediates and the final products on a 4% agarose gel. The final product (Fig. 2, lane 4) is a high-purity, high-complexity, double-strand-

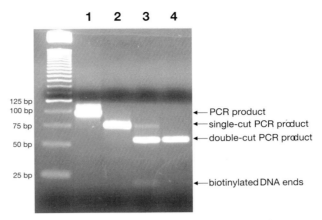

FIGURE 2. PCR synthesis of DNA encoding random peptides. PCR products were generated with the primers and template described in Fig. 1 (lane *1*); digested with *Not*I (lane *2*); digested with *Not*I, followed by *Cla*I (lane *3*); or digested with both enzymes, then incubated with streptavidin beads to remove any remaining uncut DNA, single-cut DNA, and biotinylated DNA ends (lane *4*). Samples were electrophoresed on a 4% agarose gel. A 25-bp DNA ladder is included.

ed library DNA with overhanging ends that is now ready for directional ligation into a suitable expression system.

PROTOCOL 2 USE OF PCR TO ANALYZE LIGATION SUBSTRATES AND PRODUCTS

The restriction digest products of the PCR-generated synthetic DNA library are next ligated into a linearized expression vector system. In the example protocol presented below, different restriction enzyme sites flank the 5′ (*Not*I) and 3′ (*Cla*I) sides of the library, and the same restriction sites are used for cloning into the vector. A starting vector that lacks library inserts can significantly contaminate the final library if the restriction enzyme reactions are incomplete. This contamination can occur from either self-ligation of single-cut vector or transformation of starting, uncut circular plasmid. PCR products generated with primers that flank the library cloning sites are larger from plasmids that contain library inserts than from starting vector. PCR, therefore, can be diagnostic of the efficiency of the linearization and ligation reactions.

MATERIALS

▼ CAUTION

See Appendix for appropriate handling of materials marked with <!>.

BUFFERS, SOLUTIONS, AND REAGENTS
  ddH$_2$O

ENZYMES AND ENZYME BUFFERS
  10x Vent buffer
  Vent DNA polymerase

NUCLEIC ACIDS AND OLIGONUCLEOTIDES
  dNTPs, 20 mM stocks of each
  PCR primers that anneal to the vector sequence that flanks the cloning sites of the synthetic DNA library
  Plasmid DNA (see step 1 below)

SPECIAL EQUIPMENT

Agarose gel electrophoresis supplies, including ethidium bromide<!> (see step 5)
Thermal cycler

METHOD

1. Assemble the following reaction:
   10–50 ng of plasmid DNA
   50 pmoles of each primer
   5 μl of 10x Vent buffer
   0.5 μl each of dATP, dCTP, dTTP and dGTP
   1 unit of Vent DNA polymerase
   Bring to a final volume of 50 μl with ddH₂O.
   Separate reactions should be assembled for each of the following plasmid templates:
   undigested vector
   digested vector, no ligation
   digested vector ligated in the absence of library DNA
   digested vector ligated with library DNA

2. Preheat the mixtures in a thermal cycler for 20 seconds at 94°C.

3. Perform 30 cycles of PCR with the following parameters:
   94°C for 10 seconds
   55°C for 10 seconds (or appropriate annealing temperature for primers being used)
   72°C for 15 seconds
   Increase the extension time of the final cycle to 30 seconds.

4. Cool the reactions to 4°C.

5. Run 25 μl of each reaction on a 4% agarose gel, and stain with ethidium bromide for analysis (see Sambrook and Russell 2001).

ANALYSIS

In the example shown in Figure 3, PCR analysis of the linearized vector before ligation demonstrates that it is completely linearized, as no PCR product corresponding to starting vector is obtained (Fig. 3, lane 4). The PCR product from vector ligated in the absence of library insert indicates that a small amount of self-ligation occurred (Fig. 3, lane 2) and sug-

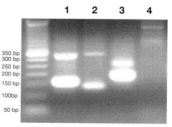

FIGURE 3. PCR analysis of random peptide library ligation. Restriction endonuclease-digested random peptide library DNA was ligated with similarly digested vector DNA. PCR was performed with primers flanking the cloning site on undigested vector (lane *1*), digested vector ligated in the absence of library DNA (lane *2*), digested vector ligated with library DNA (lane *3*), and digested vector (lane *4*). Samples were electrophoresed on a 4% agarose gel. A 50-bp DNA ladder is included.

gests that a small amount of single-cut vector is present. However, when ligation occurs with vector plus library insert, the predominant DNA product corresponds to correctly ligated library DNA with no detectable PCR product corresponding to starting vector (Fig. 3, lane 3). The lack of detectable PCR product corresponding to starting vector when ligation occurs with vector plus insert, but its synthesis using vector alone in the ligation reaction (compare Fig. 3, lanes 2 vs. 3), can be explained by ligation of library DNA onto single-cut vector. This side-reaction product will not form a closed circular plasmid, will not clone, and will not yield PCR product.

In this example, PCR provides an important analytical tool to verify that the restriction endonuclease and ligation reactions occurred with high fidelity. Thus, the investigator can proceed to the next steps of library construction with well-characterized DNA, increasing the likelihood that the final library will be of high quality. The final library is obtained through transformation and amplification in bacteria, followed by purification of the plasmid DNA using standard protocols (see Sambrook and Russell 2001).

PROTOCOL 3   USE OF PCR FOR QUALITY CONTROL OF A RANDOM PEPTIDE DNA LIBRARY

The use of PCR to characterize two different types of DNA libraries and subject them to quality control is described in this and the following protocol. In this protocol, the random peptide library encoded in the plasmid vector is analyzed with PCR. The library was obtained after transformation of bacteria with the ligation reaction described in Protocol 2, and expansion and purification of the plasmid using standard protocols.

MATERIALS

....

BUFFERS, SOLUTIONS, AND REAGENTS
    ddH_2O

ENZYMES AND ENZYME BUFFERS
    10x Vent buffer
    Vent DNA polymerase

NUCLEIC ACIDS AND OLIGONUCLEOTIDES
    dNTPs, 20 mM stocks of each
    PCR primers that anneal to the vector sequence that flanks the cloning sites of the synthetic DNA library
    Plasmid DNA (see step 1 below)

SPECIAL EQUIPMENT
    Agarose gel electrophoresis supplies, including ethidium bromide<!> (see step 5)
    Thermal cycler

▼ CAUTION

See Appendix for appropriate handling of materials marked with <!>.

METHOD

....

1. Use the plasmid DNA prepared from the random peptide library after amplification in bacteria to assemble the following reaction:
    10–50 ng of plasmid DNA
    50 pmoles of each primer

5 μl of 10x Vent buffer
0.5 μl each of dATP, dCTP, dTTP, and dGTP
1 unit of Vent DNA polymerase
Bring to a final volume of 50 μl with ddH$_2$O.

Separate reactions should be assembled for each of the following plasmid templates:

Expression vector without library DNA insert
DNA library mixture
Miniprep DNA prepared from individual bacterial clones from DNA library

2. Preheat the mixtures in a thermal cycler for 20 seconds at 94°C.
3. Perform 30 cycles of PCR with the following parameters:
   94°C for 10 seconds
   55°C for 10 seconds (or appropriate annealing temperature for primers being used)
   72°C for 15 seconds
   Increase the extension time of the final cycle to 30 seconds.
4. Cool the reactions to 4°C.
5. Run 25-μl aliquots of the reactions on a 4% agarose gel and stain with ethidium bromide for analysis (see Sambrook and Russell 2001).

ANALYSIS

In the example shown in Figure 4, analysis of the PCR products using the random 12-mer library mixture as the template shows that most of the library contains inserts of the correct size (Fig. 4, lane 2). Furthermore, there is no detectable PCR product corresponding to starting vector (compare Fig. 4, lanes 1 and 2). PCR analysis of individual, randomly chosen clones contained within the library confirms that the majority of inserts are of correct size and that no detectable starting vector contaminates the library (Fig. 4, lanes 3–20). DNA sequence analysis verifies that these PCR products represent correct random 12-mer

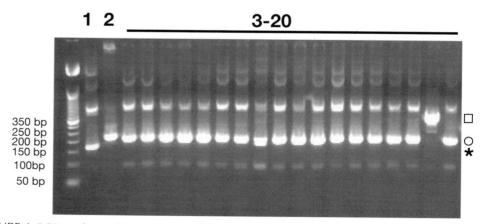

FIGURE 4. PCR analysis of a random peptide library. PCR was performed with primers flanking the cloning site on vector (lane *1*), amplified vector ligated with library DNA (lane *2*), and miniprep DNA prepared from individual bacterial clones of amplified vector ligated with library DNA (lanes *3–20*). Samples were electrophoresed on a 4% agarose gel. The size of the PCR product expected for vector (*) and vector containing the random peptide library DNA (○) are indicated. An incorrect-sized PCR product containing multimeric insert is also shown (□). A 50-bp DNA ladder is included.

library members (data not shown). Thus, PCR analysis of the intermediate reaction products (Protocol 2, Fig. 3) was predictive of obtaining a final library of high quality (Fig. 4).

PROTOCOL 4   USE OF PCR FOR QUALITY CONTROL OF A GENOMIC DNA LIBRARY

In a second example demonstrating the use of PCR to analyze production of a DNA library and subject it to quality control, analysis of a human genomic DNA fragment library is described. In this protocol, human genomic DNA is digested with a cocktail of 4-bp recognition restriction endonucleases to generate fragments ranging from ~50 to 500 bp (Fig. 5). Although typical genomic DNA libraries contain much larger inserts, the ultimate utility of this type of expression library is to isolate gene fragments that encode functional peptides or protein fragments using phenotypic and/or biochemical screens. Genomic DNA fragments are blunt-ended using Klenow DNA polymerase, ligated into a vector linearized with *Pml*I (an enzyme that yields blunt-ended product), transformed into bacteria, and amplified to make the library.

MATERIALS

BUFFERS, SOLUTIONS, AND REAGENTS
    ddH$_2$O

ENZYMES AND ENZYME BUFFERS
    10x Vent buffer
    Vent DNA polymerase

NUCLEIC ACIDS AND OLIGONUCLEOTIDES
    dNTPs, 20 mM stocks of each
    PCR primers that anneal to the vector sequence that flanks the cloning sites of the genomic DNA library
    Plasmid DNA (see step 1 below)

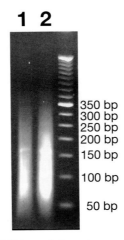

FIGURE 5. Preparation of blunt-ended DNA used for a human genomic DNA fragment library. Male human genomic DNA was digested with a combination of restriction endonucleases (*Aci*I, *Hin*P1I, *Hpa*II, HpyCH4IV, *Bfa*I, *Mse*I, *Nla*III, *Rsa*I, and *Sau*3AI) and analyzed by electrophoresis on a 4% agarose gel. A 50-bp DNA ladder is included.

SPECIAL EQUIPMENT

Agarose gel electrophoresis supplies, including ethidium bromide<!> (see step 5)
Thermal cycler

METHOD

1. Use plasmid DNA prepared from a genomic library after amplification in bacteria to assemble the following reaction:
   10–50 ng of plasmid DNA
   50 pmoles of each primer
   5 μl of 10x Vent buffer
   0.5 μl each of dATP, dCTP, dTTP, and dGTP
   1 unit of Vent DNA polymerase
   Bring to a final volume of 50 μl with ddH$_2$O.
   Separate reactions should be assembled for each of the following plasmid templates:
   Expression vector without library DNA insert
   Genomic DNA library mixture
   Miniprep DNA prepared from individual bacterial clones from genomic DNA library

2. Preheat the mixture in a thermal cycler for 20 seconds at 94°C.

3. Perform 30 cycles of PCR with the following parameters:
   94°C for 15 seconds
   55°C for 10 seconds (or appropriate annealing temperature for primers being used)
   72°C for 40 seconds
   Increase the extension time of the final cycle to 60 seconds.

4. Cool the reactions to 4°C.

5. Run 25-μl aliquots of the reactions on a 4% agarose gel and stain with ethidium bromide for analysis (see Sambrook and Russell 2001).

ANALYSIS

In the example shown in Figure 6, PCR analysis of the library mix using primers that flank the *Pml*I site indicates that the library contains inserts of the desired size range, but it is contaminated by a significant amount of starting vector (Fig. 6A, lane 2). PCR analysis of individual clones picked from this library confirms the results from the use of the library mixture as template, and indicates that starting vector comprises ~40% of this library (10 of 24 clones analyzed, Fig. 6A, lanes 3–26). To increase the quality of this library, the library was digested with *Pml*I and reamplified in bacteria with the goal of decreasing the frequency of contaminating starting vector. PCR analysis of the library mixture after digestion with *Pml*I demonstrates that the amount of contaminating, insert-minus circular vector is greatly diminished by the restriction endonuclease reaction (Fig. 6B, lanes 2 vs. 3). PCR analysis of individual clones from the library after *Pml*I digestion and reamplification in bacteria confirms that the occurrence of insert-minus DNA is greatly diminished compared to the original library (0 of 20, <5%, Fig. 6B, lanes 4–23) and that the clones contain inserts of the desired size range. DNA sequence analysis of clones from the genomic fragment library demonstrates that the library contains human genomic DNA fragments of the desired size (data not shown). Thus, PCR provides a rapid, sensitive, and convenient method to analyze the original and improved versions of this library.

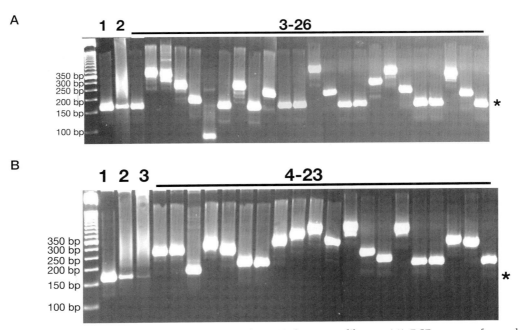

FIGURE 6. PCR analysis of a human genomic DNA fragment library. (A) PCR was performed with primers flanking the cloning site on vector (lane *1*), amplified vector ligated with digested human genomic DNA (lane *2*), and miniprep DNA prepared from individual bacterial clones of amplified vector ligated with digested human genomic DNA (lanes *3–26*). Samples were electrophoresed on a 4% agarose gel. The size of the PCR product expected for vector (*) is indicated. A 50-bp DNA ladder is included. (B) The human genomic DNA library depicted in *A* was digested with *Pml*I and reamplified in bacteria. PCR was performed with primers flanking the cloning site on vector (lane *1*), the human genomic DNA library shown in *A* (lane *2*), *Pml*I-cut and reamplified human genomic DNA library (lane *3*), and miniprep DNA prepared from individual bacterial clones of *Pml*I-cut and reamplified human genomic DNA library (lanes *4–23*). Samples were electrophoresed on a 4% agarose gel. The size of the PCR product expected for vector (*) is indicated. A 50-bp DNA ladder is included.

TROUBLESHOOTING AND SUMMARY

Together, the PCR and DNA sequence analyses described in the examples above validate the utility of PCR in the synthesis and analysis of DNA libraries. The PCR synthetic and analytical methods described in this chapter can be employed to generate a variety of peptide- or genomic-based libraries for expression and selection in numerous systems either as free peptides or as components of fixed protein scaffolds. Care must be taken to assure that the final product of the reaction that generates the random amino-acid-encoding library is full-length, double-stranded DNA. If gel analysis indicates the presence of a robust product, but a low level of transformants is observed after restriction digestion, purification, and ligation, then the starting material was most likely not fully double stranded. Lengthen the final extension step to 2 minutes.

The PCR analytical protocols described for the genomic fragment library can readily be adapted for libraries containing larger genomic clones by using polymerases and cycle conditions optimized for larger PCR products. PCR is, therefore, a powerful tool for synthesis and/or analysis of a variety of types of gene libraries.

REFERENCES

Chien C., Bartel P.L., Sternglanz R., and Fields S. 1991. The two-hybrid system: A method to identify and clone genes for protein that interact with a protein of interest. *Proc. Natl. Acad. Sci.* **88:** 9578–9582.

Fearon E.R., Finkel T., Gillison M.L., Kennedy S.P., Casella J.F., Tomaselli G.F., Morrow J.S., and Dang C.V. 1992. Karyoplasmic interaction selection strategy: A general strategy to detect protein-protein interactions in mammalian cells. *Proc. Natl. Acad. Sci.* **89:** 7958–7962.

Korn B., Sedlacek Z., Manca A., Kioschis P., Konecki D., Lehrach H., and Poustka A. 1992. A strategy for the selection of transcribed sequences in the Xq28 region. *Hum. Mol. Genet.* **1:** 235–242.

Pelletier J.N., Ardnt K.M., Pluckthun A., and Michnick S.W. 1999. An in vivo library-versus-library selection of optimized protein-protein interactions. *Nat. Biotechnol.* **17:** 683–690.

Sambrook J. and Russell D.W. 2001. *Molecular cloning: A laboratory manual*, 3rd edition. Cold Spring Harbor Laboratory Press, Cold Spring Harbor, New York.

Sato A.K., Sexton D.J., Morganelli L.A., Cohen E.H., Wu Q.L., Conley G.P., Streltsova Z., Lee S.W., Devlin M., DeOliveira D.B., Enright J., Kent R.B., Wescott C.R., Ransohoff T.C., Ley A.C., and Ladner R.C. 2002. Development of mammalian serum albumin affinity purification media by peptide phage display. *Biotechnol. Prog.* **18:** 182–192.

Scott J.K. and Smith G.P. 1990. Searching for peptide ligands with an epitope library. *Science* **249:** 386–390.

Stemmer W.P.C. 1994. DNA shuffling by random fragmentation and reassembly: *In vitro* recombination for molecular evolution. *Proc. Natl. Acad. Sci.* **91:** 10747–10751.

Velculescu V.E., Zhang L., Vogelstein B., and Kinzler K.W. 1995. Serial analysis of gene expression. *Science* **270:** 484–487.

Xu X., Leo C., Jang Y., Chan E., Padilla D., Huang B.C., Lin T., Gururaja T., Hitoshi Y., Lorens J.B., Anderson D.C., Sikic B., Luo Y., Payan D.G., and Nolan G.P. 2001. Dominant effector genetics in mammalian cells. *Nat. Genet.* **27:** 23–29.

CLONING OF PCR PRODUCTS

Following PCR, it is sometimes necessary to clone the amplified fragment into a plasmid. Once the PCR product is subcloned into a vector, bacteria containing the plasmid can be frozen, allowing an almost indefinite availability of the amplified material. Examples of times when cloning is advantageous include

- when the template is in very limited supply and further manipulations of the obtained product are desired

- when the product is difficult to obtain because of size or GC content

- when constructing a molecular standard

In addition, cloning of the product may facilitate sequencing allowing the use of primers in the vector. Due to the variety of available plasmids with different promoters and selectable markers, cloning of the PCR product is also useful when expression is desired, when mutations are to be introduced into the fragment prior to expression, or when sequence tags encoded in the vector are to be added in frame. Other examples in which cloning of the amplified product is necessary include the last step in the "Fluorescent mRNA Differential Display in Gene Expression Analysis" (p. 327) and in the "Beyond Classic RACE (Rapid Amplification of cDNA Ends)" (p. 359) in Sections 5 and 6. In this section, a number of strategies are presented.

CHOICE OF CLONING STRATEGY

The cloning strategy to be used should be chosen prior to beginning the experiment. Several of the methods require the addition of extra nucleotides to the primers and, in some instances, use of particular thermostable DNA polymerases is recommended. As discussed in "Strategies for Cloning PCR Products" (p. 405), the amplicon can be engineered through the addition of defined 5′ and 3′ functional elements for direct mammalian cell expression or in vitro transcription/translation. If accuracy is essential, one of the enzymes with proofreading activity should be used, provided that a uracil DNA glycosylase contamination control system is not in use. Non-proofreading enzymes add a non-template deoxyadenosine to the 3′ ends of the product, which can be exploited for cloning. Most of the commercially available methods are discussed in "Strategies for Cloning PCR Products" (p. 405). These include the enzyme-dependent methods, such as DNA ligase and topoisomerase I; enzyme-independent methods; and systems for in vitro and in vivo recombination.

"Bidirectional and Directional Cloning of PCR Products" (p. 421) provides a comprehensive set of protocols for blunt-end cloning, easy determination of insert orientation, and suggestions on choices of plasmid vectors. This method has some distinct advantages because any PCR product can be rapidly ligated into a plasmid using these methods.

The final chapter in the section, "Ribocloning: DNA Cloning and Gene Construction Using PCR Primers Terminated with a Ribonucleotide" (p. 441), provides a unique method for the generation of sticky ends to facilitate the rapid cloning of an amplicon. The essential part of this method is the specificity of RNase A, which can detect and cut the nucleic acid strand on the 3′ side of a cytosine or uracil ribose base, leaving a 3′ overhang. Two PCR primer pairs are used in this method—one pair for the insert and the second for the vector. All vectors have a rC substitution in their sequence so that 3′ sticky ends can be generated. The ends of the primers are designed to anneal to bring the vector and insert together. Because the length of the complementary region is quite large, up to 20 nucleotides, very stable complexes are formed that have a high efficiency of transformation into competent *E. coli*. This method can be coupled to any of the procedures in this book that recommend the cloning of a PCR product.

In some instances, when undesirable products are present in the reaction after visualization on a gel, purification of the desired fragment is recommended prior to cloning. "DNA Purification" (p. 87) in Section 2 describes a reliable method for PCR amplicon purification. Finally, extra care must be taken with any cloned amplicon so that it does not become a major source of PCR contamination in the laboratory.

27 Strategies for Cloning PCR Products

Peter D'Arpa

Department of Biochemistry and Molecular Biology, F. Edward Hébert School of Medicine, Uniformed Services University of the Health Sciences, Bethesda, Maryland 20814-4799

The ease of adding nucleotide sequences to the termini of PCR products has led to the development of a variety of strategies for their cloning (Aslanidis and de Jong 1990; Jung et al. 1990; Haun et al. 1992; Smith et al. 1993; Weiner 1993; Aslanidis et al. 1994; Testori et al. 1994; Chuang et al. 1995; Ailenberg and Silverman 1996; Padgett and Sorge 1996; Bielefeldt-Ohmann and Fitzpatrick 1997; Delidow 1997; Gabant et al. 1997; Horton et al. 1997; Ido and Hayami 1997; Zeng et al. 1997; Arashi-Heese et al. 1999; Gal et al. 1999; Tillett and Neilan 1999; Geiser et al. 2001). Because the cloning of PCR products is typically the first step for generating a reagent that will be used to achieve an experimental goal, the efficiency of the cloning procedure is usually the primary consideration, and cloning strategies should be simple in design and execution, requiring a minimum of enzymatic steps. Toward this goal, companies market and continue to develop reagent kits that improve the ease and rapidity of cloning PCR products. With these considerations in mind, this chapter focuses on the more common and efficient strategies in use today for cloning PCR products. Less widely employed strategies and those for more specialized applications have been reviewed elsewhere (Guo and Bi 2002).

The strategies covered in this chapter are categorized into those that employ (1) DNA ligase, (2) vaccinia virus topoisomerase I (TOPO), (3) ligase-independent cloning (LIC) using uracil DNA glycosylase in which single-strand overhangs produced on the PCR product and vector anneal in vitro and nick-sealing occurs in *Escherichia coli*, (4) in vitro recombination of a vector and a PCR product that have homologous duplex ends, (5) in vivo recombination in *E. coli* following transformation of a mixture of linear vector and PCR product that have homologous duplex ends. Also covered is the production of linear PCR products having defined 5′ and 3′ functional elements enabling direct mammalian cell expression or in vitro transcription/translation.

In large part, the technical details of the procedures have been omitted because they are already detailed in manufacturers' literature (Invitrogen 2001, 2002a,b; Qiagen 2001) or protocol manuals (Sambrook et al. 1989; Ausubel et al. 2003); this chapter focuses on an overview of the strategies, their molecular basis, and their advantages and disadvantages for specific applications.

PCR products are cloned for a variety of applications, and the choice of the strategy depends on the application. When cloning PCR products for the purpose of synthesizing the nucleic acid or the protein encoded by the PCR product, the DNA must be cloned into the vector in the proper orientation with respect to the promoter. This is typically accomplished by adding different terminal sequences to the PCR product via the primer tails. In other applications, such as the shotgun cloning of genomic DNA, directional cloning is not required.

Essential to the success of the cloning of PCR products is the quality of their ends. The coupling efficiency of the synthesis of oligonucleotide primers is typically ~98–99%, and therefore 39–63% of a 50-mer oligonucleotide is truncated on the 5′ terminus before purification (LifeTechnologies 1999). Subsequent purification methods differ in their ability to remove truncated products; therefore, for certain applications, especially when longer primers are used, high-performance liquid chromatography (HPLC) or polyacrylamide gel electrophoresis (PAGE) purification may be necessary (consult manufacturers' literature to determine the level of purity required for each method).

LIGASE-DEPENDENT CLONING

DNA Ligase

Currently, the most common method for cloning PCR products employs DNA ligase to join covalently a DNA duplex end having a 5′ phosphate with a duplex having a 3′ hydroxyl (OH) such as is produced by restriction enzyme cleavage. The reaction proceeds via a phospho-AMP intermediate and occurs only when the 5′ phosphate and 3′ OH are properly juxtaposed by base-pairing of overhangs or if both molecules have blunt termini. PCR products typically lack 5′ phosphates because they are generated using synthetic oligonucleotide primers that lack 5′ phosphates. PCR products without terminal phosphates can be directly ligated into a vector having 5′-terminal phosphates, such as is the case for certain types of TA and blunt-end cloning vectors (see below). Alternatively, PCR products having 5′-terminal phosphates (produced either by amplification with kinased primers, kinase treatment of the PCR product, or restriction enzyme cleavage of the PCR product) can be ligated to vectors without 5′ phosphates that result from phosphatase treatment. Phosphatase treatment of the vector is used in strategies where the termini of the vector are compatible for ligation (e.g., a vector cut with a single restriction enzyme) to prevent recircularization without insertion of the PCR product.

In cloning reactions, DNA ligase must join both termini of a linear PCR product to a linear vector to form a circular recombinant molecule. As discussed above, this can be accomplished if either the PCR product or the vector has termini with 5′ phosphates, or if the vector and the insert both have termini with 5′ phosphates. The recombinant molecule resulting when both the insert and vector have 5′-terminal phosphates can be closed circular; when only one of the reactant DNAs has 5′-terminal phosphates, the doubly nicked circular form results. Both types of recombinants transform *E. coli*, and cloning strategies producing both types are commonly employed.

Directional Cloning of a PCR Product Having Noncompatible Termini

In this strategy, the termini of the PCR product have different restriction enzyme cleavage sites that are added to the DNA of interest via the tails of the synthetic oligonucleotide PCR primers (neither restriction enzyme should be present within the sequence to be amplified). Following PCR, about one-tenth of the reaction mix is electrophoresed on an agarose gel to check whether the expected size has been obtained. The PCR product is then typically purified out of the PCR mix using a silica membrane kit (e.g., QIAquick, Qiagen; NucleoSpin, BD Biosciences Clontech) and then cut with restriction enzymes, gel-purified, and placed into a ligase reaction mixture with the vector having compatible ends. As both the PCR product and the vector are cut with restriction enzymes, both the insert and the vector contain 5′ phosphates on both termini, and thus, closed circular recombinants can be produced. In "sticky end" ligations, the overhangs anneal at the ligation temperature of 16°C, which stabilizes the juxtaposition of the 5′ phosphate and 3′ hydroxyl to provide an improved efficiency of ligation compared to "blunt-end" cloning.

Sticky-end ligations with a vector having noncompatible termini are preferred because self-ligation, which can cause a high background of wrong colonies, is prevented. However, there are pitfalls to the effective execution of this strategy.

1. Failure can result from incomplete cutting of one of the restriction sites on the vector; the singly cleaved vector has compatible sticky ends that can join during the ligation reaction, producing an unacceptable high background of colonies without the inserted PCR product.

2. Partial cutting of the PCR product, although less deleterious than partial cutting of the vector, can adversely impact on cloning efficiency because of erroneous estimation of the molar quantity of doubly cut insert (because it is typically assumed to be 100%), and because intermediates consisting of vector linked to a singly cut PCR product are dead ends.

3. The cutting of PCR products can be incomplete because the restriction sites are too close to the ends (Zimmermann et al. 1998). Efficiency of cleavage increases with an increase in the number of nucleotides flanking the restriction enzyme recognition site. Therefore, it is advisable to consult tables listing the efficiency of cutting close to DNA ends (see New England Biolabs catalog or Web site), to design cloning strategies with enzymes that cut efficiently close to the ends, to add more (up to 6 are required) rather than fewer nucleotides to the ends of PCR primers flanking restriction sites, and to use sufficient amounts of restriction enzymes and longer cutting times.

4. The presence of 5′-truncated primers can also result in a PCR product cut on only one terminus, which can ligate to the vector to produce a dead-end product.

5. A high background of the wrong colonies can also be due to carryover of the super-coiled plasmid that was used as the template for the PCR. Small amounts of the super-coiled PCR template may co-migrate with the linear PCR product, and as a result co-purify. This is more likely to occur if the PCR product is large or if the plasmid preparation contains a low level of deleted plasmids (not detectable by ethidium bromide staining) that retain the origin of replication and the selectable marker and can thus transform E. coli. One solution to this problem is to cut the template prior to PCR to produce fragments that cannot easily self-ligate and have different ends from the PCR product (and of course using enzymes that do not cut within the insert). Another way to avoid this problem is to use a template containing a different antibiotic-resistance gene than is used for selecting the recombinants. Some vectors have two selectable markers (e.g., pCR-Blunt II-TOPO contains kanamycin- and zeocin-resistance genes), and thus if the PCR template plasmid contains a kanamycin-resistance gene, zeocin can be used for selection of the correct recombinants.

Nondirectional Ligase-dependent Cloning: Sticky-end Cloning Using a Singly Cut Vector and a Compatible PCR Product with Identical Termini

When only a single restriction site is available on a cloning vector, an additional step, alkaline phosphatase treatment of the vector, is usually employed to reduce vector self-ligation. Although this strategy is commonly used and can be successful, it is more prone to failure, and therefore, the preferred method is noncompatible, sticky-end cloning, which should be used wherever possible.

Nondirectional Ligase-dependent Cloning: Blunt-end Cloning

Any DNA with a blunt terminus can be joined to any other DNA with a blunt terminus, regardless of the nucleotide sequence, provided that one has a 5′ phosphate and the other a 3′ hydroxyl. PCR products generated with proofreading polymerases (i.e., those having 3′→5′ exonuclease activity) are mostly blunt (Lohff and Cease 1992) and can be ligated directly to a blunt-ended vector. However, the stabilization of the juxtaposition of the 5′ phosphate with the 3′ hydroxyl that is provided by the annealing of overhangs in sticky-end cloning is lacking in blunt-end cloning. As compared to sticky-end cloning, blunt-end cloning is less efficient, and 10-fold greater concentrations of ligase are used. Reduced efficiency also results from the potential of the vector to self-ligate, and thus, the need to remove the phosphate groups with alkaline phosphatase.

A great improvement in the efficiency of blunt-end cloning can be achieved by performing the ligation reaction in the presence of the restriction enzyme used to linearize the vector. In this strategy, self-ligated vectors are continually reopened, which drives the reaction toward the desired product. The PCR-Script Cloning Kits (Stratagene) use this strategy by employing the blunt-cutting, 8-base cutter *Srf*I. In this system, neither terminus of the PCR product can have the sequence 5′-GCCC-3′ (the *Srf*I half site), and the *Srf*I cleavage site cannot be present within the amplified region. Stratagene offers vectors with sequencing primer sites and RNA polymerase promoters flanking the insert, and a vector for mammalian expression via a cytomegalovirus (CMV) promoter.

Nondirectional Ligase-dependent Cloning: TA Cloning

PCR products generated with a nonproofreading polymerase contain deoxadenosine (dA) overhangs and can be cloned directly (without post-PCR enzymatic steps) into vectors having deoxythymidine (dT) overhangs. Vectors for this purpose can be generated by cutting two different restriction sites in order to leave a 3′ dT overhang on each terminus of the linearized vector (the recessed strands have 5′ phosphate) (Mead et al. 1991), or dT can be added with a nonproofreading polymerase in the presence of the deoxynucleotide triphosphate dTTP. The vectors cannot religate because dT cannot pair with dT. Several TA cloning vectors are commercially available and supplied in linearized form with 3′ dT overhangs. The TA cloning strategy is an efficient one. The PCR products lack 5′ phosphates and thus do not ligate with each other. Nonproofreading polymerases such as *Taq* generally add a dA overhang, although the nucleotide added depends on the preceding nucleotide, and the highest cloning efficiencies have been obtained using PCR primers with a 5′-terminal dA (Hu 1993). For efficient addition of dA overhangs to the PCR product, a final extension step of 10 minutes and 72°C is recommended.

TOPO CLONING

In the TOPO cloning method, vaccinia virus topoisomerase I is used instead of DNA ligase to join DNA molecules (Shuman 1994). Vaccinia TOPO specifically recognizes the duplex sequence 5′-(C/T)CCTT↓NNN-3′ and cleaves after the 3′ dT in one strand (the arrow indicates the cleavage site). This cleaved intermediate consists of Tyr-274 of topoisomerase I (TOPO) covalently attached to the dT via a 3′ phosphodiester bond (i.e., (C/T)CCTT-3′-**TOPO**), and a free 5′ OH on the other side of the break (i.e., **OH**-5′-NNN). This 5′ OH can attack the phosphotyrosyl bond between the DNA and TOPO, which results in the religation of DNA and the release of TOPO (reversal of the cleavage reaction). However, if the 5′-(C/T)CCTT-3′ is located close to the 3′ terminus, the single-stranded **OH**-5′-NNN dissociates from the duplex as a result of insufficient base-pairing, which leaves TOPO covalently linked to the 3′ end of the duplex. This TOPO-DNA molecule is said to be "activated," because it is susceptible to attack by the 5′-OH of a heterologous duplex DNA having a complementary 5′ protrusion. Thus, a TOPO-DNA adduct can be produced that is highly recombinogenic with DNAs having complementary single-strand protrusions (Fig. 1). TOPO cloning (Invitrogen) takes advantage of this property for producing cloning vectors and tools (Shuman 1994). A variety of TOPO cloning vectors are available that are shipped as linearized plasmids with TOPO covalently linked to the 3′ termini. These include vectors for blunt T-A and directional cloning. Because a free 5′ OH is required to attack the phosphotyrosyl bond between the vector and TOPO, PCR products to be cloned must contain 5′ OH,

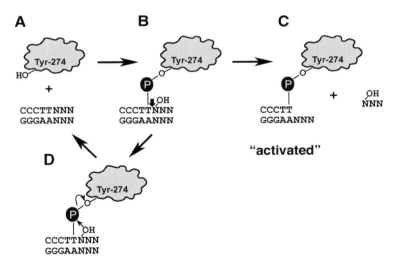

FIGURE 1. Production of TOPO-activated DNA fragments. (*A*) Vaccinia virus TOPO with the OH of Tyr-274 is shown together with a DNA fragment containing a TOPO cleavage site. (*B*) TOPO cleaves after the second deoxythymidine in the sequence CCCTT (see solid downward arrow), and in so doing is linked to the DNA via a 3′ phosphotyrosyl bond, which leaves a free 5′ OH on the other side of the break. (*C*) When the cleavage site is located close to the 3′ end of the duplex DNA, the single stranded **OH**-5′-NNN can dissociate from the duplex because of weak base-pairing energy, leaving a 5′ overhang and TOPO covalently linked to the 3′ end of the duplex. (*D*) The free 5′ OH can attack the phosphotyrosyl bond, which results in the religation of DNA and the release of TOPO (reversal of the cleavage reaction). The TOPO-activated molecule (shown in *C*) can combine with a heterologous DNA duplex having a complementary 5′ overhang. (Adapted from Invitrogen 2002b [©Invitrogen Corporation. All Rights Reserved].)

and therefore must not be phosphorylated (i.e., unphosphorylated primers should be used for PCR and the products must not be kinased), and restriction fragments must be treated with alkaline phosphatase before TOPO cloning in order to generate free 5′ OH termini.

The efficiency of TOPO cloning is due to a lack of a requirement for post-PCR restriction digestion, the inability of the vector to circularize without insert, and the fact that the TOPO reaction is rapid, requiring only 5 minutes for most applications. In addition, gel purification is not required if a single dominant PCR product is obtained; however, gel purification is recommended for PCR products greater than 1 kb because TOPO cloning is very efficient for small fragments that contaminate certain PCRs (Invitrogen 2002a). Further efficiency is obtained through selection strategies that are positive (blue/white test) and negative (disruption of the lethal *E. coli* gene *ccd*B). Some of the cloning vectors contain two resistance markers, so that a marker other than the marker on the template can be used for selecting the recombinants, which can eliminate a high background due to colonies containing the plasmid that was used as the template for generating the PCR product.

Invitrogen sells TOPO cloning vectors with a variety of DNA functional elements for applications including generic cloning of PCR products and the directional cloning of PCR products for a variety of applications. Furthermore, once a PCR product is TOPO-cloned into their pENTR system, a one-step, in vitro reaction using lambda phage recombination proteins (Gateway) can transfer the insert in the desired orientation into a large number of destination vectors designed for a variety of applications.

TOPO TA Cloning

As with TA vectors used for ligase-dependent cloning, the TOPO TA cloning vectors have termini with single 3′ dT overhangs, and are used to clone PCR products having a single, overhanging 3′ dA produced by the non-template-dependent terminal transferase activity of polymerases such as *Taq*, which lack proofreading. As supplied by Invitrogen, the 3′ termini of these vectors are linked to vaccinia TOPO. Self-closure of the vector is prevented because the 5′-overhanging dT of one terminus cannot pair with the 3′-overhanging dT of the other terminus, and thus the 5′ OH of one terminus is blocked from attacking the phosphotyrosyl bond of the other terminus. The PCR product with dA overhangs can pair with dT overhangs of the vector, juxtaposing the 5′ OH groups of the PCR product to attack of the phosphotyrosyl bonds of the vector (Fig. 2).

PCR products amplified with polymerase mixtures (*Taq* plus a proofreading polymerase) should contain a 10-fold excess of *Taq* to ensure the presence of 3′ dA-overhangs on the PCR product (Invitrogen 2002a). PCR products amplified solely with a proofreading polymerase lack 3′ dA overhangs, which must be added in a post-amplification reaction using *Taq* polymerase.

TOPO Blunt-end Cloning

As with conventional, ligase-mediated cloning of blunt-ended DNA fragments, recircularization of the vector is a potential problem. In TOPO cloning of blunt-ended PCR products, the 5′ OH of each blunt end is not sterically hindered from attacking the phosphotyrosyl bond on the other end to produce a recircularized vector that lacks the insert. To overcome this, Invitrogen has designed TOPO cloning vectors in which ligation of the insert disrupts the lethal *E. coli* gene *ccd*B (which is toxic because it traps gyrase; Van Melderen 2002). Thus, cells having plasmids that self-ligate without insert are killed upon plating.

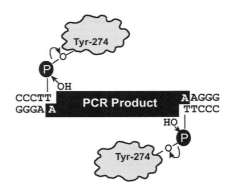

FIGURE 2. TOPO TA cloning. The nontemplate-dependent terminal transferase activity of *Taq* polymerase adds a single dA to the 3′ ends of PCR products. The linear Topo TA cloning vector contains overhanging 3′ deoxythymidine (dT) residues, and is "TOPO-activated" (TOPO is attached via a 3′ phosphotyrosyl linkage), which enable it to efficiently ligate with PCR products having single 3′ dA overhangs on both sides. The 5′ OH of each end of the PCR product is shown attacking the phosphotyrosyl bond between the vector DNA and topoisomerase I, which results both in the release of topoisomerase I molecules and the production of the doubly nicked, circular, recombinant molecule (not shown). (Reprinted, with permission, from Invitrogen 2002a [©Invitrogen Corporation. All Rights Reserved].)

Directional Cloning

The TOPO cloning vectors for directional cloning have one blunt terminus and one terminus with a 4-bp, 5′ protrusion. The PCR product can be cloned directionally into the vector by virtue of 4 bases added to the forward primer that are complementary to the vector's 5′ protrusion. The vector's 5′ protrusion (GTGG) invades the blunt-ended PCR product, annealing with the added bases (CACC), which correctly orients the PCR product and positions its 5′ OH for attack of the phosphotyrosyl bond (Fig. 3). Invitrogen's literature reports that inserts in the correct orientation are obtained in >90% of recombinant clones (Invitrogen 2002a).

Production of Linear Constructs with Defined 5′ and 3′ Functional Elements That Are Used Directly for Mammalian Cell Expression or In Vitro Transcription/Translation

TOPO Tools

The TOPO Tools kit enables the TOPO-mediated joining of functional elements, such as promoters, tags, and terminator regions, to PCR-amplified DNA. After adding the 5′ and 3′ functional elements, the construct is PCR-amplified, and the linear product is used directly for in vitro transcription (RNA probe and antisense) and transcription/translation (protein analyis), or is transfected for in vivo expression and protein analysis (protein–protein interaction analyses by coprecipitation and mammalian two-hybrid).

To create a construct having specific 5′-side and 3′-side functional elements, different 11-bp sequences are incorporated into the primers for amplifying a DNA of interest (Fig. 4). This creates a PCR product having one terminus with complementarity to a 5′-TOPO-adapted functional element, and the other terminus with complementarity to a

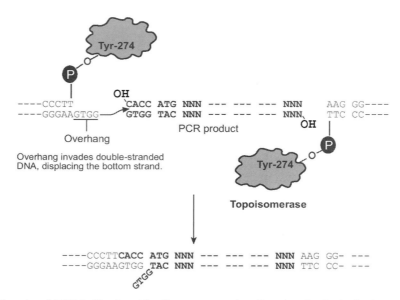

FIGURE 3. Directional TOPO Cloning. The linear vector for directional TOPO cloning (pENTR in this example) is shown as two independent TOPO-linked DNA molecules. The vector terminus on the left has a 5′ protrusion (GTGG), and the vector terminus on the right is blunt. Directional insertion of the PCR product into the vector occurs via complementarity between the 5′-protruding bases of the vector and the 5′-terminal 4 bases of the duplex PCR product that were added onto the forward primer (CACC). The 5′-protruding GTGG of the vector accesses the CACC of the PCR product via strand invasion. The annealing of the 4 bp juxtaposes the 5′ OH on the PCR product to attack the phosphotyrosyl bond. The other terminus of the PCR product is blunt, and its 5′ OH is not hindered from attacking the phosphotyrosyl bond on the blunt-ended side of the vector (*right*). (Reprinted, with permission, from Invitrogen 2002a [©Invitrogen Corporation. All Rights Reserved].)

3′-TOPO-adapted functional element. The 5′ terminal 6 bases of the primers' 11-base sequence are complementary to the 5′ overhang sequence of the TOPO-adapted functional element, and the internal 5 bases comprise the TOPO recognition sequence. In the TOPO Tools reaction, cleavage of the TOPO recognition sequences on the PCR product creates ends with a 3′-phosphotyrosyl-linked TOPO and a 5′, 6-base overhang complementary to the overhang of a functional element (either 5′ side or 3′ side). Thus, both termini of the PCR product and both functional elements (5′ side and 3′ side) have TOPO linked to the 3′ strand and have a 6-base overhanging, 5′ strand terminating with an OH. When the 6-bp overhang of the PCR product anneals with the 6-bp overhang of a functional element, each 5′ OH attacks the other molecule's phosphotyrosyl bond. Thus, both strands of the functional element ligate to both strands of the PCR product, and two TOPO molecules are released on each functional side (this is in contrast to TOPO cloning where only a single strand on each side of the insert is ligated to the vector, producing the doubly nicked, circular form when transformed). The addition of the functional elements is directional due to the different 6 terminal nucleotides added onto the primers incorporated into the PCR product, and different 5′ functional elements such as promoters, and 3′ functional elements such as epitopes, purification tags and polyadenylation sequences can be added to the proper sides of the PCR product in a single reaction. After the 5′- and 3′-functional elements are added to the PCR product, another round of PCR is used to produce enough of the linear construct for in vitro reactions or in vivo transfection. Overall, production of the construct requires one PCR step for adding different 11-bp sides to the gene of interest, a TOPO reac-

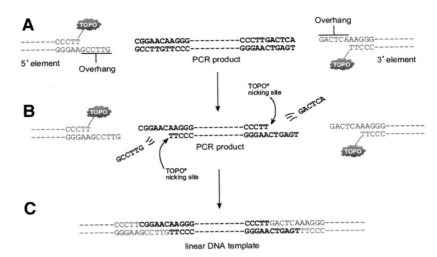

FIGURE 4. TOPO Tools for producing linear DNA constructs with defined 5′ and 3′ functional elements for direct use in vitro and in vivo. (*A*) The 5′ and 3′ functional elements and the PCR product to which they are to be added are shown in linear order from 5′ to 3′. Different 11-bp sequences are incorporated onto the termini of the PCR product via the forward and reverse primers (shown in bold). (*B*) In the TOPO Tools reaction, TOPO cleaves the PCR product, creating TOPO-activated overhangs (and the release of single-stranded 6-mers) complementary to overhangs of the TOPO-activated functional elements. (*C*) The annealing of the two overhangs juxtaposes the 5′ OH of each to attack the phosphotyrosyl bond of the other, resulting in two TOPO molecules released, and both strands of each functional element ligated to both strands of the PCR product, which creates the linear recombinant template. (Reprinted, with permission, from Invitrogen 2002b [©Invitrogen Corporation. All Rights Reserved].)

tion to simultaneously add the 5′- and 3′- functional elements, and another PCR for amplification.

These steps can be accomplished in less than a single day as compared to the several days often required for ligation of insert with vector, transformation, and the testing of colonies for the correct recombinants (Invitrogen 2002b). A major advantage of the system is its adaptability to high-throughput analysis of multiple DNAs of interest. HPLC purification of PCR primers is recommended to ensure that they are full length.

TAP Express Rapid Gene Expression Kit

In this system, a promoter and a terminator are added to the respective sides of a PCR product using two PCR steps. The first step amplifies the gene of interest using (1) a 5′-side primer with a tail complementary to the promoter and (2) a 3′-side primer with a tail complementary to the terminator. In the second PCR step, each strand of the PCR product of the first reaction becomes a primer for extension on the promoter or the terminator. Also included in this reaction is one primer that is identical to the very 5′ end of the promoter and another that is complementary to the very 3′ end of the terminator. These primers have modified 5′-end bases, and the resulting linear PCR products are said to exhibit improved expression after transfection as compared to PCR products not having the 5′-modified bases.

The advantages of the TOPO Tools (Invitrogen) and TAP Express (Gene Therapy Systems) systems is that the linear constructs can be produced in a single day. For quick

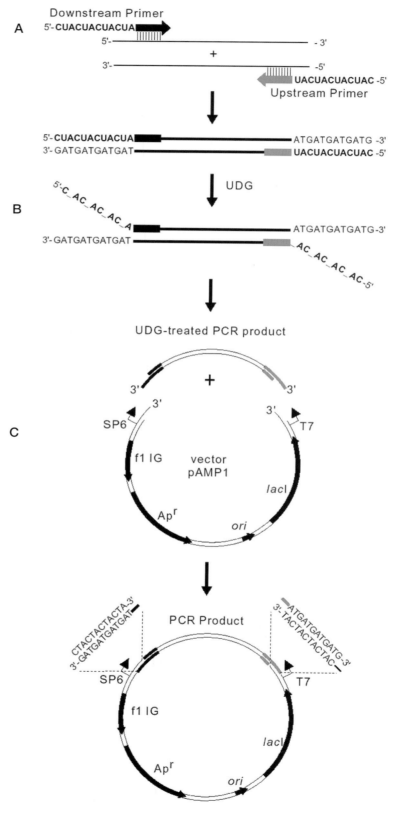

FIGURE 5. (*See facing page for legend.*)

tests, they may be preferable to making plasmid clones in certain circumstances, especially in high-throughput situations.

LIGASE-INDEPENDENT CLONING

Uracil DNA Glycosylase

In the uracil DNA glycosylase (UDG) method (Nisson et al. 1991; Varshney and van de Sande 1991; Rashtchian et al. 1992), specific primer tails are synthesized with dUMP instead of dTMP. The resulting PCR products contain dUMP residues in the primer region, which are susceptible to deglycosylation by UDG, rendering the dUMP residues abasic and unable to base-pair. This creates 3′ overhangs on the PCR product that anneal with complementary 3′ overhangs, which have been designed on the commercially supplied vector. The deglycosylation and the annealing with the vector occur simultaneously in a 30-minute reaction, which is then transformed into *E. coli*, where the insert–vector junctions are repaired. Directional cloning is accomplished by using different 5′-side and 3′-side PCR primers to create different 3′ overhangs on each side of the PCR product (Fig. 5). The advantages of the system are the elimination of the time-consuming tasks associated with some other systems for cloning PCR products (restriction endonuclease digestion, PCR product purification, ligation, or end-polishing). The disadvantage is the limited number of vectors available that use this cloning system (e.g., CloneAmp, Invitrogen).

BD In-Fusion

In the BD In-Fusion system (BD Biosciences Clontech), the PCR product is recombined with a linearized vector of your choice in an in vitro reaction catalyzed by a proprietary enzyme (Fig. 6). Directionality of cloning is achieved by making each side of the PCR product homologous to each side of the linearized vector by adding different 15-base tails to each primer (dA overhangs do not affect the reaction, so a proofreading or a nonproofreading polymerase can be used). The linearized vector and the PCR product are mixed with the proprietary BD In-Fusion enzyme, and in a single 30-minute reaction, the enzyme catalyzes the alignment and strand displacement of the homologous ends of the PCR product with the vector, while a 3′-exonuclease activity removes the single-stranded region; the nicks are repaired after transformation of *E. coli*.

The advantages of this system are its simplicity (it requires only the design of 15 nucleotide tails for each primer) and its universal applicability to any vector and any restriction site within a vector that can be used for linearization. Restriction enzyme digestion of

FIGURE 5. UDG for creating single-stranded overhangs on the PCR product for annealing with complementary overhangs on the vector; repair and covalent joining of the vector–insert junctions occur in vivo. The UDG method relies on the incorporation of dUMP residues in place of dTMP in the 5′ end of each amplification primer. (*A*) Target DNA is amplified using primer tails synthesized with dUMP residues instead of dTMP; the resulting PCR products have dUMP-containing sequence at their 5′ termini. (*B*) UDG treatment of the PCR product renders dUMP residues abasic and unable to base-pair, producing 3′ overhangs. (*C*) The single-stranded 3′ overhangs of the PCR product are annealed with complementary 3′ overhangs of the commercially supplied, linearized vector (pAMP1, Invitrogen). The creation of single-strand tails on the PCR product (via deglycosylation of dUMP residues by UDG) and the annealing of the PCR product to the vector occur in a single 30-minute reaction. (Reprinted, with permission, from Invitrogen 2002a [©Invitrogen Corporation. All Rights Reserved].)

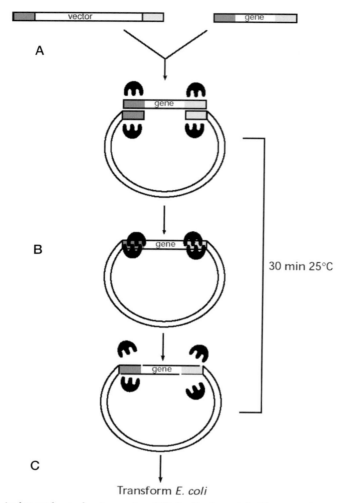

FIGURE 6. Ligation-independent cloning using BD In-Fusion. (*A*) The reaction mixture contains a vector of your choice linearized by a restriction cut (**vector**), a PCR product generated with primers containing 15-bp 5′ termini homologous to the ends of the linear vector (**gene**), and the proprietary BD In-Fusion enzyme. (*B*) The In-Fusion enzyme catalyzes the alignment and strand displacement of the homologous ends of the PCR product with the vector, while a 3′-exonuclease activity removes single-stranded regions. (*C*) The nicks are sealed after transformation of *E. coli*.

the PCR product is not required, and when a single predominant PCR product is obtained, only a cleanup with a silica membrane kit is required. If only a single restriction site is available for insertion of a PCR product, the system is advantageous because it nonetheless provides an efficient method for directional cloning. For expression cloning, no additional amino acids are added to the expressed protein, and epitope tags can be added via PCR. It is especially useful for high-throughput applications because once multiple PCR products are in hand, all subsequent steps are the same (i.e., different combinations of restriction enzymes are not required to cut different PCR products). Another advantage is that linearized vectors are commercially available in a kit, and these vectors are adapted for the direct transfer of the insert to a variety of other types of functional plasmids via *Cre-loxP*-based in vitro recombination (BD In-Fusion PCR Cloning Kit, BD Biosciences Clontech).

Although 15-nucleotide tails are required on the PCR primers, these are smaller than the 25-nucleotide tails required for either Xi-cloning (Gene Therapy Systems) or BP Clonase (Invitrogen).

Gateway BP Clonase System

In this in vitro site-specific recombination system from Invitrogen, specific sequences are directionally added to the 5′ termini of PCR primers. These sequences are homologous to sequences on the vector, and recombination between the PCR product and the vector is mediated by specific recombinases in vitro. In the Gateway BP Clonase system (Invitrogen), recombination is based on the phage-lambda site-specific recombination system (Landy 1989) and uses the bacteriophage lambda recombination protein (Int) and the *E. coli*-encoded protein, integration host factor (IHF). The sequences added to the PCR product are the 25-bp *att*B sequences (+ 4 terminal Gs); directionality is provided by different *att*B sequences, *att*B1 and *att*B2, that are added onto the 5′ side and 3′ side of the PCR product (Fig. 7). The in vitro recombination of the *att*B sequences flanking the PCR product with the *att*P sequences flanking the *ccd*B gene (negative selection markers; Bernard and Couturier 1992) on the "donor" vector causes excision of the *ccd*B gene and insertion of the PCR product to generate an "entry" clone. This recombinant entry clone is flanked by ~100-bp *att*L sequences, which result from recombination of *att*B with *att*P, and thus is not used for protein expression. Instead, the original PCR product can be transferred from this entry clone into a variety of expression vectors using another in vitro recombination reaction where the *att*L-flanked PCR product recombines with *att*R on the "destination" expression vector, resulting in an expression clone in which the PCR product is again flanked by *att*B sites, which are 25 bp. A potential disadvantage of the system is that the peptide encod-

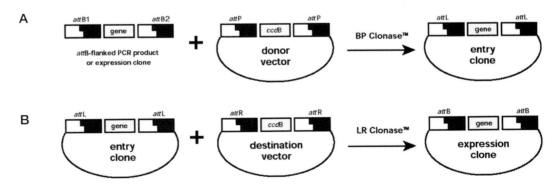

FIGURE 7. Cloning PCR products by in vitro recombination using the Gateway system (Invitrogen). (*A*) The target sequence (gene) is amplified by PCR with primers containing the 25-bp *att*B sequences (+ 4 terminal Gs); these are directional, with *att*B1 on the 5′ side, and *att*B2 on the 3′ side of the gene. The resulting PCR product is mixed with a donor vector in the presence of BP Clonase (bacteriophage lambda recombination protein, Int, and the *E. coli*-encoded protein, IHF). BP Clonase catalyzes recombination between the *att*B sites of the PCR product (gene) and the *att*P sites of the donor vector. The *E. coli* gene, *ccd*B (prevents growth of *E. coli*) in the donor vector is replaced with the PCR product (gene) to create the entry clone in which the PCR product (gene) is flanked by the *att*L sequences that were created by the recombination. (*B*) From the entry clone construct, the PCR product can be transferred to a variety of expression vectors by in vitro recombination with the destination vector that is mediated by LR Clonase (mix of lambda phage Int and Xis proteins, and *E. coli*-encoded IHF), to produce an expression clone. (Reprinted, with permission, from Invitrogen 2002a [© Invitrogen Corporation. All Rights Reserved].)

ed by an *att*B sequence (8 amino acids) is added onto the expressed protein; however, insertion of a protease cleavage sequence (e.g., TEV protease) can permit removal of amino-terminal peptides from the expressed protein. Advantages of the system include the short reaction time of the recombination reaction (1 hour), which is not affected by a 3′-dA overhang so a proofreading or a nonproofreading polymerase can be used. Digestion with restriction enzymes and ligation are not required.

The addition of the *att*B sequences by the first in vitro recombination reaction, which is the first step toward getting a PCR product into the Gateway system, can also be accomplished by cloning of a PCR product between two *att*B sites in a vector by traditional restriction cleavage/ligation, or by directional TOPO cloning.

RECOMBINATION IN VIVO

Xi-Clone In Vivo Cloning System

In this system from Gene Therapy Systems, a linear vector and a PCR product with homologous termini to the vector are transformed together into *E. coli* and the two linear molecules recombine in vivo. The PCR product is generated using PCR primers having 25-nucleotide tails that are homologous to the termini of the vector. The two linear molecules with 25-bp homology on their termini are then transformed into a proprietary recombinase-positive strain of *E. coli* (SmartCells) where they join, producing a circular plasmid.

The advantage of the system is the savings in time because restriction digestion and ligation are not required. Also, additional amino acids are not added onto expressed proteins. Linear vectors are commercially available in kit form with instructions on primer design. A kit is also available for adapting your vector of choice for Xi-cloning. A disadvantage is the 25 nucleotides that need to be added to PCR primers, which are an added expense and a potential source of amplification difficulties.

SUMMARY AND PERSPECTIVE

Cloning PCR products is typically the first step to generate a reagent for an experiment, and therefore efficiency is the goal and has been the impetus for the development of new ligase-independent methods for the efficient cloning of PCR products. Traditional cloning of PCR products involves restriction digestion of the PCR product and the vector to make their termini compatible to enable ligase-mediated production of the desired recombinant. Such traditional methods are effective and are the most commonly employed means of generating recombinant molecules today, although they require multiple post-PCR enzymatic steps (i.e., are less efficient), are less applicable to high-throughput cloning, and cause some strategies to be limited by the presence of restriction sites within the PCR product. Recently developed ligase-independent methods have the advantages of amenability to high-throughput cloning and the ability to insert the PCR product into any restriction site on a plasmid that can be used for linearization; the presence of restriction sites within the PCR product is not limiting because the product is not digested. Most importantly, these methods can be more efficient than traditional methods due to fewer enzymatic steps.

TOPO-activated vectors (Invitrogen) are provided with vaccinia topoisomerase I covalently linked via a phosphotyrosyl bond to each 3′ end, and are used to ligate PCR products directly without restriction digestion. Cloning PCR products into TOPO vectors in our experience is highly efficient, and a variety of vectors are available for cloning PCR products that have blunt termini or dT overhangs. Directional cloning is also available and, once cloned directionally into the so-called "entry" vector, the PCR product can be efficiently transferred via in vitro recombination to a variety of "destination" vectors for expression in a variety of systems.

Other methods involve the addition of termini to the PCR product (15–29 bp) that are used for ligase-independent recombination with the vector in vitro (BD Biosciences Clontech or Invitrogen). The BD Infusion system (BD Biosciences Clontech) uses a proprietary enzyme for recombining PCR products with termini homologous to the termini of any linearized vector; and no additional bases need to be added to the PCR product beyond the region of homology with the vector. Vectors are also available that can transfer PCR products cloned in this way into a variety of expression systems via additional in vitro recombination reactions.

Traditional methods of cloning PCR products continue to be used extensively for historical reasons, because they are tried and true, and because most molecular biology laboratories have large inventories of restriction enzymes. Newer ligase-independent methods, including TOPO cloning and in vitro recombination, are being increasingly adopted due to their efficiency.

REFERENCES

Ailenberg M. and Silverman M. 1996. Description of a one step staggered reannealing method for directional cloning of PCR-generated DNA using sticky-end ligation without employing restriction enzymes. *Biochem. Mol. Biol. Int.* **39:** 771–779.

Arashi-Heese N., Miwa M., and Shibata H. 1999. XcmI site-containing vector for direct cloning and in vitro transcription of PCR product. *Mol. Biotechnol.* **12:** 281–283.

Aslanidis C. and de Jong P.J. 1990. Ligation-independent cloning of PCR products (LIC-PCR). *Nucleic Acids Res.* **18:** 6069–6074.

Aslanidis C., de Jong P.J., and Schmitz G. 1994. Minimal length requirement of the single-stranded tails for ligation-independent cloning (LIC) of PCR products. *PCR Methods Appl.* **4:** 172–177.

Ausubel F.M., Brent R., Kingston R.E., Moore D.D., Seidman J.G., Smith J.A., and Struhl K. 2003. *Current protocols in molecular biology.* Wiley, New York.

Bernard P. and Couturier M. 1992. Cell killing by the F plasmid CcdB protein involves poisoning of DNA-topoisomerase II complexes. *J. Mol. Biol.* **226:** 735–745.

Bielefeldt-Ohmann H. and Fitzpatrick D.R. 1997. High-efficiency T-vector cloning of PCR products by forced A tagging and post-ligation restriction enzyme digestion. *BioTechniques* **23:** 822–826.

Chuang S.E., Wang K.C., and Cheng A.L. 1995. Single-step direct cloning of PCR products. *Trends Genet.* **11:** 7–8.

Delidow B.C. 1997. Molecular cloning of PCR fragments with cohesive ends. *Mol. Biotechnol.* **8:** 53–60.

Gabant P., Dreze P.L., Van Reeth T., Szpirer J., and Szpirer C. 1997. Bifunctional lacZ alpha-ccdB genes for selective cloning of PCR products. *BioTechniques* **23:** 938–941.

Gal J., Schnell R., Szekeres S., and Kalman M. 1999. Directional cloning of native PCR products with preformed sticky ends (autosticky PCR). *Mol. Gen. Genet.* **260:** 569–573.

Geiser M., Cebe R., Drewello D., and Schmitz R. 2001. Integration of PCR fragments at any specific site within cloning vectors without the use of restriction enzymes and DNA ligase. *BioTechniques* **31:** 88–92.

Guo B. and Bi Y. 2002. Cloning PCR products. *Methods Mol. Biol.* **192:** 111–120.

Haun R.S., Serventi I.M., and Moss J. 1992. Rapid, reliable ligation-independent cloning of PCR products using modified plasmid vectors. *BioTechniques* **13:** 515–518.

Horton R.M., Raju R., and Conti-Fine B.M. 1997. A T-linker strategy for modification and directional cloning of PCR products. *Methods Mol. Biol.* **67:** 101–110.

Hu G. 1993. DNA polymerase-catalyzed addition of nontemplated extra nucleotides to the 3′ end of a DNA fragment. *DNA Cell Biol.* **12:** 763–770.

Ido E. and Hayami M. 1997. Construction of T-tailed vectors derived from a pUC plasmid: A rapid system for direct cloning of unmodified PCR products. *Biosci. Biotechnol. Biochem.* **61:** 1766–1767.

Invitrogen. 2001. Five-minute cloning of Taq polymerase-amplified, PCR products. TOPO TA Cloning® (version N, 072601, 25-0184). Invitrogen Life Technologies, Inc., Rockville, Maryland.

———. 2002a. pENTR Directional TOPO® cloning kits (version B, July 9, 2002, 25-0434). Invitrogen Life Technologies, Inc., Rockville, Maryland.

———. 2002b. TOPO® tools technology. For generating functional constructs containing your PCR products and a choice of TOPO®-adapted elements (version B, 072501, 25-0413). Invitrogen Life Technologies, Inc., Rockville, Maryland.

Jung, V., Pestka S.B., and Pestka S. 1990. Efficient cloning of PCR generated DNA containing terminal restriction endonuclease recognition sites. *Nucleic Acids Res.* **18:** 6156.

Landy A. 1989. Dynamic, structural, and regulatory aspects of lambda site-specific recombination. *Annu. Rev. Biochem.* **58:** 913–949.

Life Technologies. 1999. Helpful tips for custom primers. *Focus* **21:** 69–71.

Lohff C.J. and Cease K.B. 1992. PCR using a thermostable polymerase with 3´ to 5´ exonuclease activity generates blunt products suitable for direct cloning. *Nucleic Acids Res.* **20:** 144.

Mead D.A., Pey N.K., Herrnstadt C., Marcil R.A., and Smith L.M. 1991. A universal method for the direct cloning of PCR amplified nucleic acid. *Bio/Technology* **9:** 657–663.

Nisson P.E., Rashtchian A., and Watkins P.C. 1991. Rapid and efficient cloning of Alu-PCR products using uracil DNA glycosylase. *PCR Methods Appl.* **1:** 120–123.

Padgett K.A. and Sorge J.A. 1996. Creating seamless junctions independent of restriction sites in PCR cloning. *Gene* **168:** 31–35.

Qiagen. 2001. *PCR cloning handbook,* April. Qiagen.

Rashtchian A., Buchman G.W., Schuster D.M., and Berninger M.S. 1992. Uracil DNA glycosylase-mediated cloning of polymerase chain reaction-amplified DNA: Application to genomic and cDNA cloning. *Anal. Biochem.* **206:** 91–97.

Sambrook, J., Fritsch E.F., and Maniatis T. 1989. *Molecular cloning: A Laboratory manual.* Cold Spring Harbor Laboratory Press, Cold Spring Harbor, New York.

Shuman S. 1994. Novel approach to molecular cloning and polynucleotide synthesis using vaccinia DNA topoisomerase. *J. Biol. Chem.* **269:** 32678–32684.

Smith C., Day P.J., and Walker M.R. 1993. Generation of cohesive ends on PCR products by UDG-mediated excision of dU, and application for cloning into restriction digest-linearized vectors. *PCR Methods Appl.* **2:** 328–332.

Testori A., Listowsky I., and Sollitti P. 1994. Direct cloning of unmodified PCR products by exploiting an engineered restriction site. *Gene* **143:** 151–152.

Tillett D. and Neilan B.A. 1999. Enzyme-free cloning: a rapid method to clone PCR products independent of vector restriction enzyme sites. *Nucleic Acids Res.* **27:** e26.

Van Melderen L. 2002. Molecular interactions of the CcdB poison with its bacterial target, the DNA gyrase. *Int. J. Med. Microbiol.* **291:** 537–544.

Varshney U. and van de Sande J.H. 1991. Specificities and kinetics of uracil excision from uracil-containing DNA oligomers by *Escherichia coli* uracil DNA glycosylase. *Biochemistry* **30:** 4055–4061.

Weiner M.P. 1993. Directional cloning of blunt-ended PCR products. *BioTechniques* **15:** 502–505.

Zeng Q., Eidsness M.K., and Summers A.O. 1997. Near-zero background cloning of PCR products. *BioTechniques* **23:** 412–418.

Zimmermann K., Schogl D., and Mannhalter J.W. 1998. Digestion of terminal restriction endonuclease recognition sites on PCR products. *BioTechniques* **24:** 582–584.

28 Bidirectional and Directional Cloning of PCR Products

Gina L. Costa and Michael P. Weiner

454 Life Sciences, Molecular Sciences, Branford, Connecticut 06405

This chapter presents methods for improved blunt-end cloning of PCR-generated DNA fragments (Costa et al. 1994a). "Polishing" *Taq* DNA polymerase-generated PCR fragments with *Pfu* DNA polymerase increases the yield and efficiency of cloning. Using a triple primer set consisting of two outside, asymmetrically distanced, primers and one fragment-specific primer, it is possible to determine both the presence and the orientation of cloned inserts. Application of these methods allows fragments to be generated and cloned in one day, and putative clones to be analyzed the next, thereby saving a substantial amount of time and effort. Cloning kits available for PCR product cloning include the PCR-Script Cloning Kit (bidirectional) (Stratagene Cloning Systems) and PCR-Script Direct Cloning Kit (directional) (Stratagene Cloning Systems).

PROTOCOL 1 PREPARATION OF CLONING VECTORS

Several vectors and vector derivatives have been created for cloning PCR products. These include the standard pBluescript-type with multiple cloning sites and abbreviated multiple cloning sites, as contained in the PCR-Script Direct plasmids (see Fig. 1) (Bauer et al. 1992; Costa and Weiner 1994a,b; Costa et al. 1994b,c). Abbreviated multiple cloning sites allow the user to incorporate common restriction enzyme sites into the PCR primer sets without the problem of having the same target sequence occurring in the plasmid vector. Chloramphenicol (cam)-resistant plasmids should be used when subcloning DNA fragments from ampicillin (amp)-resistance-encoding plasmids. This allows the use of antibiotic selection against parental plasmids after bacterial transformation.

To increase the efficiency of blunt-end cloning of PCR-generated fragments, it was found that a restriction enzyme, added in a functional-unit excess relative to the units of T4 DNA ligase, increases the efficiency of the ligation reaction (Liu and Schwartz 1992). This simultaneous restriction digestion and ligation reaction results in an increased efficiency of the blunt-end cloning by two mechanisms. First, as long as the PCR fragment does not, when ligated with the vector, create a restriction enzyme target site, the available circular vector is removed from the overall reaction by recombinant insertion. An increased amount of linear vector is made available during the ligation reaction by the restriction enzyme on self-ligated vector molecules. Second, because linear DNA molecules transform *E. coli* at a greatly reduced efficiency, they do not significantly contribute to the number of colonies observed after transformation. Both of these mechanisms result in a reduced overall transformation efficiency, but because only the linearized, nonrecombinant plasmids are reduced, the overall recombinant efficiency actually increases.

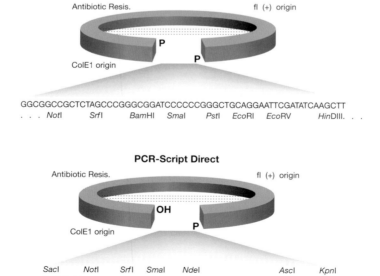

FIGURE 1. Several vectors have been developed for directional and bidirectional cloning. These include derivatives that encode either chloramphenicol or ampicillin resistance with the modified multiple cloning sites optimized for specific cloning operations (e.g., general subcloning or protein expression).

The PCR-Script method uses the restriction enzyme *Srf* I (Simcox et al. 1991). *Srf* I has an octanucleotide recognition sequence (5´-GCCC|GGGC-3´) that is rare and would occur on an average of 1 in 65,000 bp. (Because of the bias against CpG sequences in some DNA, its actual occurrence in mammalian DNA is closer to 1 in 100,000 bp.) The target site is blunt-ended and contains an internal 6-base recognition site (5´-CCC|GGG-3´) that can be recognized by another blunt-end restriction enzyme (*Sma*I). This was important to the development of the PCR-Script Direct method because the actual PCR cloning with directionality occurs in a reaction identical to that described for PCR-Script.

The addition of a restriction endonuclease to the ligation reaction allows an overall fourfold increase in cloning efficiency, along with a greatly reduced background. For the bidirectional cloning of PCR-generated DNA fragments, it is recommended to use a PCR-Script-type reaction containing a predigested vector DNA and *Pfu* DNA polymerase-generated, or *Pfu*-polished, inserts (Costa and Weiner 1994c,d). For the directional cloning of PCR-generated DNA fragments, it is recommended to use a PCR-Script Direct-type cloning reaction with a *Pfu* DNA polymerase-generated, or *Pfu*-polished monophosphorylated, insert. The procedure is the same for both the PCR-Script bidirectional and the PCR-Script Direct directional cloning.

MATERIALS

> **▼ CAUTION**
>
> *See Appendix for appropriate handling of materials marked with <!>.*

BUFFERS, SOLUTIONS, AND REAGENTS

Chloroform-isoamyl alcohol (24:1)<!>
Phenol:chloroform-isoamyl alcohol (25:24:1)<!>
Lithium chloride<!> (LiCl; 10 M)
Phenol (Tris-buffered)<!>, pH 8.0
TE buffer (5 mM Tris-HCl<!>, pH 8.0, 0.1 mM EDTA)
Ethanol<!>, 100%

ENZYMES AND ENZYME BUFFERS

Alkaline phosphatase (0.1–0.2 units) (for use in Method B, directional cloning)

> Commercially available molecular-biology-grade alkaline phosphatase often contains nuclease contamination. We recommend the use of bacterial alkaline phosphatase that has been purified devoid of contaminating nucleases and specifically quality-controlled for use in the PCR-Script assay.

Blunt-end restriction endonuclease (10–20 units) and reaction buffer, e.g., *Srf*I or *Sma*I
Universal buffer, 10x (for use in Method B, directional cloning, to be compatible with the restriction enzyme and the alkaline phosphatase used) (1 M potassium acetate<!>, 250 mM Tris-acetate<!>, pH 7.6, 100 mM magnesium acetate<!>, 5 mM β-mercaptoethanol<!>, 100 μg/ml BSA)

NUCLEIC ACIDS AND OLIGONUCLEOTIDES

Cloning vector (1 μg)

SPECIAL EQUIPMENT

Water baths preset to the appropriate temperature for restriction digestion, alkaline phosphatase treatments (Method B), and 65°C
Vacuum desiccator

ADDITIONAL ITEMS

Optional: Equipment and reagents for agarose gel electrophoresis, including ethidium bromide<!>

METHOD A: PREPARATION OF BIDIRECTIONAL CLONING VECTORS

Blunt-end cloning procedures utilize cloning vectors with blunt ends to capture DNA fragments for bidirectional insertion. Therefore, blunt-end cloning vectors do not require nucleotide overhangs for clonal insertion. Subsequently, blunt DNA fragments, such as PCR products, may be cloned directly in these vectors. Blunt-end cloning is an inherently inefficient method, with recombinant insertion generally accounting for less than 10% of total transformants (see Fig. 2A). Increased efficiency can be achieved by including a restriction enzyme in the ligation reaction, as in the PCR-Script method (see Fig. 2B) (Liu and Schwartz 1992).

METHOD

1. Digest the 1-μg vector DNA with an appropriate blunt-end restriction endonuclease in a 20-μl reaction mixture as follows;

Vector DNA (1 μg/μl)	1 μl
Restriction enzyme (10 units/μl)	1–2 μl
Reaction buffer, 10x	2 μl
H₂O	to 20 μl

2. Incubate the digestion at the appropriate temperature for 1 hour.

   Optional: A 1-μl aliquot of the digestion can be run on an agarose gel to check for linearization of the vector DNA.

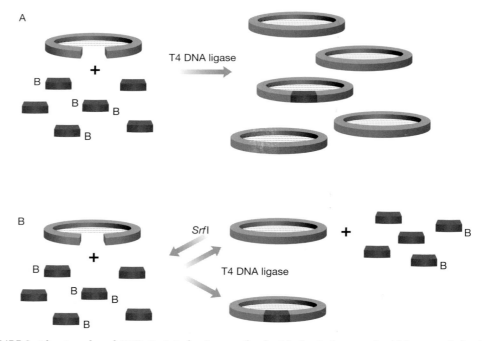

FIGURE 2. Blunt-end and PCR-Script cloning methods. Methods for standard blunt-end cloning (*A*) include incubation of the PCR product with predigested vector DNA and T4 DNA ligase. More efficient methods (*B*) include the addition of the restriction enzyme (in this example, *Srf*I endonuclease) to regenerate the linearized vector from the self-ligated vector during the ligation reaction. B denotes extended nucleotide bases.

3. Extract the digestion with phenol/chloroform and add an equal volume of Tris-buffered phenol (pH 8.0), vortex, and transfer the aqueous top phase to a new tube. Add an equal volume of chloroform to the tube and vortex. Centrifuge briefly, and carefully transfer the aqueous phase to a new tube.

4. Heat-treat the extracted DNA for 20 minutes at 65°C to remove any remaining chloroform (the boiling point of chloroform is 55°C).

5. Precipitate the DNA extracted in step 2 with 0.1 volume of 10 M LiCl and 2.5 volumes of ice-cold 100% ethanol. Mix gently and centrifuge at room temperature for 10 minutes at 12,000g.

6. Following centrifugation, carefully decant the supernatant. Dry the DNA pellet under vacuum for 10 minutes.

7. Redissolve the DNA in 50 μl of TE buffer. Store the predigested bidirectional cloning vector at –20°C until use.

> When redissolved in 50 μl of TE, the final concentration of the digested DNA should be ~10 ng/μl.

METHOD B: DIRECTIONAL CLONING VECTORS

Known characteristics of T4 DNA ligase and *E. coli* transformation have been used to create a directional cloning method that does not require the addition of extra bases to the primers. First, T4 DNA ligase requires both a 5′-phosphate and a 3′-hydroxyl group to efficiently ligate two strands of DNA. Second, linear DNA transforms *recBC*-proficient hosts of *E. coli* at a greatly reduced efficiency (it is decreased by approximately four orders of magnitude). Therefore, it was reasoned that directional cloning could be achieved by creating a monophosphorylated vector and a monophosphorylated insert. In the desired orientation, the ligation would result in a single-nicked, circular molecule. In the undesired, opposite orientation, the ligation would result in a linear molecule that would transform *E. coli* with a drastically reduced efficiency. A monophosphorylated vector is created by enzymatically treating the vector with a restriction endonuclease, removing the exposed 5′ phosphates with an alkaline phosphatase, and subsequently digesting the vector with a second restriction endonuclease. Degenerate restriction endonucleases may also be used for this procedure.

Proper DNA sequence manipulation will enable the enzymatically processed vector to be used in a PCR-Script-type reaction, whereby a self-ligated vector is susceptible to restriction by the endonuclease present in the ligation reaction, and the reading frame of the reporter gene is conserved. Owing to the importance of recreating a restriction enzyme site following vector self-ligation, the necessity of using highly purified enzymes for performing the directional and bidirectional cloning protocols as described cannot be overstated. Nuclease contamination must be determined and eliminated prior to performing the described experiments.

In a specific example, using the PCR-Script Direct directional cloning method, we enzymatically processed an SK(+) multiple cloning site that was engineered to contain both a *Srf*I (5′-GCCC|GGGC-3′) site and a *Sma*I (5′-T<u>CCC|GGG</u>C-3′; where the *Sma*I target sequence is underlined) site (Weiner 1993). The vector was first digested with *Srf*I, followed by removal of the 5′ phosphates with alkaline phosphatase and a second digestion with *Sma*I. Removal of the short DNA fragment after the *Srf*I–*Sma*I digestions results in the retention of an *Srf*I site (see Fig. 3). Phenotypic selection can still be used, since the reading frame is conserved. The monophosphorylated vector is produced by the general guidelines below.

METHOD

1. Digest the appropriate vector DNA with the first blunt-end restriction endonuclease (*Srf*I) as follows:

Vector DNA (1 μg/μl)	1 μg/μl
Restriction enzyme (10 units/μl)	1–2 μl
Universal buffer,* 10x	2 μl
H_2O	to 50 μl

2. Incubate at the appropriate temperature for 1 hour.

   Optional: A 1-μl aliquot of the reaction can be run on an agarose gel to check for linearization of the vector DNA.

   *A buffer that is compatible with the first restriction endonuclease digestion as well as the alkaline phosphatase dephosphorylation should be used to optimize the enzymatic processing of the vector DNA.

2. Inactivate the restriction enzyme by incubating the reaction for 20 minutes at 65°C. Remove to ice.

3. Add the alkaline phosphatase enzyme (0.1–0.2 units) directly to the heat-treated reaction mixture and incubate according to the manufacturer's guidelines.

4. Phenol/chloroform-extract the restriction-digested, alkaline-phosphatase-treated plasmid DNA. Add an equal volume of Tris-buffered phenol, vortex, and transfer the aqueous phase to a new tube. Add an equal volume of chloroform to the tube and vortex. Transfer the aqueous top phase to a new tube. Heat-treat the extracted DNA for 20 minutes at 65°C to remove any remaining chloroform.

5. Set up a second 30-μl restriction enzyme digestion containing the processed alkaline-phosphatase-treated vector DNA with the downstream blunt-end restriction endonuclease by adding a 15-μl aliquot of phenol/chloroform-extracted DNA, H_2O, 1x Universal buffer, and enzyme (10–20 units). Allow this digestion to incubate at the recommended temperature for 1 hour.

6. Inactivate the second restriction enzyme by incubating the reaction for 20 minutes at 65°C. Remove to ice.

7. Precipitate the monophosphorylated DNA with 0.1 volume of 10 M LiCl and 2.5 volumes of ice-cold 100% ethanol. Mix gently and centrifuge at room temperature at 12,000*g* for 10 minutes.

8. Following centrifugation, decant the supernatant. Dry the DNA pellet in vacuo for 10 minutes.

9. Redissolve the DNA in 25 μl of TE buffer.

   When redissolved into 25 μl of TE, the final concentration of the monophosphorylated DNA should be ~10 ng/μl. The monophosphorylated directional cloning vector can be stored at –20°C until use.

PROTOCOL 2   PREPARATION OF INSERT

STAGE 1: PRIMER-KINASING TREATMENT OF A DNA PRIMER

Studies have shown that many species of DNA polymerases (e.g., T7, modified T7, *Taq*, Vent, *Tth*, and Klenow) exhibit terminal deoxynucleotidyl transferase (TdT) activity (Clark

1988; Hu 1993). The 3´-end nucleotide extension of PCR products by DNA polymerases has been found to be both nucleotide- and polymerase-specific. For example, *Taq* DNA polymerase-generated PCR products would be preferentially modified as follows (+ for extension; – for nonaddition).

3´ NUCLEOTIDE EXTENSIONS ASSOCIATED WITH *TAQ*-GENERATED PCR PRODUCTS (HU 1993)

3´-End nucleotide	3´-End extension
A	+A (at very low efficiency)
C	+A > +C
G	+G > +A > +C
T	–T > +A

There appears to be no consistent pattern by which bases are added to templates by polymerases. Therefore, it cannot be assumed that all DNA polymerases can be used to create blunt-end DNA fragments. However, for certain DNA polymerases, the expected 3´-end nucleotide of a PCR product can be controlled by the 5´-end nucleotide of the PCR primer (Hu 1993; Costa and Weiner 1994d).

For directional cloning using a monophosphorylated vector, insert monophosphorylation can be achieved by kinase-treating one primer prior to the PCR. Preferably, this could be achieved by synthesizing a PCR primer with a 5´ phosphate group chemically attached. Synthesis of a PCR primer with a 5´-terminal phosphate group ensures that all single-stranded DNA has been monophosphorylated. An advantage to kinase treatment is that all preexisting primer sets can be modified for directional cloning using a monophosphorylated vector. T4 polynucleotide kinase treatment is a simple and rapid technique.

MATERIALS

BUFFERS, SOLUTIONS, AND REAGENTS
ATP (10 mM)

ENZYMES AND ENZYME BUFFERS
Kinase buffer (10 mM $MgCl_2$<!>, 100 mM Tris-HCl<!>, pH 7.5, 5 mM dithiothreitol<!>)
T4 polynucleotide kinase (10 units)

NUCLEIC ACIDS AND OLIGONUCLEOTIDES
Oligonucleotide primers (1 µg/µl)

SPECIAL EQUIPMENT
Water baths, preset to 37°C and 95°C

METHOD

1. Add the following to a microcentrifuge tube:

Kinase buffer, 10x	3 µl
ATP, 10 mM	0.5 µl
T4 DNA kinase (10 units/µl)	1 µl
Chosen primer	5 µg
H_2O	to 30 µl

2. Incubate for 1 hour at 37°C.

3. Boil the reaction at 95°C to inactivate the T4 polynucleotide kinase.

STAGE 2: PCR AMPLIFICATION

Because no specific guidelines exist for choosing which buffer conditions to use for the various types of DNA primer-template systems, it is often advantageous to test a range of PCR buffers. A number of PCR optimization kits have been created that enable one to test several different buffer compositions (e.g., Opti-Prime PCR Optimization Kit [Stratagene] and The PCR Optimizer [Invitrogen]). By modifying specific buffer components of a PCR and by using hot-start enzymes for PCR, it is possible to improve the yield and specificity of the desired PCR products. Also, the addition of betaine at a final concentration of 0.8–1.6 M improves the amplification of DNA by reducing the formation of secondary structure in GC-rich regions.

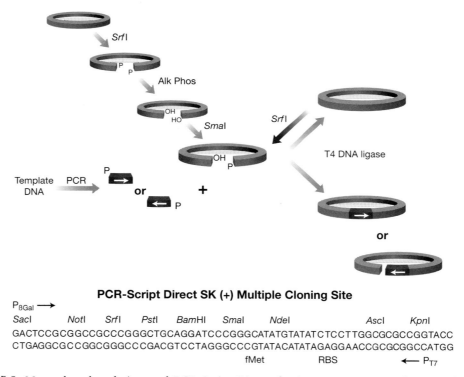

PCR-Script Direct SK (+) Multiple Cloning Site

P$_{\beta Gal}$ ⟶

Sac I   Not I   Srf I   Pst I   BamH I   Sma I   Nde I   Asc I   Kpn I
GACTCCGCGGCCGCCCGGGCTGCAGGATCCCGGGCATATGTATATCTCCTTGGCGCGCCGGTACC
CTGAGGCGCCGGCGGGCCCGACGTCCTAGGGCCCGTATACATATAGAGGAACCGCGCGGCCATGG
                                   fMet   RBS   ⟵ P$_{T7}$

FIGURE 3. Monophosphorylation and PCR-Script Direct cloning. To generate a directional cloning vector, the PCR-Script plasmid is digested with the restriction enzyme SrfI, treated with alkaline phosphatase to remove the 5′ phosphate groups, and then digested with the second restriction enzyme SmaI. Ethanol precipitation is used to remove the small (15-bp) linker. The insert fragment is created using either a machine-synthesized 5′-phosphorylated or kinase-treated primer. The monophosphorylated primer and vector are incubated in the presence of both SrfI and T4 DNA ligase. After room temperature incubation, the DNA is used to transform E. coli. P denotes 5′ phosphate group; OH denotes 3′ hydroxyl group.

MATERIALS

BUFFERS, SOLUTIONS, AND REAGENTS

Betaine solution, PCR-grade (Sigma)

dNTP stock solution (containing all four dNTPs, each at 100 mM)

TE buffer (10 mM Tris-HCl<!>, pH 7.5, 1 mM EDTA, pH 8.0)

ENZYMES AND ENZYME BUFFERS

Cloned *Pfu* DNA polymerase (Stratagene)

PCR buffer, 10x (as supplied by enzyme manufacturer) or PCR optimization buffer; e.g., Opti-Prime PCR optimization kit (Stratagene) or The PCR Optimizer (Invitrogen)

Thermostable DNA polymerase (5 units); e.g., *Taq* DNA polymerase, TaqBead Hot Start Polymerase (Promega), Platinum *Taq* DNA Polymerase High Fidelity (Invitrogen), AmpliTaq Gold (Applied Biosystems)

NUCLEIC ACIDS AND OLIGONUCLEOTIDES

Oligonucleotide primers

Template DNA (1–50 ng of plasmid)

SPECIAL EQUIPMENT

Thermal cycler

ADDITIONAL ITEMS

Equipment and reagents for agarose gel electrophoresis, including ethidium bromide<!>

Optional: PCR enhancer; e.g., *Taq* Extender PCR additive (Stratagene), PCRx Enhancer System (Invitrogen)

METHOD

1. Set up a 25-μl reaction in a 0.5-μl sterile, microcentrifuge tube placed on ice. Add the following reagents in order:

10x DNA polymerase buffer (for a final 1x volume)	2.5 μl
Template DNA (1–50 ng of plasmid DNA or 10^5–10^6 target molecules*)	1–10 μl
dNTP stock solution (containing all four dNTPs, each at 10 mM)	0.5 μl
Betaine solution (5 M)	2.5 μl
Upstream primer (4 μg/μl)	0.25 μl
Downstream primer (4 μg/μl)	0.25 μl
Thermostable DNA polymerase (5 units/μl)	0.25 μl
Optional: Taq Extender PCR additive (5 units) (additional)	0.25 μl
H₂O	to 25 μl

*For 3 x10^5 targets:

1 μg of human single-copy genomic DNA
10 ng of yeast DNA
1% of an M13 plaque

2. Mix well. PCR amplification should be conducted immediately.

3. Perform PCR using a thermocycler with a heated lid (in the absence of a heated lid, reactions can be overlaid with mineral oil) following the suggested temperature profile:

Cycle number	Denaturation	Annealing	Polymerization/Extension
1	4 min at 95°C	2 min at 50°C	2 min at 72°C
25–30	1 min at 94°C	2 min at 54°C	1 min at 72°C
last cycle			10 min at 72°C, then cool to 10°C temperature

4. Following thermal cycling, check the PCR products for fidelity and yield by agarose gel analysis. A 1- to 5-μl aliquot of PCR product can be monitored by ethidium bromide staining of the DNA fragments following agarose gel electrophoresis. Known amounts of control DNAs should be run as markers for PCR product size and concentration.

STAGE 3: END-POLISHING PCR PRODUCTS WITH *Pfu* DNA POLYMERASE

Optimizing primer design in accordance with the specific DNA polymerase used can only minimally increase the number of blunt-end fragments produced following PCR. The traditional Klenow polymerase should be *absolutely avoided* for end-polishing because it retains a substantial amount of extendase activity. Fortunately, T4 and *Pfu* DNA polymerases were found not to exhibit any DNA extendase or TdT activity and can be used to create blunt-end fragments following PCR (see Fig. 4) (Hu 1993; Costa and Weiner 1994c,d). PCR polishing is used to remove the 3′-end nucleotide extensions placed on completed PCR products by DNA polymerases. The resulting *Pfu*-polished molecules will ligate into blunt-end cloning vectors at high efficiency in the presence of T4 DNA ligase. End-polishing of PCR products prior to ligation has been shown to increase overall recombinant cloning efficiencies (Costa and Weiner 1994c,d; Weiner et al. 1994).

Pfu DNA polymerase is essentially inactive at temperatures below 50°C. This allows ligation reactions to be done at 4–25°C and to be set up directly from the 72°C *Pfu* polishing step. This eliminates the need to extract the enzyme prior to ligation, as would be required if T4 DNA polymerase were used for polishing. *Pfu* polishing of PCR products generates high-fidelity, blunt-end DNA fragments in 30 minutes using only a small aliquot of the PCR product. *Pfu* polishing is outlined below and can be performed directly following the PCR or following the purification of the desired PCR product.

> Before PCR polishing, it may be advantageous to verify the PCR products by agarose gel analysis to estimate the approximate concentration and to ensure that the correct PCR products have been created following thermal amplification. PCR polishing is conducted using an aliquot of the PCR amplification reaction and will therefore polish the ends of all DNA fragments present. In a typical 25-μl PCR amplification reaction, 5-10 μl of product can be used for PCR polishing.

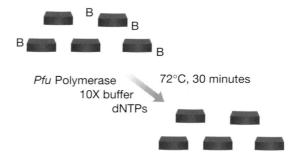

FIGURE 4. End-polishing of PCR-generated DNA fragments with *Pfu* DNA polymerase is used for increasing the amount of blunt-end DNA available for cloning. See text for protocol. B denotes extended nucleotide bases.

Because routine PCR cloning procedures require the use of a small amount of DNA insert (1–4 µl), end-polishing reactions can be set up directly from the amplification using only 5–10 µl of PCR product. When end-polishing directly from the PCR, the remaining dNTPs and reaction buffer following thermal cycling are adequate for the polishing reaction (see below). Precipitation or gel isolation of the PCR products from thermocycled reactions may also be used to increase efficiency of end-polishing; however, use of purified PCR products necessitates the addition of reaction buffer and dNTP mixtures.

MATERIALS

BUFFERS, SOLUTIONS, AND REAGENTS
dNTP stock solution (containing all four dNTPs, each at 100 mM), Method B

ENZYMES AND ENZYME BUFFERS
Cloned *Pfu* DNA polymerase (2.5 units/µl)
Cloned *Pfu* buffer, 10x (as supplied by manufacturer)

NUCLEIC ACIDS AND OLIGONUCLEOTIDES
PCR product, purified (Method B) or unpurified (Method A)

SPECIAL EQUIPMENT
Water bath, preset to 72°C

METHOD A: *Pfu* Polishing of an Unpurified PCR Product

1. For polishing PCR-generated DNA fragments, transfer an aliquot of the PCR product directly from the reaction tube into a sterile 0.5-ml microcentrifuge tube and add the following reagents, in order:

   5–10 µl of PCR product
   1 µl of cloned *Pfu* DNA polymerase (2.5 units)
   ddH$_2$O to a final volume of 10 µl
   Gently mix the components and incubate with a heated lid. (In the absence of incubation with a heated lid, reactions can be overlaid with mineral oil.)

2. Incubate the polishing reaction for 30 minutes at 72°C.

3. Following the 30-minute incubation, remove the reaction to ice.

4. End-polished DNA fragments may be added directly to a ligation reaction.

METHOD B: *Pfu* Polishing of a Purified PCR Product

1. For polishing purified PCR product, transfer an aliquot of the precipitated PCR product into a sterile 0.5-ml microcentrifuge tube and add the following reagents, in order:

   5–10 µl of precipitated PCR product
   1 µl of 10x cloned *Pfu* DNA polymerase buffer
   1 µl of dNTP mix (10 mM total; 2.5 mM each nucleotide triphosphate)
   1 µl of cloned *Pfu* DNA polymerase (2.5 units)
   ddH$_2$O to a final volume of 10 µl
   Gently mix the components and incubate with a heated lid. (In the absence of incubation with a heated lid, reactions can be overlaid with mineral oil.)

2. Incubate the polishing reaction for 30 minutes at 72°C.

3. Following the 30-minute incubation, remove the reaction to ice.

4. End-polished DNA fragments may be added directly to a ligation reaction.

OPTIONAL PCR PRODUCT PURIFICATION

The removal of excess PCR primers with selective ammonium acetate precipitation before proceeding with cloning protocols has been shown to increase the percentage of recombinants. An aliquot of the PCR product can be salted out of solution by the following protocol.

MATERIALS

BUFFERS, SOLUTIONS, AND REAGENTS

Ammonium acetate<!>, 4 M
Ethanol<!>, 100% and 70%
STE buffer, 10x (1 M NaCl, 200 mM Tris-HCl<!>, pH 7.5, 100 mM EDTA)
TE (10 mM Tris-HCl<!>, pH 7.5, 1 mM EDTA, pH 8.0)

SPECIAL EQUIPMENT

Vacuum desiccator

METHOD

1. Add 0.1 volume of 10x STE buffer.

2. Add an equal volume of 4 M ammonium acetate to the sample.

3. Add 2.5 volumes of room-temperature-equilibrated 100% ethanol.

4. Immediately spin in a centrifuge at 12,000g for 10 minutes at room temperature to pellet the DNA. *Carefully* decant the supernatant.

5. Add 200 µl of 70% (v/v) ethanol.

6. Spin in a centrifuge at 12,000g for 10 minutes at room temperature. *Carefully* decant the supernatant. Dry the pellet in vacuo.

7. Resuspend the DNA in the original volume using TE buffer. Store at 4°C until use.

PROTOCOL 3   GENERATING CLONES

E. coli Transformation

Competent cells are very sensitive to even small variations in temperature and must be stored at –80°C. Repetitive freeze-thawing will result in a loss of efficiency and should be avoided. It is important to use FALCON 2059 tubes for the transformation procedure, because the critical incubation period during the heat-pulse step described below is calculated for the thickness and shape of the FALCON 2059 tube (if different tubes are used, the conditions should be altered accordingly). In addition, β-mercaptoethanol has been shown to increase transformation efficiencies two- to threefold. Upon transformation, there seems to be a defined "window" of highest efficiency resulting from the heat pulse. Optimal efficiencies are observed when cells are heat-pulsed for 30–45 seconds. Supercompetent cells can be purchased commercially that yield extremely high efficiencies upon transformation.

β-Galactosidase Color Selection

Phenotypic selection by disruption of the β-galactosidase (β-Gal) gene is often used to detect recombinants (Sambrook et al. 2001). Such phenotypic selection is monitored by the appearance of recombinants as white colonies and nonrecombinants as blue transformant colonies on X-Gal-containing agar plates. IPTG is often used as an inducer in conjunction with X-Gal.

PCR Parameters

Depending on the needs of the investigator and the performance characteristics of the thermal cycler, sensitivity can be altered by changing both the number of cycles and the annealing temperature of segment 2. Depending on the oligonucleotide primer that is used, it may also be advantageous to calculate its optimal annealing temperature. Several resources for oligonucleotide primer design are available on-line that accurately calculate primer melting temperatures:

http://www.brinkmann.com/support_practical-pcr.asp
http://www.promega.com/biomath/calc11.htm
http://www.appliedbiosystems.com/support/techtools/calc/
http://www-genome.wi.mit.edu/cgi-bin/primer/primer3_www.cgi

Once a melting temperature is determined, a revised segment 2 can be constructed. Alternatively, the equation below can be used for calculating melting temperatures for oligonucleotides shorter than 21 nucleotides.

$$T_m = 2°C \ (A+T) + 4°C \ (G+C)$$

The following PCR program has been successfully used with 20- to 30-base oligonucleotide primers. The sensitivity of the program is determined by segment 2 and may need to be reoptimized when using a third, insert-specific oligonucleotide primer that is <20 bases.

Cycle number	Denaturation	Annealing	Polymerization/Extension
1	4 min at 94°C	2 min at 50°C	2 min at 72°C
30	1 min at 94°C	2 min at 56°C	1 min at 72°C
1			5 min at 72°C, then cool to room temperature

Kits available for cloning PCR products include:

PCR-Script SK(+) Amp cloning system (Stratagene, 211190)
PCR-Script SK(+) Cam cloning system (Stratagene, 211192)
PCR-Script Direct SK(+) cloning system (Stratagene, 211194)

MATERIALS

BUFFERS, SOLUTIONS, AND REAGENTS

ATP (10 mM)

β-mercaptoethanol<!> (*optional*)

IPTG<!> (isopropyl-β-D-thio-galactopyranoside; 100 mM in H_2O)

X-Gal (5-bromo-4-chloro-3-indoyl-β-D-galactopyranoside; 100 mg/ml in dimethyl formamide<!>)

ENZYMES AND ENZYME BUFFERS

Blunt-end restriction endonuclease (5 units) e.g., *Srf*I restriction endonuclease (Stratagene)

Ligation buffer, 10x (250 mM Tris-HCl<!>, pH 7.5, 100 mM $MgCl_2$<!>, 100 mM DTT<!>, 200 µg/ml BSA)

T4 DNA ligase (4 units)

NUCLEIC ACIDS AND OLIGONUCLEOTIDES

Blunt-end insert (100 ng/µl)

Cloning vector (10 ng/µl)

MEDIA

SOC medium (for 1 liter: 20 g of tryptone, 5 g of yeast extract, 0.5 g of NaCl. H_2O to 900 ml. Autoclave. Mix the following separately: 2.03 g of $MgCl_2$, 1.2 g of $MgSO_4$<!>, 3.6 g of glucose in H_2O to a final volume of 100 ml. Filter-sterilize and add to cooled, autoclaved medium.

LB (Luria Broth) agar plates (per liter: 10 g of NaCl, 10 g of Bacto-tryptone, 5 g of Bacto-yeast extract, 20 g of Bacto-agar. Adjust pH to 7.0 with 5 N NaOH<!>. Add deionized H_2O to a final volume of 1 liter. Autoclave and pour into petri dishes (~25 ml/100-mm plate)

LB plates containing ampicillin-methicillin (20 µg/ml ampicillin<!>, 80 µg/ml methicillin). Use for reduced satellite colony formation.

Cool autoclaved medium to 55°C before adding filter-sterilized antibiotic.

LB plates containing chloramphenicol<!> (30 µg/ml).

Cool autoclaved media to 55°C before adding filter-sterilized antibiotic.

Optional: X-gal/IPTG plates: For phenotypic selection of transformants, add a 20-µl aliquot of X-gal and a 20-µl aliquot of IPTG solution to agar plates containing the appropriate antibiotic(s). Spread immediately in an evenly distributed manner (a slight precipitate may be apparent).

Avoid mixing X-Gal and IPTG, because these chemicals will precipitate.

Important: Allow the plates to dry for 15–30 minutes before spreading transformants.

SPECIAL EQUIPMENT

FALCON 2059 polypropylene tubes

Incubator, preset to 37°C

Water baths, preset to 37°C, 42°C, 65°C

CELLS AND TISSUES

E. coli cells of the appropriate strain, e.g., XL1-Blue (Stratagene), rendered competent according to Sambrook and Russell (2001), or obtained ready-competent

METHOD

Ligation

1. In an autoclaved, sterile 1.5-ml tube, set up the PCR-Script reaction by adding the following reagents in order:

Cloning vector (10 ng/μl)	1 μl
Ligation buffer, 10x	1 μl
ATP (10 mM)	0.5 μl
Pfu-polished PCR product insert*	1–4 μl
*Srf*I restriction endonuclease (5 units)	1 μl
T4 DNA ligase (4 units)	1 μl
H$_2$O	to 10 μl

   For ligation, the ideal ratio of insert-to-vector DNA is variable. For sample DNA, a range from 5:1 (when using polished inserts) to 100:1 (when using unpolished inserts) may be necessary. A greater insert-to-vector ratio is necessary for unpolished inserts because there will be a decreased occurrence of PCR fragments with both ends blunted. It may be advantageous to optimize conditions for a particular insert using the following equation:

   $$\text{pmole ends/μg of DNA} = 2 \times 10^6/\text{number of bp}$$

2. Mix gently and incubate for 1–2 hours at room temperature.

3. Heat-treat the sample for 10 minutes at 65°C.

4. Store the sample on ice until ready to transform competent *E. coli*.

Transformation

1. Thaw competent cells on ice.

2. Gently mix the cells by swirling. Aliquot 40 μl of cells into a prechilled 15-ml FALCON 2059 tube.

3. *Optional:* Add β-mercaptoethanol (for a final 25 mM concentration) to the 40 μl of bacteria.

4. Swirl gently. Place on ice for 10 minutes; swirl gently every 2 minutes.

5. Add 2 μl of DNA from the heat-treated ligation reaction (step 4, Cloning Procedure).

6. Place on ice for 30 minutes.

7. Heat-pulse for 30 seconds in a 42°C water bath. The length of the heat pulse is critical for the highest efficiencies.

8. Place the transformation mixture on ice for 2 minutes.

9. Add 450 μl of preheated (42°C) SOC medium and incubate with shaking at 225–250 rpm for 1 hour at 37°C.

10. Plate 50–200 μl of the transformation mixture (100 μl is standard) using a sterile spreader to place the mixture onto the appropriate antibiotic-containing agar plates.

    Optional: A chromogenic substrate may be added to the LB plates to detect recombinants; see also β-Galactosidase Color Selection, below.

    If plating = 100 μl, the cells can be spread directly onto the plates. If plating <100 μl of the transformation mixture, increase the volume of the transformation mixture to be plated to a total volume of 200 μl using SOC medium.

11. Incubate the plates overnight (~16 hours) at 37°C.

12. Choose white colonies for examination, avoiding colonies with a light-blue appearance or colonies with a blue center.

    Colonies containing inserts that were initially white may turn very light blue after 2–5 days on the plate.

PROTOCOL 4 ANALYSIS OF CLONED RECOMBINANTS BY COLONY-PCR

Recombinant insert analysis of colonies resulting from transformed cells can be performed in 1 day using colony-PCR (Costa and Weiner 1994e,f). Recombinant PCR screening allows the rapid and efficient detection of cloned inserts from most ColE1-based plasmids. By using primers asymmetrically distanced from the clonal insertion site, it is possible to discern both insert presence and orientation from the resulting PCR product (see Fig. 5). One can also conduct PCR using a triple primer set containing the two asymmetric primers and an additional, fragment-specific primer from the set used to generate the original fragment. Agarose gel analysis of the PCR using such a three-primer set confirms both the presence and the orientation of the cloned insert without the need for further restriction enzyme digestion.

Further characterization of the cloned inserts can be done using restriction enzyme analysis of the colony-PCR product. Restriction enzyme digestion of the recombinant-screen PCR products can be performed directly from the amplification reaction (Costa and Weiner 1994f). In addition to restriction analysis, the recombinant-screen PCR products can be further characterized by cycle sequencing (Hedden et al. 1992; Costa and Weiner 1994f; Kretz et al. 1994). Because the primer set is designed to flank the polylinker by a distance of ~500 bases on either side of the multiple cloning site, there is a retention of common priming sites used for DNA sequencing. For sequence verification, colony–PCR procedures that result in a single product (no spurious bands or primer–dimers) may be diluted and used in a cycle-sequencing reaction. High-resolution sequences have been consistently obtained using an aliquot of a 1:50 dilution of the PCR mixtures.

PCR-mediated clonal analysis allows one to screen numerous clones in a simple, rapid, and highly efficient manner. The procedure for recombinant screening by colony-PCR is outlined below. It may be beneficial to evaluate the considerations in the troubleshooting section as a precautionary measure before proceeding with the PCR-based screening protocol.

As a control for this method, use nonrecombinant plasmid DNA (vector that does not contain insert). This will provide negative internal control colony-PCR. However, one may also transform the nonrecombinant vector DNA and inoculate colonies from the transfor-

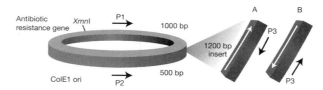

FIGURE 5. Recombinant screening method. Shown on the plasmid map are the PCR primers (P1 and P2) used in a colony-PCR screening procedure. PCR products produced using the colony-PCR method are analyzed by agarose gel electrophoresis. In a representative experiment in which a 1.2-kb insert is cloned into a plasmid, colonies selected directly from the transformation plates are inoculated into colony-PCR reaction mixtures, and PCR is conducted followed by agarose gel analysis. Nonrecombinants, which do not contain the insert, exhibit a 1.5-kb PCR product. Recombinants, which do contain the 1.2-kb insert, exhibit a 2.7-kb PCR product. Directionality of the cloned insert can be determined in a separate reaction by adding a third, insert-specific primer (P3) to the colony-PCR reaction mixture. Bidirectional cloning produces recombinants that contain inserts cloned in both directions, and the use of a third primer in the reaction mixture confirms the orientation of the cloned fragment (orientation A or B). An example of the recombinant screening of clones with a 1.2-kb insert by colony-PCR in the presence of a third primer indicates the orientation of the cloned insert after agarose gel analysis with a constitutive 2.7-kb PCR product and either a 1.7-kb PCR product (orientation A) or a 2.2-kb PCR product (orientation B).

mation plate into the standard reaction mixture to serve as a positive control for the colony-PCR. Kits available for recombinant screening include ScreenTest recombinant screening kit (Stratagene). Kits available for DNA minipreparation include QIAGEN Plasmid Mini kit (QIAGEN) and Wizard Plus SV Minipreps DNA Purification System (Promega).

MATERIALS

BUFFERS, SOLUTIONS, AND REAGENTS

Ethidium bromide<!>
Taq Extender PCR additive (Stratagene)
Tween 20, 10% solution

ENZYMES AND ENZYME BUFFERS

dNTP solution (containing all four dNTPs, each at 25 mM)
PCR buffer, 10x, as supplied by manufacturer of the chosen thermostable polymerase, or PCR optimization buffers, e.g., Opti-Prime PCR optimization kit (Stratagene), and The PCR Optimizer (Invitrogen)
Thermostable DNA polymerase (5–10 units), e.g., *Taq* DNA polymerase, cloned *Pfu* DNA polymerase (Stratagene)

NUCLEIC ACIDS AND OLIGONUCLEOTIDES

Optional: insert-specific primer (for use in bidirectional cloning to define orientation of insert)
Recombinant screening primer set

SPECIAL EQUIPMENT

Sterile pipette tips for picking colonies and inoculating PCR

Pipette tips are used in the place of sterile toothpicks because toothpicks can "wick" liquid and may therefore alter the concentration of the reaction mixture.

ADDITIONAL ITEMS

Equipment and reagents for agarose gel electrophoresis, including ethidium bromide<!>

METHOD

1. According to the number of reactions or multiples of reactions needed, prepare the PCR cocktail master mix in a single microcentrifuge tube on ice by adding the components in the order indicated below.

H_2O	to a total volume of 25.0 µl
Taq Buffer, 10X	2.5 µl
Tween 20 (10% [w/v] solution)	1.0 µl
dNTP mix (25 mM of each dNTP)	0.2 µl
Recombinant screening primer set* (100 ng/µl)	1 µl
Taq Extender PCR additive (5 units/µl)	0.25 µl
Taq DNA polymerase (5 units/µl)	0.25 µl

2. Aliquot 25 µl of the PCR cocktail master mix into each PCR tube on ice.

3. For control reaction(s), add 1 µl of the nonrecombinant DNA.

Optional: For insert orientation reaction(s), add a third, insert-specific primer to the appropriate reaction tubes.

4. For recombinant screening, stab the transformed colonies with a sterile pipette tip and swirl the colony material into the appropriate reaction tube. Immediately following inoculation into each reaction mixture, remove the pipette tip and score onto antibiotic-containing patch plates for future reference.

See also Troubleshooting, below, for PCR-mediated recombinant screening.

2. Mix each reaction gently.

3. Perform PCR using the recommended cycling parameters.

4. Analyze the PCR products using standard agarose gel electrophoresis. It is recommended to use a 1.0–1.5% (w/v) agarose gel for optimal resolution of the expected 500- to 6000-bp PCR products. Typically, 1–5 µl of each PCR is analyzed utilizing ethidium bromide staining. Images may be archived using conventional instant photography or computer-based imaging software.

◗ TROUBLESHOOTING

Under optimal conditions, colony-PCR provides an adequate amount of DNA template that will yield maximum signal in the PCR. Undoubtedly, there will be variations in thermal cyclers and reagents that may contribute to signal differences in the experiments. The following are guidelines for troubleshooting these variations in PCR-mediated recombinant screening.

- *Low signal with control DNA.* Suboptimal reagents (e.g., *Taq* DNA polymerase) and/or the thermal cycler used in conducting the assay may account for the results. The positive control is a good indicator of amplification efficiency and, when using 1–5 ng of DNA according to the specified guidelines, has been calculated to yield amounts of PCR product approaching plateau levels.

- *Low signal in the screening samples.* PCR inhibition may result in the presence of excess colony material in the colony-PCR. It is important to note that only a small amount of colony material is necessary to perform the recombinant screening method. When performing the recombinant screening method from patch plates, only "touch" onto the patch and inoculate directly into the reaction tubes.

- *Loss of sample volume.* Reduced sample volume may result from inadequate capping of the reaction tubes prior to cycling, causing evaporation of reaction mixtures. Also, the use of toothpicks in place of pipette tips for colony inoculation may act to "wick" the solution out of the reaction tube. If toothpicks are used, removal of the toothpick immediately after the PCR cocktail master mix inoculation is strongly recommended.

- *Excessive signals in the samples.* This PCR-mediated screening method has been optimized on thermal cyclers whose temperature profiles are very exact and reproducible. Thermal cyclers whose transition times are very long inadvertently add time to the PCR program and may result in excessive signals in both test samples and controls. In an attempt to reduce the signal, it may be advantageous to reoptimize segment 2 in the PCR program.

- *Multiple banding patterns.* This screening method has been designed with parameters optimized for use in colony-PCR where limited amounts of colony material are present. In the schematic representation of the recombinant screening method (see Fig. 5), two potential PCR products can be produced in the presence of a third, insert-specific oligonucleotide primer (P3). One PCR product is generated by P3+P1 or by P3+P2, and a second constitutive PCR product is generated by P1+P2. The method relies on the fact that, when limited amounts of template DNA are available, the smallest PCR product will be preferentially amplified. In cases in which miniprep or purified DNA is used, such purified DNA (as compared to colony-derived DNA) provides an optimally accessible template in cyclic amplification procedures, thereby producing both "expected" PCR products. Therefore, it is very important to calculate the expected PCR products when using a third, insert-specific oligonucleotide primer in directionality studies for the determination of insert orientation.

CONCLUSION

PCR has both simplified and accelerated the process of cloning DNA fragments. It is now possible, when appropriate primers have been obtained, to perform the PCR, clone, and transform reactions in a single day. The analysis of putative clones by colony-PCR and subsequent nucleotide verification by cycle sequencing can be completed the following day. The methods presented allow PCR cloning operations to exhibit more than 50% recombinant efficiency and to facilitate PCR screening methods that can be completed in a rapid and highly efficient manner.

ACKNOWLEDGMENTS

The authors thank John Bauer, Steve Wells, Tim Sanchez, Mark Kaderli, and Bruce Jerpseth for their substantial contributions in experimental design.

REFERENCES

Bauer J., Deely D., Braman J., Viola J., and Weiner M.P. 1992. pCR-Script SK(+) cloning system: A simple and fast method for PCR cloning. *Strategies* **5:** 62–64.

Clark J.M. 1988. Novel non-templated nucleotide addition reactions catalyzed by procaryotic and eucaryotic DNA polymerases. *Nucleic Acids Res.* **16:** 9677–9686.

Costa G.L. and Weiner M.P. 1994a. Improved PCR cloning. *Strategies* **7:** 8.

———. 1994b. pCR-Script SK(+) cloning system: Questions and answers. *Strategies* **7:** 53–54.

———. 1994c. Increased cloning efficiency with the PCR polishing kit. *Strategies* **7:** 47–48.

———. 1994d. Polishing with T4 or *Pfu* polymerase increases the efficiency of cloning PCR fragments. *Nucleic Acids Res.* **22:** 2423.

———. 1994e. ScreenTest recombinant screening in one day. *Strategies* **7:** 35–37.

———. 1994f. ScreenTest colony-PCR screening: Questions and answers. *Strategies* **7:** 80–82.

Costa G.L., Grafsky A., and Weiner M.P. 1994a. Cloning and analysis of PCR-generated DNA fragments. *PCR Methods Appl.* **3:** 338–345.

Costa G.L., Sanchez T.R., and Weiner M.P. 1994b. pCR-Script Direct SK(+) vector for directional cloning of blunt-ended PCR products. *Strategies* **7:** 5–7.

———. 1994c. New chloramphenicol-resistant version of pCR-Script vector. *Strategies* **7:** 52.

Hedden V., Simcox M., Callen W., Scott B., Cline J., Nielson K., Mathur E., and Kretz K. 1992. Superior sequencing: Cyclist Exo–*Pfu* DNA sequencing kit. *Strategies* **5:** 79.

Hu G. 1993. DNA polymerase-catalyzed addition of nontemplated extra nucleotides to the 3′ end of a DNA fragment. *DNA Cell Biol.* **12:** 763–770.

Kretz K., Callen W., and Hedden V. 1994. Cycle sequencing. *PCR Methods Appl.* **3:** S107–112.

Liu Z. and Schwartz L. 1992. An efficient method for blunt-end ligation of PCR product. *BioTechniques* **12:** 28–30.

Sambrook J. and Russell D.W. 2001. *Molecular cloning: A laboratory manual,* 3rd edition. Cold Spring Harbor Laboratory Press, Cold Spring Harbor, New York.

Simcox T., Marsh S., Gross E., Lernhardt W., Davis S., and Simcox M. 1991. *Srf* I, a new type-II restriction endonuclease that recognizes the octanucleo-tide sequence, 5′-GCCCGGGC-3′. *Gene* **109:** 121–123.

Weiner M.P. 1993. Directional cloning of blunt-ended PCR products. *BioTechniques* **15:** 502–505.

Weiner M.P., Costa G.L., Schoettlin W., Cline J., Mathur E., and Bauer J.C. 1994. Site-directed mutagenesis of double-stranded DNA by the polymerase chain reaction. *Gene* **151:** 119–123.

WWW RESOURCES

http://www.brinkmann.com/support_practical-pcr.asp, Brinkmann, Instruments, Inc. Application Support.

http://www.promega.com/biomath/calc11.htm, Promega Corporation, T_m (Melting Temperature) Calculations for Oligos.

http://www.appliedbiosystems.com/support/techtools/calc/, Applied Biosystems T_m Calculator.

http://www-genome.wi.mit.edu/cgi-bin/primer/primer3_www.cgi, Primer3 Software, Whitehead Institute, Center for Genome Research, Massachusetts Institute of Technology.

29 Ribocloning: DNA Cloning and Gene Construction Using PCR Primers Terminated with a Ribonucleotide

Wayne M. Barnes

Department of Biochemistry and Molecular Biophysics, Washington University School of Medicine, St. Louis, Missouri 63110

This lab has found that pancreatic RNase A can cleave efficiently at single rC or rU bases embedded in double-stranded DNA. This effect has been exploited in a method for cloning PCR products in which ~25-base complementary (sticky) ends for target and vector are arranged. Entire plasmid vectors are amplified using long, high-fidelity PCR with DNA "riboprimers" ending in a single rC (at their 3′ ends). Target DNA is similarly amplified with primers complementary to the vector primers and also ending in rC. After treatment with RNase (55°C), and removal of RNase with proteinase K (65°C), heating (75–80°C) melts off the primers. For optimal efficiency, loose primers and DNA molecules that retain their primers may be removed by streptavidin beads if 5′-biotinylated primers are used. Vector and target are then annealed together (65–52°C) and electroporated into bacteria having any recombination system or no recombination system, without the need for any further enzyme treatments such as restriction enzyme, kinase, phosphatase, ligase, or topoisomerase. High efficiencies of plasmid construction are achievable, exceeding 1000 transformant colonies (80–100% desired recombinants) per nanogram of target PCR DNA. No particular plasmid vector or cloning sequence location is required. No regard for restriction sites is necessary, except optionally to reduce carryover of PCR template, such as post-PCR *Dpn*I treatment.

Figure 1 illustrates the basic principle and the desired structure of the target and vector fragments right before they self-assemble. In the case of perfect complementarity between the target primers and the vector primers, a 3′-rC will always match a 5′-dG, so each primer starts with a G (see Fig. 1). This initial G is not always necessary, however, since rU ends and partial complementarity are also accommodated by the method (W. Barnes and E. Oates, in prep.). Because the target and the vector do not normally have any overlapping sequence in common, some special strategy must be used to add the overlap to either the target or the vector, or to one end of each.

Figure 2 illustrates one strategy for arranging the terminal 25-bp self-complementarity between target and vector. In this strategy, one of the PCR reactions includes, in addition to the riboprimers, two extra oligonucleotides (50-mers), called "band-aids," which

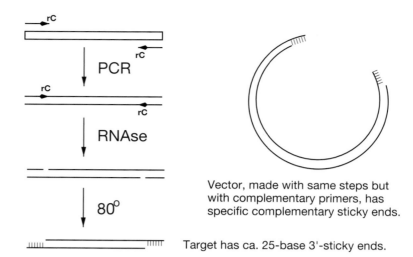

Vector, made with same steps but with complementary primers, has specific complementary sticky ends.

Target has ca. 25-base 3'-sticky ends.

FIGURE 1. The sequences of the riboprimers used in the example ribocloning are as follows, where the uppercase "c" represents the 3'-ribo base. Target primer on the left (from Shine-Dalgarno site of phage T7 gene 10) is T7gen10rC = GTT TAA CTT TAA GAA GGA GAT ATA c; vector primer (exact complement to T7gen10rC) is V01neg7TrC = GTA TAT CTC CTT CTT AAA GTT AAA c; target primer in the right (from the phage fd origin of replication present on many phagemid vectors) is T41 = GTG GCG AGA AAG GAA GGG AAG AAA Gc; vector primer V41 (exact complement to T41) is GCT TTC TTC CCT TCC TTT CTC GCC Ac.

are specified to overlap the target and vector sequence by 25 bp. As shown by their alignment in the figure, the target DNA to be amplified initially has no homology with the riboprimers (labeled "rC"). In the initial cycles of PCR, the riboprimers prime on the band-aids as template. In subsequent cycles of the PCR amplification, the extended riboprimers can prime on the target, and the full-length target template is synthesized and amplifiable directly by the riboprimers at later cycles. Two further changes to the standard PCR should be noted, however. (1) A low level of only 0.25–1 pmole of the band-aid oligonucleotides is recommended per 100-μl reaction volume. (2) Because of the low concentration of the band-aids and/or their copy, which is primed by the riboprimers, a long extension/annealing step is needed at each cycle of the PCR: 20 minutes at 62°C. This long step corresponds to the C_0t (Britten and Kohne 1968) for the band-aid oligonucleotides.

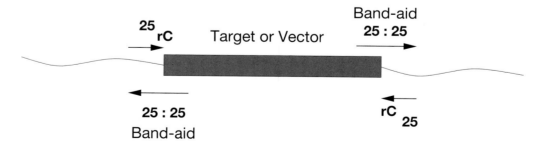

FIGURE 2. 50-mer "band-aids" (at only 1 pmole) allow riboprimers (20 pmoles) to pick up the target with vector homology. If the band-aids are on the other strand (as shown here), the PCR product will all have the C near each end. Suboptimally, the band-aids can be on the same strand as the riboprimer, but then the product will contain up to 5% nonribosylated ends.

PROTOCOL 1 RIBOCLONING

This protocol is just one way to work up the PCR products with RNase treatment. There are alternate and effective ways to purify PCR products free of primers, treat them with enzyme(s), and remove the enzyme(s). Note that it is convenient to program the incubations into a PCR machine. The warm steps are optionally followed with "chill," which is a hold on a cold block (4–6°C) for 5 minutes to overnight.

MATERIALS

BUFFERS, SOLUTIONS, AND REAGENTS

10x ATEN buffer

  0.5 M Tris-HCl<!>, pH 7.9

  2.5 M NaCl

  0.25 M Na$_4$EDTA, pH 7.9

Betaine (Sigma B-2629)

Blue dextran (Sigma D-5751)

DNA buffer (TEN)

  10 mM Tris-HCl <!>, pH 7.9

  10 mM NaCl

  0.1 mM EDTA

Ethanol<!>, 75% with DNA buffer 25%

10x KLA buffer

  500 mM Tris base<!> (Sigma; Trizma Base)

  160 mM ammonium sulfate<!>

  1% Tween 20

  25 mM MgCl$_2$<!>

  The pH comes out to 9.2 without adjustment. Add HCl dropwise to adjust pH to 7.9, testing the pH only on separate aliquots so as not to contaminate the buffer with DNA on the pH probe.

30% Polyethylene glycol 3350 (PEG)<!>, 1.5 M NaCl

Sodium acetate, 3 M, pH 5.6

T5E5

  5 mM Tris-HCl<!>, pH 7.9

  5 mM Na$_4$EDTA, pH 7.9

ENZYMES AND ENZYME BUFFERS

*Dpn*I

Klentaq LA (Barnes 1994; Clontech 8421-1)

Proteinase K<!> (Roche 1373 196)

Proteinase K storage buffer

  100 μM β-mercaptoethanol<!>

  20 mM Tris-HCl<!>, pH 7.9

  1 mM CaCl$_2$<!>

  50% (v/v) glycerol (63% by weight/volume)

RNase A (Ambion 2270 or 2272)

NUCLEIC ACIDS AND OLIGONUCLEOTIDES

Desired vector

Primers with 5′-biotin-TEG modification AND 3′-rC- or rU-terminated

ANTIBIOTICS

Kanamycin<!> (Sigma), final 25 µg/ml

Tetracycline<!> (Calbiochem), final 12 µg/ml

Ticarcillin (SmithKline Beecham), final 100 µg/ml

SPECIAL EQUIPMENT

Electroporation equipment

PCR (thin-walled) tubes; regular-walled tubes, 0.5-ml or 1.7-ml

Streptavidin beads

CELLS AND TISSUES

Electrocompetent bacteria (Dower et al. 1988; Invitrogen 18290-015)

METHOD

1. Amplify the target with 1.3 M betaine and vector with 1.9 M betaine using rC- or rU-terminated primers and KLA buffer, pH 7.9. Some vectors' plasmid replicons amplify better with the standard long PCR pH of 9.2, but the colE1 replicon amplifies better at pH 7.9.

    For best cloning efficiency later, use primers with 5′-Biotin-TEG modification and use streptavidin beads as described below. Compared to other DNA polymerases, enzyme mixture KlentaqLA (Barnes 1994) gives a higher yield and is more resistant to betaine (Baskaran et al. 1996).

    PCR from plasmid templates is a little more efficient, and the final background of template transformation is lower, if the template is linearized with a restriction enzyme before the PCR.

    *Dpn*I selection is always a good idea if the template for the vector or the target came from dam+ Escherichia coli and could be expected to contribute to an unwanted background later at the cloning step. *Dpn*I cannot cut PCR product, because it has unmethylated GATC sites (Weiner et al. 1994).

2. To each 100 µl of PCR reaction, add 1 µl (10 units) of *Dpn*I, and incubate at 37°C for 2 hours.

    This reduces the background of vector-only or target-only controls to 0–20 colonies.

3. *PEG-precipitate step:* The purpose of this step is removal of the primers and the salt. Transfer the PCR reaction to thick-walled (regular-walled) tubes, either 0.5 ml or 1.7 ml, for this step, because thin-walled PCR tubes cannot reliably withstand the centrifuge steps. Then add 5 µl of 1 mg/ml blue dextran (this is used as carrier to make the pellet easier to see) and 1/2 volume 30% PEG 3350, 1.5 M NaCl (to make final 10% PEG, 0.5 M NaCl). Wait 30 minutes at room temperature or overnight at 4°C. Centrifuge 15 minutes. Watch blue pellet while removing the supernatant, so as not to accidentally discard the pellet. If the pellet is thin and invisible, just avoid it, but save the supernatant just in case. Rinse the pellet with 75% ethanol and centrifuge for 8 minutes. Dry the pellet for 10–20 minutes in air. Resuspend the pellet in 100 µl of T5E5 buffer for at least 10 minutes on ice, with occasional vortexing.

4. *Pancreatic RNase A step:* Dilute 200 µg of RNase A to a concentration of 200 µg/ml in T5E5 buffer. DNA buffer also works (E. Oates, pers. comm.). Add 6 µg (30 µl) of RNase to each 100 µl of PCR product. Vortex and centrifuge briefly. Then incubate the PCR products with RNase for 30 minutes at 55°C. Chill.

5. *Proteinase K step:* The stock enzyme should be stored at –20°C in proteinase K storage buffer (see Materials list). Add 5 µg of proteinase K or about equal the weight (6 µg) of the RNase used above. Mix thoroughly and centrifuge briefly. Incubate for 30 minutes at 65°C. Chill.

If using biotin primers, go now to the bead purification steps below. Otherwise the DNA is now ready.

6. *Cloning step*: Double-check the target and vector DNA concentrations by loading 2–8 μl of each vector or target sample onto an agarose gel alongside a concentration standard. The exact concentration is not as important as getting the ratios to be equimolar.

> 1–10 μl of vector = 30–100 ng, 0.02 pmole
> 1–10 μl of target = 5–25 ng (equimolar target :: vector)
> 1x ATEN buffer to 40 μl

7. Use 20-μl aliquots of the cloning mix for various controls, such as before heating, no heating, gel samples, and so forth. Make up more of the cloning mix as necessary, in proportion. The minimum is 40 μl: 20 μl for gel and 20 μl for transformation of cells.

CONTROLS

The best control is a 20-μl aliquot with the two vector (or the two target) primers added—say 0.5 pmole of each (0.5 μl of 10 μM). This will "poison" the sticky ends and make the annealing impossible. This gel (and/or transformation) sample will show you what the input DNA looks like with no annealing and is analogous to a no-ligase control (see Fig. 3, lane 2).

Another good control is to include only one poison primer. This control (see Fig. 3, lane 1) shows what partial success with the annealing looks like (see below), although sometimes when it looks like this there are still hundreds to thousands of good clones. Partial annealing could be due to contaminating (cleaved or original) PCR primers. Other possible causes of poor annealing include incomplete cleavage at one or more of the ends and degradation of the single-stranded sticky ends.

SUCCESSFUL ANNEALING

Partial success with the annealing gives rise to a new band just above the vector band. This band is identified as linear DNA that is a vector and target joined at one point.

Good, successful annealing shows a third band above the partially annealed linear (vector + target) band. This is identified as the relaxed circle of vector + target. Very good cloning (almost achieved in the leftmost lane of Fig. 3) shows only the circular band.

---

8. Annealing may not be necessary for bead-purified DNA, because it seems to self-assemble at room temperature (data not shown), but this heat step is very helpful for non-biotin primers.

> Incubate for 2 minutes at 78°C, then 30 minutes at 52°C. Alternatively, incubate for 2 minutes at 78°C, 10 minutes at 65°C, and then cool slowly over 30 minutes to 52°C or lower. Chill. Save a 20-μl aliquot for an after-annealing gel sample.

> These gel samples are for monitoring and improving the recombination. They may be skipped for routine cloning, but, as for any DNA cloning experiment, if something goes wrong, the experimenter will not know what it was without the gel analysis.

9. Add 1/9 or 1/6 volume of 3 M sodium acetate, pH 5.6, and 2 mg of blue dextran. Add 2–3 volumes of ethanol to precipitate, chill to –20°C for >30 minutes, and centrifuge for 10–15 minutes. Rinse the visibly blue pellet with 75% ethanol:25% DNA buffer and then centrifuge for 8 minutes. At this 75% ethanol rinse, we include a small amount of salt in the DNA buffer so that the 25-bp sticky ends do not melt apart in the distilled H_2O during the next step.

10. Resuspend the dry pellets in 22 μl of H_2O on ice. The H_2O should turn slightly blue from the blue dextran carrier. Save 7 μl for a gel sample or a backup transformation.

11. Electroporate 15 μl of DNA into 70 μl of electrocompetent bacteria, and add 1 ml of rich medium within 5 seconds. Plate on an appropriate antibiotic medium: 100 μg/ml ticar-

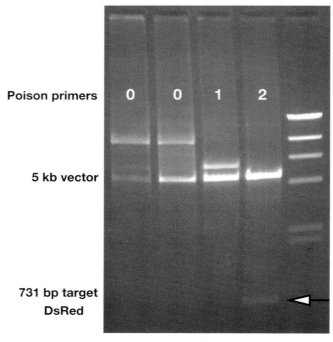

Poison primers 0 0 1 2

5 kb vector

**731 bp target
DsRed**

FIGURE 3. Some gel samples from a ribocloning are shown. These PCR products have all been cleaved with RNase and processed according to the protocols described in this chapter, including paramagnetic streptavidin beads. The first two lanes (labeled *0*) show the product vector + target circle as the most prominent band. To illustrate partial annealing (linear target + vector), one of the PCR primers was reintroduced (lane *1*). To analyze the input DNA molecules separately, two PCR primers were introduced (lane *2*) to prevent annealing. The leftmost lane used a separate preparation of vector, and its concentration was a little lower, more ideal, and more equimolar than the other lanes, which have a little too much vector.

cillin after 10 minutes, or on 25 µg/ml kanamycin after 5–16 hours at 30°C, or on 12 µg/ml tetracycline after 2 hours at 30°C. These longer post-shock incubation times are necessary to allow some antibiotic resistances to be expressed.

PROTOCOL 2 PARAMAGNETIC STREPTAVIDIN BEAD PURIFICATION

MATERIALS

BUFFERS, SOLUTIONS, AND REAGENTS
  1x ATEN buffer (see step 1)
  10x ATEN buffer (see Protocol 1 for recipe)
  Polyethylene glycol <!> (30% PEG, 1.5 M NaCl)
  T5E5 buffer (see Protocol 1 for recipe)

ENZYMES AND ENYZME BUFFERS
  *Dpn*I
  Proteinase K<!>
  RNase A

NUCLEIC ACIDS AND OLIGONUCLEOTIDES
  5′-Bio-TEG and 3′-rC primers
  Proteinase K-treated PCR product

SPECIAL EQUIPMENT
    Magnet
    Streptavidin beads (Genovision strongly recommended)

METHOD

1. Amplify the target with 5′-Bio-TEG and 3′-rC primers After the PCR, *Dpn*I, PEG-pre-cipitate, RNase, and proteinase K in T5E5 as in Protocol 1. Add 1/9 volume of 10x ATEN buffer to make the samples 1x ATEN.

2. Pre-preparation of beads: Withdraw some streptavidin beads from their storage tube, magnetize them, and discard the storage buffer. Wash the beads twice (including once at 80°C with a 5- to 10-minute soak) with 10 "volumes" of 1x ATEN buffer. Resuspend the beads in the original volume of 1x ATEN buffer for use, and store at 4°C for up to at least 2 weeks.

    One purpose of this washing is to remove loose streptavidin. The proteinase K, if left over, does not hurt the streptavidin beads.

3. To each 130 µl of proteinase-K-treated PCR product, add 15 µl of 10x ATEN and incu-bate at 52°C for 30 minutes. Sometimes the previous RNase step or the 65°C proteinase K step has melted off some of the primers. This high-salt incubation is to make sure the cleaved primers are (temporarily) back in place on the sticky ends, so that their 5′-biotin can hold them onto the beads. Without this reannealing, half of the product sometimes does not stick. Add 60–100 µl of streptavidin beads resuspended in 1x ATEN buffer as prepared in step 2. Incubate the beads for 30 minutes at 25°C. Magnetize the beads. Save the bead supernatant and run some on a gel to monitor efficiency of adsorption to beads.

4. Wash the beads once or twice quickly with 200–500 µl of 1x ATEN to remove more ini-tial PCR template and RNase.

5. Heat-elute the beads at 80°C two or three times with 100 µl of 1x ATEN for 3–5 min-utes each time. Move the tubes to 80°C hot block holes containing a magnet for anoth-er few minutes.

6. One at a time, hold the warm tubes next to a magnet in the hand, or in a magnet rack, and then withdraw the supernatant while warm. This is the ribocloning DNA with the 3′-sticky ends. Magnetize this DNA-containing supernatant, or centrifuge it, one last time to remove all traces of beads, and then transfer it to a fresh tube. Store frozen for up to 2 weeks.

7. The DNA is ready for cloning. The DNA is now in 1x ATEN, and make sure final ATEN level is 1x during annealing. Clone as described above in Protocol 1 starting with step 5.

▶ VARIATIONS AND TROUBLESHOOTING

- *The minimum enzyme incubation times.* These have not been established. This lab keeps making them shorter with no ill effect. Who knows how short they can be?

- *Some changes are not recommended, from experience.* Do not try to add PEG to the cloning step. Without the proteinase K step, RNase is carried into the cells with lethal effect and near-zero yield of transformants with some vectors. Do not try 37°C for the RNase step; either RNase does not work as well at 37°C, or DNase contaminants work better at 37°C.

- *Yield.* With *E. coli* X7029 (F⁻, ΔlacX74), which is not particularly electrocompetent (3 × 10⁸ per μg of plasmid), 1000–3000 clones are obtained per μl of original target PCR reaction. The current record for this lab is 300,000 clones, using the amount of DNA shown in the first two lanes in Figure 3, with an average of about 20,000 clones.

- *X7029 has wild-type recombination.* With JC8679, which is more electrocompetent and has high RecET activity, there are 10× more colonies, but an unacceptable portion of these are unwanted recombinants or rearrangements between repeated spans of the vector and/or target (data not shown). RecA strains such as DH5α (Invitrogen electrocompetent cells; E. Oates, pers. comm.) work well with the method given here.

EXPRESSION AND YIELD

Some 90% of the clones express the gene activity well (compare red to white colonies in Fig. 4). This is to say that, at present, a full 10% do not express the product protein, and it is not known what causes this problem. The level of PCR (i.e., DNA polymerase) -induced errors is expected to be well below 1% (about 0.2%; Barnes 1994), so PCR fidelity is not the issue.

There are other published methods that use terminal homology of target and vector PCR products and have the same freedom from restriction sites, kinase, and ligase, and the same freedom of host, cloning vector, and cloning sites. When the yield of clones is reported for these methods, it is under 100 (Tillet and Neilan 1999). Uracil glycosylase-mediated cloning (Booth et al. 1994) unfortunately is limited by the sensitivity of long and accurate PCR enzyme mixtures to inhibition by dU (Lasken et al. 1996). The method of Chen et al. (2002) uses primers with one ribo base, but kinase and ligase are still used, and the yield for this method is under 100 clones.

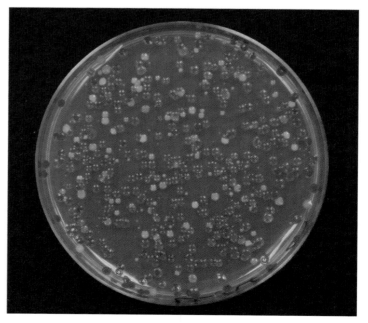

FIGURE 4. Plating of *E. coli* transformants from cloning the Ds-Red gene as in Fig. 3 (although this is not from the same cloning experiment). This plate demonstrates a current problem with the method: Some 10% of the clones do not express the Ds-red open reading frame (ORF) properly. This occurs with other target ORFs also, in the ribocloning of *lacZ* and *Taq* DNA polymerase (data not shown). The nature and cause of the cloning errors are not yet known.

ACKNOWLEDGMENTS

I thank Ed Oates, Anatoly Tzekov, Matt Smith, Ying Cui, and Elaine Frawley for crucible use of the method to improve its reliability.

REFERENCES

Barnes W.M. 1994. PCR amplification of up to 35 kb DNA with high fidelity and high yield from λ bacteriophage templates. *Proc. Natl. Acad. Sci.* **91**: 2216–2220.

Baskaran N., Kandpal R.P., Bhargava A.K., Glynn M.W., Bale A., and Weissman S.M. 1996. Uniform amplification of a mixture of deoxyribonucleic acids with varying GC content. *Genome Res.* **6**: 633–638.

Booth P.M., Buchman G.W., and Rashtchian A. 1994. Assembly and cloning of coding sequences for neurotrophic factors directly from genomic DNA using polymerase chain reaction and uracil DNA glyclosylase. *Gene* **146**: 303–308.

Britten R.J. and Kohne D.E. 1968. Repeated sequences in DNA. Hundreds of thousands of copies of DNA sequences have been incorporated into the genomes of higher organisms. *Science* **161**: 529–540.

Chen G.J., Qiu N., and Page M.G.P. 2002. Universal restriction site-free cloning method using chimeric primers. *BioTechniques* **32**: 518–524.

Dower W.J., Miller J.F., and Ragsdale C.W. 1988. High efficiency transformation of *E. coli* by high voltage electroporation. *Nucleic Acids Res.* **16**: 6127–6145.

Lasken R.S., Schuster D.M., and Rashtchian A. 1996. Archaebacterial DNA polymerases tightly bind uracil-containing DNA. *J. Biol. Chem.* **271**: 17692–17696.

Tillet D. and Neilan B.A. 1999. Enzyme-free cloning: A rapid method to clone PCR products independent of vector restriction enzyme sites. *Nucleic Acids Res.* **27**: e26.

Weiner M.P., Costa G.L, Schoettlin W., Cline J., Mathur E., and Bauer J.C. 1994. Site-directed mutagenesis of double-stranded DNA by the polymerase chine reaction. *Gene* **151**: 119–123.

MUTAGENESIS BY PCR

An important technique in molecular biology is the introduction of nucleotide base changes, deletions, or insertions into a gene to determine how the changes alter the structure or function of the product. PCR-based techniques for mutagenesis have vastly increased the number of methods to accomplish this type of research, as well as the overall efficacy of the process. There are three basic approaches to altering the nucleic acid sequence by PCR:

- random methods, exploiting the error-prone nature of *Taq* DNA polymerase
- oligonucleotide primers that are selected by the researcher to incorporate new mutations
- the gene shuffling approach, where mixtures of overlapping oligonucleotides are mixed to make new synthetic genes

All three methods are presented in this section.

RANDOM METHODS

"Mutagenic PCR" (p. 453) describes the methods for optimized incorporation of mismatched bases using thermostable polymerases without 3´→5´ exonuclease activity, also known as proofreading activity. These enzymes are biased toward incorporation of GC pairs in a reaction product when reading AT pairs within the template. Under certain reaction conditions, the introduction of bases in the product that are not present in the original template is facilitated. Optimal conditions for mismatch require adjustment of nearly all the components of the reaction, including the concentrations of nucleotides, enzyme, and magnesium, as well as the addition of manganese chloride. A protocol is given for the generation of a library of mutated amplicons that do not exhibit the GC-base sequence bias. For successful implementation of this method, a rapid and efficient method for selecting for the desired mutants within the generated library is required.

INCORPORATION OF NEW MUTATIONS

"Rapid PCR Site-directed Mutagenesis" (p. 459) presents a simple method that allows incorporation of any single-nucleotide change selected by the investigator in any plasmid, eliminating the need for subcloning and additional processing. This method is particularly useful when constructs for expression of the product are available and can be subjected to simple procedures to generate a panel of selectively modified clones. Using a single mutagenic primer pair, yields of 50% mutants can be easily obtained. This procedure is limited by the number of mismatches that can be incorporated into any one primer due to the need for perfect annealing between the 3´ end of the primer and the template for successful PCR. With all the primer-directed mutagenesis methods, care must be taken to prevent the incorporation of additional mutations, generated by *Taq* polymerase errors, from confusing the results. In this protocol, the number of PCR cycles is limited to minimize the occurrence and propagation of undesired base changes.

CHIMERIC OR HYBRID GENES

In addition to introducing point mutations, small deletions, or insertions, PCR can be used to make a chimeric or hybrid gene by splicing fragments together at a precise point. This method, described in "Mutagenesis and Synthesis of Novel Recombinant Genes Using PCR" (p. 467), requires the engineering of a region of overlap between the two fragments so that there is homology between the 3´ end of one section and the 5´ end of the second. When annealed, the region of homology serves to direct the synthesis of the hybrid molecule. Ultimately, PCR primers complementary to the 5´- and 3´-most ends of the combined fragment are used to amplify the hybrid. To be successful, this protocol must be completely mapped out. The reading frames of the molecules being spliced must be carefully considered to avoid production of an out-of-frame hybrid. In addition, the final size of the PCR product may require the adjustment of the extension phase of the final amplification reaction.

The final chapter in this section, "Streamlined Gene Assembly PCR" (p.475), describes a method for the complete synthesis of DNA fragments using overlapping 20-nucleotide primers. This method is useful for the development of codon-optimized genes and for engineering the secondary structure out of RNA molecules. Additionally, this method is amenable to generation of hybrid gene libraries using mixtures of primers with limited sequence variation. Once synthesis is complete, the gene library can be screened for an altered or improved biologic function.

Regardless of the technique used to generate the new set of mutants, a critical step—not related to PCR—is the availability of a reliable screening method to determine how the incorporated changes affect the function of the molecule under study. Is important to keep in mind that certain mutations are likely to affect one or more features of a protein, including its folding, localization, half-life, ability to interact with other molecules, and so forth. Several examples can be found in the literature, which highlight the importance of mutagenesis as a research tool.

30   Mutagenic PCR

R. Craig Cadwell[1] and Gerald F. Joyce

Departments of Chemistry and Molecular Biology and the Skaggs Institute for Chemical Biology, The Scripps Research Institute, La Jolla, California 92037

Most practitioners of PCR prefer to carry out DNA amplification in an accurate manner, introducing as few base substitutions as possible. This is especially critical when one is studying clonal isolates and must distinguish natural variation from artifactual variation that is introduced by polymerase error. Fortunately, thermostable DNA polymerases are available that operate with high fidelity due to an intrinsic $3' \rightarrow 5'$ exonuclease activity (for review, see Cha and Thilly 1993). Manipulation of the PCR conditions can lead to further improvement of copying accuracy (Ling et al. 1991; Cline et al. 1996).

Here we consider the other side of the fidelity issue—those instances where promiscuity is a virtue. Often, in probing the structure or function of a protein or nucleic acid, one wishes to generate a library of mutants and apply a screening method to isolate individuals that exhibit a particular property. For mutations over a short stretch of nucleotides within a cloned gene, it is appropriate to replace a portion of the gene with a synthetic DNA fragment that contains random or partially randomized nucleotides (Matteucci and Heyneker 1983; Wells et al. 1985; Oliphant et al. 1986; Ner et al. 1988). For mutations over a longer segment, up to the size of an entire gene, it may be preferable to scatter random mutations over the entire sequence, typically at a frequency of one or a few mutations per molecule. In such cases, it is most convenient to introduce random mutations through inaccurate copying by a DNA polymerase, especially if the polymerase is a thermostable enzyme that can operate in the context of PCR. Each pass of the polymerase during PCR allows the possibility of mutation, so that the cumulative error rate can become substantial.

The error rate of *Taq* DNA polymerase is the highest of the known thermostable DNA polymerases, in the range of 0.08×10^{-4} to 2×10^{-4} per nucleotide per pass of the polymerase, depending on the reaction conditions (Keohavong and Thilly 1989; Eckert and Kunkel 1990; Ling et al. 1991; Cline et al. 1996). Over the course of PCR, in which the polymerase makes an average of 20–25 passes, the cumulative error rate is $\sim 10^{-3}$ per nucleotide. In most cases, this is insufficient to generate a diverse library of variant sequences, especially over a region shorter than 1000 nucleotides. A further drawback is that the errors made by *Taq* DNA polymerase under standard PCR conditions are heavily biased toward $AT \rightarrow GC$ changes (Keohavong and Thilly 1989). We have devised a mutagenic PCR that has an overall error rate of $\sim 7 \times 10^{-3}$ per nucleotide and does not exhibit substantial sequence bias (Cadwell and Joyce 1992).

---

[1]Present address: Oklahoma Medical Research Foundation, Oklahoma City, Oklahoma 73104.

| PROTOCOL | ## MUTAGENIC PCR |

The top priority of mutagenic PCR is to introduce the various types of mutations in an unbiased fashion rather than to achieve a high overall level of amplification. The DNA input in a 100-μl reaction mixture consists of 10^{10} molecules (20 fmoles), which are amplified about 1000-fold to yield 10^{13} molecules (20 pmoles). This modest amplification requires an average of 10 passes of the polymerase. However, 30 cycles of PCR are carried out to ensure that mismatched termini have ample opportunity to become extended to produce complete copies. The large input prevents the PCR products from being influenced by the effects of clonal expansion. Even if a mutation occurs in the first pass of the polymerase and is passed along to all of the descendant molecules, there is very little chance that any two molecules isolated from the final population will carry the same mutation as a consequence of their being derived from a common ancestor. The protocol for mutagenic PCR is derived from "standard" PCR conditions (Coen 1991): 1.5 mM $MgCl_2$; 50 mM KCl; 10 mM Tris-HCl, pH 8.3 at 25°C; 0.2 mM each dNTP; 0.3 μM each primer; and 2.5 units of *Taq* DNA polymerase in a 100-μl volume. Incubate for 30 cycles for 1 minute at 94°C, 1 minute at 45°C, and 1 minute at 72°C in a conventional thermal cycler. The following changes are made to enhance the mutation rate:

- The $MgCl_2$ concentration is increased to 7 mM to stabilize noncomplementary pairs (Eckert and Kunkel 1991; Ling et al. 1991).

- 0.5 mM $MnCl_2$ is added to diminish the template specificity of the polymerase (Beckman et al. 1985; Leung et al. 1989).

- The concentration of dCTP and dTTP is increased to 1 mM to promote misincorporation (Leung et al. 1989; Cadwell and Joyce 1992).

- The amount of *Taq* DNA polymerase is increased to 5 units to promote chain extension beyond positions of base mismatch (Gelfand and White 1990).

MATERIALS

▼ CAUTION

See Appendix for appropriate handling of materials marked with <!>.

BUFFERS, SOLUTIONS, AND REAGENTS
Chloroform/isoamyl:alcohol <!> (24:1, v/v)
Ethanol <!>100%
Mineral oil or a wax bead
3 M NaOAc, pH 5.2
5 mM $MnCl_2$ <!>
10x mutagenic PCR buffer
  70 mM $MgCl_2$ <!>
  500 mM KCl
  100 mM Tris-HCl <!>, pH 8.3 at 25°C
  0.1% (w/v) gelatin

ENZYMES AND ENYZME BUFFERS
DNA polymerase
Native *Taq* or AmpliTaq DNA polymerase (Applied Biosystems or licensed supplier); do not substitute any other thermostable DNA polymerase

NUCLEIC ACIDS AND OLIGONUCLEOTIDES

High-purity deoxynucleoside 5′-triphosphates (Pharmacia, USB/Amersham)

10x dNTP mix containing 2 mM dGTP, 2 mM dATP, 10 mM dCTP, and 10 mM dTTP

Input DNA

Primers

GELS

Agarose gel stained with ethidium bromide<!>

SPECIAL EQUIPMENT

Thermal cycler

METHOD

1. Prepare the 10x mutagenic PCR buffer.

2. Prepare the 10x dNTP mix.

3. Prepare a solution of 5 mM $MnCl_2$. DO NOT combine with the 10x PCR buffer, which would result in the formation of a precipitate that disrupts PCR amplification.

4. Combine 10 μl of 10x mutagenic PCR buffer, 10 μl of 10x dNTP mix, 30 pmoles of each primer, 20 fmoles of input DNA, and an amount of H_2O that brings the total volume to 88 μl. Mix well.

5. Add 10 μl of 5 mM $MnCl_2$. Mix well and confirm that a precipitate has not formed.

6. Add 5 units (2 μl) of *Taq* DNA polymerase, and bring the final volume to 100 μl. Mix gently. Cover with mineral oil or a wax bead, if desired.

7. Incubate for 30 cycles for 1 minute at 94°C, 1 minute at 45°C, and 1 minute at 72°C. Do not employ a hot-start procedure or a prolonged extension time at the end of the last cycle.

8. Purify the reaction products by extraction with chloroform/isoamyl alcohol (24/1, v/v) and subsequent precipitation with 0.25 M NaOAc and 2.5 volumes of 100% ethanol.

9. Run a small portion of the purified products on an agarose gel stained with ethidium bromide to confirm a satisfactory yield of full-length material. Mutagenic PCR should be carried out in parallel with standard PCR (omitting the four changes listed above); the yields of full-length DNA should be comparable.

RESULTS

By employing a DNA of ordinary nucleotide composition, mutagenic PCR introduces errors at a frequency of 0.66% ± 0.13% per position over the course of the PCR (95% confidence interval) (Cadwell and Joyce 1992). Nearly all of these changes are base substitutions. The combined frequency of insertions and deletions is less than 0.05% (one-tailed test, 95% confidence interval). The number of mutations per DNA copy follows a Poisson distribution. The probability of mutating each of the four bases is approximately equal except for a 1.5-fold enhanced probability of mutating T residues, which is significant at the 99% confidence level. The most common specific mutations are A→T and T→A changes, which we attribute to TT mismatches that manifest as either A→T changes in the same strand or T→A changes in the opposing strand. The least common mutations are G→C and C→G changes, which presumably reflects the difficulty in forming and extending GG and CC mismatches.

Summing up all types of mutations and correcting for the base composition of the mutated gene, the ratio of AT→GC to GC→AT changes is 1.0 (0.6–1.7, 95% confidence interval).

◗ TROUBLESHOOTING

- *Artifacts.* The most common difficulty with mutagenic PCR stems from the fact that 30 temperature cycles are employed, even though *Taq* DNA polymerase makes an average of only 10 passes along the DNA. As noted above, this provides ample opportunity for extension of mismatched termini, which is necessary to lock in mutations. However, it also favors the occurrence of amplification artifacts (Mullis 1991). Compounding the problem is the markedly elevated $MgCl_2$ concentration, which lowers the stringency of primer hybridization, thereby promoting the formation of nonspecific amplification products. As a general rule, one should begin mutagenic PCR with either cDNA or a double-stranded DNA fragment that encompasses only the region of interest. It is risky to employ plasmid DNA and hopeless to begin with a genomic library. We limit the use of mutagenic PCR to DNAs no longer than about 1000 nucleotides. For longer target sequences, the DNA can be divided into two or more fragments that are mutagenized separately.

  Because mutagenic PCR enhances primer mishybridization, certain combinations of primers and target sequences will inevitably give rise to short amplification products that out-compete the full-length DNA. These artifacts are best seen by carrying out the reaction with a radiolabeled primer and separating the products on a nondenaturing polyacrylamide gel. On the basis of their size and (if necessary) sequence, it should be possible to discern the site of primer mishybridization and redesign the primers accordingly. Alternatively, it may be preferable to attach "well-behaved" primer-binding sites to the ends of the DNA and carry out PCR amplification using these handles.

- *Protocol modifications.* Another source of difficulty is the urge to make slight modifications of the protocol without evaluating their consequences. If, for example, C→G changes are less frequent than T→A changes, then why not double the concentration of dGTP to 0.4 mM? Doing so, it turns out, results in a fourfold increase in the ratio of AT→GC to GC→AT changes (Cadwell and Joyce 1992). We encourage other investigators to explore alternative reaction conditions that may lead to an improved PCR mutagenesis procedure. However, in view of the extreme sensitivity of *Taq* DNA polymerase to dNTP concentrations and other aspects of the reaction conditions, general users are encouraged to follow the protocol to the letter.

DISCUSSION

The distribution of variants that results from mutagenic PCR depends on the error rate and the length of the sequence that is being randomized. The probability P of having k mutations in a sequence of length n is given by

$$P(k,n,\varepsilon) = \{n! \ / \ [(n-k)! \ k!]\} \ \varepsilon^k \ (1-\varepsilon)^{n-k}$$

where ε is the error rate per position. In the present case, the error rate is 0.66% per position over the course of the PCR ($\varepsilon = 0.0066$). Thus, for a target sequence of 500 nucleotides, the resulting population of variants would consist of about 4% wild type, 12% one-error mutants, 20% two-error mutants, 22% three-error mutants, 18% four-error mutants, 12%

five-error mutants, and 12% mutants with six or more errors. The number of distinct sequences with k errors, N_k, increases exponentially with increasing k:

$$N_k = \{n! \; / \; [(n - k)! \; k!]\} \; 3^k$$

Thus, 20 pmoles of material resulting from the mutagenesis of a 500-nucleotide target sequence would contain all possible one-, two-, three-, and four-error mutants, but only about 2% of the possible five-error mutants and a progressively sparser sampling of higher-error mutants. These calculations refer to the composition of the DNA; for the corresponding protein they must be modified to take into account the degeneracy of the genetic code. For some purposes, an error rate of 0.66% per position will be insufficient. It is possible to carry out successive rounds of mutagenic PCR to double or even triple the overall error rate. However, two potential pitfalls must be avoided. First, if a small aliquot of one reaction mixture is used to seed the next, there is an increased chance that molecules isolated from the final pool will be related by descent. Taking one-thousandth of the products from a first mutagenic PCR to seed a second should not be a problem, but taking one-thousandth of the second to seed a third would reduce diversity to an unacceptably low level. This problem could be remedied by scaling up the third reaction mixture to a 10-ml volume, preferably in multiple reaction vessels containing 100 μl each. A second potential pitfall is the risk of generating nonspecific amplification products, made more likely by the increased number of temperature cycles. It may be necessary to gel-purify full-length DNA after the first mutagenic PCR before proceeding with the second.

A set of alternative protocols for mutagenic PCR employing *Taq* polymerase have been reported, also relying on unbalanced concentrations of the four dNTPS with the optional addition of $MnCl_2$ (Vartanian et al. 1996). Most of these protocols result in an overall error rate of 10^{-3} to 10^{-2} per nucleotide position, with substantial bias toward transition mutations. One protocol, however, deserves special attention because it provides an error rate of $\sim 10^{-1}$ per position with only modest sequence bias. This "hypermutagenic PCR" procedure requires 50 temperature cycles to obtain a good yield of PCR product; the cycles are carried out in the presence of 1 mM dGTP, 0.03 mM dATP, 0.03 mM dCTP, and 1 mM TTP, with the addition of 0.5 mM $MnCl_2$. Our laboratory has used this protocol extensively, and, even though it is prone to failure due to the highly aggressive reaction conditions, it is an excellent way to introduce random mutations at high frequency within a stretch of 20–200 nucleotides.

Recently, Stratagene developed a proprietary thermostable DNA polymerase (termed Mutazyme) that provides an error rate of 0.1–0.7% per position over the course of the PCR, depending on the amount of input DNA (Cline and Hogrefe 2000). It generates a broad spectrum of mutations under standard PCR conditions, albeit with significant bias toward GC→AT changes. Because PCR mutagenesis with Mutazyme employs standard concentrations of the four dNTPs and no added $MnCl_2$, it is less prone to generating amplification artifacts compared to our PCR mutagenesis procedure and even more so compared to hypermutagenic PCR.

The mutagenic PCR protocol described here, employing *Taq* polymerase, provides an effective way to generate a library of DNAs that contain random mutations over a stretch of 100–1000 nucleotides. If one is interested in a library of RNAs, then a promoter sequence for T7 RNA polymerase can be included near the 5′ end of the appropriate PCR primer, allowing the DNA products to serve as templates in an in vitro transcription reaction (Chamberlin and Ryan 1982). If one is interested in a library of proteins, then the PCR primers can be designed to include either restriction sites for cloning into a suitable expression vector or a ribosome-binding site and start codon for in vitro translation.

An important advantage of mutagenic PCR is that it allows repeated randomization of a population of nucleic acids without isolating clones and obtaining sequence information. After one has generated a library of mutants and applied a screening method to obtain individuals that exhibit a particular property, the selected individuals can then be used directly as input for a second mutagenic PCR. Repeating the cycle of selection and mutagenic amplification allows one to carry out in vitro evolution of nucleic acids, including those that have catalytic function (Beaudry and Joyce 1992; Bartel and Szostak 1993). Similarly, a population of protein-encoding DNAs, harvested from a selected subset of cells or viral particles, can be treated as an ensemble and subjected to mutagenic PCR to produce variants of the selected variants.

REFERENCES

Bartel D.P. and Szostak J.W. 1993. Isolation of new ribozymes from a large pool of random sequences. *Science* **261:** 1411–1417.

Beaudry A.A. and Joyce G.F. 1992. Directed evolution of an RNA enzyme. *Science* **257:** 635–641.

Beckman R.A., Mildvan A.S., and Loeb L.A. 1985. On the fidelity of DNA replication: Manganese mutagenesis in vitro. *Biochemistry* **24:** 5810–5817.

Cadwell C. and Joyce G.F. 1992. Randomization of genes by PCR mutagenesis. *PCR Methods Appl.* **2:** 28–33.

Cha R.S. and Thilly W.G. 1993. Specificity, efficiency, and fidelity of PCR. *PCR Methods Appl.* **5:** S18–S29.

Chamberlin M. and Ryan T. 1982. Bacteriophage DNA-dependent RNA polymerases. *Enzymes* **15:** 85–108.

Cline J. and Hogrefe H. 2000. Randomize gene sequences with new PCR mutagenesis kit. *Strategies* **13:** 157–162.

Cline J., Braman J.C., and Hogrefe H.H. 1996. PCR fidelity of *Pfu* DNA polymerase and other thermostable polymerases. *Nucleic Acids Res.* **24:** 3546–3551.

Coen D.M. 1991. The polymerase chain reaction. In *Current protocols in molecular biology* (ed. F.M. Ausubel et al.), pp. 151.1–151.7. Wiley Interscience, New York.

Eckert K.A. and Kunkel T.A. 1990. High fidelity DNA synthesis by the *Thermus aquaticus* DNA polymerase. *Nucleic Acids Res.* **18:** 3739–3744.

———. 1991. DNA polymerase fidelity and the polymerase chain reaction. PCR *Methods App.* **1:** 17–24.

Gelfand D.H. and White T.J. 1990. Thermostable DNA polymerases. In *PCR protocols: A guide to methods and applications* (ed. M.A. Innis et al.), pp. 129–141. Academic Press, San Diego, California.

Keohavong P. and Thilly W.G. 1989. Fidelity of DNA polymerases in DNA amplification. *Proc. Natl. Acad. Sci.* **86:** 9253–9257.

Leung D.W., Chen E., and Goeddel D.V. 1989. A method for random mutagenesis of a defined DNA segment using a modified polymerase chain reaction. *Technique* **1:** 11–15.

Ling L.L., Keohavong P., Dias C., and Thilly W.G. 1991. Optimization of the polymerase chain reaction with regard to fidelity: Modified T7, *Taq*, and Vent polymerases. *PCR Methods Appl.* **1:** 63–69.

Matteucci M.D. and Heyneker H.L. 1983. Targeted random mutagenesis: The use of ambiguously synthesized oligonucleotides to mutagenize sequences immediately 5′ of an ATG initiation codon. *Nucleic Acids Res.* **11:** 3113–3121.

Mullis K.B. 1991. The polymerase chain reaction in an anemic mode: How to avoid cold oligodeoxyribonuclear fusion. *PCR Methods Appl.* **1:** 1–4.

Ner S.S., Goodin D.B., and Smith M. 1988. A simple and efficient procedure for generating random point mutations and for codon replacements using mixed oglinucleotides. *DNA* **7:** 127–134.

Oliphant A.R., Nussbaum A.L., and Struhl K. 1986. Cloning of random-sequence oligodeoxynucleotides. *Gene* **44:** 177–183.

Vartanian J.-P., Henry M., and Wain-Hobson S. 1996. Hypermutagenic PCR involving all four transitions and a sizeable proportion of transversions. *Nucleic Acids Res.* **24:** 2627–2631.

Wells J.A., Vasser M., and Powers D.B. 1985. Cassette mutagenesis: An efficient method for generation of multiple mutations at defined sites. *Gene* **54:** 31–323.

Rapid PCR Site-directed Mutagenesis

Michael P. Weiner and Gina L. Costa

454 Life Sciences, Branford, Connecticut 06405

In vitro site-directed mutagenesis is an invaluable technique for studying protein structure–function relationships, gene expression, and vector modification. Several methods for performing site-directed mutagenesis have appeared in the literature, but these methods generally require single-stranded DNA (ssDNA) as the template (Kunkel 1985; Taylor et al. 1985; Vandeyar et al. 1988; Sugimoto et al. 1989). PCR-mediated methods have been developed that allow the use of double-stranded DNA (dsDNA) templates. These methods use high temperature to denature the double-stranded molecules (Jones and Winistorfer 1992; Watkins et al. 1993; Picard et al. 1994), but often require the use of multiple pairs of primers. To circumvent such disadvantages, we have developed a PCR-based method of site-directed mutagenesis (PCR–SDM) that allows site-specific mutation in virtually any double-stranded plasmid, thus eliminating the need for subcloning fragments into M13-based bacteriophage vectors and for ssDNA rescue (Weiner et al. 1994). The PCR-SDM system uses a single mutagenesis primer set and generates mutants with >50% efficiency in ~3 hours. The protocol is simple to perform and uses either mini- or maxi-prepared DNA.

The PCR-SDM system (see Fig. 1) uses increased template concentration and reduced cycling number to decrease potential second-site mutations during the PCR. The polymerase adjunct, *Taq* Extender PCR additive, is included in the PCR to increase yield and reliability of amplification from long templates (Nielson et al. 1994). The *Dpn*I endonuclease (target sequence: 5′-G^{m6}ATC-3′) is specific for methylated and hemimethylated DNA and is used to digest parental DNA, thus enhancing selection for mutation-containing, amplified DNA. DNA isolated from almost all *Escherichia coli* strains is Dam-methylated and therefore susceptible to *Dpn*I digestion. DNA isolated from *Dam*-deficient *E. coli* or other host organisms can be methylated in vitro using Dam methylase. Cloned *Pfu* DNA polymerase is used prior to end-to-end ligation of the linear template to remove any bases extended onto the 3′ ends of the product by *Taq* DNA polymerase (Costa and Weiner 1994a,b). The recircularized vector DNA incorporating the desired mutations is then ligated and used to transform *E. coli*. A single buffer (SDM buffer) has been developed that can be used for all of the steps involved in the procedure.

MUTAGENIC PRIMERS

Oligonucleotides for use in site-directed mutagenesis must be designed individually, according to the desired mutation. The following considerations should be taken into account when designing mutagenesis primers:

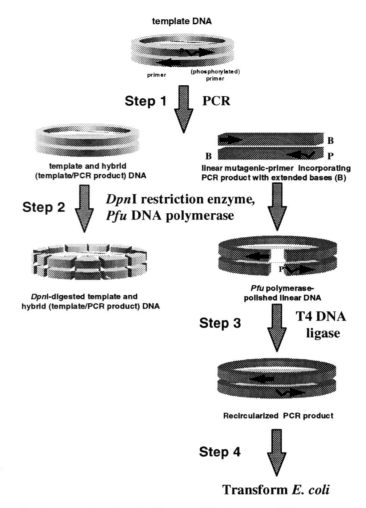

FIGURE 1. Schematic of the method for SDM using PCR. Template DNA is treated by the PCR-SDM protocol for a limited number of PCR cycles. The resulting mixture of template, newly synthesized, and hybrid parental/newly synthesized DNA is treated with *Dpn*I (target site 5′G^{m6}ATC) and *Pfu* DNA polymerase. (P) 5′ Phosphate; (B) 3′-terminal extended base(s). The end-polished PCR product is then intramolecularly ligated together and used to transform *E. coli*.

1. Both the mutagenic primers must anneal to different strands of the plasmid.

2. The distance between the primers is not crucial, but the primers should not overlap.

3. Generally, primers should be >20 bases in length (shorter primers may be used; see point 4, below).

4. The mismatched sequence portions should be at the 5′ end of one or both of the primers with 15 bases of correct sequence on the 3′ side.

5. The primers should be capable of synthesizing a PCR product (see PCR Considerations, below).

6. One or both of the primers must be 5′ phosphorylated (Sambrook and Russell 2001).

PCR CONSIDERATIONS

It is desirable to optimize the PCR before beginning the PCR-SDM procedure. The conditions needed to synthesize full-length products should be established. *Taq* Extender

(Stratagene) is a PCR adjunct that increases the yield, efficiency, and reliability of *Taq* DNA polymerase-driven PCR. Pilot PCRs should be performed with reduced (10–50 ng) template concentrations and increased cycle numbers (30–40 cycles). Other reagent concentrations (including the primer set) should be kept constant, as described for the SDM protocol. These concentrations should also be used during optimization of the denaturing, annealing, and extension times. The parameters that generate full-length, linear template molecules as the major amplification product should be used in the PCR-SDM method. After these conditions have been established, the template concentration should be increased, the cycle number reduced, and the PCR-SDM protocol followed as outlined below.

TABLE 1. Introducing restriction endonuclease sites by silent mutations

R.E.[a]	Target sequence	Reading frame 1 (amino acids)		Reading frame 2 (amino acids)			Reading frame 3 (amino acids)		
		1	2	1	2	3	1	2	3
*Alw*441	GTGCAC	V	H	CRSG	A	LPHQR	LSXWPQRMTKVAEG	C	T
*Apa*I	GGGCCC	G	P	WRG	A	LPHQR	LSXWPQRMTKVAEG	G	P
*Bam*HI	GGATCC	G	S	WRG	I	LPHQR	LSXWPQRMTKVAEG	D	P
*Bcl*I	TGATCA	X	S	LMV	I	IMTNKSR	FSYCLPHRITNVADG	D	HQ
*Bgl*II	AGATCT	R	S	XQKE	I	FLSYXCW	LSXPQRITKVAEG	D	L
*Bsp*MII	TCCGGA	S	G	FLIV	R	IMTNKSR	FSYCLPHRITNVADG	P	DE
*Cla*I	ATCGAT	I	D	YHND	R	FLSYXCW	LSXPQRITKVAEG	S	IM
*Eco*RI	GAATTC	E	F	XRG	I	LPHQR	LSXWPQRMTKVAEG	N	S
*Eco*RV	GATATC	D	I	XRG	Y	LPHQR	LSXWPQRMTKVAEG	I	S
*Hind*III	AAGCTT	K	L	XQKE	A	FLSYXCW	LSXPQRITKVAEG	S	FL
*Hpa*I	GTTAAC	V	N	CRSG	X	LPHQR	LSXWPQRMTKVAEG	L	T
*Kpn*I	GGTACC	G	T	WRG	Y	LPHQR	LSXWPQRMTKVAEG	V	P
*Mlu*I	ACGCGT	T	R	YHND	A	FLSYXCW	LSXPQRITKVAEG	R	V
*Msc*I	TGGCCA	W	P	LMV	A	IMTNKSR	FSYCLPHRITNVADG	G	HQ
*Nae*I	GCCGGC	A	G	CRSG	R	LPHQR	LSXWPQRMTKVAEG	P	A
*Nar*I	GGCGCC	G	A	WRG	R	LPHQR	LSXWPQRMTKVAEG	A	P
*Nco*I	CCATGG	P	W	SPTA	M	VADEG	FSYCLPHRITNVADG	H	G
*Nde*I	CATATG	H	M	SPTA	Y	VADEG	FSYCLPHRITNVADG	I	CXW
*Nhe*I	GCTAGC	A	S	CRSG	X	LPHQR	LSXWPQRMTKVAEG	L	A
*Nru*I	TCGCGA	S	R	FLIV	A	IMTNKSR	FSYCLPHRITNVADG	R	DE
*Pst*I	CTGCAG	L	Q	SPTA	A	VADEG	FSYCLPHRITNVADG	C	SR
*Pvu*I	CGATCG	R	S	SPTA	I	VADEG	FSYCLPHRITNVADG	D	R
*Pvu*II	CAGCTG	Q	L	SPTA	A	VADEG	FSYCLPHRITNVADG	S	CXW
*Sal*I	GTCGAC	V	D	CRSG	R	LPHQR	LSXWPQRMTKVAEG	S	T
*Sma*I	CCCGGG	P	G	SPTA	R	VADEG	FSYCLPHRITNVADG	P	G
*Spe*I	ACTAGT	T	S	YHND	X	FLSYXCW	LSXPQRITKVAEG	L	V
*Sph*I	GCATGC	A	C	CRSG	M	LPHQR	LSXWPQRMTKVAEG	H	A
*Sst*I	GAGCTC	E	L	XRG	A	LPHQR	LSXWPQRMTKVAEG	S	S
*Sst*II	CCGCGG	P	R	SPTA	A	VADEG	FSYCLPHRITNVADG	R	G
*Stu*I	AGGCCT	R	P	XQKE	A	FLSYXCW	LSXPQRITKVAEG	G	L
*Xba*I	TCTAGA	S	R	FLIV	X	IMTNKSR	FSYCLPHRITNVADG	L	DE
*Xho*I	CTCGAG	L	E	SPTA	R	VADEG	FSYCLPHRITNVADG	S	SR
*Xma*III	CGGCCG	R	P	SPTA	A	VADEG	FSYCLPHRITNVADG	G	R

Adapted from Weiner and Scheraga (1989) (for Apple-based computers) and Shankarappa et al. (1992) (for PC-based computers). Supplementary material containing these computer programs written for Apple MacIntosh computers is available from the Quantum Chemistry Program Exchange (QCPE), Department of Chemistry, Indiana University, Bloomington, Indiana 47405 (Program No. QMAC006).
[a](R.E.) Restriction endonuclease.

TABLE 2. The genetic code

Amino acid	Abbreviation 1	Abbreviation 3	Codon[a]
Alanine	A	Ala	(GCN)
Arginine	R	Arg	(CGN) or (AGR)
Asparagine	N	Asn	(AAY)
Aspartic acid	D	Asp	(GAY)
Cysteine	C	Cys	(TGY)
Glutamine	Q	Gln	(CAR)
Glutamic acid	E	Glu	(GAR)
Glycine	G	Gly	(GGN)
Histidine	H	His	(CAY)
Isoleucine	I	Ile	(ATH)
Leucine	L	Leu	(CTN) or (TTR)
Lysine	K	Lys	(AAR)
Methionine	M	Met	(ATG)
Phenylalanine	F	Phe	(TTY)
Proline	P	Pro	(CCN)
Serine	S	Ser	(TCN) or (AGY)
Threonine	T	Thr	(ACN)
Tryptophan	W	Trp	(TGG)
Tyrosine	Y	Tyr	(TAY)
Valine	V	Val	(GTN)

[a](N) Any base; (R) purine; (Y) pyrimidine; (H) A, C, or T.

INTRODUCTION OF TRANSLATIONALLY SILENT MUTATIONS

For easy analysis and subsequent manipulation, it is often desirable to incorporate translationally silent restriction sites into the mutagenesis primer during site-specific mutagenesis (Weiner and Scheraga 1989; Shankarappa et al. 1992). Tables 1 and 2 can be used for reverse translation of protein-coding regions to determine where, in a particular sequence, a translationally silent restriction site can be inserted.

PROTOCOL   RAPID PCR SITE-DIRECTED MUTAGENESIS

The advantages of the PCR-SDM method include the following:

- Use of increased template concentration, which allows fewer cycles to be used, thus reducing the potential for amplifying nonspecific products

- Inclusion of *Taq* Extender, which provides increased yield and reliability in the amplification of longer PCR products

- Use of *Dpn*I restriction endonuclease, which reduces the number of parental molecules

- Use of *Pfu* DNA polymerase to remove undesired base extensions, which improves the efficiency of blunt-end ligation

The complete protocol can be performed rapidly and does not require purification or precipitation procedures between steps. PCR-SDM can be used for both large deletions and insertions.

MATERIALS

BUFFERS, REAGENTS, AND SOLUTIONS

ATP (10 mM)

dNTP solution (containing all four dNTPs, each at 6.25 mM)

SDM buffer 10x (20 mM Tris-HCl<!>, pH 7.5, 8 mM $MgCl_2$<!>, 40 µg/ml BSA)

ENZYMES AND ENZYME BUFFERS

Taq DNA polymerase
Cloned *Pfu* DNA polymerase (Stratagene)
Taq Extender PCR additive (Stratagene)
*Dpn*I restriction endonuclease
T4 DNA ligase

NUCLEIC ACIDS AND OLIGONUCLEOTIDES MUTAGENESIS PRIMERS

Template DNA, dsDNA plasmid, mini- or maxi-preparation (0.5 pmole template = 0.33 µg/kb x size of template [kb])

Template DNA must contain methylated $G^{m6}ATC$ sequences; if not, in vitro methylation with Dam methylase must be done prior to the initiation of PCR-SDM.

Only plasmids harboring the bacteriostatic antibiotic resitance genes, such as ampicillin, tetracycline, and chloramphenicol, are suitable for use with this rapid protocol. Plasmids harboring bacteriocidal antibiotic resistance genes, such as kanamycin and streptomycin, can be used, but only if transformed cells are subjected to a growth period prior to the application of antibiotic selection. See note to step 14.

Mutagenic primers (15 pmoles of primer = [5 ng/base] x size of primer [base])

The presence of a 5´ phosphate on one or both primers is critical. The primer(s) can either be phosphorylated with T4 polynucleotide kinase or synthesized with a 5´-terminal phosphate.

MEDIA

LB agar plates (10 g of Bacto-tryptone, 5 g of Bacto-yeast extract, 10 g of NaCl, H_2O to 1 liter, final pH 7.0, with 15 g of agar per liter)

SPECIAL EQUIPMENT

FALCON 2059 tubes or equivalent
Thermal cycler
Water baths or appropriate heating blocks, preset to 37°C, 42°C, 72°C
Incubator, preset to 37°C
Microcentrifuge tubes

VECTORS AND BACTERIAL STRAINS

Heat-shock-competent *E. coli* cells (e.g., XL1-blue, Stratagene)

ADDITIONAL

Optional: Equipment and reagents for agarose gel electrophoresis, including ethidium bromide<!> (see step 9).

METHOD

PCR

1. In a sterile microfuge tube on ice, assemble the PCR-SDM reaction:

Template DNA	0.5 pmole
SDM buffer, 10x	2.5 µl
dNTP solution (6.5 mM)	1 µl
Mutagenic primers	15 pmoles of each
H_2O	to 24 µl

2. Add 2.5 units of *Taq* DNA polymerase and 2.5 units of *Taq* Extender PCR additive.

> These enzymes can be mixed together and stored as a 1:1 (v/v) mixture at –20°C for at least 3 months.

3. Carry out 7–12 cycles of amplification as follows; the PCR conditions are:

Cycle number	Denaturation	Annealing	Polymerization/Extension
1	4 min at 94°C	2 min at 50°C	2 min at 72°C
5–10	1 min at 94°C	2 min at 56°C	1 min at 72°C, then cool to room temperature
1	1 min at 94°C	2 min at 56°C	5 min at 72°C, then cool to room temperature

> Times and temperatures may need to be adapted to suit different types of equipment and reaction volumes. See PCR considerations above.

> If the thermal cycler is not equipped with a heated lid, use either mineral oil or paraffin wax to prevent evaporation of liquid from the reaction mixture during PCR.

Digesting and Polishing the PCR-SDM Product

4. Following the PCR, cool the reaction by placing it on ice for 2 minutes.

5. Add the following components directly to the 25-µl amplification reaction.

*Dpn*I restriction enzyme (10 units/1 µl)	1 µl
Pfu DNA polymerase (2.5 units/µl)	1 µl

> If mineral oil is used to overlay reactions during cycling, it is important to insert the pipette tip below the mineral oil when adding additional components to the reaction tube in the digestion, polishing, and ligation steps.

6. Mix gently and then centrifuge the reaction in a microcentrifuge for 1 minute. Immediately incubate the reaction for 30 minutes at 37°C.

7. Incubate the reaction for an additional 30 minutes at 72°C.

Ligating the PCR-SDM Product

8. Add the following components to the *Dpn*I, cloned *Pfu* DNA polymerase-treated product:

H_2O	100 µl
SDM buffer 10x	10 µl
ATP, 10 mM	5 µl

9. Mix gently, then centrifuge the reaction in a microcentrifuge for 1 minute.

> *Optional:* To verify the integrity of the PCR-SDM product, remove a 5-µl aliquot of the sample and analyze by standard agarose gel electrophoresis. A single band should be apparent.

10. Remove 10 µl of the above reaction to a sterile microcentrifuge tube and add 4 units of T4 DNA ligase (4 units/1 µl).

> There seems to be considerable variation in the ability of different batches of T4 DNA ligase to ligate blunt-ended DNA molecules. This leads to variations in mutagenesis efficiency, which can be anywhere from 30% to 70% in this assay. Once a good batch of enzyme has been identified, it is recommended that it be held in reserve for use in PCR-SDM.

11. Incubate the reaction for 1 hour at 37°C.

Transforming into Competent Cells (Rapid Transformation Protocol)

12. Gently thaw the heat-shock-competent cells on ice, and aliquot 40 µl of the cells to a prechilled FALCON 2059 polypropylene tube.

13. Add 1 µl of the ligase-treated DNA to the cells, swirl gently, and incubate for 30 minutes on ice.

14. Heat-shock for 30 seconds at 42°C and place on ice for 2 minutes.

    This heat pulse has been optimized for the FALCON 2059 tubes. If different tubes are used, conditions should be optimized accordingly.

    If plasmids harboring bacteriocidal antibiotic resistance genes are used, after 2 minutes on ice, add 260 µl of LB and incubate at 37°C, shaking at 250 rpm, for 30 minutes. Then continue with step 15.

15. Immediately plate the entire volume of transformed cells onto LB agar plates containing the appropriate antibody selection. Incubate the plates overnight at 37°C.

ACKNOWLEDGMENTS

The authors thank Dr. Joseph Sorge, Jr. and John Bauer for their support with this research. We thank Daniel McMullan, Barbara McGowan, and Mila Angert for their substantial contributions in experimental design.

REFERENCES

Costa, G.L. and Weiner M.P. 1994a. Protocols for cloning and analysis of blunt-ended PCR generated DNA fragments. *PCR Methods Appl.* **3:** S95–S106.

———. 1994b. Polishing with T4 or Pfu polymerase increases the efficiency of cloning PCR fragments. *Nucleic Acids Res.* **22:** 2423.

Jones, D.H. and Winistorfer S.C. 1992. Recombinant circle PCR and recombination PCR for site-directed mutagenesis without PCR product purification. *BioTechniques* **12:** 528–533.

Kunkel T. 1985. Rapid and efficient site-specific mutagenesis without phenotypic selection. *Proc. Natl. Acad. Sci.* **82:** 488–492.

Nielson K.B., Schoettlin W., Bauer J.C., and Mathur E. 1994. *Taq* Extender PCR additive for improved length, yield and reliability of PCR products. *Strategies Mol. Biol.* **7:** 27.

Picard V., Ersdal-Badju E., Lu A., and Bock S.C. 1994. A rapid and efficient one-tube PCR-based mutagenesis technique using Pfu DNA polymerase. *Nucleic Acids Res.* **22:** 2587–2591.

Sambrook J. and Russell D.W. 2001. *Molecular cloning: A laboratory manual,* 3rd edition. Cold Spring Harbor Laboratory Press, Cold Spring Harbor, New York.

Shankarappa B., Vijayananda K., and Ehrlich G.D. 1992. SILMUT: A computer program for the identification of regions suitable for silent mutagenesis to introduce restriction enzyme recognition sequences. *BioTechniques* **12:** 882–884.

Sugimoto M., Esaki N., Tanaka H., and Soda K. 1989. A simple and efficient method for oligonucleotide-directed mutagenesis using plasmid DNA template and phosphorothioate-modified nucleotide. *Anal. Biochem.* **179:** 309–311.

Taylor J.W., Ott J., and Eckstein F. 1985. The rapid generation of oligonucleotide-directed mutations at high frequency using phosphorothioate-modified DNA. *Nucleic Acids Res.* **13:** 8765–8785.

Vandeyar M., Weiner M.P., Hutton C., and Batt C. 1988. A simple and rapid method for the selection of oligodeoxynucleotide-directed mutants. *Gene* **65:** 129–133.

Watkins B., Davis A.E., Cocchi F., and Reitz Jr., M.S. 1993. A rapid method for site-specific mutagenesis using larger plasmids as templates. *BioTechniques* **15:** 700–704.

Weiner M.P. and Scheraga H.A. 1989. A set of Macintosh programs for the design of synthetic genes. *Comput. Appl. Biosci.* **5:** 191–198.

Weiner M.P., Costa G.L., Schoettlin W., Cline J., Mathur E., and Bauer J.C. 1994. Site-directed mutagenesis of double-stranded DNA by the polymerase chain reaction. *Gene* **15:** 119–123.

32 Mutagenesis and Synthesis of Novel Recombinant Genes Using PCR

Abbe N. Vallejo, Robert J. Pogulis, and Larry R. Pease

Department of Immunology, Mayo Clinic Rochester, Rochester, Minnesota 55905

Since its original description, PCR (Mullis et al. 1986) has become a fundamental analytical tool in cellular and molecular biology. The literature is replete with the use of this technique in the identification of new genes or members of gene families, even those from distantly related species. PCR has permitted the introduction of mutations into DNA sequences (Higuchi et al. 1988) to allow assessment of the biological function of genes. PCR has also been used to simplify site-directed mutagenesis and to generate recombinant or chimeric gene constructs (Ho et al. 1989; Horton et al. 1990).

Mutagenesis by PCR is accomplished by incorporating the desired genetic changes in custom-made primers used in amplification reactions. Because these mutagenizing primers have terminal complementarity, two separate DNA fragments amplified from a target gene may be fused into a single product (see below) by primer extension without relying on restriction endonuclease sites or ligation reactions.

The relative ease of rapidly combining two DNA fragments by overlap extension has led to its application in the production of chimeric genes. As long as pairs of primers used in amplification reactions have complementary terminal regions, overlapping strands of PCR-amplified DNA fragments from various sources may be spliced together by the extension of overlapping strands (hence, the term *s*plicing by *o*verlap *e*xtension or gene SOEing) in a subsequent reaction. The ability to generate recombinants by SOEing is limited only by the knowledge of the mechanisms of folding and/or assembly of the encoded novel protein when transfected into cells. SOE has been used routinely to generate mutant molecules and complex hybrid genes, to engineer proteins, and to create mutant libraries (Hunt et al. 1990a; Daugherty et al. 1991; Davis et al. 1991; Pullen et al. 1991; Cai and Pease 1992; Gobius et al. 1992; Bobek et al. 1993; Kirchoff and Desrosiers 1993; Pease et al. 1993; Yun et al. 1994).

CONCEPT OF OVERLAP EXTENSION

Overlap extension uses PCR both for introducing site-specific mutations and for generating recombinant gene constructs. Details of the strategy have been described previously (Ho et al. 1989; Horton et al. 1990). Briefly, mutagenesis is achieved by PCR with the use of specially designed oligonucleotide primers that include in their sequence the desired changes (i.e., substitutions, insertions, or deletions) to be incorporated in the gene construct (Fig. 1). Because the mutagenizing primers also have terminal complementarity, two overlapping fragments can be fused together in a subsequent extension reaction. The inclusion of outside primers in the extension reaction amplifies the fused product by PCR. In principle, the primers may be moved anywhere along the targeted gene to introduce mutations. Thus, the regions of the gene containing the introduced mutations may be lengthened in a single reaction.

The ability to fuse two DNA fragments by overlap extension can be exploited further to splice (or SOE together) two or more DNA fragments from different genes to generate a chimeric product. Like the mutagenizing primers, SOEing primers have terminal complementarity. Thus, DNA fragments generated by PCR can be spliced together by primer extension and amplified to yield a recombinant product (Fig. 2).

A limitation of SOE, however, is the difficulty of manipulating large DNA segments, i.e., those greater than 1–2 kb. To circumvent this difficulty, a cassette system can be easily targeted, modified by SOE, and reinserted using restriction endonuclease sites designed into the cassette structure (Horton et al. 1990; Hunt et al. 1990b). Thus, specific segments of genes can be manipulated at will. This cassette approach also allows easy shuffling or replacement of gene segments. Such a cassette system has been a very valuable tool in dissecting the structure–function relationship of class I major histocompatibility complex molecules (Horton et al. 1990; Hunt et al. 1990a; Cai and Pease 1992).

DESIGN OF OLIGONUCLEOTIDE PRIMERS

SOEing oligomers have two sequence regions, namely a "priming" region and an "overlap" region. The priming region is the 3′ end of the oligomer, which serves as the PCR primer, and the overlap region is the 5′ end of the oligomer, which is complementary to a sequence of a DNA fragment that will be fused with it in the overlap extension reaction. As depict-

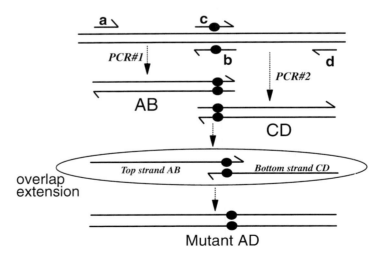

FIGURE 1. Mutagenesis by overlap extension. Two segments of a gene are PCR-amplified independently and then fused together in a subsequent reaction. Mutations are introduced into a targeted region with the use of specially designed mutagenizing primers (b and c), which contain nucleotide mismatches (represented by the solid circles) in the center of the primers. Because these primers are complementary, strands of PCR products generated independently with these primers will have an overlap that can be extended in a subsequent reaction to form the mutant product. When outside primers (a and d) are added to this latter reaction, the fused mutant product is amplified as soon as it is formed.

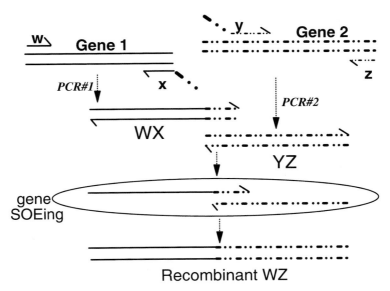

FIGURE 2. Gene SOEing. The reactions involved in SOEing are similar to those depicted in Fig. 1, with two exceptions. The first is that the PCR products to be fused are derived from two different genes. Second, the fusion is mediated by an overlap of the two strands of PCR products that were created with the use of SOEing primers (x and y). In this case, the 5′ region of primer x, used in the amplification of gene 1, is complementary to a segment of gene 2 (i.e., the 5′ region of primer y). A recombinant product is therefore formed when this overlap is extended in a subsequent reaction, and this recombined product may be amplified with the inclusion of outside primers (w and z) in this latter extension reaction.

ed in Figure 2, primer x contains a priming sequence for the amplification of gene 1 and also has an overlap sequence at its 5′ end that is complementary to a segment of gene 2.

Mutagenizing oligomers can have their priming and overlap regions completely coinciding with each other. The centers of the oligomers contain the mismatches or deletions that will be incorporated by overlap extension (Fig. 1).

In designing the oligomers, the actual lengths of the priming and overlap regions may depend on the particular situation. One approach is to determine the length by a simple estimate of the melting temperature (Suggs et al. 1981; Horton and Pease 1991), which is calculated as follows:

$$T_{m} = ([G+C] \times 4) + ([A+T] \times 2)$$

This quantity is an estimate of the denaturation temperature (in degrees Centigrade) of the oligomer. As a rule of thumb, our oligomers are designed such that both their priming and overlap regions have T_{m} values around 50°C. This 50°C rule generally gives a length estimate of 13–20 bp for both the priming and the overlap regions of the oligomers. The mismatches incorporated in mutagenizing oligomers are not included in this estimation.

It must be emphasized, however, that T_{m} does not estimate accurately the annealing temperature of any specific oligomer, because it does not take into consideration the concentration of salts present in the cycling reaction. Nevertheless, our 50°C T_{m} rule has usually yielded reliable oligomers that anneal at this temperature.

The usual rules in designing PCR primers apply when synthesizing SOEing and mutagenizing primers (Horton and Pease 1991; Rychlik 1993). Oligomer sequences must always be checked for the potential to form secondary structures and for complementarity to primers in the same reaction. Proper priming is achieved when at least the five bases of the

3′-terminal end of the primer are complementary to the template. Introducing mismatches too close to the 3′ end of the primer could result in the lack of amplification.

PROTOCOL PCR MUTAGENESIS BY OVERLAP EXTENSION AND GENE SOEing

MATERIALS

BUFFERS, SOLUTIONS, AND REAGENTS
Sterile H$_2$O

ENZYMES AND ENZYME BUFFERS
10x PCR buffer (Roche) containing 100 mM Tris-HCl<!>, pH 8.3, 500 mM KCl<!>, and 15 mM MgCl$_2$<!> (see Note 4, for determining the appropriate concentration of Mg^{++})
Restriction enzymes and buffers (see step 8 and Note 6)
Taq DNA polymerase (Roche)

NUCLEIC ACIDS AND OLIGONUCLEOTIDES
10x dNTPs: Commercially available 100 μM dNTP solutions (Amersham Biosciences) come as aqueous stocks and are diluted with sterile H$_2$O.
PCR templates (see Note 3)
5′ and 3′ primers (see Note 2)

SPECIAL EQUIPMENT
DNA sequencing equipment (see Note 6)
Thermal cycler

ADDITIONAL ITEMS
Agarose: The concentration of agarose needed to resolve DNA fragments depends on the fragment size to be isolated. In our experience, 0.8–1% SeaKem GTG agarose (Cambrex) in 1x TAE is generally satisfactory in resolving DNA fragments from 0.25 to 2 kb.
Agarose gel electrophoresis units and supplies
GENECLEAN II kit (BIO 101)

VECTORS AND BACTERIAL STRAINS
Cloning vector and bacterial strain for sequencing PCR products (see step 8 and Note 6)

METHOD

PCR Mutagenesis by Overlap Extension

1. Set up the PCR assays to produce products AB and CD (refer to Fig. 1) in individual microfuge tubes. These reaction mixtures can be assembled at room temperature. See Notes 2 and 3 regarding template and primer concentrations.

Reaction No.	1	2
PCR product	AB	CD
Template	parental template	parental template
5′ primer	a	c*
3′ primer	b*	d
10x PCR buffer	10 μl	10 μl
10x dNTPs	10 μl	10 μl
Taq DNA polymerase	2.5 units	2.5 units
Sterile H$_2$O	to 100 μl	to 100 μl

* Denotes mutagenizing primers.

2. Cycle for 30 rounds (1 minute at 94°C, 1 minute at 50°C, and 2 minutes at 72°C).

3. Run the entire volume of each reaction on an agarose gel.

4. Cut out the bands of interest (i.e., products AB and CD), and recover the DNA fragments by GENECLEAN II.

5. Set up the overlap extension reaction in a new microfuge tube.

Product	AD
Template 1	AB
Template 2	CD
5′ primer	a
3′ primer	d
10x PCR buffer	10 μl
10x dNTPs	10 μl
Taq DNA polymerase	2.5 units
Sterile H$_2$O	to 100 μl

This reaction is not very sensitive to the amounts of templates added. Typically 10–100 ng of each template may be used. Primers a and d may also be added from the beginning of this PCR.

6. Amplify by PCR as in step 2 above.

7. Run the reactions on agarose gels, cut out the band of interest (i.e., product AD), and recover the DNA fragment as in step 4 above.

8. The purified mutant product AD may then be digested with the appropriate restriction enzymes and ligated to an appropriate cloning vector for subsequent transformation and sequencing (see Note 6, below).

Gene SOEing

1. Set up PCR in individual microfuge tubes to generate products WX and YZ from genes 1 and 2, respectively (Fig. 2). See Notes 2 and 3 regarding template and primer concentrations.

Reaction No.	1	2
PCR product	WX	YZ
Template	gene 1	gene 2
5′ primer	w	y*
3′ primer	x*	z
10x PCR buffer	10 µl	10 µl
10x dNTPs	10 µl	10 µl
Taq DNA polymerase	2.5 units	2.5 units
Sterile H$_2$O	to 100 µl	to 100 µl

*Denotes SOEing primers

2. Cycle for 30 rounds (1 minute at 94°C, 1 minute at 50°C, 2 minutes at 72°C).

3. Run the entire volume of each reaction on an agarose gel.

4. Cut out the bands of interest (i.e., products WX and YZ), and recover the DNA fragments by GENECLEAN II.

5. Set up the SOEing reaction in a new microfuge tube.

Product	WZ
Template 1	WX
Template 2	YZ
5′ primer	w
3′ primer	z
10x PCR buffer	10 µl
10x dNTPs	10 µl
Taq DNA polymerase	2.5 units
Sterile H$_2$O	to 100 µl

6. Amplify by PCR as in step 2 above.

7. Run the reaction on agarose gels, cut out the band of interest (i.e., product WZ), and recover the DNA fragment as in step 4 above.

8. The purified recombinant product WZ may then be digested with the appropriate restriction enzymes and ligated to an appropriate cloning vector for subsequent transformation and sequencing (see Note 6 below).

NOTES/COMMENTS

1. *Oligonucleotide purification.* Oligomers used in this laboratory are synthesized using a DNA Synthesizer (Perkin-Elmer, Applied Biosystems Division) in our institutional core facility, and the final material comes as an aqueous ammonia solution. The solution is dried under vacuum (SpeedVac SC100, Savant). The residue is resuspended in 1 ml of sterile water and desalted in Sephadex G-25 (NAP-10 columns, Amersham Biosciences). Five fractions of 500 µl are collected, and the absorbance of each fraction is determined at 260 nm. The fractions containing the first peak are pooled (typically fractions 3–4 from NAP-10 columns), and the absorbance of the pooled material is measured at 260 and 300 nm. The concentration of the purified oligomer is determined by the formula: µg/ml = $(A_{260nm} - A_{300nm})$ × dilution factor × 33, where the number 33 is the approximate concentration of oligonucleotides (µg/ml) per unit of absorbance at 260 nm (Sambrook et al. 1989). The concentration of the stock oligomer may also be expressed in molar terms as follows: µM = {$[(A_{260nm} - A_{300nm})$ × dilution factor] ÷ ε_{260}} × 10^6; where ε_{260}, the extinc-

tion coefficient, equals the number of bases times 10,000. The purified oligomer is aliquoted into smaller quantities and stored at 20°C. Another option from our core facility is to have their staff purify the oligomers and deliver them in a solution of 20% acetonitrile in H_2O.

2. *Primer concentration.* In a standard reaction of 100 μl, we find that 100 pmoles of each primer is optimal. This gives a final concentration of 1 pmole/μl (or 1 μM). It is recommended, however, that the amounts of primers be determined empirically. The important point is that large amounts of primers increase amplification errors caused by mispriming.

3. *Template concentration.* In theory, larger amounts of template reduce the number of cycles required to generate enough product and, therefore, lessen the chance of *Taq* polymerase-induced base misincorporations. However, high concentrations of template tend to impede successful amplification, and fewer rounds of amplification produce more open-ended strands (i.e., single-stranded products that extend beyond the length of the primer), which contribute to the background. Thus, it is necessary to titrate the amount of template to determine the optimal concentration. We generally find that 250–500 ng of cloned template or as much as 1–2 μg of genomic DNA yields the PCR products of interest without significant background amplification.

4. *Magnesium concentration.* Perhaps the single most important parameter for PCR is the amount of Mg^{++} present in the amplification reaction. High concentrations of Mg^{++} generally produce background amplification and increase *Taq*-induced errors that can be reduced by lowering the amount of Mg^{++} in the reaction. In most of our PCR experiments, we find that 1.5 mM Mg^{++} is optimal.

5. *Purification of PCR and SOEing products.* Agarose gel purification of the intermediate PCR products leads to cleaner SOE reactions (Ho et al. 1989). This purification step removes templates as well as open-ended strands that could generate unwanted products. Similarly, gel purification of SOE products ensures cleaner fragments for subsequent cloning.

6. *Cloning and sequencing of SOEing products.* Mutant and/or recombinant genes generated by SOE must always be cloned and sequenced to determine the accuracy of the introduced genetic changes and to ascertain that there have been no amplification errors. We routinely sequence DNA using an automated sequencer (Perkin-Elmer, Applied Biosystems Division) available in our institutional core facility.

Cloning of the PCR-generated genes could be facilitated by designing amplification primers (i.e., a-d and w-z primer pairs in Figures 1 and 2, respectively) with unique flanking restriction enzyme sites (details of the strategy are discussed elsewhere) (Kwok et al. 1994). The product of the third PCR (mutant AD or recombinant WZ) may then be digested with these enzymes and cloned into a suitable vector or exchanged with the appropriate fragment of a plasmid cassette (Horton et al. 1990; Hunt et al. 1990b).

REFERENCES

Bobek L.A., Tsai H., and Levine M.J. 1993. Expression of human salivary histatin and cystatin/histatin chimeric cDNAs in *Escherichia coli. Crit. Rev. Oral Biol. Med.* **4:** 581–590.

Cai Z. and Pease L.R. 1992. Structural and functional analysis of three D/L-like class I molecules from H-Z^v: Indications of an ancestral family of *D/L* genes. *J. Exp. Med.* **175:** 583–596.

Daugherty B.L., deMartino J.A., Law M.F., Kawka D.W., Singer I.I., and Mark G.E. 1991. Polymerase chain reaction facilitates the cloning, CDR-grafting and rapid expression of a murine monoclonal antibody directed against the CD18 component of leukocyte integrins. *Nucleic Acids Res.* **19:** 2471–2476.

Davis G.T., Bedzyk W.D., Voss E.W., and Jacobs T.A.V. 1991. Single chain antibody (SCA) encoding genes: One step construction and expression in eukaryotic cells. *BioTechnology* **9:** 165–169.

Gobius K.J., Rowlinson S.W., Barnard R., Mattick J.S., and Waters M.J. 1992. The first disulfide loop of the rabbit growth hormone receptor is required for binding to the hormone. *J. Mol. Endocrinol.* **9:** 213–220.

Higuchi R., Krummel B., and Saiki R.K. 1988. A general method of in vitro preparation and specific mutagenesis of DNA fragments: Study of protein and DNA interactions. *Nucleic Acids Res.* **16:** 7351–7367.

Ho S.N., Hunt H.D., Horton R.M., Pullen J.K., and Pease L.R. 1989. Site-directed mutagenesis by overlap extension using polymerase chain reaction. *Gene* **77:** 51–59.

Horton R.M. and Pease L.R. 1991. Recombination and mutagenesis of DNA sequences using PCR. In *Directed mutagenesis: A practical approach* (ed. M.J. McPherson), pp. 217–247. Oxford University Press, United Kingdom.

Horton R.M., Cai Z., Ho S.N., and Pease L.R. 1990. Gene splicing by overlap extension: Tailor-made genes using the polymerase chain reaction. *BioTechniques* **8:** 528–535.

Hunt H.D., Pullen J.K., Dick R.F., Bluestone J.A., and Pease L.R. 1990a. Structural basis of K^{bm8} allore-activity: Amino acid substitutions on the β-pleated floor of the antigen recognition site. *J. Immunol.* **145:** 1456–1462.

Hunt H.D., Pullen J.K., Cai Z., Horton R.M., Ho S.N., and Pease L.R. 1990b. Novel MHC variants spliced by overlap extension. In *Transgenic mice and mutants in MHC research* (ed. I.K. Egorov and C.S. David), pp. 47–55. Springer-Verlag, Berlin.

Kirchoff F. and Desrosiers B.C. 1993. A PCR-derived library of random mutations within the V3 region of simian immunodeficiency virus. *PCR Methods Appl.* **2:** 301–304.

Kwok S., Chang S.Y., Sninsky J.J., and Wang A. 1994. A guide to the design and use of mismatched and degenerate primers. *PCR Methods Appl.* **3:** S39–S47.

Mullis K., Faloona F., Scharf S., Sakki R., Horn G., and Erlich H. 1986. Specific enzymatic amplification of DNA in vitro: The polymerase chain reaction. *Cold Spring Harbor Symp. Quant. Biol.* **51:** 263–273.

Pease L.R., Horton R.M., Pullen J.K., Hunt H.D., Yun T.J., Rohren E.M., Prescott J.L., Jobe S.M., and Allen K.S. 1993. Amino acid changes in the peptide binding site have structural consequences at the surface of class I glycoproteins. *J. Immunol.* **150:** 3375–3381.

Pullen J.K., Hunt H.D., and Pease L.R. 1991. Peptide interactions with the K^b antigen recognition site. *J. Immunol.* **146:** 2145–2151.

Rychlik W. 1993. Selection of primers for polymerase chain reaction. *Methods Mol. Biol.* **15:** 31–40.

Sambrook J., Fritsch E.F., and Maniatis T. 1989. *Molecular cloning: A laboratory manual*, 2nd edition, Cold Spring Harbor Laboratory Press, Cold Spring Harbor, New York, pp. 6.28-6.29, E.10-E.15.

Suggs S.V., Hirose T., Miyaki T., Kawashima E., Johnson M.J., Itakura K., and Wallace R.B. 1981. Use of synthetic oligodeoxyribonucleotides for the isolation of cloned DNA sequences. In *Developmental biology using purified genes* (ed. D.D. Brown and C.F. Fow), pp. 683–693. Academic Press, New York.

Tautz D. and Renz M. 1983. An optimized freeze-squeeze method for the recovery of DNA fragments from agarose gels. *Anal. Biochem.* **132:** 14–19.

Yun T.J., Tallquist M.D., Rohren E.M., Sheil J.M., and Pease L.R. 1994. Minor pocket B influences peptide binding, peptide presentation and alloantigenicity of H-2K^b. *Int. Immunol.* **6:** 1037–1047.

33 Streamlined Gene Assembly PCR

Wayne M. Barnes[1,2] and Elaine R. Frawley[1]

[1]*Department of Biochemistry and Molecular Biophysics, Washington University School of Medicine, St. Louis, Missouri 63110;* [2]*DNA Polymerase Technology, Inc., St. Louis, Missouri 63108*

The temperature cycling steps of the PCR process can be used to assemble whole genes and plasmids from identically sized pieces as small as 40 nucleotides in a process for constructing genes with artificial, designed sequences (Stemmer et al. 1995). Despite its title, the original protocol entailed two sequential PCR-like reactions. The first cycling protocol (55 cycles) is called "assembly." It contains no PCR primers per se; instead, the 40-mers all prime on each other, building up the product gene by extension of 20 bp at each extension step. No actual product band is visible; instead, the result is a messy smear. In a second cycling protocol, real PCR primers are supplied, and 1 or 2 µl from the first PCR is the template for an additional 23 cycles, for a total of 78 cycles. Finally, the PCR product is digested by restriction enzymes and gel-purified for cloning.

STREAMLINED PROTOCOL

We have made a few changes to the protocol so that only one round of 20–25 cycles is necessary to obtain a single or major product band (detected by agarose gel analysis), which can be followed directly by cloning without gel-purification of the target DNA.

Compared to the original method, the suggested changes include the following:

1. **Less is more.** Twenty times less DNA is used during the assembly. In our hands, the original level of DNA always gives rise to nonspecific smears instead of a visible target band. We found that the optimal clarity of target band resulted from the use of picomole amounts of each 40-mer per 100-µl reaction. We recommend bracketing this level on either side by a factor of two, such that we routinely attempt an assembly in parallel reactions with 1 pmole (or 1/6, 1/3, 2/3 pmole) of each gel-purified 40-mer per 100 µl of reaction volume.

2. **Long extension time: 20 minutes.** The low concentration of DNA actually runs afoul of the principle of C_0t (Britten and Kohne 1968). Therefore, the span of the anneal/extend portion of each cycle must be lengthened to ~20 minutes to allow time for the complementary pieces of the DNA strands to find each other. The recommended thermal cycler parameters are 1 minute at 93°C, 20 minutes at 55°C. The latter temperature should be raised to 60°C or 62°C if possible (if product still results) to achieve the maximum possible specificity. Logically, one might expect that only 7–10 long cycles would be necessary, and that the extension/annealing step of the last 10–15

cycles could be reduced to the normal 2 minutes or so. However, this did not work for us, so we keep all of the cycles long.

3. **Hot start.** The use of a hot start, which we provide with Rockstart, the magnesium precipitate method (Barnes and Rowlyk 2002), is often crucial to achieve a single or prominent band as analyzed on an agarose gel.

4. **Betaine.** To improve specificity and to suppress misannealing of oligonucleotides that happen to have high GC content, we include betaine at a level of 1.3 or 2.0 M (Baskaran et al. 1996).

5. **PCR primers present.** We use 200 nM of PCR primers in the single PCR necessary for the gene assembly. No subsequent separate PCR is required in this protocol. In our case, the PCR primers are riboprimers T7gen10rC and T41, so that the PCR product may be processed and cloned by the method of ribocloning (see Chapter 29).

6. **Gel-purify the oligonucleotides.** Occasionally, we obtain the desired PCR product using 40-mers directly from the manufacturer. More often, we gel-purify our input 40-mers. The quality of the input 40-mers is paramount to the success of the gene assembly. A single poor or absent 40-mer prevents the appearance of the intact target product gene. A common contaminant in preparations of synthetic oligonucleotides is $n - 1$ (39-mers). Each of these 39-mers, should they be incorporated into the product, causes a frameshift mutation that is fatal to the translatability of the gene. Therefore, we carry out preparative acrylamide gel electrophoresis on each input 40-mer, using 12% acrylamide, 6 M urea, and a migration distance of about 30–35 cm. We find that similar purification, which can be included with the price of the 40-mers as supplied by commercial manufacturers, is not necessarily of sufficient quality, as judged by the continued presence of the 39-mer on our 12% gels.

The largest gene that we have attempted to construct by this method was 2187 bp, encoding a CryV ICP (insecticidal crystal toxin). Gel purification of the 40-mers was not required in this case; a 2-kb band arose in 22 cycles, using 40-mers. A bioassay showed that at least 3 of 24 clones exhibited insecticidal activity. We did not confirm the DNA sequence of these clones.

Figures 1–3 illustrate the assembly of our designed gene for Ds-Red (W.M. Barnes, unpubl.) having an open reading frame (ORF) of 678 bp, and the total PCR product size was 731 bp. Our assemblies designated T0, T3, and T4 differ by a few 40-mers each for our project to introduce various mutations into the protein (E.R. Frawley and W.M. Barnes, unpubl.). Our proportion of red colonies (expressing the Ds-Red protein) ranges from 5% to 35% with no gel purification of the target band.

Occasionally (data not shown) it is desirable (but not necessary) to do a re-PCR step of about 18 cycles, using the same PCR primers and 1 µl from the assembly PCR as template.

PROTOCOL 1 DESIGN OF CODONS

Our procedure for designing a gene for optimal codon frequency uses, as one of its inputs, a codon table based on counts from real genes to choose its frequencies. We make little or no attempt to adjust any other features of the gene sequence. Our genes have only been expressed in bacteria so far, thus we do not yet know whether our simple strategy leads to superior plant transgenes. Software programs in CAPITALS are from the DOS Compugene suite of DNALYSIS software, available from the author (W.M.B.) for computers running Windows.

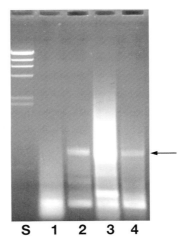

FIGURE 1. Comparison gel between our method and that of Stemmer et al. (1995). (Lane 1) Lack of specific product of the 55-cycle assembly specified by Stemmer et al. (4.5 pmoles each oligonucleotide per 100-μl reaction). Other very similar examples are shown in their paper. (Lane 2) PCR product that results from amplifying template from that assembly in a second cycling step of 23 cycles, as recommended by Stemmer et al. (Lane 3) Smear resulting from our method and too much template (4 pmoles each 40-mer). (Lane 4) Assembly performed in one reaction of 25 total cycles according to method described above, starting with about a pmole each 40-mer. The desired product (Ds-Red gene) is shown with an arrow.

1. To emulate the codon frequency of a highly expressed plant gene, we counted the codons in the small subunit of tomato Rubisco (ribulose bisphosphate carboxylase/oxygenase) using the program CODONS. Unfortunately, this single gene was perhaps a sample that was too small and extreme, because it does not have an example of every codon, so we added the codon table from another gene that is highly expressed in plants—the coat protein from tobacco mosaic virus (TMV)—and a third (but bacterial)

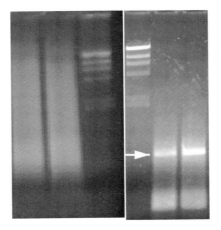

Bench Start   **Rockstart**

FIGURE 2. Rockstart is required for some assemblies. On the left, assemblies T3 and T4 were attempted using 0.6 pmoles of template oligonucleotides at 2.0 M betaine. The cycling program was 16 cycles at 93°C for 1 minute, 62°C for 20 minutes. Twenty and 24 cycles produced worse results (not shown). Only when Rockstart was used did the assemblies result in the desired product band in 25 cycles, in addition to one smaller unwanted band.

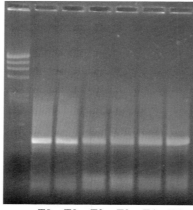

FIGURE 3. Rockstart is not always required. Pools T0, T3, and T4 were assembled in duplicate reactions for 24 cycles, according to our protocol, except that normal buffer with soluble magnesium (KLA buffer, see Chapter 6) was added during reaction setup.

gene with no known problems in plant gene expression, namely NPT-II (kanamycin resistance) from Tn5. The CODONS program accepts several genes in series, and then creates a total codon table automatically.

2. Next, we generated a listing of the amino acids of our targets Ds-Red and the CryV protein, in the single-letter code, using the program TRAFIG. We edited the output file to create a standard fasta format, which served as input to the program REVTRANS, which asks not only for amino acid sequence, but also for a codon table to emulate. REVTRANS conveniently reads in the output format from the summary table created by CODONS. The sequence-generating program runs a little slowly (no matter how fast the computer), since it only creates 18 codons per second. This is because it uses the time since the new year to the nearest 1/18 of a second to seed a pseudorandom number generator. The REVTRANS program can be run again and again, and a different (partially homologous) DNA sequence results each time, but each sequence will code for the input protein and have nearly identical frequencies in its codon table, emulating the frequencies requested by the user.

PROTOCOL 2 ORDERING THE 40-MERS

1. Print out the top strand. The program TRAFIG may be run with no gene to translate, but specifying a line length of 40 for the output. This will provide a convenient listing of the top strand of the designed sequence, at 40 per line, so that each line may be (manually) given a name or number and ordered from a commercial provider by E-mail. The last line of less than 40 is not ordered.

2. Print out the bottom strand. The program OTHSTR serves to generate the complementary (bottom) strand, since both strands must be ordered for the assembly. Before running TRAFIG to generate 40-per-line of the bottom strand, care must be taken (delete a few of the first few bases) so that this output will start exactly 20 bases after the right side of the map of the top strand (and the first one ordered ends at nucleotide 21 of the last top strand oligonucleotide). The goal is to create exact 20:20 overlaps throughout the gene assembly, as specified by Stemmer et al. (1995). If this printout is in the right phase, the last line will be only 20 bases. This 20-mer is not ordered, either.

3. Order a "touch-up" oligonucleotide. Probably, the desired gene is not exactly ($40n + 20$) in length, so an extra, offset 40-mer must be ordered that extends the gene by a few base pairs, replacing the few bases that were left out at the start of the phased bottom strand printing above. This extra oligonucleotide will be redundant to the right side of the bottom strand by more than 20, but it will participate in the assembly as template to extend the gene by the necessary few base pairs homologous to the PCR primer on that side (see Fig. 4). Actually, this touching-up may be done on either end (top strand on the left, bottom strand on the right), as long as the overlaps are at least 20, to alter or adjust the edges of the desired artificial gene.

4. PCR primers of normal size (about 25 bases) may be ordered to prime the amplification. Alternatively, and depending on the cloning strategy, the first and last 40-mers can be used as the PCR primers, if they are used at full strength (200 nM) in the reaction.

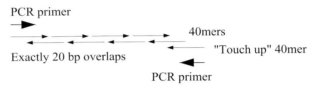

FIGURE 4. Compared to Stemmer et al. (1995), our PCR-assembly reactions include, from the beginning, the PCR primers and an additional "touch-up" 40-mer at the 5′ end of the bottom strand.

PROTOCOL 3   GEL PURIFICATION

MATERIALS

BUFFERS, SOLUTIONS, AND REAGENTS

Distilled H$_2$O

Ethanol, 95%<!>

Formamide<!> (deionize with Amberlite monobed resin MB-1 and store in the dark)

Gel buffer is the standard TBE + 6 M urea

Methylene Blue<!>, 100x, 0.05%

Sodium dodecyl sulfate stock<!>, 10%

TAE buffer

    40 m M Tris acetate<!>

    20 m M sodium acetate<!>

    0.2 m M EDTA, pH 8.3

Blue dextran, 1 mg/ml

TEN buffer

    10 mM Tris-HCl<!>, pH 7.9

    10 mM NaCl,

    0.1 mM EDTA

▼ CAUTION

See Appendix for appropriate handling of materials marked with <!>.

NUCLEIC ACIDS AND OLIGONUCLEOTIDES

40-mers

GELS

Polyacrylamide<!> gel with an acrylamide:bis ratio of 100:1 (total 12% acrylamide) and gel dimensions of 40 cm x 20 cm x 2 mm

SPECIAL EQUIPMENT
    Scalpel
    UV light<!> (UVGL-58, Ultraviolet Products)
    UV-filter safety glasses

METHOD

1. Suspend each 40-mer at 1 mM in TEN buffer and load only 1000 pmoles (1 μl + 4 μl formamide) per lane (1 cm wide).

2. After the 40-mers migrate ~30–35 cm, soak the gel in two changes of distilled H_2O to remove the urea, and then stain in 0.0005% Methylene Blue overnight. The visible blue bands do not comigrate exactly, due to sequence effects, even though they are all 40-mers. Alternatively, instead of staining, use UV shadowing to detect the gel bands as less fluorescent shadows over common printer paper when illuminated from above with a handheld UV light, while wearing safety glasses.

3. Excise each most-prominent band with a new scalpel blade and soak over one or two nights at 37°C in 3 volumes of 0.1% SDS and TAE. Recover the supernatant (avoiding the gel piece) and precipitate with 3 volumes of ethanol.

4. Estimate the yield of 40-mers for a representative number of the oligonucleotides by re-electrophoresis, staining with Methylene Blue, and comparing to a few input, unpurified 40-mers. In our experiments, the overall yield was usually about 30%, and this yield was assumed for all.

5. Pool the oligonucleotides and adjust their concentration (each) to an estimated 1/2 pmole per μl in TEN buffer.

PROTOCOL 4   PCR-ASSEMBLY REACTION

MATERIALS

▼ CAUTION

See Appendix for appropriate handling of materials marked with <!>.

BUFFERS, SOLUTIONS, AND REAGENTS
    Magnesium chloride<!>, 35 mM
    Rockstart buffer (http://klentaq.com)
    Three-reaction master mix
        82 μl of 5 M betaine
        3.2 μl of 10 mM each dNTP mix
        1 μl of KlentaqLA enzyme mixture (Clontech 8421-1)
        6.3 μl of left PCR riboprimer from 10 μM stock
        6.3 μl of right PCR riboprimer from 10 μM stock
        (no template yet)
        153.5 μl of H_2O to make 252 μl

GELS
    Agarose gel, 1.4%

SPECIAL EQUIPMENT
    PCR machine
    PCR tubes

METHOD

1. To each of three PCR tubes, add 10 µl of Rockstart buffer and 10 µl of 35 mM MgCl$_2$. Allow the ingredients to sit together for at least 15 minutes to form the magnesium precipitate (Barnes and Rowlyk 2002).

2. Prepare the three-reaction master mix.

3. After the magnesium precipitate has formed, add 80 µl of the three-reaction master mix to each reaction tube containing Rockstart precipitate.

4. Add 1/2, 1, or 2 µl of the mix of 40-mers at a pmole each per µl. Mixing is not necessary. Cycle 25 cycles at 93°C 1 for minute, 62°C for 20 minutes.

5. Analyze 15 µl of the PCR product on a 1.4% agarose gel.

ACKNOWLEDGMENTS

We thank Katherine Rowlyk and Anatoly Tzekov for technical assistance with early demonstrations of these methods. W.M.B. is affiliated with DNA Polymerase Technology, Inc., an inventor and provider of Rockstart, Klentaq1, and Compugene software.

REFERENCES

Barnes W.M. and Rowlyk K.R. 2002. Magnesium precipitate hot start method for PCR. *Mol. Cell. Probes* **16:** 167–171.

Baskaran N., Kandpal R.P., Bhargava A.K., Glynn M.W., Bale A., and Weissman S.M. 1996. Uniform amplification of a mixture of deoxyyribonucleic acids with varying GC content. *Genome Res.* **6:** 633–638.

Britten R.J. and Kohne D.E. 1968. Repeated sequences in DNA. Hundreds of thousands of copies of DNA sequences have been incorporated into the genomes of higher organisms. *Science* **161:** 529–540.

Stemmer W.P.C., Crameri A., Ha K.D., Brennan T.M., and Heyneker H.L. 1995. Single-step assembly of a gene and entire plasmid from large numbers of oligodeoxyribonucleotides. *Gene* **164:** 49–53.

WWW RESOURCE

http://klentaq.com DNA Polymerase Technology, Inc. homepage

1 Licensing

Applied Biosystems, as the exclusive licensee of Hoffmann-La Roche, owner of the basic PCR process and reagent patents (Mullis 1987; Mullis et al. 1987, 1990), oversees all license agreements for PCR. When using PCR for internal research purposes, the purchase and use of reagents and equipment that has been licensed by Applied Biosystems to other suppliers for PCR should give the end user all of the necessary coverage. Internal research does not include PCR testing for others or using the information obtained by PCR for analytical, diagnostic, or quality-control purposes. "Internal research" is defined by Applied Biosystems as follows: "For clarity, internal research does not include PCR-based testing for a third party, PCR performed under contract for a third party, or internal routine PCR-based analysis of product or process quality" (http://www.applied biosystems.com /legal/pcrlicfaqs.cfm). Much research falls in this category.

There are several situations where service rights and additional license agreements are necessary. Specifically, the third-party use of PCR, defined as "the performance or offering of commercial services of any kind using PCR, including, for example, reporting the results of purchaser's activities for a fee or other commercial consideration" (http://www.applied biosystems.com/legal/pcrlicfaqs.cfm), is granted unless the purchaser obtains an additional license. This includes providing services and the use of PCR even internally in applied fields, such as use of PCR for quality control. Under some circumstances, when using only a licensed kit with an authorized machine, all the necessary licenses may have been obtained. If using an internally developed process, additional service licenses will most likely be needed.

Furthermore, if you want to manufacture PCR-specific reagents, kits, and machines and want to be an authorized supplier, contact the Applied Biosystems Licensing Department at the address below. Additionally, questions related to necessary licenses and the establishment of third-party relationships to perform PCR and additional information about licenses can be obtained from:

Licensing Department
Applied Biosystems
850 Lincoln Centre Drive
Foster City, California 94404
Fax number: (650) 638-6071
Telephone: (650) 638-5845

For additional information see:
http://www.appliedbiosystems.com/legal/pcrlicfaqs.cfm

REFERENCES

Mullis K.B. 1987. Process for amplifying nucleic acid sequences. U.S. patent no. 4,683,202.

Mullis K.B., Erlich H.A., Gelfand D.H., Horn G., and Saiki R.K. 1990. Process for amplifying, detecting, and/or cloning nucleic acid sequences using a thermostable enzyme. U.S. patent no. 4,965,188.

Mullis K.B., Erlich H.A., Arnheim N., Horn G.T., Saiki R.K., and Scharf S.J. 1987. Process for amplifying, detecting, and/or cloning nucleic acid sequences. U.S. patent no. 4,683,195.

WWW RESOURCE

http://www.appliedbiosystems.com/legal/pcrlicfaqs.cfm PCR Licensing, Patents & Trademarks, Applied Biosystems

Modifications of Oligonucleotides for Use in PCR

The solid-phase synthesis of oligonucleotide deoxyribonucleotides, specifically using the 5′-dimethoxytrityl, β-cyanoethyl, diisopropyl phosphoramidite chemistry (Beaucage and Caruthers 1981; Sinha et al. 1984), is presently the most common method used for producing high-quality primers for PCR. Improvements to the basic b-cyanoethyl phosphoramidite chemistry include improved molecules for blocking and deprotecting guanosine during synthesis. From this core chemistry, numerous modifications can be made either during the actual synthesis or after the oligonucleotide has been prepared for use. These tables show some of the most commonly used modifications and what is necessary to produce that particular modification. By no means should these be construed as the only PCR-relevant modifications available.

TABLE 1. Additions to the 5´ end

Additions	Purpose	Potential use(s) in PCR	Necessary for incorporation
^{32}P-^{33}P	radioactively labeled oligonucleotide	detection	^{32}P or ^{33}P γ-labeled ATP with T4 polynucleotide kinase (Richardson 1965)
Primary amine group	provides site for further modification of the oligonucleotide	detection, modification	easily incorporated during synthesis. Basis for additional modifications listed below
Biotin	detection with avidin, multiple enzyme conjugates possible	detection, single-strand purification and PCR	amine linker on 5´ end of oligonucleotide and NHS-biotin (Chollet and Kawashima 1985) or can be directly added during synthesis
Digoxigenin	detection with anti-Dig Fab, multiple enzyme conjugates possible	detection	amine linker on 5´ end of oligonucleotide (Connolly 1987) and NHS-digoxigenin
Fluorescein Texas Red rhodamine	fluorescently labeled oligonucleotide, detection by fluorometer of UV light	in situ PCR, detection	amine linker on 5´ end of oligonucleotide and NHS-conjugated dye or, in the case of fluorescein, can be added during synthesis
6-FAM HEX TET	fluorescently labeled oligonucleotide, detection by fluorometer or charged-coupled device (such as the one in the ABI DNA sequencers)	in situ PCR, detection, quantitative PCR	all three dyes can be directly added during synthesis or added to a 5´ amine linker in their NHS form (Smith et al. 1985)
Ruthenium (TBR)	detection by a specific electrochemiluminescence reaction	quantitative PCR	can be directly added during synthesis (Telser et al. 1989)
Cyanine dyes	used at the 5´ end of primers	labeling PCR	incorporated by phosphoramidite

5´-End additions, in general, do not block the subsequent use of the oligonucleotide in a PCR. For use in PCR, the 3´ hydroxyl must remain unmodified. (6-FAM) 6-Carboxy-fluorescein; (HEX) hexachloro-6-carboxy-fluorescein; (TET) tetrachloro-6-carboxy-fluorescein; (Ruthenium [TBR]) tris (2,2´-bipyridine) ruthenium (II) chelate; (Dig Fab) fragment of an immunoglobulin specific for digoxigenin; (ABI) Applied Biosystems; (NHS) N-hydroxy-succinimide ester. Cyanines (Cy3-phosphoramidite), Indodicarbocyanine 3-1-O-(2-cyanoethyl)-(N-N-diisopropyl)-phosphoramidite; (Cy5-phosphoramidite), Indodicarbocyanine 5-1-O-(2-cyanoethyl)-(N,N-diisopropyl)-phosphoramidite.

TABLE 2. Additions to the 3′ end

Additions	Purpose	Potential use(s) in PCR	Necessary for incorporation
^{32}P-^{33}P	radioactively labeled oligonucleotide	detection	TdT, ^{32}P or ^{33}P-α (usually dC or dA) (Collins and Hunsaker 1985)
^{35}S	radioactively labeled oligonucleotide	detection	TdT, dATP-^{35}S
Biotin	detection with avidin, multiple enzyme conjugates possible	detection, single-strand purification and PCR	usually added during synthesis (CPG derivatized) or TdT tailed with dUTP-biotin (Kumar et al. 1988)
Digoxigenin	detection with anti-Dig Fab, multiple enzyme conjugates possible	detection	TdT tailed with dUTP-digoxigenin (BMB)
Fluorescein tetraethyl-rhodamine	fluorescently labeled oligonucleotide, detection by a fluorometer or UV light	in situ PCR, detection	usually added during synthesis (CPG derivatized) or TdT tailed (Trainor and Jensen 1988)
TAMRA	fluorescently labeled oligonucleotide, additionally used as a quencher in the TaqMan System, detection by a fluorometer or similar luminescence	quantitative PCR (when used with 5′-end-labeled HEX, TET, or 6-FAM), multiplex PCR, detection	TAMRA-CPG (linker arm nucleotide)
Dabcyl	used as a quencher in dual-labeled probes	quantitative PCR	incorporated during synthesis using dabcyl CPG
Dideoxy-nucleotide triphosphate (ddNTP)	multiple oligonucleotide species, detection by numerous methods (fluorescence, autoradiography)	detection, DNA sequencing	usually added during synthesis
Phosphate	oligonucleotide does not serve as primer in PCR	detection (i.e., TaqMan System), inhibition of extension	usually added during synthesis (CPG derivatized) but can be added enzymatically

3′-End additions, in general, prevent further use of the oligonucleotide in PCR. The 3′ hydroxyl is either blocked or missing. (TAMRA) 6-Carboxy-tetramethyl-rhodamine; (Dig Fab) fragment of an immunoglobulin specific for digoxigenin; (Dabcyl-CPG) 1-Dimethoxytrityloxy-3-[O-(N-4′-sulfonyl-4-(dimethylamino)-azobenzene)-3-aminopropyl]-propyl-2-O-succinoyl-long chain alkylamino-CPG. (HEX) hexachloro-6-carboxy-fluorescein; (TET) tetrachloro-6-carboxy-fluorescein; (6-FAM) 6-carboxy-fluorescein; (CPG) controlled pore glass; (BMB) Boehringer Mannheim. Do not use ammonium hydroxide with TAMRA; it degrades the moiety.

TABLE 3. Internal additions

Additions	Purpose	Potential use(s) in PCR	Necessary for incorporation
Biotin	multilabeled oligonucleotide, detection by avidin, multiple enzyme conjugates possible	detection	biotin phosphoramidite (Misiura et al. 1990) or modified base (Haralambidis et al. 1987)
Digoxigenin	multilabeled oligonucleotide, detection by anti-Dig Fab, multiple enzyme conjugates possible	detection	internal amine phosphoramidite (like amine-VCN from CLONTECH) plus NHS-digoxigenin (BMB)
Fluorescein Texas Red rhodamine	multilabeled oligonucleotide, detection by fluorometer or UV light	in situ PCR, detection	fluorescein phosphoramidite or an internal amine linker with an NHS form of the dye; can be easily added during synthesis
Degenerative (wobble), inosine	multiple oligonucleotide species, detection by DNA sequencing	detection, PCR for related sequences	
Thioate bond	a more nuclease-resistant oligonucleotide than the standard phosphodiester oligonucleotide linkage	probe preparation via PCR or probing in general, especially where nucleases are present (i.e., in situ)	TETD (ABI) or similar sulfurizing reagent (Hollway et al. 1993); easily performed on a DNA synthesizer
Peptide nucleic acid (PNA)	nuclease-resistant, higher annealing oligonucleotides for use in detection; cannot serve as primers, since they do not contain phosphate	detection, masking, strand displacement	synthesized using standard technology

Internal labeling may, or may not, prevent further use in a PCR. This is dependent on many factors: the number inserted, type of insertion, position of the insertions, and how many base pairs are interrupted. (Dig Fab) Fragment of an immunoglobulin specific for digoxigenin; (NHS) N-hydroxy-succinimide ester; (BMB) Boehringer Mannheim; (ABI) Applied Biosystems.

REFERENCES

Beaucage S.L. and Caruthers M.H. 1981. Deoxynucleoside phosphoramidites—A new class of key intermediates for deoxypolynucleotide synthesis. *Tetrahedron Lett.* **22:** 1859–1862.

Chollet A. and Kawashima E.H. 1985. Biotin-labeled synthetic oligodeoxyribonucleotides: Chemical synthesis and uses as hybridization probes. *Nucleic Acids Res.* **13:** 1529–1541.

Collins M.L. and Hunsaker W.R. 1985. Improved hybridization assay employing tailed oligonucleotide probes: A direct comparison with 5′-end-labeled oligonucleotide probes and nick-translated plasmid probes. *Anal. Biochem.* **151:** 211–224.

Connolly B.A. 1987. The synthesis of oligonucleotides containing a primary amino group at the 5′-terminus. *Nucleic Acids Res.* **15:** 3131–3139.

Haralambidis J., Chai M., and Chollet A. 1987. Preparation of base-modified nucleosides suitable for nonradioactive label attachment and their incorporation into synthetic oligodeoxyribonucleotides. *Nucleic Acids Res.* **15:** 4857–4876.

Hollway B., Erdman D.D., Durigon E.L., and Murtagh, Jr., J.J. 1993. An exonuclease-amplification coupled capture technique improves detection of PCR product. *Nucleic Acids Res.* **21:** 3905–3906.

Kumar A., Tchen P., Roullet F., and Cohen J. 1988. Non-radioactive labeling of synthetic oligonucleotide probes with terminal deoxynucleotidyl transferase. *Anal. Biochem.* **169:** 376–382.

Misiura K., Durrant I., Evans M.R., and Gait M.J. 1990. Biotinyl and phosphotyrosinyl phosphoramidite derivatives useful in the incorporation of multiple reporter groups on synthetic oligonucleotides. *Nucleic Acids Res.* **18:** 4345–4354.

Richardson C.C. 1965. Phosphorylation of nucleic acid by an enzyme from T4 bacteriophage-infected *Escherichia coli*. *Proc. Natl. Acad. Sci.* **54:** 158–165.

Sinha N.D., Biernat J., McManus J., and Köster H. 1984. Polymer support oligonucleotide synthesis XVIII: Use of β-cyanoethyl-N, N-dialkylamino-/N-morpholino phosphoramidite of deoxynucleosides for the synthesis of DNA fragments simplifying deprotection and isolation of the final product. *Nucleic Acids Res.* **12:** 4539–4557.

Smith L.M., Fung S., Hunkapillar M.W., Hunkapillar T.J., and Hood L.E. 1985. The synthesis of oligonucleotides containing an aliphatic amino group at the 5´ terminus: Synthesis of fluorescent DNA primers for use in DNA sequence analysis. *Nucleic Acids Res.* **13:** 2399–2419.

Telser J., Cruickshank K.A., Schanze K.S., and Netzel T.L. 1989. DNA oligomers and duplexes containing a covalently attached derivative of tris-(2,2´-bipyridine) ruthenium (II): Synthesis and characterization by thermodynamic and optical spectroscopic measurements. *J. Am. Chem. Soc.* **111:** 7221–7226.

Trainor G.L. and Jensen M.A. 1988. A procedure for the preparation of fluorescence-labeled DNA with terminal deoxynucleotidyl transferase. *Nucleic Acids Res.* **16:** 11846.

3 Cautions

The following general cautions should always be observed.

- **Become completely familiar with the properties of substances used before** beginning the procedure.

- **The absence of a warning** does not necessarily mean that the material is safe, since information may not always be complete or available.

- **If exposed** to toxic substances, contact your local safety office immediately for instructions.

- **Use proper disposal procedures** for all chemical, biological, and radioactive waste.

- **For specific guidelines on appropriate gloves,** consult your local safety office.

- **Handle concentrated acids and bases** with great care. Wear goggles and appropriate gloves. A face shield should be worn when handling large quantities. Do not mix strong acids with organic solvents as they may react. Sulfuric acid and nitric acid especially may react highly exothermically and cause fires and explosions. Do not mix strong bases with halogenated solvent as they may form reactive carbenes which can lead to explosions.

- **Handle and store pressurized gas containers** with caution as they may contain flammable, toxic, or corrosive gases; asphyxiants; or oxidizers. For proper procedures, consult the Material Safety Data Sheet that must be provided by your vendor.

- **Never pipette** solutions using mouth suction. This method is not sterile and can be dangerous. Always use a pipette aid or bulb.

- **Keep halogenated and nonhalogenated** solvents separately (e.g., mixing chloroform and acetone can cause unexpected reactions in the presence of bases). Halogenated solvents are organic solvents such as chloroform, dichloromethane, trichlorotrifluoroethane, and dichloroethane. Some nonhalogenated solvents are pentane, heptane, ethanol, methanol, benzene, toluene, *N,N*-dimethylformamide (DMF), dimethyl sulfoxide (DMSO), and acetonitrile.

- **Laser radiation,** visible or invisible, can cause severe damage to the eyes and skin. Take proper precautions to prevent exposure to direct and reflected beams. Always follow manufacturer's safety guidelines and consult your local safety office. See caution below for more detailed information.

- **Flash lamps,** due to their light intensity, can be harmful to the eyes. They also may explode on occasion. Wear appropriate eye protection and follow the manufacturer's guidelines.

- **Photographic fixatives and developers** contain chemicals that can be harmful. Handle them with care and follow manufacturer's directions.

- **Power supplies and electrophoresis equipment** pose serious fire hazard and electrical shock hazards if not used properly.

- **Microwave ovens and autoclaves** in the lab require certain precautions. Accidents have occurred involving their use (e.g., to melt agar or bacto-agar stored in bottles or to sterilize). If the screw top is not completely removed and there is not enough space for the steam to vent, the bottles can explode and cause severe injury when the containers are removed from the microwave or autoclave. Always completely remove bottle caps before microwaving or autoclaving. An alternative method for routine agarose gels that do not require sterile agar is to weigh out the agar and place the solution in a flask.

- **Ultra-sonicators** use high-frequency sound waves (16–100 kHz) for cell disruption and other purposes. This "ultrasound," conducted through air, does not pose a direct hazard to humans, but the associated high volumes of audible sound can cause a variety of effects, including headache, nausea, and tinnitus. Direct contact of the body with high-intensity ultrasound (not medical imaging equipment) should be avoided. Use appropriate ear protection and display signs on the door(s) of laboratories where the units are used.

- **Use extreme caution when handling cutting devices** such as microtome blades, scalpels, razor blades, or needles. Microtome blades are extremely sharp! Use care when sectioning. If unfamiliar with their use, have someone demonstrate proper procedures. For proper disposal, use the "sharps" disposal container in your lab. Discard used needles *unshielded*, with the syringe still attached. This prevents injuries (and possible infections; see Biological Safety) while manipulating used needles, since many accidents occur while trying to replace the needle shield. Injuries may also be caused by broken Pasteur pipettes, coverslips, or slides.

GENERAL PROPERTIES OF COMMON CHEMICALS

The hazardous materials list can be summarized in the following categories:

- Inorganic acids, such as hydrochloric, sulfuric, nitric, are phosphoric, and colorless liquids with stinging vapors. Avoid spills on skin or clothing. Spills should be diluted with large amounts of water. The concentrated forms of these acids can destroy paper, textiles, and skin, as well as cause serious injury to the eyes.

- Inorganic bases such as sodium hydroxide are white solids that dissolve in water and under heat development. Concentrated solutions will slowly dissolve skin and even fingernails.

- Salts of heavy metals are usually colored powdered solids which dissolve in water. Many of them are potent enzyme inhibitors and therefore toxic to humans and to the environment (e.g., fish and algae).

- Most organic solvents are flammable volatile liquids. Avoid breathing the vapors, which can cause nausea or dizziness. Also avoid skin contact.

- Other organic compounds, including organosulfur compounds such as mercaptoethanol and organic amines, can have very unpleasant odors. Others are highly reactive and should be handled with appropriate care.

- If improperly handled, dyes and their solutions can stain not only your sample, but also your skin and clothing. Some of them are also mutagenic (e.g., ethidium bromide), carcinogenic, and toxic.

- All names ending with "ase" (e.g., catalase, β-glucuronidase, and zymolase) refer to enzymes. There are also other enzymes with nonsystematic names like pepsin. Many of them are provided by manufacturers in preparations containing buffering substances, etc. Be aware of the individual properties of materials contained in these substances.

- Toxic compounds are often used to manipulate cells. They can be dangerous and should be handled appropriately.

- Be aware that several of the compounds listed have not been thoroughly studied with respect to their toxicological properties. Handle each chemical with the appropriate respect. Although the toxic effects of a compound can be quantified (e.g., LD_{50} values), this is not possible for carcinogens or mutagens where one single exposure can have an effect. Also realize that dangers related to a given compound may depend on its physical state (fine powder vs. large crystals/diethylether vs. glycerol/dry ice vs. carbon dioxide under pressure in a gas bomb). Anticipate under which circumstances during an experiment exposure is most likely to occur and how best to protect yourself and your environment.

HAZARDOUS MATERIALS

In general, proprietary materials are not listed here. Follow the manufacturer's safety guidelines that accompany the product.

Acetic acid (glacial) is highly corrosive and must be handled with great care. Liquid and mist cause severe burns to all body tissues. It may be harmful by inhalation, ingestion, or skin absorption. Wear appropriate gloves and goggles and use in a chemical fume hood. Keep away from heat, sparks, and open flame.

Acetone causes eye and skin irritation and is irritating to mucous membranes and upper respiratory tract. Do not breathe the vapors. It is also extremely flammable. Wear appropriate gloves and safety glasses.

Acrylamide (unpolymerized) is a potent neurotoxin and is absorbed through the skin (the effects are cumulative). Avoid breathing the dust. Wear appropriate gloves and a face mask when weighing powdered acrylamide and methylene-bisacrylamide. Use in a chemical fume hood. Polyacrylamide is considered to be nontoxic, but it should be handled with care because it might contain small quantities of unpolymerized acrylamide.

Ammonium persulfate ($[NH_4]_2S_2O_8$) is extremely destructive to tissue of the mucous membranes and upper respiratory tract, eyes, and skin. Inhalation may be fatal. Wear appropriate gloves, safety glasses, and protective clothing and use only in a chemical fume hood. Wash thoroughly after handling.

Ammonium sulfate ($[NH_4]_2SO_4$) may be harmful by inhalation, ingestion, or skin absorption. Wear appropriate gloves and safety glasses.

Bacterial strains (shipping of): The Department of Health, Education, and Welfare (HEW) has classified various bacteria into different categories with regard to shipping requirements (see Sanderson and Zeigler, *Methods Enzymol. 204: 248–264 [1991]*). Nonpathogenic strains of *E. coli* (such as K12) and *B. subtilis* are in Class 1 and are considered to present no or minimal hazard under normal shipping conditions. However, *Salmonella, Haemophilus,* and certain strains of *Streptomyces* and *Pseudomonas* are in Class 2. Class 2 bacteria are "Agents of ordinary potential hazard: agents which produce disease of varying degrees of severity...but which are contained by ordinary laboratory techniques." For detailed regulations regarding the packaging and shipping of Class 2 strains, see Sanderson and Ziegler (*Methods Enzymol. 204: 248–264 [1991]*) or the instruction brochure by Alexander and Brandon (*Packaging and Shipping of Biological Materials at ATCC* [1986]) available from the American Type Culture Collection (ATCC), Rockville, Maryland.

BCIG, *see* **5-Bromo-4-chloro-3-indolyl-β-D-galactopyranoside**

BCIP, *see* **5-Bromo-4-chloro-3-indolyl-phosphate**

Biotin may be harmful by inhalation, ingestion, or skin absorption. Wear appropriate gloves and safety glasses and use in a chemical fume hood.

Bisacrylamide is a potent neurotoxin and is absorbed through the skin (the effects are cumulative). Avoid breathing the dust. Wear appropriate gloves and a face mask when weighing powdered acrylamide and methylene-bisacrylamide.

Blood (human) and blood products and Epstein-Barr virus. Human blood, blood products, and tissues may contain occult infectious materials such as hepatitis B virus and HIV that may result in laboratory-acquired infections. Investigators working with EBV-transformed lymphoblast cell lines are also at risk of EBV infection. Any human blood, blood products, or tissues should be considered a biohazard and should be handled accordingly. Wear disposable appropriate gloves, use mechanical pipetting devices, work in a biological safety cabinet, protect against the possibility of aerosol generation, and disinfect all waste materials before disposal. Autoclave contaminated plasticware before disposal; autoclave contaminated liquids or treat with bleach (10% [v/v] final concentration) for at least 30 minutes before disposal. Consult the local institutional safety officer for specific handling and disposal procedures.

5-Bromo-4-chloro-3-indolyl-β-D-galactopyranoside (BCIG; X-gal) is toxic to the eyes and skin and may be harmful by inhalation, ingestion, or skin absorption. Wear appropriate gloves and safety goggles.

5-Bromo-4-chloro-3-indolyl-phosphate (BCIP) is toxic and may be harmful by inhalation, ingestion, or skin absorption. Wear appropriate gloves and safety glasses. Do not breathe the dust.

Bromophenol blue may be harmful by inhalation, ingestion, or skin absorption. Wear appropriate gloves and safety glasses and use in a chemical fume hood.

CaCl$_2$, *see* **Calcium chloride**

Cacodylate contains arsenic, is highly toxic, and may be fatal if inhaled, ingested, or absorbed through the skin. Wear appropriate gloves and safety glasses and use in a chemical fume hood. *See also* **Potassium cacodylate; Sodium cacodylate.**

Calcium chloride (CaCl$_2$) is hygroscopic and may cause cardiac disturbances. It may be harmful by inhalation, ingestion, or skin absorption. Do not breathe the dust. Wear appropriate gloves and safety goggles.

Carbenicillin may cause sensitization by inhalation, ingestion, or skin absorption. Wear appropriate gloves and safety glasses.

Carbon dioxide, (CO_2), in all forms may be fatal by inhalation, ingestion, or skin absorption. In high concentrations, it can paralyze the respiratory center and cause suffocation. Use only in well-ventilated areas. In the form of dry ice, contact with carbon dioxide can also cause frostbite. Do not place large quantities of dry ice in enclosed areas such as cold rooms. Wear appropriate gloves and safety goggles.

Cesium chloride, (CsCl), may be harmful by inhalation, ingestion, or skin absorption. Wear appropriate gloves and safety glasses.

Cetyltrimethylammonium bromide (CTAB) is toxic and an irritant and may be harmful by inhalation, ingestion, or skin absorption. Wear appropriate gloves and safety glasses. Avoid breathing the dust.

$CHCl_3$, *see* Chloroform

Chloramphenicol is a probable carcinogen and may be harmful by inhalation, ingestion, or skin absorption. Wear appropriate gloves and safety glasses and use in a chemical fume hood.

Chloroform ($CHCl_3$) is irritating to the skin, eyes, mucous membranes, and respiratory tract. It is a carcinogen and may damage the liver and kidneys. It is also volatile. Avoid breathing the vapors. Wear appropriate gloves and safety glasses and always use in a chemical fume hood.

Chromogenic substrates may be carcinogenic. Consult your local institutional safety officer for handling and disposal procedures.

Citric acid is an irritant and may be harmful by inhalation, ingestion, or skin absorption. It poses a risk of serious damage to the eyes. Wear appropriate gloves and safety goggles. Do not breathe the dust.

CO_2, *see* Carbon dioxide

Cobalt chloride ($CoCl_2$) may be harmful by inhalation, ingestion, or skin absorption. Wear appropriate gloves and safety glasses.

$CoCl_2$, *see* Cobalt chloride

CsCl, *see* Cesium chloride

CTAB, *see* Cetyltrimethylammonium bromide

DEPC, *see* Diethyl pyrocarbonate

Digoxigenin may be fatal if inhaled, ingested, or absorbed through the skin. Wear appropriate gloves and safety glasses and use in a chemical fume hood.

Diethylamine ($NH[C_2H_5]_2$), is corrosive, toxic, and extremely flammable. It may be harmful by inhalation, ingestion, or skin absorption. Wear appropriate gloves and safety glasses and use only in a chemical fume hood. Keep away from heat, sparks, and open flame.

Diethyl pyrocarbonate (DEPC) is a potent protein denaturant and a suspected carcinogen. Aim bottle away from you when opening it; internal pressure can lead to splattering. Wear appropriate gloves, safety goggles, and lab coat, and use in a chemical fume hood.

N,N-Dimethylformamide (DMF, HCON [CH_3])$_2$, is a possible carcinogen and is irritating to the eyes, skin, and mucous membranes. It can exert its toxic effects through inhalation, ingestion, or skin absorption. Chronic inhalation can cause liver and kidney damage. Wear appropriate gloves and safety glasses and use in a chemical fume hood.

Dimethyl sulfoxide (DMSO) may be harmful by inhalation or skin absorption. Wear appropriate gloves and safety glasses and use in a chemical fume hood. DMSO is also combustible. Store in a tightly closed container. Keep away from heat, sparks, and open flame.

Dithiothreitol (DTT) is a strong reducing agent that emits a foul odor. It may be harmful by inhalation, ingestion, or skin absorption. When working with the solid form or highly concentrated stocks, wear appropriate gloves and safety glasses and use in a chemical fume hood.

DMF, *see* **N,N-Dimethylformamide**

DMSO, *see* **Dimethyl sulfoxide**

Dry ice, *see* **Carbon dioxide**

DTT, *see* **Dithiothreitol**

Ethanol (CH₃CH₂OH) may be harmful by inhalation, ingestion, or skin absorption. Wear appropriate gloves and safety glasses.

Ethidium bromide is a powerful mutagen and is toxic. Consult the local institutional safety officer for specific handling and disposal procedures. Avoid breathing the dust. Wear appropriate gloves when working with solutions that contain this dye.

Fast Red may cause methemoglobinemia through overexposure. It may be harmful by inhalation, ingestion, or skin absorption. Wear appropriate gloves and safety glasses.

Fluorescein may be harmful by inhalation, ingestion, or skin absorption. Wear appropriate gloves and safety glasses and use in a chemical fume hood.

Formaldehyde (HCHO) is highly toxic and volatile. It is also a possible carcinogen. It is readily absorbed through the skin and is irritating or destructive to the skin, eyes, mucous membranes, and upper respiratory tract. Avoid breathing the vapors. Wear appropriate gloves and safety glasses and always use in a chemical fume hood. Keep away from heat, sparks, and open flame.

Formalin is a solution of formaldehyde in water. *See* **Formaldehyde**

Formamide is teratogenic. The vapor is irritating to the eyes, skin, mucous membranes, and upper respiratory tract. It may be harmful by inhalation, ingestion, or skin absorption. Wear appropriate gloves and safety glasses and always use a chemical fume hood when working with concentrated solutions of formamide. Keep working solutions covered as much as possible.

Guanidine thiocyanate may be harmful by inhalation, ingestion, or skin absorption. Wear appropriate gloves and safety glasses.

HCl, *see* **Hydrochloric acid**

HCHO, *see* **Formaldehyde**

Hematoxylin, Mayer's may be harmful by inhalation, ingestion, or skin absorption. Wear appropriate gloves and safety goggles.

H₂NOH, *see* **Hydroxylamine**

HOCH₂CH₂SH, *see* **β-Mercaptoethanol**

H₃PO₄, *see* **Phosphoric acid**

Hydrochloric acid, HCl, is volatile and may be fatal if inhaled, ingested, or absorbed through the skin. It is extremely destructive to mucous membranes, upper respiratory tract, eyes, and skin. Wear appropriate gloves and safety glasses and use with great care in a chemical fume hood. Wear goggles when handling large quantities.

Hydroxylamine (H_2NOH) is corrosive and toxic. It may be harmful by inhalation, ingestion, or skin absorption. Wear appropriate gloves and safety glasses and use only in a chemical fume hood.

IAA, *see* **Isoamyl alcohol**

IPTG, *see* **Isopropyl-β-D-thiogalactopyranoside**

Isoamyl alcohol (IAA) may be harmful by inhalation, ingestion, or skin absorption and presents a risk of serious damage to the eyes. Wear appropriate gloves and safety goggles. Keep away from heat, sparks, and open flame.

Isopropanol is flammable and irritating. It may be harmful by inhalation, ingestion, or skin absorption. Wear appropriate gloves and safety glasses. Do not breathe the vapor. Keep away from heat, sparks, and open flame.

Isopropyl-β-D-thiogalactopyranoside (IPTG) may be harmful by inhalation, ingestion, or skin absorption. Wear appropriate gloves and safety glasses.

Kanamycin may be harmful by inhalation, ingestion, or skin absorption. Wear appropriate gloves and safety glasses. Use only in a well-ventilated area.

KCl, *see* **Potassium chloride**

$KH_2PO_4/K_2HPO_4/K_3PO_4$, *see* **Potassium phosphate**

$KMnO_4$, *see* **Potassium permanganate**

KOH, *see* **Potassium hydroxide**

LiCl, *see* **Lithium chloride**

Liquid nitrogen (LN_2) can cause severe damage due to extreme temperature. Handle frozen samples with extreme caution. Do not breathe the vapors. Seepage of liquid nitrogen into frozen vials can result in an exploding tube upon removal from liquid nitrogen. Use vials with O-rings when possible. Wear cryo-mitts and a face mask.

Lithium chloride (LiCl) is an irritant to the eyes, skin, mucous membranes, and upper respiratory tract. It may be harmful by inhalation, ingestion, or skin absorption. Wear appropriate gloves and safety goggles, and use in a chemical fume hood. Do not breathe the dust.

Magnesium acetate may be harmful by inhalation, ingestion, or skin absorption. Wear appropriate gloves and safety glasses.

Magnesium sulfate ($MgSO_4$) may be harmful by inhalation, ingestion, or skin absorption. Wear appropriate gloves and safety glasses and use in a chemical fume hood.

Manganese chloride ($MnCl_2$) may be harmful by inhalation, ingestion, or skin absorption. Wear appropriate gloves and safety glasses and use in a chemical fume hood.

β-Mercaptoethanol (2-Mercaptoethanol, $HOCH_2CH_2SH$) may be fatal if inhaled or absorbed through the skin and is harmful if ingested. High concentrations are extremely destructive to the mucous membranes, upper respiratory tract, skin, and eyes. β-Mercaptoethanol has a very foul odor. Wear appropriate gloves and safety glasses and always use in a chemical fume hood.

MgSO$_4$, *see* **Magnesium sulfate**

MnCl$_2$, *see* **Manganese chloride**

MOPS, *see* **3-(*N*-Morpholino)-propanesulfonic acid**

Na$_2$HPO$_4$, *see* **Sodium hydrogen phosphate**

NaN$_3$, *see* **Sodium azide**

NaOAc, *see* **Sodium acetate**

NaOH, *see* **Sodium hydroxide**

NBT, *see* **4-Nitro blue tetrazolium chloride**

NH(C$_2$H$_5$)$_2$, *see* **Diethylamine**

(NH$_4$)$_2$SO$_4$, *see* **Ammonium sulfate**

(NH$_4$)$_2$S$_2$O$_8$, *see* **Ammonium persulfate**

N-**Hydroxysuccinimide** is an irritant and may be harmful by inhalation, ingestion, or skin absorption. Wear appropriate gloves and safety glasses.

4-Nitro blue tetrazolium chloride (NBT) may be harmful by inhalation, ingestion, or skin absorption. Wear appropriate gloves and safety glasses.

N-**Lauroylsarcosine** is an irritant and may be harmful by inhalation, ingestion, or skin absorption. Wear appropriate gloves and safety glasses.

3-(*N*-Morpholino)-propanesulfonic acid (MOPS) may be harmful by inhalation, ingestion, or skin absorption. It is irritating to mucous membranes and upper respiratory tract. Wear appropriate gloves and safety glasses and use in a chemical fume hood.

Osmium tetroxide (OsO$_4$) (osmic acid) is highly toxic if inhaled, ingested, or absorbed through the skin. Vapors can react with corneal tissues and cause blindness. There is a possible risk of irreversible effects. Wear appropriate gloves and safety goggles and always use in a chemical fume hood. Do not breathe the vapors.

OsO$_4$, *see* **Osmium tetroxide**

PEG, *see* **Polyethyleneglycol**

Pepsin may be harmful by inhalation, ingestion, or skin absorption. Wear appropriate gloves and safety glasses.

Permount, *see* **Toluene**

Phenol is extremely toxic, highly corrosive, and can cause severe burns. It may be harmful by inhalation, ingestion, or skin absorption. Wear appropriate gloves, goggles, protective clothing, and always use in a chemical fume hood. Rinse any areas of skin that come in contact with phenol with a large volume of water and wash with soap and water; do not use ethanol!

Phosphoric acid (H$_3$PO$_4$) is highly corrosive and is harmful by inhalation, ingestion, or skin absorption. Wear appropriate gloves and safety glasses. Do not inhale the vapor.

Piperidine is highly toxic and is corrosive to the eyes, skin, respiratory tract, and gastrointestinal tract. It reacts violently with acids and oxidizing agents and may be harmful by inhalation, ingestion, or skin absorption. Do not breathe the vapors. Keep away from heat, sparks, and open flame. Wear appropriate gloves and safety glasses and use in a chemical fume hood.

Polyacrylamide is considered to be nontoxic, but it should be treated with care because it may contain small quantities of unpolymerized material (*see* **Acrylamide**).

Polyethyleneglycol (PEG) may be harmful by inhalation, ingestion, or skin absorption. Avoid inhalation of powder. Wear appropriate gloves and safety glasses.

Polyvinylpyrrolidone may be harmful by inhalation, ingestion, or skin absorption. Wear appropriate gloves and safety glasses and use in a chemical fume hood.

Potassium cacodylate, *see* **Cacodylate**

Potassium chloride (KCl) may be harmful by inhalation, ingestion, or skin absorption. Wear appropriate gloves and safety glasses.

Potassium hydroxide (KOH) and KOH/methanol is highly toxic and may be fatal if swallowed. It may be harmful by inhalation, ingestion, or skin absorption. Solutions are corrosive and can cause severe burns. It should be handled with great care. Wear appropriate gloves and safety goggles.

Potassium permanganate ($KMnO_4$) is an irritant and a strong oxidant. It may form explosive mixtures when mixed with organics. Use all solutions in a chemical fume hood. Do not mix with hydrochloric acid.

Potassium phosphate (KH_2PO_4/K_2HPO_4/K_3PO_4) may be harmful by inhalation, ingestion, or skin absorption. Wear appropriate gloves and safety glasses. Do not breathe the dust. *$K_2HPO_4 \cdot 3H_2O$ is dibasic and KH_2PO_4 is monobasic.*

Proteinase K is an irritant and may be harmful by inhalation, ingestion, or skin absorption. Wear appropriate gloves and safety glasses.

Pyridine is highly toxic and extremely destructive to the mucous membranes, upper respiratory tract, skin, and eyes. It may be harmful by inhalation, ingestion, or skin absorption. It is a possible mutagen and may cause male infertility. Keep away from heat, sparks, and open flame. Wear appropriate gloves and safety glasses and always use in a chemical fume hood.

Radioactive substances: When planning an experiment that involves the use of radioactivity, include the physicochemical properties of the isotope (half-life, emission type, and energy), the chemical form of the radioactivity, its radioactive concentration (specific activity), total amount, and its chemical concentration. Order and use only as much as really needed. Always wear appropriate gloves, lab coat, and safety goggles when handling radioactive material. **X-rays** and **gamma rays** are electromagnetic waves of very short wavelengths either generated by technical devices or emitted by radioactive materials. They may be emitted isotropically from the source or may be focused into a beam. Their potential dangers depend on the time period of exposure, the intensity experienced, and the wavelengths used. Be aware that appropriate shielding is usually of lead or other similar material. The thickness of the shielding is determined by the energy(s) of the X-rays or gamma rays. Consult the local safety office for further guidance in the appropriate use and disposal of radioactive materials. Always monitor thoroughly after using radioisotopes. A convenient calculator to perform routine radioactivity calculations can be found at:

http://www.graphpad.com/calculators/radcalc.cfm

SDS, *see* **Sodium dodecyl sulfate**

Sodium acetate (NaOAc), *see* **Acetic acid**

Sodium azide (NaN$_3$) is highly poisonous. It blocks the cytochrome electron transport system. Solutions containing sodium azide should be clearly marked. It may be harmful by inhalation, ingestion, or skin absorption. Wear appropriate gloves and safety goggles and handle with great care. Sodium azide is an oxidizing agent and should not be stored near flammable chemicals.

Sodium citrate, *see* **Citric acid**

Sodium cacodylate may be carcinogenic and contains arsenic. It is highly toxic and may be fatal by inhalation, ingestion, or skin absorption. It also may cause harm to the unborn child. Effects of contact or inhalation may be delayed. Do not breathe the dust. Wear appropriate gloves and safety goggles and use only in a chemical fume hood. *See also* **Cacodylate.**

Sodium dodecyl sulfate (SDS) is toxic, an irritant, and poses a risk of severe damage to the eyes. It may be harmful by inhalation, ingestion, or skin absorption. Wear appropriate gloves and safety goggles. Do not breathe the dust.

Sodium hydrogen phosphate, Na$_2$HPO$_4$ (sodium phosphate, dibasic) may be harmful by inhalation, ingestion, or skin absorption. Wear appropriate gloves and safety glasses and use in a chemical fume hood.

Sodium hydroxide (NaOH) and solutions containing NaOH are highly toxic and caustic and should be handled with great care. Wear appropriate gloves and a face mask. All other concentrated bases should be handled in a similar manner.

SYBR GREEN I/GOLD is supplied by the manufacturer as a 10,000-fold concentrate in DMSO which transports chemicals across the skin and other tissues. Wear appropriate gloves and safety glasses and decontaminate according to your Safety Office guidelines. *See* **DMSO.**

TAE buffer contains **Tris** and **Glacial acetic acid.**

TEAC, *see* **Tetraethylammonium chloride**

TEMED, *see* **N,N,N´,N´-Tetramethylethylenediamine**

Tetracycline may be harmful by inhalation, ingestion, or skin absorption. Wear appropriate gloves and safety glasses and use in a chemical fume hood.

Tetraethylammonium chloride (TEAC) may cause allergic skin reaction and may be harmful by inhalation, ingestion, or skin absorption. It is irritating to the mucous membranes and upper respiratory tract. Do not breathe the dust. Wear appropriate gloves and safety glasses.

*N,N,N´,N´-***Tetramethylethylenediamine (TEMED)** is highly caustic to the eyes and mucous membranes and may be harmful by inhalation, ingestion, or skin absorption. Wear appropriate gloves and tightly sealed safety goggles.

Toluene (C$_6$H$_5$CH$_3$) vapors are irritating to the eyes, skin, mucous membranes, and upper respiratory tract. Toluene can exert harmful effects by inhalation, ingestion, or skin absorption. Do not inhale the vapors. Wear appropriate gloves and safety glasses and use in a chemical fume hood. Toluene is extremely flammable. Keep away from heat, sparks, and open flame.

Tris may be harmful by inhalation, ingestion, or skin absorption. Wear appropriate gloves and safety glasses.

Trizol may be fatal if absorbed through the skin, inhaled, or swallowed. It can also cause severe burns. Wear appropriate gloves, safety goggles, protective clothing, and always use in a chemical fume hood. Rinse any areas of skin that come in contact with trizol with a large volume of water and wash with soap and water; do not use ethanol!

Urea may be harmful by inhalation, ingestion, or skin absorption. Wear appropriate gloves and safety glasses.

UV light and/or **UV radiation** is dangerous and can damage the retina. Never look at an unshielded UV light source with naked eyes. Examples of UV light sources that are common in the laboratory include hand-held lamps and transilluminators. View only through a filter or safety glasses that absorb harmful wavelengths. UV radiation is also mutagenic and carcinogenic. To minimize exposure, make sure that the UV light source is adequately shielded. Wear protective appropriate gloves when holding materials under the UV light source.

X-gal may be toxic to the eyes and skin. Observe general cautions when handling the powder. Note that stock solutions of X-gal are prepared in DMF, an organic solvent. For details, see *N,N*-dimethylformamide (DMF). See also **5-Bromo-4-chloro-3-indolyl-β-D-galactopyranoside (BCIG)**.

Xylene is flammable and may be narcotic at high concentrations. It may be harmful by inhalation, ingestion, or skin absorption. Wear appropriate gloves and safety glasses and use only in a chemical fume hood. Keep away from heat, sparks, and open flame.

Xylene cyanol, *see* **Xylene**

Suppliers

With the exception of those suppliers listed in the text with their addresses, all suppliers mentioned in this manual can be found in the BioSupplyNet Source Book and on the Web site at:

http://www.biosupplynet.com

If a copy of the BioSupplyNet Source Book was not included with this manual, a free copy can be ordered by any of the following methods:

- Complete the Free Source Book Request Form found at the Web site at: http://www.biosupplynet.com
- E-mail a request to info@biosupplynet.com
- Fax a request to 1-919-659-2199.

Index